SEISMIC DESIGN AND RETROFIT OF BRIDGES

SEISMIC DESIGN AND RETROFIT OF BRIDGES

M. J. N. PRIESTLEY
F. SEIBLE
Department of Applied Mechanics and Engineering Sciences
University of California, San Diego, USA

G. M. CALVI
Dipartimento di Meccanica Strutturale
Università di Pava, ITALY

A Wiley-Interscience Publication
JOHN WILEY & SONS, INC.
New York · Chichester · Brisbane · Toronto · Singapore

Library of Congress Cataloging in Publication Data:
Priestley, M. J. N.
  Seismic design and retrofit of bridges / M. J. N. Priestley, Frieder
Seible, Gian Michele Calvi.
    p. cm.
  Includes bibliographical references and index.
  ISBN 0-471-57998-X (cloth : alk. paper)
  1. Bridges–Design and construction   2. Earthquake engineering.
3. Bridges–Remodeling.   I. Seible, Frieder.   II. Calvi, Gian
Michele.   III. Title.
TG300.P64   1996
624'.252–dc20   95-35406

10 9 8 7 6 5 4

CONTENTS

Preface   **xv**

1 Seismic Design Philosophy for Bridges   **1**

   1.1   Introduction, 1
   1.2   Damage to Bridges in Recent Earthquakes, 3
         1.2.1   Seismic Displacements, 5
               (a) Span Failures Due to Unseating at Movement Joints, 5
               (b) Amplification of Displacements Due to Soils Effects, 5
               (c) Pounding of Bridge Structures, 8
         1.2.2   Abutment Slumping, 9
         1.2.3   Column Failures, 12
               (a) Flexural Strength and Ductility Failures, 12
               (b) Column Shear Failures, 18
         1.2.4   Cap Beam Failures, 22
         1.2.5   Joint Failures, 23
         1.2.6   Footing Failure, 27
         1.2.7   Failures of Steel Bridge Components, 27
   1.3   Design Philosophy, 29
         1.3.1   Strength Design Versus Elastic Design, 29
         1.3.2   Aspects of Ductility and Energy Dissipation, 30

1.3.3 Capacity Design Principles, 33
1.3.4 Strength Definitions, 38
    (a) Required Strength, 38
    (b) Nominal Strength, 38
    (c) Expected Strength, 39
    (d) Dependable Strength, 39
    (e) Extreme Strength, 39
    (f) Ideal Strength, 39
1.3.5 Application of Structural Reliability Theory to Seismic Design, 40
1.3.6 Design and Response Limit States, 41
    (a) Member Limit States, 41
    (b) Structure Limit States, 42
1.4 Retrofit Philosophy, 45
1.4.1 Prioritization of Retrofit, 45
1.4.2 Assessment of Existing Structures, 46
1.4.3 Retrofit Concepts, 47
    (a) Span Unseating, 47
    (b) Liquefaction and Abutment Slumping, 49
    (c) Column Failures, 49
    (d) Cap Beam Flexural and Shear Strength, 49
    (e) Joint Failures, 51
    (f) Footing Failure, 51
    (g) Seismic Isolation as a Retrofit Approach, 53
1.5 Design Process, 54
1.5.1 Selection of Design Seismicity, 55
1.5.2 Conceptual Design, 55
1.5.3 Modeling and Analysis, 56
1.5.4 Design and Detailing, 56
1.5.5 Design Checking (Reanalysis), 57
1.5.6 Seismic Isolation, 57
1.5.7 Assessment and Retrofit of Existing Bridges, 57

2 Seismicity and Geotechnical Considerations   58

2.1 Introduction, 58
2.2 Characterization of Ground Motion, 59
2.2.1 Source Mechanism and Attenuation Laws, 59
2.2.2 Wave Propagation Effects, 61
2.2.3 Return Periods and Risk Evaluation, 65
2.2.4 Accelerograms, 67
2.2.5 Magnitude and Intensity, 69
2.2.6 Directionality Effects, 69
2.2.7 Components and Maxima Values, 70
    (a) Acceleration, 70

(b) Velocity, 71
(c) Displacement, 71
2.2.8 Duration Effects, 71
2.2.9 Energy Considerations, 72
2.2.10 Geographical Amplifications, 73
2.3 Soil Properties and Response, 74
2.3.1 Geotechnical Investigations and Prediction of Soil
Properties, 74
2.3.2 Soil Modification of Rock Accelerograms, 76
2.3.3 Dislocations, 81
2.3.4 Slumping, 81
2.3.5 Liquefaction, 88
2.4 Design Ground Motion, 94
2.4.1 Elastic Response Spectra, 94
2.4.2 Inelastic Response Spectra, 97
2.4.3 Risk–Importance–Return Period Relationships, 100
2.4.4 Design Response Spectra, 102
2.4.5 Generation of Artificial Ground Motion, 106
2.5 Effects of Nonsynchronous Input Motion, 108
2.5.1 Synchronous and Nonsynchronous Input to
Piers, 108
2.5.2 Linear Multimodal Approach, 109
2.5.3 Nonlinear Approach, 110
2.5.4 Preliminary Design Approach, 110
2.5.5 Generation of Nonsynchronous Ground Motion
Time Histories, 113

3 Conceptual Design   116

3.1 Introduction, 116
3.2 Constraints, 117
3.2.1 Functional Constraints, 117
(a) Alignment, 117
(b) Width, 120
3.2.2 Geographical Constraints, 121
3.2.3 Superstructure Categories, 124
(a) Concrete Sections, 124
(b) Steel Superstructure Sections, 129
3.3 Seismic Design Alternatives, 130
3.3.1 Superstructure–Pier Connection, 130
(a) Moment-Resisting Connection, 130
(b) Bearing-Supported Superstructures, 133
(c) Isolated Superstructures, 136
3.3.2 Superstructure–Abutment Connection, 137
(a) Monolithic Connection, 137

(b) Bearing-Supported Superstructure, 139
(c) Isolated Superstructure, 139
(d) Transverse Response, 139
(e) Approach Settlement, 141
3.3.3 Bent Configurations, 141
(a) Single-Column Bents, 141
(b) Multicolumn Bents, 143
(c) Linked-Column Bents, 143
3.3.4 Column–Pier Section Alternatives, 144
(a) Circular Sections, 144
(b) Rectangular Columns, 146
(c) Rectangular Piers, 146
(d) Hollow-Section Columns, 147
3.3.5 Bent–Foundation Connection, 148
3.3.6 Foundation System, 149
(a) Spread Footings, 149
(b) Rocking Spread Footings, 150
(c) Integral Pile-Shaft/Columns, 151
(d) Pile-Supported Footings, 151
3.4 Deep-Valley-Crossing Considerations, 152
3.5 Long-Span-Bridge Considerations, 154

4 Modeling and Analysis   **155**

4.1 Introduction: Seismic Bridge Assessment and Design
Tools, 155
4.2 Modeling and Analysis Objectives, 157
4.3 Fundamentals of Seismic Bridge Behavior: Structural
Dynamics, 160
4.3.1 Dynamic Behavior of Bridges: Equation of
Motion, 160
(a) Mass, 163
(b) Stiffness, 168
(c) Damping, 174
4.3.2 Bridge Dynamic Response Characteristics, 177
(a) Single-Degree-of-Freedom Characteristics, 177
(b) Multidegree-of-Freedom Characteristics, 181
4.3.3 Elastic Seismic Response of Bridges: Maximum
Response Values, 183
4.4 Modeling of Bridge Structures, 189
4.4.1 General Modeling Issues, 189
(a) Structural Systems, 191
(b) Individual Structural Members, 193
4.4.2 Modeling of Bridge Components, 196
(a) Superstructure, 197

(b) Single-Column Bents, 203

(c) Multicolumn Bents, 205

(d) Foundations, 210

(e) Abutments, 216

(f) Movement Joints and Restrainers, 219

4.4.3 Substitute Structure Models, 223

4.5 Methods of Analysis, 226

4.5.1 Types of Analysis Tools, 227

4.5.2 Static or Quasistatic Analysis Tools, 227

(a) Solution Strategies, 229

(b) Analysis Tools, 232

4.5.3 Response Spectrum Analyses, 241

4.5.4 Time-History Analyses, 243

4.6 Bridge Response Analysis Example: Response Assessment of a Long Regular Viaduct, 249

5 **Design**   **265**

5.1 Introduction, 265

5.2 Material Properties for Seismic Resistance, 266

5.2.1 Unconfined Concrete, 266

(a) Compression Strength f_c', 266

(b) Modulus of Elasticity, 266

(c) Tensile Strength of Concrete, 267

5.2.2 Confined Concrete, 267

(a) Confining Effect of Transverse Reinforcement, 267

(b) Compression Stress–Strain Relationships for Confined Concrete, 270

5.2.3 Reinforcing Steel, 273

(a) Monotonic Characteristics, 273

(b) Inelastic Cyclic Response, 275

(c) Temperature and Strain-Aging Effects, 276

5.2.4 Structural Steel, 277

5.2.5 Prestressing Steel, 277

5.2.6 Advanced Composite Materials, 278

5.3 Capacity Design Process, 280

5.3.1 Flexural Strength Requirements for Plastic Hinges, 280

(a) Force-Based Design (Conventional Design), 281

(b) Relationship Between Force-Reduction Factor and Displacement Ductility Factor, 285

(c) Displacement-Based Design, 289

(d) Design for Required Flexural Strength of Plastic Hinges, 293

5.3.2 Flexural Ductility and Inelastic Rotation, 307
    (a) Required Ductility, 307
    (b) Assessment of Member Inelastic Rotation and
        Ductility Capacity, 307
    (c) Confinement for Plastic Hinges, 312
    (d) Serviceability Considerations, 321
5.3.3 Required Strength of Capacity-Protected Actions, 322
    (a) Standard Prescriptive Force-Based Design, 322
    (b) Actions Determined by Displacement-Based
        Design, 326
    (c) Actions Determined by Inelastic Time-History
        Analyses, 326
5.3.4 Design Strength of Capacity-Protected Actions, 326
    (a) Flexural Strength, 327
    (b) Shear Strength, 331
    (c) Torsional Strength, 345
5.4 Design of Beam–Column Joints, 348
5.4.1 Shear Force in Beam–Column Joints, 349
    (a) Knee Joints, 352
    (b) Tee Joints, 353
5.4.2 Nominal Shear Stress, 355
5.4.3 Principal Stress Levels in Joints, 355
    (a) Principal Tension Stress, 357
    (b) Principal Compression Stress, 358
    (c) Example of Principal Stress Computation, 358
5.4.4 Mechanisms of Force Transfer in Cracked
    Joints, 360
    (a) Knee Joints, 360
    (b) Tee Joints, 377
    (c) Longitudinal Response, 384
    (d) Joint Reinforcement for Longitudinal and
        Transverse Seismic Response, 385
    (e) Example of Joint Design for Combined
        Longitudinal and Transverse Response of an
        Internal Column of a Multicolumn Bent, 386
5.5 Anchorage, Development, and Splicing, 389
5.5.1 Introduction, 389
5.5.2 Codified Development Equations, 392
5.5.3 Anchorage in Confined Conditions, 393
5.5.4 Splicing of Reinforcing Bars, 396
5.5.5 Flexural Bond, 400
5.6 Footings and Pile Caps, 400
5.6.1 Introduction, 400
5.6.2 Design for Flexure, 401
    (a) Stability, 401

(b) Design Moments, 403
(c) Flexural Strength, 406
(d) Detailing, 408
5.6.3 Design for Shear Strength, 408
(a) Design Shear Force, 408
(b) Shear Strength, 408
(c) Detailing, 409
5.6.4 Design of Column–Footing Joints, 410
(a) Joint Cracking, 410
(b) Mechanisms of Force Transfer, 411
(c) Summary of Footing Joint Design, 412
(d) Footing Joint Design Example, 413
5.6.5 Design of Piles, 415
5.7 Superstructure Longitudinal Design, 416
5.8 Movement Joints, 417
5.8.1 Seating, 418
5.8.2 Shear Keys, 421
5.8.3 Restrainer Design, 425
5.9 P–Δ Effects, 427
5.10 Design Examples:
5.10.1 Example 5.1: Cantilever Column with Curved
Continuous Box-section Superstructure, 431
5.10.2 Example 5.2: Two Column Bent Designed to
Displacement-based Consideration, 443

6 Design of Bridges Using Isolation and Dissipation Devices   457

6.1 Introduction, 457
6.1.1 Concepts of Seismic Isolation and Energy
Dissipation, 457
6.1.2 Period Shift, 458
6.1.3 Damping, 460
6.1.4 Seismicity Aspects, 461
6.1.5 Vertical Response, 461
6.1.6 Design Philosophy and Structural Aspects, 462
6.1.7 Displacement Problems, 463
6.1.8 Past Applications and Response in Past
Earthquakes, 464
6.1.9 Viewpoints on the Applicability of Seismic Isolation
to Bridges, 465
6.2 Isolation and Dissipation Devices, 466
6.2.1 Main Properties of the Most Common Devices, 466
(a) Laminated-Rubber Bearings, 466
(b) Lead–Rubber Bearings, 469
(c) Sliding Bearings, 472

(d) Steel Hysteretic Dampers, 475

(e) Hydraulic Dampers, 478

(f) Lead-Extrusion Dampers, 480

(g) Initiating and Limiting Devices, 481

6.2.2 Performance Requirements and Testing of Isolating Devices, 483

(a) Fundamental Features, 483

(b) Performance Comparison, 484

(c) Testing Specifications, 484

6.3 Modeling, Analysis, and Design, 487

6.3.1 Modeling, 487

(a) Preliminary Design, 487

(b) Models for the Design Earthquake, 488

(c) Models for the Extreme Earthquake, 488

6.3.2 Analysis, 490

(a) Static, Linear, Single-DOF Analysis, 490

(b) Dynamic Modal Analysis, 493

(c) Time-History Analysis, 498

6.3.3 Design Principles, 500

6.3.4 Capacity Design Principles and Protection Factors, 503

6.3.5 Design Methods, 504

(a) Design of I/D System and Pier Reinforcement, 505

(b) Given Reinforcement Design, 511

6.4 Foundation Rocking, 516

6.4.1 Introduction, 516

6.4.2 Rocking Response of Bridges, 519

6.4.3 Response Spectra Design Approach, 525

6.4.4 Time-History Analysis, 530

6.5 Active Control, 531

7 Seismic Assessment of Existing Bridges   **534**

7.1 Introduction: Procedures for Seismic Assessment, 534

7.1.1 Prioritization Schemes, 534

7.2 Assessment Limit States, 537

7.2.1 Serviceability Limit State, 537

7.2.2 Damage-Control Limit State, 538

7.2.3 Survival Limit State, 538

7.3 Assessment Analysis Schemes, 538

7.3.1 Capacity/Demand Ratio Analyses, 539

7.3.2 Plastic Collapse Mechanism (Pushover) Analysis, 540

7.3.3 Inelastic Time-History Analysis, 544

7.4 Assessment of Member Strength and Deformation
     Characteristics, 546
     7.4.1 Material Strengths, 546
     7.4.2 Relative Capacity Considerations, 547
     7.4.3 Elastic Stiffness, 548
     7.4.4 Flexural Strength, 549
     7.4.5 Flexural Strength of Column Sections with Lap-
            Spliced Longitudinal Reinforcement, 549
     7.4.6 Deformation Capacity of Plastic Hinges, 552
            (a) Sections Without Lap-Spliced
                Reinforcement, 552
            (b) Sections with Lap-Spliced Reinforcement, 555
     7.4.7 Flexural Strength and Ductility of Cap Beams, 557
            (a) Effects of Low Positive Moment Capacity, 557
            (b) Effects of Terminition of Longitudinal
                Reinforcement, 557
     7.4.8 Shear Strength, 560
     7.4.9 Joint Strength and Deformation Characteristics, 562
     7.4.10 Footing Strength and Deformation
            Characteristics, 570
            (a) Stability, 570
            (b) Flexural Strength, 570
            (c) Shear Strength, 573
            (d) Column–Footing Joint Strength, 574
            (e) Footing Failure as an Accepted Mechanism of
                Response, 574
            (f) Strength and Rotation Capacity of Column-
                Base Hinges, 574
            (g) Pile Capacity, 576
     7.4.11 Superstructure Strength and Deformation
            Characteristics, 577
     7.4.12 Abutment Assessment, 578
7.5 Miscellaneous Details, 578
     7.5.1 Shear Keys, 579
     7.5.2 Strut Action in Wide Bridges, 579
     7.5.3 Bearings, 580
     7.5.4 Restrainer and Movement Joint Details, 581
     7.5.5 Steel Superstructure Assessment, 582

8 Retrofit Design   585

8.1 Introduction: Retrofit Philosophy, 585
8.2 Retrofit of Concrete Columns, 586
     8.2.1 Column Retrofit Techniques, 587
            (a) Steel Jacketing, 587

(b) Concrete Jacketing, 588

(c) Composite Materials Jackets, 589

8.2.2   Column Retrofit Design Criteria, 590

(a) Confinement for Flexural Ductility
      Enhancement, 590

(b) Confinement for Flexural Integrity of Column
      Lap Splices, 598

(c) Shear Strength Enhancement, 606

(d) Column Stiffness Considerations, 614

(e) Column Repair by Jacketing, 615

(f) Example 8.1: Column Steel Jacket Retrofit
      Design, 619

(g) Example 8.2: Alternative Design with
      Composite Materials Jacket, 622

8.3   Retrofit of Cap Beams, 623

8.3.1   Reduction of Cap Beam Seismic Forces, 625

8.3.2   Enhancement of Cap Beam Strength, 625

(a) Flexural Strength, 625

(b) Shear Strength, 627

(c) Torsional Strength, 630

8.4   Retrofit of Cap Beam–Column Joint Regions, 630

8.4.1   Joint Force Reduction, 631

8.4.2   Damage Acceptance with Subsequent Repair, 631

8.4.3   Joint Prestressing, 632

8.4.4   Jacketing, 634

8.4.5   Joint Replacement, 634

8.5   Superstructure Retrofit, 638

8.5.1   Movement Joint Retrofit, 638

(a) Restrainers, 638

(b) Seat Extenders, 641

8.5.2   Superstructure Flexural Capacity, 642

(a) Strength Increase, 643

(b) Force Reduction, 644

8.6   Footing Retrofit, 646

8.6.1   Stability Enhancement, 646

8.6.2   Flexural Enhancement, 647

8.6.3   Shear Enhancement, 649

8.6.4   Footing Joint Shear Force Enhancement, 649

8.7   Seismic Isolation in Bridge Retrofit, 653

References   **654**

List of Symbols   **667**

Index   **679**

PREFACE

Recent earthquakes, particularly the 1989 Loma Prieta and 1994 Northridge earthquakes in California, and the 1995 Kobe earthquake in Japan, have caused collapse of, or severe damage to, a considerable number of major bridges that were at least nominally designed for seismic forces. This behavior led to a marked increase in discussion about seismic design philosophy and in the research activity on the seismic design of new bridges and the assessment and retrofit of existing bridges that was initiated following the 1971 San Fernando earthquake. Our involvement in the relevant experimental and analytical research, and in design philosophy development, has led us to attempt to present a unified approach for seismic design of new bridges, and the assessment and retrofit of older bridges which have substandard seismic characteristics. This book, the end product of this effort, is primarily directed towards practicing bridge designers, though the extent of new design and research information, presented here for the first time, should make it of some interest to researchers investigating seismic design of bridges. It should also be appropriate for graduate courses or upper level undergraduate course in seismic design of bridges.

The approach relies heavily on the principles of capacity design, where a strength hierarchy is established in a bridge to ensure that damage is controllable, and occurs only where the designer intends. This approach, which is well established for seismic design of buildings, has been extended and modified here to reflect the special demands and characteristics of bridges. Particular emphasis is placed on designing adequate displacement and ductility capacity into new bridges, with less significance placed on strength. The 1995 Kobe earthquake graphically illustrated that strength alone will not ensure good seismic performance. The book is developed around two alternative design strategies: the traditional force-based approach where force levels are related to acceleration spectra, with checks to ensure adequate displacement capacity exists, and the newer displacement-based design approach, where displacements are the starting point in the design. It is felt that this later approach will largely supersede force-based design in the next decade.

Introductory chapters discuss design philosophy and its impact on the performance of bridges in recent earthquakes, seismicity and soils effects, including liquefaction, in a form facilitating understanding by structural engineers, and the importance of rational consideration, from a seismic design viewpoint, of the various structural configuration possibilities in the conceptual design phase. Analysis is considered subservient to design in this book, since we have been dismayed by the end results of a number of recent bridge designs which have been driven by seismic analytical considerations. Nevertheless, extensive discussion of analysis is provided in Chapter 4, with emphasis on the importance of realistic modeling assumptions and appropriate choice of analytical tools. Chapters 5 to 8 provide detailed information on the design of new bridges and the assessment and retrofit of existing bridges. A separate chapter is devoted to design and retrofit using seismic isolation and dissipation devices. Many design and analysis examples, some quite extensive in scope, are included in these chapters. Design aids in the form of charts and tables are also provided.

Throughout the book we attempt to present design information based on rational argument, supported by the most recent experimental research results. This has led us to recommend many procedures that differ from, or deal with areas not covered by existing codified approaches, particularly for detailed assessment of displacement capacity, design for shear strength, and the behavior of column-cap beam and column-footing connections. In some cases, the information presented may be considered controversial, in that it sometimes appears to be less conservative than previous approaches, but we believe that the excellent correlation with experimental research results supports the recommended approach.

We gratefully acknowledge support from a wide range of sources, including past and present graduate researchers and technicians at the University of California, San Diego, and the Università di Pavia. We also particularly acknowledge the support of the California Department of Transportation (Caltrans) for both financial support and involvement in the extensive experimental research at UCSD, on which so much of this book is based. The list of people at Caltrans we should acknowledge is too long to reasonably include here. However, one name must be mentioned, that of James R. Roberts, for many years Chief Bridge Engineer in California and currently, Director of the Caltrans Engineering Services Center. Without his drive, vision, and sense of urgency following the Loma Prieta earthquake, much of the research which forms the basis for this book would not have been done. We also acknowledge the support of the University of California, San Diego, the Università di Pavia, and the New Zealand Earthquake Commission, through its Distinguished Research Fellowship program.

We are most grateful to Joan Welte, Maria Martin, Lesleigh Bradford-Brown, and especially Celia Villamil, who typed most of the chapters and revisions. A large number of graduate students were pressed, under veiled threats related to their research programs, to prepare the hundreds of line drawings in the book. In this regard, we gratefully acknowledge Andy Budek, Rigoberto Burgueno, Bill Bruin, Duane Gee, Jason Ingham, Danny Innamorato, Mervyn Kowalsky, Guido Magenes, Alberto Pavese, Paolo Moncecchi, Gene Smith, Sri Sritharan, and especially Marc Veletzos.

A debt of love and gratitude is due to our wives and families, for the long hours spent on preparing this book rather than spending time with them. Their patience and understanding was essential to the successful completion of this book. Finally, it will be evident to anyone familiar with capacity design that this book owes a great deal to the seismic design philosophy developed over the past 25 years in New Zealand, principally by Professors Robert Park and Tom Paulay. Without their inspiration this book would not have been written, and in consequence, we humbly dedicate it to them.

NIGEL PRIESTLEY
FRIEDER SEIBLE
MICHELE CALVI

San Diego and Pavia
October 1995

1

SEISMIC DESIGN PHILOSOPHY FOR BRIDGES

1.1 INTRODUCTION

Bridges give the impression of being rather simple structural systems. Indeed, they have always occupied a special place in the affections of structural designers because their structural form tends to be a simple expression of their functional requirement. As such, structural solutions can often be developed that are both functionally efficient and aesthetically satisfying, as evidenced by the many excellent texts on aesthetic aspects of bridge design (see, e.g., [L1,W1]).

Despite, or possibly because of their structural simplicity, bridges, in particular those constructed in reinforced or prestressed concrete, have not performed as well as might be expected under seismic attack. In recent earthquakes in California [F1,P1,E1,E2], Japan [K1,P2], and Central and South America [E2,E4], modern bridges designed specifically for seismic resistance have collapsed or have been severely damaged when subjected to ground shaking of an intensity that has frequently been less than that corresponding to current code intensities.

As will be seen subsequently, this unexpectedly poor performance can in the majority of cases be attributed to the design philosophy adopted, coupled with a lack of attention to design details. Earthquakes have a habit of identifying structural weaknesses and concentrating damage at these locations. With building structures, the consequences may not necessarily be disastrous, because of the high degree of redundancy generally inherent in building structural systems. This enables alternative load paths to be mobilized if necessary. Typically, bridges have little or no redundancy in the structural systems, and

1

failure of one structural element or connection between elements is thus more likely to result in collapse than is the case with a building. This leads to the above-mentioned warning that the structural simplicity of bridges may be a mixed blessing. While the simplicity should lead to greater confidence in the prediction of seismic response, this also results in greater sensitivity to design errors.

Bridges are typically more sensitive to soil–structure interaction effects than are buildings. Dynamic response to ground shaking may be less predictable, particularly for long bridges, as a consequence of nonsynchronous seismic input at different supports, resulting from traveling-wave effects. Further, bridges are often constructed at sites with difficult ground conditions: river and estuary crossings will not infrequently be over sandy or silty soils with a potential for liquefaction; where roads or railroads cross active faults, the local geography will typically be contorted as a result of earlier fault movements, and bridging of the fault will be necessary. As a consequence, large relative displacements of supports could occur due to fault dislocation in a future earthquake.

All the foregoing considerations indicate a need for special care in the seismic design of bridges. The main purpose of this book is to examine the seismic response of bridges in detail and to develop design strategies that make the structure as insensitive as possible to the unknown characteristics of the input seismic excitation. The basis for this will be the approach known as *capacity design,* where locations (plastic hinges) of potential inelastic flexural deformation are selected and detailed carefully while undesirable plastic hinge locations, or undesirable inelastic deformation mechanisms, such as shear, are inhibited by providing them with an appropriate strength margin above that corresponding to the designed plastic hinge strength. This approach has become well established for building structures in recent years [P3,P4]. Aspects of the philosophy have also been incorporated in a number of bridge design specifications [M1,C1,E5]. As will be seen subsequently, the procedure is generally somewhat simpler for bridges than for buildings.

Application of the capacity design philosophy developed for bridges in this book will result in structures that can confidently be expected to survive a design-level earthquake with only minor, repairable damage. It is recognized, however, that the highway systems of most countries are rather fully developed and that the number of new bridges being constructed each year is a comparatively small proportion of the total bridge inventory. Many existing bridges were constructed before seismic actions were adequately understood. As a consequence, failures of bridges in seismic regions, similar to the examples described in Section 1.2, can be expected to occur in future earthquakes unless remedial actions are taken. The design of retrofit measures for existing bridges should thus be a matter of considerable concern. Although the underlying philosophy for retrofit design should conform to that for new design, a less conservative approach might be justifiable for assessment of the expected seismic performance of existing bridges. The fact that an older bridge was designed to lower seismic standards than would be required today does not necessarily mean that it must be expected to collapse or be severely damaged in an earthquake of the intensity envisaged by current standards. Further, techniques for retrofitting bridges will often involve measures that would not be considered as appropriate for new bridge design. Consequently, methods for seismic assessment and retrofit of existing bridges are considered separately from design measures for new bridges in this book and are given considerable emphasis. Much of this work is an outcome of intensive research and design development following the 1989 Loma Prieta earthquake [E1] and the 1994 Northridge earthquake [E2] in California.

To be able to effectively design either new bridges or retrofit measures for existing bridges, a clear understanding of potential problem areas is essential. There is no better way of developing this understanding than by a systematic examination and categorization of failures and damage that have occurred to bridges in earlier earthquakes. It has often been said that those who ignore the lessons of history are doomed to repeat its mistakes. In no area is this more true than in seismic design. It is thus appropriate, before developing design philosophy or design details, to spend some time in reviewing past earthquake damage to bridges and common design deficiencies.

1.2 DAMAGE TO BRIDGES IN RECENT EARTHQUAKES

In reviewing bridge damage caused by recent earthquakes, three basic design deficiencies can be identified, although each deficiency may be manifested in different forms. All tend to be a direct consequence of the elastic design philosophy almost uniformly adopted for seismic design of bridges prior to 1970 and still used in some countries. Elastic seismic design of older bridges typically utilized comparatively low allowable stress levels, corresponding to seismic forces that were but a small fraction of the actual forces that could be developed in an elastically responding structure of unlimited strength. With the actual strength of older bridge structures typically 100 to 300% higher than the elastic design force levels, even moderate seismic excitation could be expected to develop the available capacities.

The consequences of this elastic design approach were:

1. Seismic deflections, based on the specified lateral force levels, were seriously underestimated. This was compounded by the use of gross-section rather than cracked-section member stiffness in the computation of displacements, which resulted in artificially low predictions of expected displacements.

2. Since seismic force levels were artificially low, the ratio of gravity load to seismic force adopted for design was incorrect. This led to moment patterns under combined gravity load plus seismic force that were not only low but often had the wrong shape. Points of contraflexure were seriously mislocated,

resulting in premature termination of reinforcement. Locations and moment magnitudes of critical sections were also incorrect. In many cases, when gravity load and seismic forces resulted in moments at a given section that were of opposite sign, the final design moment could even be of the incorrect sign, due to gravity-load domination.

An example of this is illustrated in Fig. 1.1, which represents moments in a two-column bridge bent under dead load D [Fig. 1.1(a)] and transverse seismic response E [Fig. 1.1(b)]. When the level of transverse seismic force corresponding to the elastic design assumption is considered, the moments resulting from the combined actions $(D + E)$ are as shown in Fig. 1.1(c) by the solid curve. However, since material stress levels for elastic seismic design were well below yield or strength values, higher lateral forces could clearly be sustained before member strengths were achieved. With increased seismic lateral force levels E corresponding to the development of the critical member flexural strength at joint A, the dashed-line moment distribution of Fig. 1.1(c) is obtained. Because of the higher level of lateral force, the point of contraflexure B predicted by the "elastic" force combination has shifted to C. Thus cap beam negative reinforcement terminated in accordance with the elastic moment distribution could result in premature failure some distance from joint A.

At the other end of the cap beam (joint D), the elastic moment distribution $(D + E)$, based on reduced seismic inertia force, predicts a small residual negative moment and only nominal positive moment reinforcement would be carried into the joint region. However, at levels of lateral force sufficient to develop the negative moment capacity at A, the dashed $(D + E)$ curve in Fig. 1.1(c) indicates that a positive moment of considerable magnitude develops at D. The reinforcement and anchorage details adopted, based on the elastic moment distribution, are thus again likely to cause premature failure.

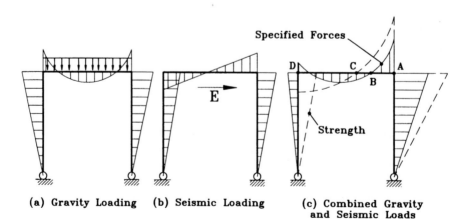

(a) Gravity Loading   (b) Seismic Loading   (c) Combined Gravity
                                                         and Seismic Loads

FIG. 1.1   Elastic moment distribution.

3. Inelastic structural actions and associated concepts of ductility and capacity design, which will be elaborated in subsequent sections, are crucial to the survival of inelastic systems under severe seismic response and were not considered in the elastic design process. Thus critical potential hinge locations were not detailed to sustain large inelastic deformations without strength degradation, and member shear strength was not set higher than flexural strength, to avoid the possibility of brittle shear failure.

As will be seen in the following examples, most of the observed damage can be attributed to one, or a combination, of the three deficiencies noted above.

1.2.1 Seismic Displacements

A direct consequence of the underestimated seismic displacements, which were based on elastic theory, gross-section stiffness, and low lateral force levels, was that seating lengths provided at movement joints were unrealistically short, and lateral separations between adjacent structures were typically inadequate, resulting in pounding.

(a)   Span Failures Due to Unseating at Movement Joints. There have been many examples of bridge failure caused by relative movement of spans in the longitudinal direction exceeding seating widths, resulting in unseating at unrestrained movement joints. This has been a particular problem for multi-span bridges with tall columns. Adjacent frames separated by movement joints may move out of phase, increasing the relative displacement across the joint. Figure 1.2 shows an example of span collapse from the 1971 San Fernando earthquake [F1]. Although linkage restrainer bolts were provided across the movement joints in this bridge, they had insufficient strength to restrain the longitudinal relative movement.

Figure 1.3 shows another example, also from the San Fernando earthquake. In this case a simply supported span has unseated due to excessive relative displacement between adjacent bents. The support lines are skewed to the bridge axis, and it has been observed that skewed spans develop larger displacements than right spans, as a consequence of a tendency for the skew span to rotate in the direction of decreasing skew, thus tending to drop off the supports at the acute corners. This behavior is due to a combination of longitudinal and transverse response and is illustrated schematically in Fig. 1.4. It will be seen that transverse response in either direction will cause binding in one obtuse corner, causing clockwise rotation. These rotations tend to accumulate under cyclic transverse response.

(b)   Amplification of Displacements Due to Soils Effects. When bridges are built on soft or liquefiable soils, the problem is compounded. Soft soils will generally result in amplification of structural vibrational response, increasing the probability of unseating. When bridge bents are supported by piles though

FIG. 1.2 Span collapse, 1971 San Fernando earthquake.

FIG. 1.3 Unseating, 1971 San Fernando earthquake (simple span, skew).

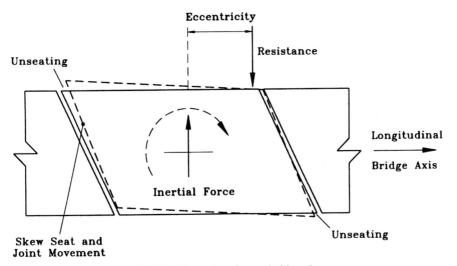

FIG. 1.4 Unseating due to bridge skew.

saturated sandy silts or silty sands, liquefaction of the ground may occur, causing a loss of support to the piles, with excessive vertical and/or lateral displacements unrelated to vibrational response. Bridges with simply supported spans are particularly susceptible to span failure resulting from liquefaction, and examples abound from most moderate to large earthquakes, including the 1964 Alaska earthquake [B9], 1985 Chile earthquake [E3], and 1990 Costa Rica earthquake [E4], to name a few. Figures 1.5 and 1.6 show span

FIG. 1.5 Liquefaction failure, Rio Viscaya bridge, 1990 Costa Rica earthquake.

FIG. 1.6 Unseating of skewed spans at a soft soil site, Rio Bananito bridge, 1990 Costa Rica earthquake.

failures of modern bridges in the Costa Rica earthquake, the first resulting from liquefaction at internal supports and the second due to relative displacement between the abutment and an internal pier at a site with soft soils. The supports of the bridge in Fig. 1.6 were skewed at about 30° to the transverse axis, and the spans were thrown off the internal support in the direction of decreasing skew, as postulated in Fig. 1.4. A third example is given by the failure of the east link span to the 250-m (820-ft) Nishinomiya-ko arch bridge of the Wangan expressway (Fig. 1.7) in the 1995 Kobe earthquake [P2]. This 50-m (164-ft) steel span was designed to be simply supported to allow adjustment of elevation of supports if soil settlements occurred. During the earthquake, the tension-link restrainers connecting the link span to the arch support failed, and large movements of the arch bridge support resulting from slumping and liquefaction at the site resulted in unseating of the span.

(c) Pounding of Bridge Structures. Underpredicted seismic displacements may lead to the provision of inadequate clearance between adjacent structures, resulting in pounding damage. This is a problem primarily when the adjacent

FIG. 1.7 Unseating of simply supported link span, Nishinomiya-ko bridge, 1995 Kobe earthquake.

structures are of different heights, resulting in impact between the superstructure of one, and the column (or columns) of another. An example occurred in the 1989 Loma Prieta earthquake at the China Basin/Southern viaduct section of I-280 in San Francisco, as shown in Fig. 1.8, where a 150-mm (6-in.) separation between the lower roadway and columns independently supporting an upper-level deck proved inadequate. Impact forces from pounding of bridge components can be very high, causing amplification of member shear forces with increased probability of brittle shear failure. Pounding for structures of unequal height must therefore be avoided through realistic assessment of deformation and provision of adequate separation. There is, however, theoretical evidence that pounding between superstructures of equal height, as might occur with parallel bridge structures carrying opposite directions of traffic, can be beneficial to seismic response [T1]. This is because the impacts between structures of different fundamental periods act to spoil the buildup of resonant response.

1.2.2   Abutment Slumping

Again, related to response of soft soils and incompletely consolidated abutment fill, slumping of abutment fill and rotation of abutments have been widespread in recent earthquakes. Under longitudinal response earth pressures on the abutment increase due to seismic accelerations. Impact of the bridge with the abutment may generate high passive pressures, which will

FIG. 1.8   Pounding of I-280 China Basin/Southern viaduct, 1989 Loma Prieta earthquake.

induce a further increase in lateral pressures at levels below the point of deck or superstructure impact. Inadequately compacted natural or fill soils tend to slump toward the bridge, pushing the lower part of the abutment inward with the moving soil. Contact between the top of the abutment and the superstructure limits the inward displacement at the top, resulting in a rotation of the abutment, as suggested in Fig. 1.9. Typical consequences are damage to the top of the abutment backwall from the superstructure impact, and damage to the pile support system if abutment rotations are large.

Figures 1.10 and 1.11 show examples of abutment failure from the 1990 Costa Rica earthquake [E4]. The bridge shown in Fig. 1.10, a three-span bridge consisting of simply supported I girders had severe slumping and a 9° rotation of one abutment, resulting in a lateral displacement of the pile tops of about 660 mm (26 in.) and in flexural and shear failure of the supporting piles, particularly the front piles, which were raked at a 1 : 5 slope. Abutment fill material slumped about 1 m (40 in.) behind the abutment.

An extreme case of abutment backwall damage caused by superstructure impact is shown in Fig. 1.11. This bridge, already depicted in Fig. 1.5, pushed

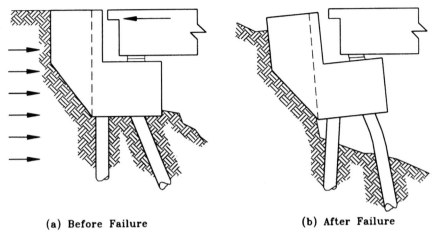

(a) Before Failure   (b) After Failure

FIG. 1.9   Abutment slumping and rotation.

FIG. 1.10   Abutment slumping and rotation failure, Rio Banano bridge, 1990 Costa Rica earthquake.

11

FIG. 1.11   Abutment failure due to passive pressure, Rio Viscaya bridge, 1990 Costa Rica earthquake.

its end span through and over the abutment when an internal pier on liquefying sand collapsed and rotated toward the abutment, pushing the span with it. The passive pressure buildup behind the abutment caused rotation of the abutment and pile failure. Although severe damage can result from abutment slumping, total collapse is uncommon and generally related to other causes, as was the case for the bridge of Fig. 1.11. It is frequently possible within hours or days to reinstate the bridge temporarily for emergency traffic, although expensive repair or replacement may be needed later.

1.2.3   Column Failures

Failure of the columns of bridge bents may result from a number of deficiencies related to the consequences of elastic design philosophy identified earlier in this section. The most common of these are discussed below.

(a)   Flexural Strength and Ductility Failures.   Until the 1970s, designers were generally unaware of the need to build ductility capacity into potential plastic hinge regions. Indeed, the concept of plastic hinging was irrelevant to the elastic design approach utilized. Four particular deficiencies can be identified.

(i)   Inadequate Flexural Strength.   Low seismic lateral force levels were typically used to characterize seismic action. For example, in California, design for lateral forces equivalent to 6% of the gravity weight was common, although it is now appreciated that elastic response levels may exceed 100% of the

gravity weight. Although the discrepancy between the design and actual elastic response levels seems large, the discrepancy between actual strength and elastic response level is typically much less, as a result of the conservative nature of the elastic analysis adopted for column flexural design. This is illustrated in Fig. 1.12, which compares typical column axial load–moment interaction diagrams for elastic and strength design.

Elastic design was typically based on a linear interaction between the moment M and axial load P, from about 45% of the flexural strength under pure bending to about 30% of the axial compression strength for pure compression. As shown in Fig. 1.12, this implies a reduction in moment capacity as the axial compression increases, whereas for the low axial load levels common for bridge columns, the moment capacity is typically enhanced by the axial compression. As a consequence, the actual flexural strength is not infrequently as much as three to four times the design elastic level, as suggested in Fig. 1.12. There may be further enhancement from material strengths, particularly reinforcement yield strength, exceeding specified values, and from strain hardening of reinforcement. As a result of this behavior, lateral flexural strengths corresponding to about 25% of the gravity weight are common in existing Californian bridges. This is of a similar magnitude to that required by current ductile design practice, but much less than possible elastic response levels, and is thus inadequate for an elastic design approach, where special ductile detailing provisions are not implemented. The importance of this is discussed further in Section 1.3.3.

(ii)   Undependable Column Flexural Strength. Column longitudinal reinforcement was often lap spliced immediately above the foundation, with a

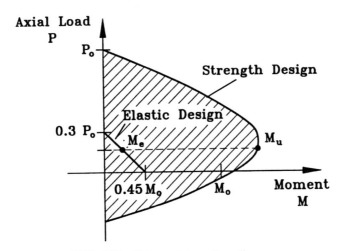

FIG. 1.12   Column interaction diagram.

splice length inadequate to develop the strength of the bars. Lap-splice lengths as short as 20 bar diameters were commonly provided at the base of Californian bridge columns designed prior to 1971. Tests indicate that this is insufficient to enable the flexural strength of the column to develop [C2]. It should be pointed out that even much longer lap splices, satisfying current code requirements, are likely to fail if the column is subject to moderate ductility levels. This is discussed further in Section 5.5.4. Figure 1.13 shows damage to the base of a column, attributable to lap-splice bond failure, which occurred in the 1989 Loma Prieta earthquake [E1]. Inadequate flexural strength may also result from butt welding of longitudinal reinforcement close to maximum moment locations. It appears difficult to ensure the strength and ductility of welds, and indeed, these may not seem critical if an elastic design approach is used. Figure 1.14 shows an example of flexural failure of a column of the Hanshin expressway in the 1995 Kobe earthquake [P2], caused by failure of a large number of butt welds at the same location, close to the column base. Tension shift effects [see Section 5.3.4(*a*)] result in peak reinforcement strains being almost constant for a height above the column base equal to half the column diameter. The failure depicted in Fig. 1.14 was one of a large number

FIG. 1.13   Bond failure of lap splices at column base, 1989 Loma Prieta earthquake.

FIG. 1.14   Weld failure of column longitudinal reinforcement, 1995 Kobe earthquake.

of columns (at least 50) in the Hanshin expressway, where weld failure contributed to column failure.

(iii)   Inadequate Flexural Ductility.   Despite the higher-than-anticipated flexural strength of existing reinforced concrete bridge columns, this strength is still frequently very much less than that required for elastic response to expected seismic intensities, as noted above. The consequence is that to survive intense seismic attack, structures must possess *ductility.* Ductility, sometimes termed *toughness,* is the property of being able to deform through several cycles of displacements much larger than the yield displacement, without significant strength degradation. The definition and importance of ductility are discussed in more detail in Section 1.3. Displacement ductility factors (multiples of yield displacement) as high as $\mu_\Delta = 6$ or 8 may be needed in some cases.

At displacement ductility levels of about 2 to 3, concrete compression strains in the plastic hinge regions exceed the unconfined compression strain capacity, and spalling of the cover concrete occurs. Unless the core concrete is well confined by close-spaced transverse hoops or spirals, crushing rapidly extends into the core, the longitudinal reinforcement buckles, and rapid strength degradation occurs, resulting eventually in an inability to support gravity load. This behavior may be accelerated when transverse reinforcement is lap spliced in the cover concrete, as is common with older bridges. When the cover concrete spalls, the hoops lose effectiveness at the lap locations.

Figures 1.15 and 1.16 show examples of flexural plastic hinge failures in recent earthquakes. The low levels of transverse reinforcement present in these hinges should be noted. In the example of Fig. 1.16 from the 1994

FIG. 1.15 Confinement failure at column top, 1971 San Fernando earthquake.

FIG. 1.16 Flexural plastic hinges in columns captured by a connecting wall, Bull Creek Canyon Channel bridge, 1994 Northridge earthquake.

Northridge earthquake, close-spaced transverse reinforcement had been placed over a distance equal to the column diameter from the column base, but a wall connecting the columns over the bottom region of the columns forced the plastic hinge to develop immediately above the wall, where vertical spacing of transverse reinforcement had been increased.

(iv) Premature Termination of Column Reinforcement. In Japan, a number of bridge columns developed flexure–shear failures at column midheight during the 1982 Urahawa-ohi [K1] and 1995 Kobe [P2] earthquakes, as a consequence of premature termination of the column longitudinal reinforcement. An example is shown in Fig. 1.17, where the flexure–shear failure apparent corresponds to the bar cutoff location at column midheight. Bar termination was based on the design moment envelope, without accounting for the effects of tension shift due to diagonal shear cracking [see Section 5.3.4(*a*)]. This deficiency was exacerbated by the provision of a short development length of bars lap spliced at this location. The effects of rotational inertia increasing the moment at column midheight (see Section 4.3.1(*a*)] may also have been significant in this case. Failure of the 18 columns of the collapsed section of the Hanshin expressway in the 1995 Kobe earthquake was also initiated by premature termination of 33% of the column longitudinal reinforcement at 20% of the column height. This forced the plastic hinge to form above the base, where it could not benefit from confinement provided by the strong footing, which was essential for survival of the columns, since very little

FIG. 1.17 Flexural–shear failure at pier midheight of Route 43/2 overpass, due to premature termination of longitudinal reinforcement; 1995 Kobe earthquake.

confinement reinforcement was provided. This dramatic failure is shown in Fig. 1.18.

(b)   Column Shear Failures.   The shear strength of concrete column sections results from a combination of mechanisms involving concrete compression shear transfer, aggregate interlock along inclined flexure–shear cracks, arching action sustained by axial forces, and truss mechanisms utilizing horizontal ties provided by transverse reinforcement. Shear resistance is discussed in some detail in Section 5.3.4(*b*).

Shear mechanisms interact in a complex fashion. If the transverse reinforcement of the truss mechanism yields, flexure–shear crack widths increase rapidly, reducing the strength of the concrete shear-resisting mechanisms utilizing aggregate interlock. As a consequence, shear failure is brittle and involves rapid strength degradation. Inelastic shear deformation is thus unsuitable for ductile seismic response.

Short columns are particularly susceptible to shear failure as a consequence of the high shear/moment ratio and conservatism in the flexural strength design of older columns. As noted above with reference to Fig. 1.12, actual flexural strength of existing columns will be many times the design strength if designed on the basis of elastic theory. Shear strength equations for column design, when utilized at all, were generally less conservative than flexural design equations in older bridges. In California it is common to find the

FIG. 1.18   Flexural failure above column base of columns of the Hanshin expressway, due to premature termination of longitudinal reinforcement and inadequate confinement; 1995 Kobe earthquake.

transverse reinforcement of older bridge columns to consist of No. 4 bars (12.7 mm diameter) spaced vertically at 12-in. (305-mm) centers, regardless of column size or shear force, leading to the suspicion that shear design was not considered essential. Prior to 1970, there was also a lack of appreciation of the need to ensure that actual shear strength exceeded actual flexural strength (one of the tenets of the capacity design philosophy discussed subsequently in Section 1.3), and as a consequence it is not uncommon to find bridge columns where flexural strength may be two to three times the shear strength.

An example is provided by the I-5/I-605 separator, a major freeway bridge structure severely damaged in the 1987 Whittier earthquake [P1]. Analysis of the columns that failed in shear indicated that shear strengths were only about 30% of the flexural strength. It is important to realize that it is the actual strength that will be developed under seismic attack unless the actual strength exceeds the elastic response level. Figure 1.19 shows damage to one of the columns of this bridge. Several transverse ties fractured, the shear crack widths exceeded 25 mm (1 in.), and virtually no shear strength remained after the earthquake. Bowing of the longitudinal column reinforcement like the staves of a barrel, apparent in Fig. 1.19, indicated that axial shortening of the column occurred and collapse was imminent.

Shear failures also occurred extensively in the 1971 San Fernando [C1], 1994 Northridge [E2], and 1995 Kobe [P2] earthquakes. Figure 1.20, from the

FIG. 1.19 Brittle shear failure of column of the I-5/I-605 separator, 1987 Whittier earthquake.

FIG. 1.20   Shear failure outside plastic hinge region, 1971 San Fernando earthquake.

San Fernando earthquake, is typical of brittle shear failure where flexural strength exceeds shear strength. There is no indication that plastic hinging developed at the member ends. In contrast, the column in Fig. 1.21, also from the San Fernando earthquake, has no apparent damage in the midregion, but a plastic hinge has clearly formed at the top of the column, with subsequent shear failure within the hinge region. This leads to an important observation: shear strength in plastic hinge regions is less than in nonhinging regions. This is a consequence of a reduction in aggregate interlock shear transfer in plastic hinges as flexure–shear cracks increase in width under the action of flexural ductility. Design equations should thus recognize an influence of flexural ductility demand on shear strength, an approach developed in Section 5.3.4(*b*)(iii).

The cause of failure of six of the seven bridge structures that failed in 1994 Northridge earthquake has been attributed to column shear failures [P5]. Because of failure of transverse reinforcement, column shear failure often results in a loss of structural integrity of the column, with subsequent failure under gravity loads. Examples of column collapse caused by shear failures are shown in Fig. 1.22. Because of the sudden and brittle nature of shear failure, special efforts must be taken in new and retrofit designs to guard against it.

FIG. 1.21 Shear failure within plastic hinge region, 1971 San Fernando earthquake.

(a)   (b)

FIG. 1.22 Examples of column shear failures, 1994 Northridge earthquake. (a) I-10 Freeway at Fairfax/Washington undercrossing; (b) I-118 Mission/Gothic undercrossing.

1.2.4 Cap Beam Failures

Cap beam failures caused by the 1989 Loma Prieta earthquake and subsequent seismic assessments of existing concrete bents indicate significant deficiencies in three areas: (1) shear capacity, particularly where seismic and gravity shears are additive; (2) premature termination of cap beam negative moment (top) reinforcement; and (3) insufficient anchorage of cap beam reinforcement into the end regions. A further deficiency, involving the shear strength of the joint regions, is discussed in Section 1.2.5. While the first two problems were encountered primarily in outrigger bent caps, the third deficiency has been found to be widespread in many older multicolumn bents.

An example of damage to outrigger cap beams as a consequence of the mechanisms illustrated in Fig. 1.1 is shown in Fig. 1.23 and refers to a bent of the China Basin viaduct on I-280 in San Francisco, damaged in the 1989 Loma Prieta earthquake [E1]. Negative moments under combined gravity load and seismic force extended beyond the region where all cap beam top reinforcement was terminated. The resulting flexural cracks, which developed at the location of the final top bar termination, were steeply inclined under the influence of shear, as is evident in Fig. 1.23. The shear reinforcement was inadequate to control the flexure–shear crack, and very large crack widths developed, nearly resulting in collapse of the bent.

FIG. 1.23 Cap beam flexure and shear and joint failures, I-280 China Basin viaduct, 1989 Loma Prieta earthquake.

Damage resulting from inadequate and poorly developed cap beam bottom reinforcement, also discussed in reference to Fig. 1.1, was evident in a number of cases in the Loma Prieta earthquake. Figure 1.24 shows an example from an uncollapsed portion of the Cypress viaduct, where a wide flexural crack has formed at the inner face of a column as a result of the inadequate anchorage provided by the 900-mm (3-ft) development length of No. 18 bars (57 mm diameter) into the joint region. While there exists a large discrepancy between different design codes for required development length, the 15 bar diameters provided was clearly insufficient. Insufficient anchorage of cap beam reinforcement was also clearly contributory to the failure of the Cypress viaduct, as evidenced by Fig. 1.25, which shows top and bottom reinforcement of the upper cap beam exposed after the column has been wrenched free of the reinforcement.

1.2.5 Joint Failures

An important consequence of the 1989 Loma Prieta earthquake was the focus on joint shear problems in the connection between cap beams and columns. Transfer of member forces through connections result in horizontal and vertical joint shear forces that may be many times the shears in the connected members. Analytical aspects related to determination of, and design for, joint shears are considered in some detail in Section 5.4. It has been uncommon

FIG. 1.24 Cap beam positive moment cracks at inner column face, 1989 Loma Prieta earthquake.

FIG. 1.25   Anchorage failure for cap beam reinforcement, Cypress viaduct, 1989 Loma Prieta earthquake.

for these shear forces to be considered in bridge design, and until recently, properly designed joint shear reinforcement was almost never provided. Without this reinforcement, joint shear failure may occur, as evidenced by the diagonal cracking of the joint region of a Cypress viaduct bent shown in Fig. 1.26(a) and by the crack pattern in the joint region of a bent of the Embarcadero viaduct (I-480) depicted in Fig. 1.26(b). Joint shear failure is generally accepted as the major contributor to the collapse of a 1.6-km (1-mile) length of the Cypress viaduct during the 1989 Loma Prieta earthquake, with the tragic loss of 43 lives.

Although double-deck viaducts such as the Cypress viaduct are especially susceptible to joint shear problems, single-level multicolumn bents may also be at risk. A review of damage in the 1971 San Fernando earthquake [F1] reveals clear evidence of joint shear failure in at least one case [P6]. Several cases of joint shear failure of outrigger bent knee joints occurred in the Loma Prieta earthquake. The example of cap beam flexure–shear failure of Fig. 1.23 also shows clear evidence of joint shear failure, with a diagonal crack propagating from the outside top corner of the joint.

A further example of knee-joint shear failure, of bent 38 of the I-980 connector in Oakland California, damaged in the Loma Prieta earthquake, is shown in Fig. 1.27. Although this bridge was designed and constructed in the 1980s in accordance with current bridge design requirements, several undesirable features are apparent. Joint reinforcement consists of very light transverse spirals, which had the capacity to resist less than 10% of the horizontal joint shear force. Column reinforcement was anchored by straight-bar

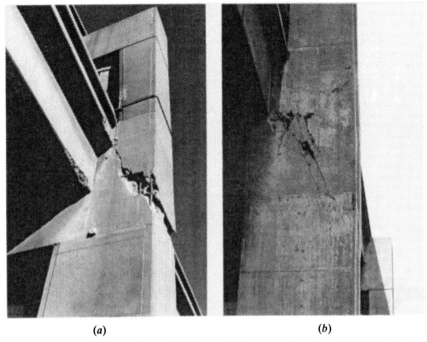

(a)   (b)

FIG. 1.26 Joint shear failure, 1989 Loma Prieta earthquake. (a) I-880 viaduct; (b) Embarcadero viaduct.

extensions into the joint, and the mechanism for transfer of moment from column to cap beam is not obvious, particularly under moments that tend to open the joint. Under negative (closing) moments, cap beam top reinforcement can be expected to yield back to the start of the 90° hook down into the column, because of the lack of horizontal or vertical shear reinforcement near the top of the joint region, thus transferring the full development force to the hook. The vertical extensions of these hooks are inadequately restrained back into the joint to enable this transfer to take place without the hooks tending to straighten.

Under seismic attack, diagonal shear cracks developed corresponding to both opening and closing joint moments. The bent top beam bars in the cap beam attempted to straighten, as noted above, and one bar, visible in Fig. 1.27(a), fractured at the start of the bend. Testing of an exact scale model of the joint duplicated the joint damage but not the bar fracture, and indicated that the joint probably resisted a maximum moment of less than 60% of the nominal strength of the connecting members [I2]. It is felt that the bar fracture was the result of a number of contributing factors, including the high plastic strain (about 6%) induced by bending the 57-mm (No. 18) bars to a 450-mm radius, strain-aging effects [P8], and

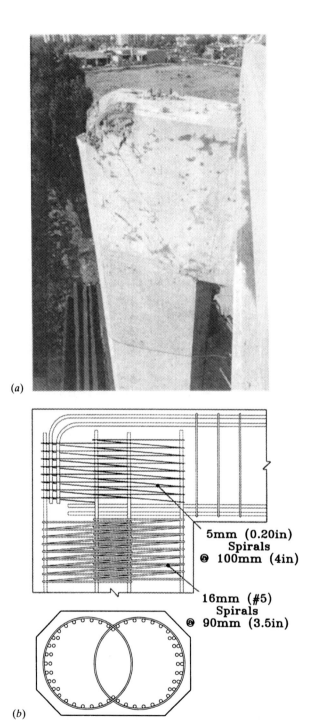

(a)

(b)

5mm (0.20in)
Spirals
@ 100mm (4in)

16mm (#5)
Spirals
@ 90mm (3.5in)

FIG. 1.27 I-980 joint failure (bent 38), 1989 Loma Prieta earthquake. (a) General view; (b) Reinforcement configuration.

increased strain in the edge bar, compared with interior bars, as a result of its hook location (see Fig. 1.27) and the effects of cap beam flexure and torsion under longitudinal response.

1.2.6 Footing Failures

There have been comparatively few reported incidences of footing failures caused by seismic actions. Several reasons for this are possible: maximum feasible footing forces may not have developed as a consequence of premature failure of columns in flexure or shear; other mechanisms, particularly rocking of the footing, may have protected the footing from excessive seismic forces; footings are often situated some distance subgrade and may not have been adequately inspected for damage after earthquakes. Despite the lack of reported damage, analyses of typical details common in earlier footing designs reveal deficiencies in (1) footing flexural strength (particularly due to the common omission of a top reinforcement mat); (2) footing shear strength, since footing shear reinforcement was rarely provided; (3) joint shear strength in the region immediately below the column, which is subjected to high shear forces, as with the cap beam–column connections discussed above; (4) anchorage and development of column reinforcement, which in the past has typically been provided by straight-bar extensions down into the footing or by 90° hooks bent out away from the column axis, a practice that exacerbates joint shear problems; and (5) inadequate connection between tension piles and footing. A dramatic example of the consequences of inadequate development of column longitudinal reinforcement is provided by Fig. 1.28. This bridge bent, which failed in the 1971 San Fernando earthquake, had straight-bar anchorage of column reinforcement in a footing reinforced solely with a bottom two-way flexural mat of reinforcement. Flexural failure of the essentially unreinforced footing is likely to have contributed to this anchorage failure. Design of footings to avoid these deficiencies is covered in Section 5.6.

1.2.7 Failures of Steel Bridge Components

There is a general perception that steel bridge components are less susceptible to bridge damage than are equivalent concrete components. Although it is true that steel superstructures will be lighter than equivalent concrete superstructures (typically, by about 30%), this does not mean that damage will not occur to the superstructure or its supporting members. In the 1995 Kobe earthquake there were many examples of buckling of steel I-beam bridge girders as a result of inadequate bracing (e.g., Fig. 1.29). Steel columns have often been suggested as a preferred alternative to concrete columns. However, the complete failure of a number of steel columns (e.g., Fig. 1.30) and the

FIG. 1.28 Pullout failure, 1971 San Fernando earthquake.

FIG. 1.29 I-beam buckling, 1995 Kobe earthquake.

FIG. 1.30   Steel column collapse, 1995 Kobe earthquake.

buckling of many more in the 1995 Kobe earthquake confirm experimental evidence that such columns have little ductility capacity.

1.3   DESIGN PHILOSOPHY

1.3.1   Strength Design Versus Elastic Design

Strength design philosophy has largely superseded elastic design to specified stress levels as the preferred basis for the design of reinforced concrete buildings [P4]. The change to strength design has been less complete for bridges, where for service loads, design to stress limits is still the most common design basis in most countries.

The relative merits of elastic and strength design for service loads on bridge superstructures can, perhaps, be argued without a clear conclusion. It is evident, however, from the discussion in Section 1.2 that the use of elastic design promotes a false sense of the response levels to be expected under seismic attack and typically results in severely underestimated displacements. It also encourages designers to ignore aspects of ductility and the provision of a rational hierarchy of strengths. Further, the moment patterns resulting from the combination of reduced seismic and full gravity actions can be serious distortions of the actual patterns that can be expected to develop at full strength.

It is thus essential that strength design be adopted as the basis for all aspects of seismic design. It will be assumed in this book that the reader is familiar with the basics of strength design for flexure. Some review of the appropriate material is included in Chapter 5, but readers who need more detailed background are encouraged to refer to one of the many basic texts covering the topic (e.g, [P3,M2]).

1.3.2 Aspects of Ductility and Energy Dissipation

It is now well known that structures with less strength than that corresponding to elastic response to a given earthquake can nevertheless survive with little or no apparent damage. This behavior is generally attributed to the ability of the structures to deform inelastically without significant strength loss, through a number of cycles of displacement response. If the level of inelastic deformation required is sufficiently low, inelastic strains induced in critical parts of the structure (plastic hinges) may be such that the damage is insignificant and appears to be representative more of elastic than of inelastic response. For example, the reinforced concrete bridge pier shown in Fig. 1.31(a) may respond transversely under a moderate earthquake, achieving its expected strength. As shown in Fig. 1.31(b), elastic response to the excitation would require a base shear strength of V_{me}. The actual response corresponds to ductile response with base shear force V_{md} and a peak displacement Δ_M at the center mass which may be rather similar to that for the equivalent elastic response, in

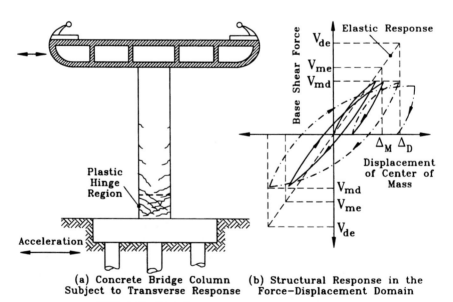

(a) Concrete Bridge Column   (b) Structural Response in the
Subject to Transverse Response   Force–Displacement Domain

FIG. 1.31 Ductile response of a concrete bridge.

accordance with the *equal-displacement approximation*. Inspection of the pier after the earthquake reveals only hairline cracks. This is because the peak compression strains in the plastic hinge region at the base of the column were smaller than the crushing strain. Although some of the longitudinal reinforcement in the plastic hinge region may have sustained peak strains many times the yield strain, the residual cracks are small after seismic excitation ceases because of the influence of the gravity axial compression load tending to close the cracks and the softening of the yielded reinforcement for the reversed response direction as a result of Bauschinger effects.

Although not elastic, the response has satisfied performance criteria for what might be termed a serviceability-level earthquake, one that might be expected to occur once or more within the expected design life of the bridge. Within these performance criteria, it is thus possible to design for a base shear strength that is perhaps less than 50% of the true linear elastic response level.

Under more intensive excitation, the bridge pier responds at essentially the same base shear strength since the flexural capacity has already been achieved but is required to sustain larger displacements as shown in Fig. 1.31(b) by the linked response curve. At these larger displacements, concrete compression strains within the plastic hinge region exceed the crushing strain, resulting in spalling, and steel strains are of sufficient magnitude so that the flexural cracks within the hinge region remain comparatively wide after excitation ceases. Provided that the reinforcement of the hinge region is designed properly, the core concrete remains competent, and post-earthquake repair is possible by reinstating the spalled cover concrete and perhaps sealing the flexural cracks by epoxy injection grouting.

This level of response is visibly inelastic but will be acceptable for many bridges, provided that the probability of occurrence during the design life of the bridge is sufficiently low. In this case the equivalent elastic strength required would be V_{de} [see Fig. 1.31(b)], and the reduction in the design level of base shear force given by the ratio V_{de}/V_{md} may be as large as 5 or so. The penalty in not designing to the elastic response level is that the possibility of minor, repairable damage must be accepted during the design life of the bridge. The benefit is clearly reduced cost.

It is obvious that if sufficient strength is built into the pier and into the supporting pile structure of Fig. 1.31(a), the bridge could be designed to respond elastically. The penalty for this would be the high initial cost associated with providing the extra strength. A more economical solution might be to design for a lesser strength than V_{de} but with strains corresponding to the serviceability limit discussed earlier. The choice between designing to force levels corresponding to serviceabililty limits and force levels corresponding to ductile response with controlled damage will be a matter of economics and the importance of keeping the bridge open immediately after a major earthquake. For critical lifeline structures, where alternative routes are not available into an earthquake-affected region for rescue operations or to key hospitals that will be needed to treat injuries after an earthquake, the importance of serviceability will be extremely high.

It should be noted that there may be additional problems associated with elastic response as well as the high initial cost. Design elastic lateral response levels may exceed the acceleration of gravity ($1.0g$), resulting in extremely hazardous driving conditions. Coupled with the high levels of response displacement possible, severe accidents and loss of life may result that would not eventuate in a bridge designed for lower lateral acceleration levels with ductile response. In a long elevated viaduct, this could also result in traffic disruption that could take hours or even days to clear, thus eliminating the advantage of structural serviceability after the design-level earthquake.

Satisfactory response of the bridge in Fig. 1.31(a) relies on the capacity of the column to displace inelastically through several cycles of response without significant degradation of strength or stiffness, a quality termed *ductility*. If the strength or stiffness degrade excessively, the response displacements increase significantly beyond those corresponding to elastic response, and the structure may collapse. Relationships between strength, stiffness, energy, dissipation, and deformation are discussed in some detail in Section 5.3.1.

Ductility is often defined mathematically as the ratio of deformation at a given response level to deformation at yield response. Thus in relation to the base shear:displacement relationship of Fig. 1.32, response is idealized by an equivalent bilinear curve by extrapolating the elastic response up to the expected strength V_d to obtain the yield displacement Δ_y. If the maximum

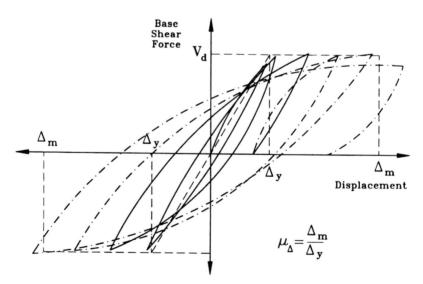

FIG. 1.32 Definition of displacement ductility.

displacement expected during the design-level earthquake is Δ_m, the maximum expected displacement ductility factor is defined as

$$\mu_\Delta = \frac{\Delta_m}{\Delta_y} \tag{1.1}$$

Although the ductility ratio defined in Eq. (1.1) refers to displacements, curvature ductility ratios relating to maximum and yield curvatures at critical sections are also frequently defined, and other measures of ductility, including rotational ductility and strain ductility, are sometimes useful indicators of the level of inelastic response. As discussed in some detail in Section 5.3.2, curvature ductility relates to the response of an individual section and depends on its reinforcement and axial load level. Displacement ductility relates to overall structural response. The relationship between the curvature and displacement ductility factors depends on structural geometry and is of considerable importance in determining safe levels of inelastic displacement for the structure as a whole.

The area within the inelastic base-shear:displacement response curve of Fig. 1.32 is a measure of the energy-dissipating capacity of the structure. Sometimes termed *hysteretic damping,* this can in turn be related to the effective damping of the response (see Section 4.3.1(c). For short-period structures, high hysteretic damping is very important, since the maximum response displacements are very sensitive to the damping. For longer-period structures, the energy-dissipating characteristics are less important.

An example of ductile flexural response obtained from a well-confined and detailed bridge column tested in laboratory conditions is shown in Fig. 1.33(a) in terms of a lateral force–displacement hysteresis plot. In Fig. 1.33, lateral force levels V_y and V_i correspond to the first yield of longitudinal reinforcement and ideal flexural strength, which is the flexural strength calculated based on measured material properties. The excess strength apparent in Fig. 1.33(a) over V_i is a result of strain hardening of longitudinal reinforcement at high ductility levels.

1.3.3 Capacity Design Principles

The performance illustrated in Figs 1.31(b), 1.32, and 1.33(a) is characteristic of structural response dominated by inelastic flexural yielding of carefully detailed plastic hinges. This requires significant amounts of closely spaced and well-anchored transverse reinforcement to enable high compression strains to be developed within the core concrete after spalling of the cover concrete. The transverse reinforcement also restrains the longitudinal reinforcement against buckling. Aspects related to the design of reinforcement for plastic hinges are covered in some detail in Section 5.3.2.

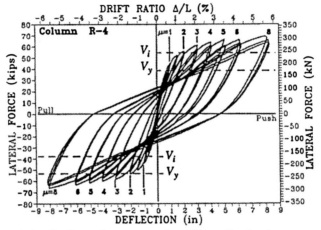

(a) ductile flexural response of a well-confined column

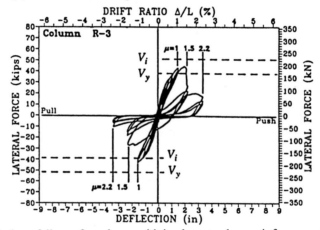

(b) shear failure of a column with inadequate shear reinforcement

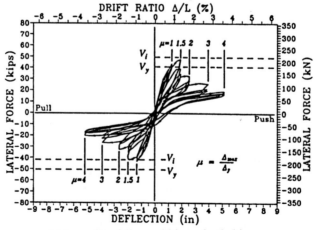

(c) lap splice failure within a plastic hinge

FIG. 1.33 Various modes for inelastic response of a bridge column.

34

With seismic design of building frames it is generally required that plastic hinges be located at the ends of beams and that columns be protected against inelastic response. This is primarily to ensure against the formation of soft-story sway mechanisms in multistory frames [P4]. In bridges it is generally neither practical nor desirable to provide for plastic hinge formation in the superstructure, and hence column hinges are typically chosen as the site for inelastic deformation.

To ensure that ductile inelastic flexural response is achieved, it is essential that nonductile deformation modes be inhibited. For example, if the shear strength of a bridge column is developed at less than the flexural strength, very brittle response can result, as indicated by Figs. 1.19 to 1.22 and the hysteresis loops of Fig. 1.33(b), where strength and stiffness can be seen to degrade rapidly once the initial shear strength has been developed, and the ideal flexural strength is not achieved. Bond failure of reinforcing bar lap splices in critical regions can also cause rapid strength degradation, as seen in Fig. 1.33(c). If regions of the bridge that have not been designed and detailed for flexural ductility unexpectedly become plastic hinge locations, the structural ductility capacity will be limited by the reduced integrity of the actual plastic hinges compared with the intended plastic hinges.

It is thus necessary to ensure an adequate margin of strength between nonductile failure modes and the designated ductile mode of deformation. This is the basis of capacity design. As will be clear from examination of Fig. 1.31(b), the consequence of the flexural strength of the plastic hinge regions exceeding the design strength is that the higher, actual strength will be attained, unless this exceeds the equivalent elastic response level. Since this is typically many times the design strength, it is uncommon for the latter limit to be achieved. Design of the flexural plastic hinges is typically based on conservative estimates of material properties and allowable strain limits, according to a relationship of the form

$$\phi_f M_n \geq M_r \qquad (1.2)$$

where M_r in the required flexural strength, M_n the calculated (nominal) flexural strength based on low estimates of expected material strengths, and ϕ_f a flexural strength reduction factor. As discussed in Section 5.3.1, use of a flexural strength reduction factor is of doubtful relevance to ductile seismic design. The actual response level will typically exceed M_r by a significant margin as a result of material strengths exceeding the nominal strengths, use of the flexural strength reduction factor, and the effects of increases in reinforcement stress above the design limit (normally yield stress), at high levels of ductility, as a consequence of strain hardening.

In systems where the dynamic response may include several modes of vibration, it is also possible that the forces may be amplified above those resulting from the design assumption of the seismic force distribution, and the actual strength of the plastic hinge region, as a consequence of higher-mode response effects. This effect is generally more significant in buildings than in bridges. A rather complete review of higher-mode effects is given in [P4].

A general relationship between the strength of a member or action that must be protected against undesirable inelastic action is given by the basic capacity design equation, which may be written as

$$\phi_s S_n \geq S_r = \omega \phi^\circ S_m \qquad (1.3)$$

where S_m is the design required strength of the action to be protected, corresponding to the nominal strength of the plastic hinge or hinges; ϕ° is a strength amplification factor relating the maximum feasible flexural capacity of the plastic hinge to the nominal strength; ω is a dynamic amplification factor accounting for increases in the design action as a result of higher-mode dynamic effects; and ϕ_s, the strength-reduction factor appropriate for action S, reflects the probability and consequences of the strength of action S being less than the nominal strength.

Discussion of the appropriate numeric values for the variables in Eqs. (1.2) and (1.3) is included in Chapters 5 and 6. However, the process may be clarified by consideration of an illustrative example, based on the structure of Fig. 1.34, which consists of regular spans of circular columns with a monolithic connection to the superstructure. Thus under transverse response the bridge acts as a simple vertical cantilever, while under longitudinal response, the column is subjected to double bending. After initial sizing of the column and determining mass, stiffness, and hence the natural response periods for transverse and longitudinal directions, the required seismic design column

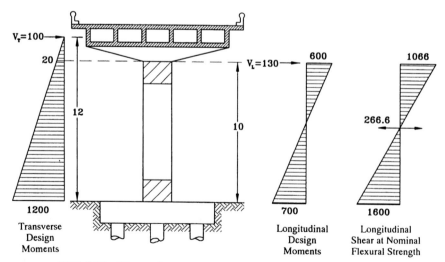

FIG. 1.34   Illustrative example for the capacity design process.

shear forces $V_T = 100$ and $V_L = 130$ are calculated from a response spectrum analysis, using techniques described in Section 4.3.3, and are reduced by a factor of 5 from the elastic response levels, to account for expected ductility capacity. The different shears for the two directions result from increased relative stiffness in the longitudinal direction. These numbers have no units in this example, which is purely to illustrate the capacity design process.

Moments induced by these column shears are shown in Fig. 1.34. The required capacity of $M_{br} = 1200$ at the column base is governed by transverse response, while at the top of the column longitudinal capacity governs, with $M_{tr} = 600$. Actual flexural capacity provided at the column base is calculated from Eq. (1.2) using a flexural strength reduction factor $\phi_f = 0.75$ and nominal material strengths. We will assume that an exact balance is achieved between the required strength and the dependable strength. At the top of the column the flexural capacity has been reduced by terminating 50% of the column reinforcement at about two-thrids of the column height. This was the maximum that could be terminated without violating the minimum reinforcement requirements for the column. As a consequence of the influence of the axial compression on the column, the flexural strength is not reduced in proportion to the reduction in longitudinal reinforcement, and a nominal capacity of 1066 units results. Using $\phi_f = 0.75$, this corresponds to a dependable capacity of 800 units, which is 33% more than the value required. The nominal flexural strengths at the top and bottom of the column, also shown in Fig. 1.34, are thus 1066 and $1200/0.75 = 1600$ units, respectively.

The column must now be designed for shear, using capacity design principles. It will be obvious that the critical direction will be longitudinal response, which has so far not governed the column design. The longitudinal shear corresponding to development of nominal flexural strength at column top and bottom is found to be $V_{ld} = 266.6$ units, or more than twice the value calculated from the response spectrum analysis but still less than the elastic response value of $5 \times 130 = 650$ units.

Equation (1.3) must now be utilized to account for the possibility of material overstrength and strain hardening. This is related primarily to reinforcing steel characteristics, described in Section 5.2.2. The ratio of maximum feasible yield strength to the design nominal strength of the reinforcement is 1.3. This corresponds, for example, to U.S. grade 60 reinforcement, which has a nominal yield strength of 60 ksi (414 MPa) and a maximum permitted yield strength of 78 ksi (538 MPa). A further allowance of 10% is made for strain hardening and the influence of enhanced concrete compression strength, giving $\phi° = 1.43$. This allowance is lower than might often be appropriate, as a consequence of the rather low ductility levels actually expected in the longitudinal direction.

Referring again to Eq. (1.3), there is no dynamic amplification factor in this case, since the structure essentially responds as a single-degree-of-freedom system in the longitudinal direction. Hence $\omega = 1.0$. Assuming a shear strength-reduction factor of $\phi_s = 0.85$, the required nominal shear strength V_n is found from

$$0.85V_n > 1.0 \times 1.43 \times 266.6$$

Hence the minimum value for nominal shear strength is 448.5 units. This is still less than the elastic response level but is 3.45 times the initial calculated required shear strength for the longitudinal direction. Failure to provide this level of shear strength could, in theory, result in brittle shear failure under the design level of seismic response.

The example just completed represents a strict interpretation of the capacity design approach, in conjunction with typical strength design principles, as currently adopted in the United States and in many other countries. It will be seen that the results appear to be very conservative. A less conservative approach will be recommended in this book, where the full protection of capacity design is provided but relaxations in the conservatism of strength design are adopted.

1.3.4 Strength Definitions

It is apparent from the example in the preceding section that several different measures of strength must be considered in the design process. The number of possibilities is further compounded by the use of different approaches for strength design used in different countries. It is therefore necessary to define the appropriate terms and to form relationships between them. This is carried out below using the example of Fig. 1.34 to illustrate the relationships.

(a) Required Strength. Required strength is the strength S necessary to satisfy levels of the appropriate design action (e.g., flexure, shear, etc.) based on structural analysis. When the action considered is the flexural strength of the designated plastic hinges, the required strength results directly from the seismic analyses, using the appropriate representation of the design earthquake, in conjunction with the gravity loads. Thus in the example, the required strength for the column base moment is 1200 units.

When the action considered is to be protected against inelastic response by capacity principles, the required strength is the end result of the capacity analysis process, represented by the right-hand side of Eq. (1.3). Thus for shear design of the column, the required strength is $1.0 \times 1.43 \times 266.6 = 381.2$ units. The symbol S_r is used for required strength throughout this book. In terms of moments or shears, this appears as M_r or V_r, respectively.

(b) Nominal Strength. Nominal strength, sometimes termed *characteristic strength,* is the strength of action S based on section or member analysis using conservative estimates of material properties. The symbol S_n will be used. The relationship between nominal strength and required strength is expressed by

$$\phi_s S_n > S_r \tag{1.4}$$

The value of material properties adopted in calculating S_n varies among countries and design codes, as do the equations for predicting strength from the material properties. As a consequence, the appropriate values for ϕ_s are also variable. In Europe and New Zealand it is common to base S_n on characteristic material strengths, which represent the lower 5 percentile estimate of actual strength based on specified strength. In Japan and on the American continent it is common to use a lower 5 percentile estimate for concrete compression strength, while the reinforcement yield strength is taken as the absolute minimum permitted for the grade. In the example of Fig. 1.34, the minimum nominal flexural strength at the top of the column is found by inverting Eq. (1.4) as $M_n = 600/0.75 = 800$ units. In fact, a rather higher design strength was provided, as noted earlier.

(c) **Expected Strength.** It is also possible to design on the basis of expected or mean strength. There is some justification for adopting this for seismic design, since better estimates of ductility capacity and deformations will result. Lower values of the strength-reduction factor might then be appropriate. Appropriate values of material strength for seismic design are discussed in more detail in Section 5.3.1.

(d) **Dependable Strength.** Dependable strength is the strength of the action that has a sufficiently high probability of being exceeded in the design earthquake to satisfy performance criteria and is given by the left-hand side of Eq. (1.4). In Europe this is generally termed the *design strength.*

(e) **Extreme Strength.** Extreme strength is the strength of the action that has a sufficiently low probability of being exceeded in the design earthquake to satisfy the performance criteria and is sometimes termed the *overstrength* of the action. This normally relates to the maximum feasible flexural strength of the critical section in a potential plastic hinge. Extreme strength is related to the nominal strength by the relationship

$$S^\circ = \phi^\circ S_n \tag{1.5}$$

Thus the extreme strength of the column top moment capacity in the example of Fig. 1.34 is $M^\circ = 1.43 \times 1066 = 1524$ units. Clarification of the relationship between the various strength measures is provided by the typical frequency-strength distribution of Fig. 1.35.

(f) **Ideal Strength.** Ideal strength relates to experimental research results and is the best prediction of strength of a particular test unit using measured material properties. It thus corresponds to nominal strength but using measured rather than nominal material strengths. It is used primarily to assess the validity of equations used for predicting strength [see, e.g., Fig. 1.33(*a*)].

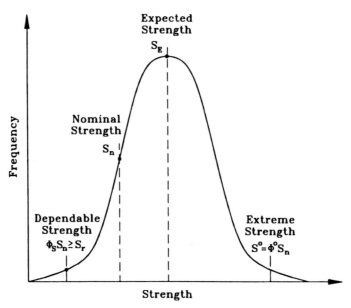

FIG. 1.35 Typical frequency–strength distribution and definition of measures of strength.

1.3.5 Application of Structural Reliability Theory to Seismic Design

It would appear that structural reliability theory should have significant application to seismic design because of the probabilistic nature of seismic action. In fact, although the method has promise, it has not yet been developed sufficiently to be fully incorporated into design methodology. Seismic design differs from design to almost any other action or load case in that the strength of the structure is expected to be developed in less than the design-level event. Thus the probabilistic base for structural reliability theory has little applicability to strength design for seismic action. In fact, it can be seen from examination of the force–deformation characteristics of Figs. 1.31 and 1.32 that the choice of design strength may be somewhat arbitrary, in that there is a trade-off between required strength and required ductility. This is expressed in different seismic design codes for bridges in widely varying ductility factors. It is however, at least philosophically possible as well as desirable to apply reliability theory to determining displacement design limits, but the uncertain nature and distribution of seismic excitation and the lack of precision in current methods for determining the displacement capacities of structures convince us that it will be some years before it can be applied realistically to seismic design. Nevertheless, it will be clear from the formulation of the various strength levels defined in the preceding section that reliability theory has an important role to play in calibrating the various factors involved, particularly the overstrength factor ϕ°.

1.3.6 Design and Response Limit States

There are a number of limit states of structural response that it is useful to consider in the design process. It has already been noted in Section 1.3.2 that a serviceability limit state may be defined which may correspond to less than the full design response level and for which no repair is expected. The term *limit state* is applied to both member and structure response in somewhat different fashions. A more complete description of member and structure limit states is provided in the following.

(a) Member Limit States

(i) Cracking Limit State. The onset of cracking generally marks the point for a significant change of stiffness, as shown in the typical moment–curvature relationship of Fig. 1.36(*a*). For critical members expected to respond in the inelastic range to the design-level earthquake, this limit state has little significance, as it is likely to be exceeded in even rather minor seismic excitation. The limit state may, however, be important for members that are expected to respond essentially elastically to the design-level earthquake. For example, the appropriate stiffness to be used for a prestressed superstructure under longitudinal response will depend on whether or not the cracking limit state is exceeded.

(ii) First-Yield Limit State. A second significant change in stiffness occurs at the onset of yield in the extreme tension reinforcement. This limit state is useful for defining the appropriate stiffness to be used in elastic analyses of ductile systems using simplified hysteresis rules such as elastoplastic or bilinear response.

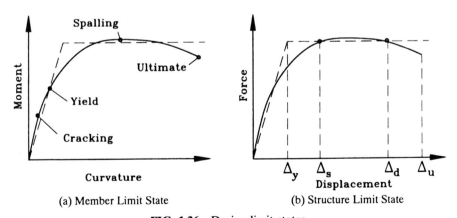

(a) Member Limit State   (b) Structure Limit State

FIG. 1.36 Design limit states.

(iii) *Spalling Limit State.* With unconfined sections, the onset of spalling of cover concrete may be a significant limit state, particularly for sections subjected to high levels of axial compressions, where spalling is typically associated with onset of negative stiffness and possibly sudden strength loss. Excedance of this state represents a local condition that can be expected to require remedial action. For well-confined sections this is likely to be the only significance of exceeding this limit state, since the section can be expected to support much larger deformations without excessive distress, and resistance may in fact continue to increase.

(iv) *Ultimate Limit State.* Definition of the ultimate limit state is somewhat subjective. It is sometimes taken to correspond to a critical physical event, such as fracture of confinement reinforcement in a potential plastic hinge zone, which results in a sudden drop in resistance and obvious physical deterioration. Another common definition is the state existing when the lateral resistance has decreased by a specified amount (20% is commonly used) from the maximum attained strength. Neither definition truly corresponds to an ultimate condition, since at least some residual strength is maintained for further increase of displacements. However, the occurrence of negative stiffness of the moment–curvature characteristic is cause for concern under dynamic response, since it implies redistribution of strain energy from elastically responding portions of the structure into the section with negative stiffness. Analyses and shake-table testing have shown this to be a potentially disastrous condition [A1].

(b) Structure Limit States

(i) *Serviceability Limit State.* As noted earlier, relatively frequently occurring earthquakes should not impair the functionality of the bridge. This condition may be taken as that occurring when traffic is stopped, or restricted after an earthquake, while remedial measures are undertaken. Spalling of concrete should not have occurred, and crack widths should be sufficiently small so that injection grouting is not needed. The displacement ductility factor at which this occurs depends on section properties, axial load level, and structure ductility, but an average value of about $\mu_\Delta = 2$ is typical, although lower values are considered appropriate in many countries. The acceptable probability for the occurrence of an earthquake corresponding to the serviceability limit state will depend on the importance of preserving functionality of the bridge. As a consequence, this limit state is sometimes referred to as the *functionality limit state.* Thus for standard highway bridges of no special importance, the serviceability limit state may be chosen to correspond to a level of shaking likely to occur on average once every 50 years. For a critical lifeline bridge, which requires a high

degree of protection to preserve functionality during the emergency period following an earthquake, an earthquake with a much lower probability of occurrence will be appropriate. It is not currently common practice to require specific evaluation of the serviceability limit state. It is clear, nevertheless, that this is an important design function for special bridges. As a consequence, specific recommendations are included in Section 5.3.2(d).

(ii)   Damage-Control Limit State. Under a design-level earthquake of reduced probability of occurrence compared to the serviceability limit state, a certain amount of repairable damage may be permissible. The damage may include spalling of cover concrete requiring cover replacement, and the formation of wide flexural cracks requiring injection grouting to avoid later corrosion problems. However, the essential aspect of response to this limit state is that the required repair should essentially be superficial. Fracture of transverse reinforcement or buckling of longitudinal reinforcement should not occur, and the core concrete in plastic hinge regions should not need replacement. Ground shaking of intensity likely to induce response corresponding to the damage-control limit state should have a low probability of occurrence during the expected life of the structure. With well-designed bridges, this limit state generally corresponds to displacement ductility factors in the range of $3 \leq \mu_\Delta \leq 6$, although an upper limit of $\mu_\Delta = 4$ would be felt to be appropriate in Europe.

(iii)   Survival Limit State. It is important that a reserve of capacity exists above that corresponding to the damage control limit state, to ensure that during the strongest ground shaking considered feasible for the site, collapse of the bridge does not take place. Protection against loss of life is the prime concern here and must be accorded high priority in the overall design philosophy. Extensive damage may have to be accepted under an earthquake corresponding to the survival limit state, to the extent that it may not be economically or technically feasible to repair the bridge after the earthquake. Demolition and replacement may be required.

Although the survival limit state is of critical concern, its determination has received comparatively little attention. It is almost a truism to state that this limit state corresponds to the condition when the bridge is no longer able to support its gravity loads and therefore collapses, but this is a very valuable and effective way of defining the limit state. Even when the lateral resistance of a critical section of the bridge has been substantially reduced, the structure may still be stable, since lateral response displacements essentially have an upper bound for any given seismic event. For example, the five columns of a critical bent of the I-5/I-605 separator, one of which is shown in Fig. 1.19, suffered shear failure during the 1987 Whittier earthquake [P1] and are esti-

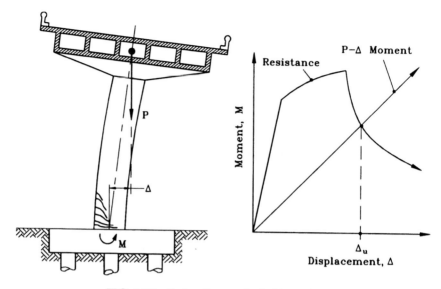

FIG. 1.37 P–Δ collapse of a bridge column.

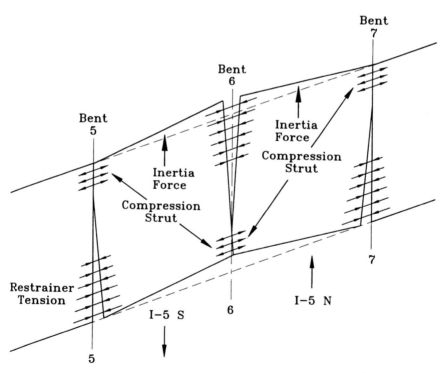

FIG. 1.38 Lateral resistance mechanism after shear failure of columns of the I-5/I-605 separator, 1987 Whittier earthquake.

44

mated to have had negligible residual strength after the earthquake. Despite this and the simple support provided for the two spans supported by the bent, failure did not occur, and the bridge was capable of withstanding an unusually strong aftershock two days after the main event.

Collapse of a structure will occur when gravity-load capacity is reduced below the level of existing gravity loads as a result of (say) total shear failure or disintegration of a column plastic hinge. Alternatively, collapse results from a stability failure, where the $P-\Delta$ moments exceed the residual capacity of the bridge columns, as shown in Fig. 1.37. If the ultimate displacement capacity assessed from the intersection of the resistance and $P-\Delta$ curves exceeds the maximum expected in the survival-level earthquake, collapse should not occur. In the case of the I-5/I-605 separator, the gravity-load capacity was just maintained and sufficient lateral displacement capacity was provided by a mechanism involving membrane action of the deck slab and cable restrainers across movement joints, as shown in Fig. 1.38, despite the total lack of resistance of the columns to lateral forces.

1.4   RETROFIT PHILOSOPHY

1.4.1   Prioritization of Retrofit

Seismic performance of bridges designed before modern bridge design codes were developed was described in Section 1.2. An understanding of the reasons leading to damage and collapse of these bridges under strong seismic excitation has led to greatly improved design methods, which can be expected to reduce damage of recent bridges to acceptable levels. However, the problem remains of the performance of existing bridges, designed by codes that are now known to provide inadequate safety, which may yet be subjected to intense shaking within their design life.

In many seismic regions, California in particular, there was a great expansion of freeway and highway systems in the 1950s and 1960s, before modern bridge seismic design codes had been developed. As a consequence, the stock of existing bridges designed to substandard seismic codes will exceed the number of new, well-designed bridges for many years. It can thus be argued that it is more important to develop methods for assessing and retrofitting existing bridges than to develop refinements to new codes, which may not be perfect but which provide reasonable safety. In fact, until recently, the amount of research and design effort put into the general area of seismic retrofit of bridges was insignificant. However, over the past 5 to 10 years, and in particular since the 1989 Loma Prieta earthquake [E1], very substantial advances have been made. As a consequence, seismic retrofit methods are considered in considerable detail in this book.

Given the large number of existing substandard bridges, it is clear that seismic upgrading cannot be advanced simultaneously in a uniform fashion over the entire bridge stock. It is thus important that a method for prioritization of existing bridges for retrofit be employed to determine which bridges are at greatest risk, and hence must be upgraded immediately, and which bridges may with little risk be left until later. The prioritization scheme should include consideration of a number of seismic, structural, and social issues. In general terms, these can be categorized as follows:

- *Site Seismicity.* The probability of occurrence of strong ground shaking is perhaps the single most important consideration.
- *Structural Vulnerability.* Different structural systems can be considered to be at higher risk of collapse than others, and thus deserve earlier attention.
- *Social Consequences of Failure.* In this category of factors, traffic density (which is an indicator of number of people at risk), importance of the structure to postearthquake rescue and emergency services, and availability of alternative routes, as a measure of economic impact of traffic disruption, need to be considered.

A prioritization scheme needs to be simple and should ideally not require structural calculations. Inevitably, such a scheme can only provide a rather coarse sieve, enabling sorting of those bridges that merit more detailed immediate attention. Prioritization models are discussed in detail in Section 7.1.

1.4.2 Assessment of Existing Structures

When, as a consequence of the prioritization process described above, a bridge is accorded high priority for retrofit considerations, the next process is a detailed evaluation of the expected performance under seismic attack. In this respect, comparing strength and reinforcement details with requirements of modern design codes is not likely to be very illuminating. Codes are by nature simple, conservative documents intended to provide adequate safety under a wide range of design conditions. The level of safety provided to new structures is expected to be adequate, but not uniform, in all cases. Assessment in accordance with such codes is thus not likely to provide definitive answers about structural safety. Further, design codes are based on a pass/fail approach. For example, the code may require a transverse volumetric ratio of confinement of 1.2% in a particular plastic hinge. If the bridge being assessed has 0.9% transverse reinforcement, it does not satisfy the code but may have adequate deformation capacity for the design-level earthquake.

The assessment procedure should thus be more precise than a code evaluation exercise. It should be capable of tracing the expected force–deformation response, noting when section or connection deterioration occurs, and determining the significance of this deterioration to safety. An early attempt at such a process was the assessment methodology developed by the Applied Technology Council [A2]. More recent assessment methodologies based on a more extensive analytical and experimental data base [P6] form the basis for the detailed procedures suggested in Chapter 7. Application of these procedures, which are based on a plastic collapse mechanism analysis, typically results in less conservative decisions about the need for retrofit for a given bridge than will result from application of earlier approaches based on the capacity/demand ratio approach or on code compliance evaluation.

Because of greater precision in analysis techniques used and greater certainty of member strengths resulting from materials testing, it will often be appropriate in the assessment procedures to accept reductions in capacity protection coefficients used to set the margins of strength of undesirable (i.e., shear failure) mechanisms over preferred (i.e., flexural hinging) mechanisms. There may also be a case for accepting higher probability of responding at the damage control limit state than would be the case for a new bridge. However, the probability of attaining the survival limit state should not exceed that considered appropriate for new bridges.

1.4.3 Retrofit Concepts

In recent years retrofit procedures have been developed to mitigate most of the problem areas identified in Section 1.2. Design approaches for the retrofit measures are presented in Chapter 8. Next we review briefly some of the more common methods for enhancing seismic performance.

(a) Span Unseating. In Section 1.2.1(*a*), span unseating resulting from insufficient seating length at movement joints was identified as one of the more common problems with existing bridges. It was recognized in the early 1970s that it was comparatively straightforward and inexpensive to retrofit this deficiency. The most common approach has been to provide cable or bar restrainers between the elements on either side of the movement joint, or between the span and a pier in some cases. Extensive retrofit programs adding restrainers to deficient bridges have been carried out in California [G1] and in Japan [K1]. Typical examples are shown in Fig. 1.39.

Further details to improve safety at movement joints include the increase of seating length for spans simply supported on piers or abutments by the addition of bolsters to the supporting element or by the use of strong pipes through diaphragms on either side of a movement joint internal in the span. These strong pipes are designed to have sufficient strength to

FIG. 1.39 Retrofit against span unseating by installation of cable restrainers across a movement joint.

support the entire dead-load shear if the seismic displacement is such that the seating support is lost. Details and appropriate design methods are given in Section 8.5.1. Placing hydraulic dampers in movement joints, which are effective in reducing displacements under high-velocity input, have also been utilized.

(b) *Liquefaction and Abutment Slumping.* Some of the most difficult and costly retrofit problems relate to reducing risks associated with soils failure. Risk of liquefaction can be reduced or eliminated by vibroflotation of the site [W2], by placing stone columns, or by other means of site densification or reduction of pore-water pressure buildup. Structural consequences of liquefaction can be reduced by increasing the redundancy of the bridge. Thus the retrofit measures should attempt to make the superstructure continuous across supports, and where possible, monolithic with the piers.

(c) *Column Failures.* The method used most extensively to date for enhancing the flexural ductility, shear strength, or lap-splice performance of bridge columns has been the use of steel jacketing, a technique developed at the University of California–San Diego [C1,P6]. For circular columns two steel half-shells are rolled to a radius slightly larger than the column, placed around the column, and site welded up the vertical seams. The small gap between the column and the jacket is then grouted. Ratios of column diameter to shell thickness in the range 100:1 to 200:1 are common. The steel jacket acts as an extremely efficient transverse reinforcement to enhance confinement of plastic hinges and the shear strength of truss shear mechanisms. Rectangular or other noncircular sections have also been retrofitted successfully using steel jackets. In these cases the jacket is rolled to a circular or elliptical shape and the gap between the jacket and column is filled with concrete.

Figure 1.40 shows two examples of steel jacket retrofit in the field. Several hundred bridges have now been retrofitted with steel column jackets. Other techniques providing confinement and shear strength using concrete jackets or composite materials (fiberglass–epoxy; carbon fiber–epoxy) have been verified by tests [P7] and used in the field to a minor extent. Design methods for different column retrofit techniques are given in detail in Section 8.2.

(d) *Cap Beam Flexural and Shear Strength.* Generally, the requirement for retrofit of cap beam deficiencies, which are discussed in Section 1.2.4, is to enhance the flexural and shear strength above the levels corresponding to formation of column plastic hinges. One of the most common retrofit measures involves prestressing the cap beam, using external tendons anchored against end blocks, as shown in Fig. 1.41, or tendons placed in ducts cored through the length of the cap beam. Prestressing can provide the necessary flexural

(a)

(b)

FIG. 1.40 Examples of column retrofit using steel jackets. (a) Full retrofit with elliptical jacket; (b) Partial retrofit with elliptical jacket.

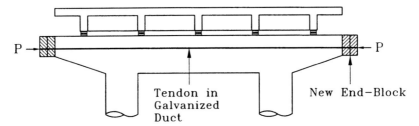

FIG. 1.41 Cap beam retrofit using external prestressing.

and shear strength enhancement for very little cost in many cases and also enhances joint performance. This and other methods of cap beam retrofit are discussed in Section 8.3.

(e) Joint Failures. Column–cap beam or column–footing deficiencies as a result of inadequate joint shear strength or anchorage of column reinforcement generally require elaborate retrofit measures to ensure performance comparable to that of new designs. In the most extreme case it may be necessary to remove the joint concrete and place additional joint shear reinforcement as shown in Fig. 1.42, which shows repair and retrofit of the knee joint of the I-980 bent 38, which was damaged in the 1989 Loma Prieta earthquake and is shown in damaged condition in Fig. 1.27. It is, however, also possible to apply external retrofit of knee joints using heavily reinforced concrete jackets bonded to the existing concrete. In either of the methods, the size of the joint may be increased both in thickness and in vertical extent to reduce the joint shear stresses, on one hand, and to provide increased anchorage length for column reinforcement, on the other hand. As noted above, prestressing of the cap beam can also be used to reduce principal tensile stresses in column–beam joint regions, thus reducing propensity for joint shear failure. Column–footing joints are more difficult to retrofit. Increasing the footing depth by an overlay may in some cases reduce the joint shear stresses to acceptable levels.

The consequences of joint shear distress may need careful evaluation in the assessment process, since the costs associated with retrofit may be unacceptable. In many cases it may be shown that a bridge bent with adequate redundancy may be able to sustain the expected level of displacement response without collapse, even after severe degradation of one or more column–footing or column–cap beam connections. In such cases it may be sufficient to establish that the gravity-load capacity of the bent is not impaired by the joint degradation, provided that loss of serviceability can be tolerated. Suggestions are given in Section 7.4.10.

(f) Footing Failure. As with joint problems, retrofitting to enhance footing performance can be very costly and should first be considered in the context of the ability of the bent to survive the design level of seismic response with

(a)

(b)

FIG. 1.42 Joint repair and reinstatement after damage in the Loma Prieta earthquake of 1989. (a) Reinforcement details; (b) Repaired joint.

footing deterioration. For example, the inability of a column footing to resist the moment capacity of the column base plastic hinge without uplifting may not represent undesirable response. Rocking of footings can in many cases be considered an acceptable form of seismic isolation, with large displacement capacity. The result of rocking may be to limit force levels in parts of the structure to levels that can be tolerated, whereas a full-strength retrofit of the footing may increase force levels elsewhere to the extent that additional retrofit is needed. Assessment of rocking response as a means of seismic isolation is considered in detail in Section 6.4.

Where footing retrofit cannot be avoided, it is common to increase the plan dimensions of the footing to increase rocking stability and to place additional peripheral piles if required by ground conditions. A reinforced overlay is often provided on top of the existing footing, with close-spaced dowels connecting new and existing concrete. An example is shown in Fig. 1.43. As noted above, the overlay will also reduce the shear stress levels in the column–footing joint and will provide better anchorage conditions for the column reinforcement in the footing. Aspects of footing retrofit design are considered in Section 8.6.

(g) Seismic Isolation as a Retrofit Approach. Introduction of special seismic isolation or damping devices into existing structures may be used to improve seismic performance. Where superstructures are supported on cap beams by bearings, these will often be found on assessment and site examination to be inadequate, requiring replacement. It may then be appropriate to replace the

FIG. 1.43 Footing retrofit, involving increase plan dimensions, reinforced overlay, and additional piles.

FIG. 1.44 Seismic isolation bearings placed in the Sierra Point rail overcrossing prior to the 1989 Loma Prieta earthquake (Courtesy DIS).

bearings with special isolating bearings, to reduce forces in bents to levels that can be resisted within the elastic range of response. An example of this approach is afforded by the Sierra Point bridge, shown in Fig. 1.44, where bearings between columns and superstructure were replaced with lead/rubber bearings [M3]. This bridge survived moderately intense shaking in the 1989 Loma Prieta earthquake without damage to the columns.

By adjusting deformation and strength characteristics of bearings placed at different bents, it is also possible to regularize seismic response and avoid force and damage concentration in short stiff bents. This approach is considered in some detail in Section 6.3.5.

1.5 DESIGN PROCESS

Seismic design is only a part of the entire design process. The main functional requirement will be providing a bridge for some form of traffic or material flow. For the superstructure the controlling load case will often be self-weight, while for bridges subjected to significant seismic actions, design of the substructure, including columns, cap beams, and piles, will be governed by seismic lateral forces and displacements. Thus seismic design cannot be considered in isolation from the complete design process. Nevertheless, some of the more pertinent aspects of seismic design follow a particular sequence of design steps. Since this book is structured to follow the same sequence, a brief review of the design process is warranted.

1.5.1 Selection of Design Seismicity

Before even the sructural form of the bridge is determined, the designer should consider the constraints imposed by site seismicity and the required seismic performance criteria for the bridge. Potential for liquefaction, slumping of existing or modified slopes, ground dislocation if the bridge crosses an existing fault, geographical amplification of ground motion, and potential for nonsynchronous motion should be determined. If the site has soft soils capable of resonant response within a narrow frequency band, this must be identified before the structural concept is identified, so that the structure can be "detuned" from this frequency.

There is a simple rule of thumb for seismic design that has been known for many years but still bears repeating. On stiff soils and rock, the seismic energy is likely to be concentrated in the short-period range, and flexible, long-period structures may be most appropriate. On soft flexible soils, there is likely to be much more energy concentrated in the longer-period range, and stiff short-period structures may avoid the greatest intensity of response. However, a word of caution about the latter case is appropriate. Stiff short-period bridges on flexible soils may not have the displacement capacity to cope with nonsynchronous displacements occurring at different supports. Also, high local ductility demands may occur in ductile piers consisting of short stiff piers on highly deformable foundations. This aspect is considered in Section 5.3.2. Specific aspects of seismicity that should be considered by the designer are covered in some detail in Chapter 2.

1.5.2 Conceptual Design

A key element in the success or otherwise of a bridge structure in meeting the required performance specification for the design-level earthquake is the designer's choice of structural system. Most of the parameters considered in choosing such fundamental aspects as width, alignment, length, and number of spans will be governed by nonseismic constraints, with geometric constraints resulting from traffic flow and ground terrain dominating. Nevertheless, some flexibility is generally possible, and seismic aspects should be considered in such aspects as whether the bridge will consist of a series of simple spans, or if continuous, whether it should be constructed in monolithic connection with internal piers or end abutments, or supported on bearings. Position and number of internal movement joints should be considered.

Even more flexibility will normally be possible in the choice of the bent type. The choice between single-column and multicolumn supports, section shape for columns and piers, and the type of foundation support (spread footings, pile groups, single piles) all affect the seismic response and should thus not be chosen based solely on the bridge's functional requirement. In Chapter 3 we review the conceptual design phase, comparing options and discussing merits of different configurations.

1.5.3 Modeling and Analysis

With the site seismicity established, and one or more structural options chosen for more detailed evaluation, analyses are needed to establish design forces and displacements. A number of levels of analytical approximation are possible for seismic actions, ranging from simple analysis under equivalent lateral forces representing the seismic actions, to three-dimensional inelastic time-history analyses under earthquake records representing the three-dimensional ground excitation.

Generally, comparatively simple models are most appropriate at this stage, because the element strengths and stiffnesses will not yet be known and because of the uncertainties associated with the ground motion simulation. A highly sophisticated analysis at this stage can give a false sense of confidence in the analytical results, particularly among inexperienced designers. It must be appreciated that even the most sophisticated analysis makes rather gross approximations about structural properties and boundary conditions, such as behavior of movement joints and soil–structure interaction effects. Even more approximate is the assumption of the spatial distribution of the input ground motion. The common assumption that all supports of a multispan bridge are displacing synchronously and with full coherence is nonsense for a long multispan bridge.

At this stage of the design, it is more important to model the stiffness of the structure correctly, to ensure that the displacement levels predicted are of the correct order. This involves a careful assessment of the state of cracking at high levels of response, since the response in the uncracked state is of little significance. Simple plastic collapse analyses, or analyses where the stiffness of yielding elements is reduced to represent the effective final stiffness at maximum response (the substitute structure approach [G2]), are likely to be the most useful.

Various analytical models and a brief review of the fundamentals of dynamic response are considered in Chapter 4. The more sophisticated models are included primarily for their relevance to the design checking stage.

1.5.4 Design and Detailing

The most important aspect of the design is the careful consideration of the relative strengths and deformation capacities of various elements of the lateral force resisting system of the bridge. Plastic hinge locations must be detailed so that they are capable of sustaining the levels of inelastic deformation required under the design-level earthquake within the performance criteria for acceptable damage. Because of uncertainty of the characteristics of the ground motion to which the bridge will be exposed during its design life, the estimates of deformation capacity should be conservative. The emphasis in design of plastic hinge regions is strongly on deformation capacity, with strength being of some importance but definitely playing a secondary role.

Elements of the bridge to be protected from inelastic response are designed by strength considerations, using capacity design approaches, to ensure that the designated plastic hinge mechanism can develop and that all nonductile deformation modes are inhibited. Despite uncertainties about the design-level excitation, this aim can be achieved with a high degree of certainty.

The capacity design approach is developed in considerable detail in Chapter 5. The emphasis is on basic understanding of performance and section design based on first principles. The process will not follow any specific code approach but refers to various code provisions where appropriate, primarily in the comparative sense, related to proposed design approaches. A great deal of emphasis is placed on the estimation of displacement and rotation capacity. This is based on well-established and verified analytical methods that are simple to apply but are not yet part of codified design processes.

1.5.5 Design Checking (Reanalysis)

Having completed the structural design, it may be appropriate to reanalyze the bridge under the seismic excitation to confirm that the performance criteria are being achieved. This will be the case particularly for major bridges serving critical lifeline functions. Reanalysis should be carried out using a more sophisticated analytical method than that used for the basic design, since member properties and expected deformation modes, including the extent of cracking, should be well defined by this stage. The primary purpose of the analysis will be to ensure that individual member deformations, particularly plastic hinge rotations, do not exceed the design assumptions. Small modifications of the design may result from this stage of the design, but major modifications are not expected. Aspects of this phase of the design process are included in Chapter 4.

1.5.6 Seismic Isolation

One of the design tools available to the designer is to isolate the bridge superstructure from the strongest components of ground shaking. A considerable amount of research effort has been directed toward developing effective isolation techniques for both new and existing bridges. Seismic isolation provides the designer with an alternative to the ductile design philosophy. Because of its growing importance and novelty, it is considered separately in Chapter 6. As discussed in Section 1.4.3(f), foundation rocking can be considered as a seismic isolation mechanism, and the fundamentals of rocking response are included in Chapter 6.

1.5.7 Assessment and Retrofit of Existing Bridges

As discussed earlier, the assessment and retrofit of existing bridges requires less conservative approaches than for new bridges. Although many of the fundamentals and design methods are the same as those for new bridges, the procedures for assessing and retrofitting existing bridges are considered in detail in Chapters 7 and 8, respectively.

2

SEISMICITY AND
GEOTECHNICAL CONSIDERATIONS

2.1 INTRODUCTION

Most books related to seismic design of structures begin with a summary of
the theories of earthquake generation in relation to tectonic mechanisms,
faults, and plates and then present the scales and units used to measure the
energy released by an earthquake, its potential for damage, wave propagation
theories, and attenuation laws. In this book only what is considered of specific
interest for bridge design is discussed, on the assumption that the interested
reader will refer elsewhere for more complete summaries [W6,R1,P4,N1].

The damage suffered by bridges in recent earthquakes, presented in Section
1.2, can be used as a guide to identify the most relevant aspects related to
the input motion that have to be discussed. Consider how the problem of
unseating must include consideration of the potential importance of vertical
ground motion and nonsynchronous input to piers, bearing and restrainer
failures related to maxima values for displacements and possible dislocations,
and slumping and liquefaction problems requiring assessment of the interaction between ground motion characteristics and soil conditions. It is also worth
noting that the particular nature of a bridge as an articulated and long structure
can contribute to the importance of aspects that are negligible for other kinds
of structures. For example, geographical amplification is likely to be significant,
ground dislocations within the length of the bridge are likely to take place
since bridges may cross faults, response spectra are of particular interest in
the range where they are less known and less used because of the long periods
of vibration of many bridge structures, and construction at sites with a high
potential for liquefaction may be unavoidable. Finally, the importance of the

functionality of bridges immediately after an earthquake gives a different flavor to the discussion of risk–return period relationships, which in principle should be related to network theories to examine the alternative possibilities of reaching a given area.

An essential purpose of this chapter is to improve the ability of the designer to interpret the predicted or expected ground motion at a given site or at least to better understand the geotechnical and seismological input provided by geologists and seismologists. The basic procedure to obtain a reliable input at a given site could be described in the following steps:

- Identification of the active faults and of their potential of energy release (Section 2.2.1)
- Use of attenuation laws to define the expected acceleration at the site (Section 2.2.1)
- Characterization of the expected ground motion at the bedrock as a function of source mechanism, travel path, and return period (Sections 2.2.2 and 2.2.3)
- Identification of maxima components, frequency content, and strong-motion duration (Sections 2.2.4 to 2.2.9)
- Modification of the ground motion due to local soil conditions and potential for geographical amplification (Sections 2.2.10 and 2.3.2).

Specific aspects related to soil properties, dislocations, slumping, and liquefaction are presented in Section 2.3. Comments on design ground motion in terms of response spectra, return period, site specific accelerograms, and appropriate use of strong-motion records are discussed in Section 2.4. The effects and appropriate consideration of nonsynchronous input motion are addressed in Section 2.5.

2.2 CHARACTERIZATION OF GROUND MOTION

2.2.1 Source Mechanism and Attenuation Laws

Although earthquakes can be originated by a number of natural or human-induced events, such as meteoric impact, volcanic activity, or underground nuclear explosion, the vast majority of earthquakes originate in the vicinity of the boundary between crustal tectonic plates by the release of shear stress and deformation accumulated as a consequence of relative movement of the plates. Typically, the boundaries between plates consist of complex systems of faults, with the main faults essentially parallel to the boundary and a large number of minor faults transverse to the plate boundary. Rates of average relative displacement along faults can vary from a few millimeters a year to about 100 mm/yr (4 in./yr). The magnitude of dislocation caused by an earthquake may be very small or up to several meters, depending essentially on the rate of relative displacement and on the average return period of the earthquakes generated in the fault. Larger dislocations often correspond to larger energy release and to larger ground accelerations.

The ground acceleration experienced at a site lessens with distance from the epicenter, because of the increase in the surface of the wavefronts, the damping properties of the transmitting medium, and the wave scattering at the interfaces between the layers of material. For small to moderate-sized earthquakes the source could reasonably be considered as a point source and spherically radiating waves should characterize the attenuation, while for large earthquakes a line source and cylindrical waves would seem more appropriate. Despite this argument, most attenuation laws have been developed from data on small earthquake and extrapolated to larger seismic events because of the limited data available on large earthquakes.

Typical attenuation laws take the form

$$a_0 = C_1 e^{C_2 M} (R_e + C_3)^{C_4} \tag{2.1}$$

where a_0 is the peak ground acceleration, M the Richter magnitude of the earthquake, R_e the epicentral distance, and $C_1, \ldots, C_4$ are constants (which may depend on local ground conditions). Several relationships of the general form of Eq. (2.1) have been proposed, resulting in a wide scatter of predicted values.

The Kobe earthquake (January 17, 1995) offers the possibility of analyzing a large amount of data, since a large earthquake originated in a region where many seismological stations were operating. The moment magnitude of the earthquake was about 6.9, with a rupture length inferred to be in the range of 30 to 50 km [E11]. The location of the fault rupture is evident in Fig. 2.1, where epicenter and aftershock locations are shown. Some of the recorded ground accelerations are depicted in Fig. 2.2 as a function of the distance from the epicenter [Fig. 2.2(a)] and from the fault [Fig. 2.2(b)] assumed at the centerline of the aftershocks. The open circles indicate the peak ground acceleration of each orthogonal horizontal component and the filled circles represent the maximum values obtained rotating the orthogonal components to the critical angle. In the same figures a proposed attenuation law of the form of Eq. (2.1) [K5] is also shown (soil type 1 is stiffer, type 3 is softer). It is evident that the attenuation of recorded peak acceleration depends on the fault distance rather than on the epicentral distance.

It has been proposed to use the same form of Eq. (2.1) to predict displacement and velocity as a function of magnitude and epicentral distance, calibrating the appropriate parameters C_i. The resulting attenuation laws are shown in Fig. 2.3 for magnitudes 6 and 8 and for three soil conditions. The effects of various ground conditions seem to have little impact on peak ground acceleration but are significant for velocity and displacement.

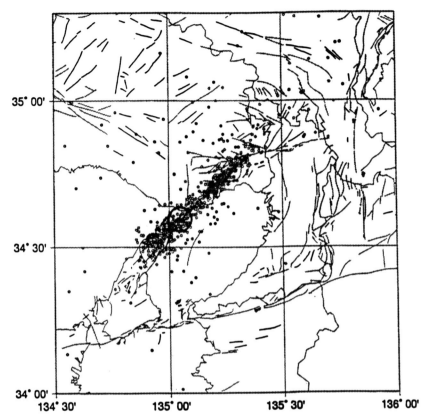

FIG. 2.1 Hypocenters of the Hyogoken–Nambu (Kobe) earthquake (1995): location of aftershocks. The mainshock epicenter is also shown. (Courtesy of Disaster Prevention Research Institute, Kyoto University [E11].)

2.2.2 Wave Propagation Effects

The energy released by earthquakes is propagated by different types of waves. Body waves, originating at the rupture zone, include P waves (primary or pressure waves), which involve particle movement parallel to the direction of propagation, and S waves (secondary or shear waves), which involve particle movement perpendicular to the direction of propagation. When body waves reach the ground surface they are reflected but also generate surface waves, which include Rayleigh (R) and Love (L) waves. Love waves produce horizontal motion transverse to the direction of propagation; Rayleigh waves produce a circular motion analogous to the motion of ocean waves. In both cases the amplitude of these waves reduces with depth from the surface.

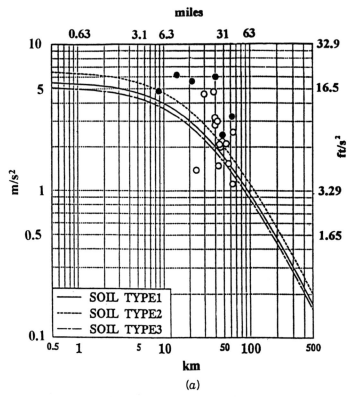

FIG. 2.2 Kobe earthquake (1995) attenuation of recorded peak acceleration at soil sites versus epicentral distance (*a*) and fault distance (*b*) compared with an attenuation law of the form of Eq. (2.1) [H7,K5].

Velocities of P (v_p) and S (v_s) waves in elastic materials are frequency independent, being, respectively,

$$v_p = \left[\frac{E(1 - v)}{(1 + v)(1 - 2v)\rho} \right]^{\frac{1}{2}} \tag{2.2}$$

$$v_s = \left[\frac{G}{\rho} \right]^{\frac{1}{2}} \tag{2.3}$$

where E is the modulus of elasticity of soil, v Poisson's ratio, G the shear modulus, and ρ the density. Considering that $G = E/2(1 + v)$, the ratio between the velocities is

$$\frac{v_p}{v_s} = \left[\frac{2(1 - v)}{1 - 2v} \right]^{\frac{1}{2}} \tag{2.4}$$

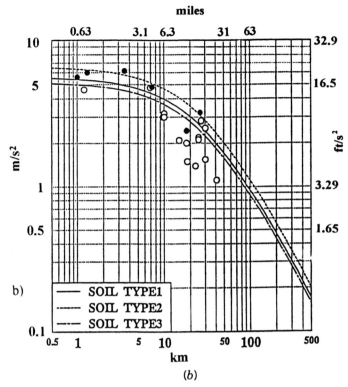

(b)

FIG. 2.2. *(Continued).*

The ratio between the velocities is therefore equal to $\sqrt{3}$ if Poisson's ratio is assumed as 0.25; 1.5 and 2 can be assumed as limit values ($v = 0.1$ and 0.5). Recording of P-S time of arrival interval at three or more noncollinear sites thus enables the epicentral position to be estimated [P4].

From the foregoing relations it is possible to evaluate the range of velocities for S waves near the earth surface as 3 to 4 km/s. This velocity can decrease to values on the order of 100 to 600 m/s in local soils above bedrock. The velocity of propagation of seismic waves is of interest for bridge design because of potentially relevant nonsynchronous input motion to piers (see Section 2.5). Soil modification of rock spectra is discussed in some detail in Section 2.3.2.

As distance from the epicenter increases, the duration of shaking at a given site increases and becomes more complex. This is because of the increase in time between the arrival of P and S waves and to the scattering effects resulting from reflection of P and S waves from the surface. Duration of shaking also increases with earthquake magnitude, due to the time difference between arrival of energy from different points on the longer rupture surface.

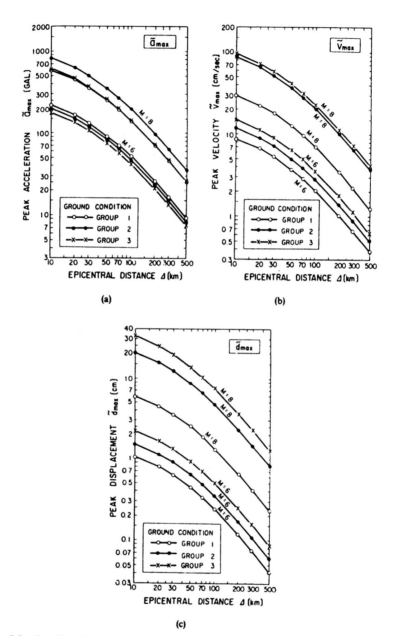

FIG. 2.3 Predicted attenuation of peak ground acceleration (*a*), velocity (*b*), and displacement (*c*) for earthquake magnitudes of 8 and 8 [K5].

2.2.3 Return Periods and Risk Evaluation

To assess the seismic risk associated with a given site, it is necessary to know not only the characteristics of strong ground shaking that are feasible for the site but also the frequency with which such events are expected. It is common to express this by the return period of an earthquake of given magnitude, which is the average recurrence interval for earthquakes of equal or larger magnitude. Large earthquakes occur less frequently than small ones. Over much of the range of possible earthquake magnitudes the probability of occurrence (effectively, the inverse of the return period) of earthquakes of different magnitudes M are well represented by a Gumbel extreme type 1 distribution, implying that

$$\lambda(M) = \alpha V e^{-\beta M} \qquad (2.5)$$

where $\lambda(M)$ is the probability of an earthquake of magnitude M or greater occurring in a given volume V of the earth's crust per unit time and α and β are constants related to the location of the given volume. Figure 2.4 shows data for different tectonic zones compared with predictions of Eq. (2.5) calibrated to the data presented in [E10]. It is seen that the form of the recurrence relationships does indeed agree well with extreme type 1 distributions, except at high magnitudes, where the probability of occurrence is overpredicted. Equation (2.5) gives poor agreement for small earthquakes, since it predicts effectively continuous slip for very small intensities, assuming a stochastic process, where

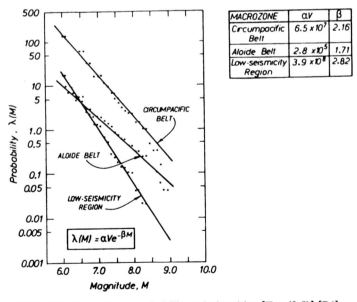

MACROZONE	αV	β
Circumpacific Belt	6.5×10^7	2.16
Aloide Belt	2.8×10^5	1.71
Low-seismicity Region	3.9×10^8	2.82

FIG. 2.4 Magnitude probability relationships [Eq. (2.5)] [P4].

the value of $\lambda(M)$ is constant with time regardless of recent earthquake activity. Also, as a stochastic process, it is assumed that the occurrence of an M8 earthquake in one year does not reduce the probability of a similar event occurring in the next few years. However, it is clear that a slip at portions of major tectonic boundaries occurs at comparatively regular intervals and generates earthquakes of comparatively uniform size. An example is the region of the San Andreas fault east of Los Angeles. Cross-fault trenching and carbon dating of organic deposits have enabled the year and magnitude of successive earthquakes to be estimated, leading to the conclusion that M7.5+ earthquakes are generated with a return period between 130 and 200 years. Even more regular behavior has been noted on the Nazca/South American plate boundary at Valparaiso in Chile. Table 2.1 lists dates and approximate magnitudes of earthquake in this region since the arrival of the Spanish in the early sixteenth century led to reliable records being kept. It will be seen that the average return period of 82 years has a standard deviation of only 7 years. In an area such as Valparaiso, the annual risk of strong ground motion is low immediately after a major earthquake and its associated aftershocks but increases to very high levels 70 years after the last major shake. Although this information can be relevant to design of structures with limited design life, it is usually less significant in bridge design.

Small earthquakes occur more frequently than large earthquakes; they can, nevertheless, generate peak ground accelerations of similar magnitude, but over a much smaller area. The quantification of seismic risk at a site thus involves assessing the probability of occurrence of ground shaking of a given intensity as a result of the combined effects of frequent moderate earthquakes occurring close to a site and infrequent larger earthquakes occurring at greater distances.

TABLE 2.1   Historical Seismicity of Valparaiso [P4]

Year	Interval (yrs)	Magnitude (approx.)
1575		7.0–7.5
	72	
1647		8.0–8.5
	83	
1730		8.7
	92	
1822		8.5
	84	
1906		8.2–8.6
	79	
1985		7.8
	Ave: 82 ± 10	Ave: 8.1 ± 0.6

2.2.4 Accelerograms

Our understanding of seismically induced forces or deformations in structures has developed, to a considerable extent, as a consequence of earthquake ground motions or structural response recorded in the form of accelerograms by strong-motion accelerographs. The time-history record of acceleration is recorded in optical or digital form. Integration of acceleration records enables velocities and displacements to be estimated. Although many useful data have recently been recorded in major earthquakes, there is still a paucity of information about the characteristics of strong ground motion, particularly for large earthquakes. Of particular concern for bridges is the lack of definitive information on long-period components present in different accelerograms.

The nature of the accelerograms depends on a number of factors, including magnitude of the earthquake, distance from the source of energy release, geologic characteristics of the rock along the wave transmission path, source mechanism, and local soil conditions. Despite the number of these factors and their variable nature it is often possible to predict the kind of motion to be expected at a certain site. In fact, the influences of some of these factors are better understood than those of others. For example, the influence of source mechanism and path geology may be understood only in a general way, while the effects of local soil conditions have been studied analytically in many cases (see Section 2.3.2).

Typical earthquake motions had been classified in the past into four different types [N1]:

1. Single-shock motions (e.g., Port Hueneme, 1957; Libya, 1963; Skopje, 1963; San Salvador, 1986)
2. Moderately long, extremely irregular motions (e.g., El Centro, 1940; Chile, 1985; Loma Prieta, 1989; Northridge, 1994)
3. Long ground motions exhibiting pronounced prevailing periods of vibration (e.g., Mexico, 1964; Bucharest, 1977; Mexico, 1985)
4. Ground motions involving large-scale permanent deformation of the ground (e.g., Anchorage, 1964; Niigata, 1964)

With regards to the factors listed previously, it can be noted that the prevailing periods of vibrations exhibited in type 3 motions can be regarded as the effects of soft layers filtering the waves; type 4 is also the result of particular soil conditions, such as presence of saturated sands; while types 1 and 2 are more likely to be registered on a hard soil [W6,X4]. It is therefore possible to record earthquake ground motions of different type produced by the same earthquake at different sites. An example of different ground motion generated by the same earthquake is again offered by the 1995 Kobe earthquake. Some near-field records can be classified as almost single-shock motion, but several records also showed long irregular motions [type 2, Fig. 2.5(a)] or prevailing long periods of vibration [type 3, Fig. 2.5(b)], and in several

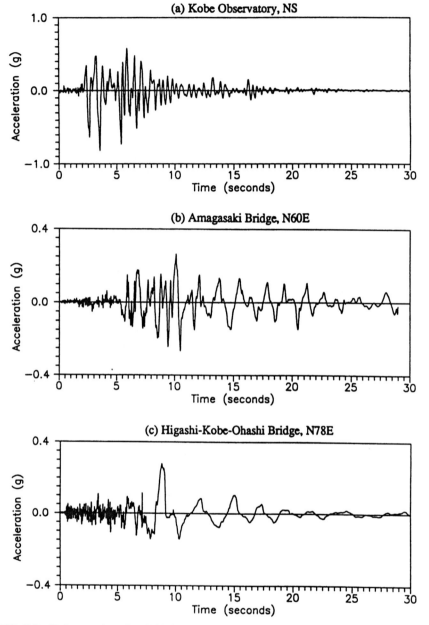

FIG. 2.5 Kobe earthquake (1995): examples of recorded acceleration ground motions [K9].

locations large-scale permanent deformation of the ground was induced [type 4, Fig. 2.5(c)].

2.2.5 Magnitude and Intensity

Earthquakes have traditionally been measured in terms of magnitude or intensity. The *magnitude* of an earthquake is a measure of the energy released at the source; *intensity* is a measure of the effects of the ground motion at a given location. The accepted measure of magnitude is based on the Richter scale and will not be discussed here, since it has little to do with seismic bridge design or the assessment of seismic response of a particular bridge. The scale historically used to measure earthquake intensity is the modified Mercalli scale, originally developed by Mercalli in 1902, modified by Wood and Neumann in 1931, and refined by Richter in 1958. This scale is based on subjective perceptions of motion and damage to buildings, particularly of unreinforced masonry structures. It is therefore only tenuously related to the performance of well-designed modern structures and particularly to the performance of bridges. It has been shown that earthquakes with significantly different magnitude (and therefore releasing very different amounts of energy) may induce similar intensities, as measured by local peak ground accelerations. For example, peak ground accelerations recorded in both the 1986 M5.4 San Salvador earthquake and the 1985 M7.8 Chilean earthquake where approximately 0.7g (while the difference in energy released was about 4000 times).

As pointed out earlier, the most important differences between moderate and large earthquakes are:

- The area subjected to strong ground motion
- The duration of strong ground motion
- The frequency composition of the ground motion (with accelerograms from large earthquakes being typically richer in long-period components)

Because of the inadequacy of either magnitude and Mercalli intensity in describing the damage potential of an earthquake at a given site, other definitions of intensity have been proposed; for example, a well-known definition, a function of both duration and acceleration, is described in [A9].

2.2.6 Directionality Effects

The surface fracture initiates at some point of a fault and propagates in one or both directions along the fault (the same is true vertically, but this is of less interest). If the fracture develops predominantly in one direction only, the ground motion at a given site can be significantly influenced by the location with respect to the direction of rupture propagation. A point located "downstream" of the rupture propagation is likely to experience enhanced peak accelerations (due to interaction between traveling waves and new waves released downstream as the fault propagates), higher-frequency components (enhanced by a kind of Doppler shift), and a shorter duration of shaking. A point located "upstream" should see reduced intensity of ground motion but with increased duration and energy content shifted toward the long-period range.

The considerations above could be of some importance for long multispan bridges, which could well cross faults and experience significantly different ground motion at different pier bases. (See Section 2.3 for a discussion of potential asynchronous input motion to piers.) The propagation of the fault fracture can be made evident by the records of the aftershock epicenters. An example is shown in Fig. 2.1, where the epicenter of the main earthquake and the aftershock hypocenters of the 1995 Kobe earthquake are depicted.

2.2.7 Components and Maxima Values

(a) Acceleration. It has been discussed in Section 2.2.1 how the peak ground (horizontal) acceleration is basically a function of the energy released and of the distance from the zone of energy release. Qualitatively, it could be considered that deeper-focus events produce lower accelerations at closer distances and lead to higher accelerations at larger distances, due to the slower attenuation rate for the travel paths; accelerations on deposits of any depth that do not involve soft clays are likely to be not more than 25% less than those on rocks, and for most practical purposes the peak acceleration values can therefore be considered the same. This is consistent with Fig. 2.3, where the peak ground acceleration is only slightly affected by the soil type. It is worth noting that earlier theoretical predictions that peak ground acceleration could not exceed $0.5g$, which were widely accepted in the 1960s and early 1970s, have recently been proven to be low; several accelerograms with peak ground acceleration components exceeding $1.0g$ have now been recorded [E2,K5,H7].

Vertical accelerations recorded by accelerographs are generally lower than corresponding horizontal components and frequently are richer in short-period components. It is often assumed that the peak vertical acceleration is two-thirds of the peak horizontal value. Although this appears reasonable for accelerograms at some distance from the epicenter, there is increasing evidence that it is nonconservative for near-field records, where peak horizontal and vertical components are often of similar magnitude. The vertical components of earthquakes are not of great significance to structural design of most kinds of structures, but this is not always the case for bridges, where potential deck unseating has to be taken into consideration: The vertical components can contribute to unseating when combined with horizontal components. Vertical accelerations may also induce significant bending moments in long prestressed spans (see Section 3.2.2) and should be considered when eccentricity of support results in column bending moments being induced by vertical seismic response.

Vertical accelerations may assume increased relevance in bridges with seismic isolation systems (see Section 6.1.5).

(b) Velocity. The considerations about the factors affecting the variability of the maximum ground velocity are almost exactly the opposite of those related to acceleration: The nature of the soil has a more significant effect on the peak ground velocity than the distance from epicenter. Typically, peak velocity values on soft soil deposits are about twice those recorded on rock sites. Again this is consistent with Fig. 2.3. It has been noted [N3] that the parameter $v_{g,max}/a_{g,max}$ can conveniently be used, since it remains approximately constant for given types of local soil. For distances of less than about 50 km from the zone of energy release and magnitude of 6.5 or larger, the following values for this ratio were suggested as being approximately correct:

$$\frac{v_{g,max}}{a_{g,max}} = \begin{cases} 0.55 \text{ m/g} \cdot \text{s} & \text{(rock and stiff soil)} \\ 1.10 \text{ m/g} \cdot \text{s} & \text{(deep medium-dense soil)} \\ 1.35 \text{ m/g} \cdot \text{s} & \text{(soft soil)} \end{cases}$$

Limiting values of peak ground velocity of 1 and 0.5 m/s have, until recently, been considered appropriate for soft soils and rock or stiff soil. However, with fault mechanisms involving directional release of energy, velocity pulses may occur which greatly exceed these values. For example, peak ground velocities of 1.3 and 1.7 m/s have been inferred from integration of accelerographs recorded in the 1994 Northridge and 1995 Kobe earthquakes. Since peak velocity has significant relevance for damage potential, particularly in degrading-strength situations, these high velocities must be viewed with some alarm.

(c) Displacement. Peak ground displacements are even more affected by local soil conditions than are peak ground velocity, with values generally below 100 mm in the case of firm soil and possibly as large as 500 mm in the case of soft soil. It has to be considered, however, that peak ground displacements are usually of only moderate interest for design, except in the case of nonsynchronous input motion or of soil failure, in which case the permanent displacement is relevant. It also has to be recognized that displacements are normally obtained from a double integration of filtered acceleration records—therefore, with less accuracy and reliability than obtained for acceleration. Permanent displacements as large as 2500 mm (8 ft) have been measured in the case of soil failure or liquefaction.

2.2.8 Duration Effects

The damage potential of an earthquake at a given site results from a combination of factors, including peak ground acceleration, frequency content, and strong-motion duration. It has been demonstrated by several seismic events (e.g., Parkfield, 1966; Ancona, 1970) that high peak ground acceleration may produce very little damage in the case of very short records with only high-frequency content, whereas a moderate peak ground acceleration combined with a large number of cycles and a lower frequency content may have a large damage potential.

The expected duration of the strong motion is therefore of some interest for the designer. It has been proposed [E8] that strong-motion duration be estimated using the following expression:

$$t = 0.02e^{0.74M} + 0.3R_e \qquad (2.6)$$

where t is the duration in seconds, M the magnitude, and R_e the epicentral distance in kilometers. The main merit of Eq. (2.6) is to show a linear increase with epicentral distance and an exponential increase with magnitude, allowing a preliminary evaluation. Actually, the effects of local soil conditions can be of fundamental importance, since the filtering offered by alluvial soil layers may produce longer records with a precise dominant period of vibration (see Section 2.3.2) corresponding to the period of vibration of the soil.

2.2.9 Energy Considerations

A comprehensive measure of the damage potential can also be obtained estimating the input energy of the expected ground motion, which unfortunately is not independent from the structural response, as shown below. The equation of motion of a linear equivalent single-degree-of-freedom structure excited by a ground acceleration record can be rewritten in energy form by multiplying each term by the response velocity (v_s) of the structure and integrating over time; as shown below, the terms of the resulting equation can be regarded as potential, kinematic, damping, hysteretic, and input energy. The following energy balance is obtained:

$$E_k + E_d + E_s + E_h = E_i \qquad (2.7)$$

where

$$E_k = \frac{mv_s^2}{2} \qquad \text{(kinetic energy)}$$

$$E_d = c \int v_s^2 \, dt \qquad \text{(damping energy)}$$

$$E_s = \frac{F^2}{2K} \qquad \text{(elastic energy)}$$

$$E_h = F \int v_s \, dt - E_s \qquad \text{(hysteretic energy)}$$

$$E_i = m \int a_g v_s \, dt \qquad \text{(input energy)}$$

where m is the mass, v_s the structure velocity, c the viscous damping, F the structure reaction force, K the elastic stiffness, a_g the ground acceleration. Note that after the end of the seismic event, when the system can be assumed to have again reached static equilibrium, the kinetic and elastic energies vanish so that the total dissipated energy is constituted only by the damping and the hysteretic terms. These two terms must therefore equate to the input energy, and the hysteretic term is usually of much greater importance for ground shaking involving significant inelastic deformations [B7].

It has been proposed to use the energy dissipated through hysteresis as a damage indicator, to use the linear elastic pseudovelocity (which is proportional to the maximum elastic energy stored in the system) as the fundamental parameter in a limit design method, and to generate input energy spectra to be used in the design process [U1]. Nevertheless, a possible practical implementation of a design criterion based on an energy equation is far from being developed, and the input energy depends on the integral of the ground acceleration time history multiplied by the structure velocity response time history, therefore being of difficult practical use [C8].

2.2.10 Geographical Amplifications

Geographical features may have a significant influence on local ground motion. In particular, steep ridges may amplify and filter the base rock acceleration in a fashion similar to structural resonance of buildings. An abutment built on top of a ridge may thus be subjected to a shaking completely different from that exciting the base of a pier in the valley. An example of geographical amplification was made evident during the 1985 Chilean earthquake (M7.8), at the Canal Beagle site near Viña del Mar. Planned housing development resulted in identical four- and five-story reinforced concrete frame apartment buildings with masonry-infilled panels being constructed by the same contractor along two ridges and in a valley immediately adjacent to one of the ridges [P4]. While the earthquake caused extensive damage to the buildings along both ridges, the buildings in the valley site escaped unscathed. Simultaneous recording of aftershock activity [C9] at the ridge and valley sites indicated intensive and consistent amplification of motion at the ridge site. A typical transfer function for aftershock activity is shown in Fig. 2.6; the amplification is particularly noticeable for frequencies of about 4 and 8 Hz. It should be recognized, however, that Fig. 2.6 represents a probable overestimation of the actual amplification that occurred during the main event, as a result of increased material damping and other nonlinear effects at the higher levels of excitation.

A second example is available [S14,I8] from strong motions records obtained during the 1957 San Francisco earthquake, which exhibit mixed effects of soil and geographical amplification. The variations in soil conditions along a 6-km section are illustrated in Fig. 2.7 together with the velocity and acceleration spectra recorded at different locations along the section. The maximum

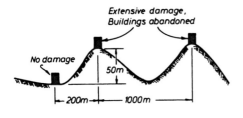
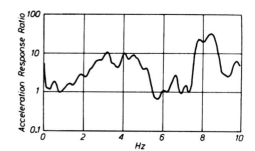

FIG. 2.6   Ridge–valley site acceleration transfer function, Chile earthquake, 1985, Viña del Mar [C9].

ground acceleration varies between 0.05 and 0.12*g*, but the differences in frequency content apparent in Fig. 2.7 are even more significant, particularly if the structures to be analyzed should have a relatively long period of vibration.

2.3   SOIL PROPERTIES AND RESPONSE

2.3.1   Geotechnical Investigations and Prediction of Soil Properties

In Section 2.2 we discussed how an earthquake is generated in a specific area of the earth crust, energy is released, and seismic waves of different nature travel in the bedrock and reach the site where a bridge is to be or has been constructed. The acceleration, velocity, and displacement time histories that will excite the bridge foundation will then be filtered by the local soil before attacking the bridge foundation system. An appropriate characterization of the local soil therefore has two objectives:

- To provide the parameters required for the definition of the ground motion (filtering effect)
- To provide the parameters required to model the soil–foundation interaction (soil stiffness) and to avoid soil failure (soil strength)

A complete presentation of techniques for soil investigations and methods for their interpretation is beyond the scope of this book. Attention will be therefore paid to the main parameters and to the assessment of the most important and potentially dangerous situations. A first basic parameter to be evaluated is the velocity of the S waves, which allow a synthetical definition of the soil type (see Section 2.3.2) and an estimation of the shear modulus of the soil, from Eq. (2.3).

A measure of the velocity of S waves can be obtained directly from cross-hole, down-hole, seismic cone [R2], or forced vibration measures [C15]. For example, in the case of cross-hole testing, the time interval between the emission of a sonic wave from a hole in the soil and the registration of the

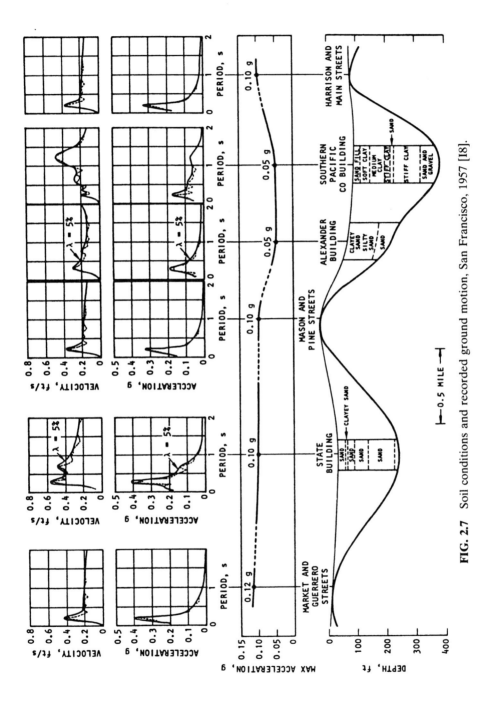

FIG. 2.7 Soil conditions and recorded ground motion, San Francisco, 1957 [18].

wave in another hole is measured. Both P and S wave velocities can be measured, with the warning that in the presence of water, P waves tend to propagate in the liquid phase, with a velocity of approximately 1500 m/s (4920 ft/s).

Wave velocities and soil damping can also be measured in the laboratory, but laboratory tests are more commonly used to obtain direct measures of shear and compression stiffness and strength. Cyclic triaxial, torsional, and shear tests can be performed for this purpose. Laboratory tests are required when potentially liquefiable (see Section 2.3.5) or soft soil is encountered. For liquefiable saturated nonplastic silts, cyclic triaxial or cyclic simple shear tests should be conducted on high-quality thin-walled tubes. With saturated soft cohesive soils the peak and the residual undrained shear strength should be measured [C13].

In situ standard penetration tests (SPT) are also used for empirical approximate evaluations of the initial shear modulus (G_0), or, with more confidence, to evaluate the liquefaction strength of cohesionless saturated soils. A comprehensive study on the relation between velocity of shear waves and SPT tests is presented in [O2], with consideration of the soil nature, soil age, and depth of testing. Similar relationships have recently been proposed to relate shear modulus and static penetration tests. It appears that the ratio between shear modulus and static strength varies in the range 1 to 10 for cohesionless soils and up to 20 for cohesive soils.

2.3.2 Soil Modification of Rock Accelerograms

The importance of local soil conditions for the modification of the accelerograms at a given site has been partially addressed in Section 2.2.10, in relation to the response spectra obtained in different locations during the San Francisco earthquake of 1957. The spectra resulted from a combination of geographical amplification and local soil conditions. Evidence of the effect of local soil filtering has again been evident in the Marina District in San Francisco during the 1989 Loma Prieta earthquake [E1]. Other examples have been made evident during several seismic events, for example:

- **Caracas, 1967.** A detailed analysis of the relationship between structural damage, depth of soil, and height of buildings led to the conclusion that for low-rise buildings, which have shorter periods of vibration, the damage was greater for shallow soil depth, while the opposite was true for high-rise buildings [S13].
- **Mexico, 1957, 1985.** The buildings constructed in the area of Mexico City, which overlays an old lake bed, seemed to suffer more damage during the 1957 earthquake, than those on adjacent rock sites. During the 1985 event a large number of strong-motion records were obtained, and the amplification and filtering produced by the alluvial deposits have become

quantitatively evident [O3,S17,S18]. In Fig. 2.8 the response spectra obtained at the epicenter (at a distance of approximately 400 km), and in Mexico City are compared, showing a tremendous amplification and a shifting of the peak to smaller frequencies (or longer periods).

- **Kalamata, Greece, 1986.** Similar to that described above for the Caracas earthquake, the map of the damage is consistent with the variation of the properties of the local soil, with severe damage to stiff buildings farther from the coast, where the alluvial deposits are thinner.

The importance of local soil conditions has been explicitly recognized in most codes of practice, where different spectra were defined as a function of local conditions. As an example, consider Eurocode 8 [E5], which recognizes three subsoil classes to be accounted for in the definition of the shape of response spectra, as follows:

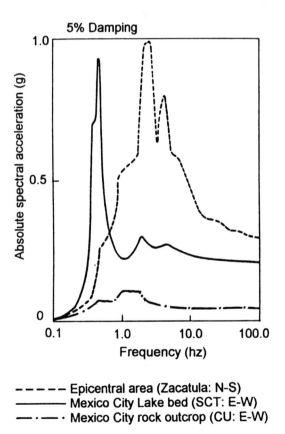

FIG. 2.8 Response spectra obtained from the ground motion histories recorded during the Mexico earthquake, 1985. The amplification and filtering effects of the alluvial deposits are evident.

A   Rock or other geological formation characterized by a shear wave
velocity (v_s) of at least 800 m/s (2625 ft/s), including at most
5 m (16 ft) of weaker material at the surface

Stiff deposits of sand, gravel, or overconsolidated clay, up to
several tens of meters thick, characterized by a gradual increase
of the mechanical properties with depth and by v_s values of at
least 400 m/s (1312 ft/s) at a depth of 10 m (33 ft)

B   Deep deposits of medium-dense sand, gravel, or medium-stiff clays
with thickness from several tens to many hundreds of meters,
characterized by minimum values of v_s increasing from 200 m/s
(656 ft/s) at a depth of 10 m (33 ft), to 350 m/s (1148 ft/s) at a
depth of 50 m (165 ft)

C   Loose cohesionless soil deposits with or without some soft
cohesive layers, characterized by v_s values below 200 m/s
(656 ft/s) in the uppermost 20 m (66 ft)

Deposits with predominant soft to medium-stiff cohesive soils,
characterized by v_s values below 200 m/s (656 ft/s) in the
uppermost 20 m (66 ft)

Typical N values of standard penetration test blows are $N > 35$ for subsoil
A, $15 < N < 35$ for subsoil B, and $N < 20$ for subsoil C. Typical unconfined
compressive strength values are above 0.4 MPa (57 psi), above 0.15 MPa
(21 psi), and below 0.15 MPa (21 psi) for subsoils A, B, and C, respectively.
Typical relative density values are, respectively, above 90%, between 60 and
80%, and between 30 and 50%.

The EC8 recommended response spectra are shown in Fig. 2.9. A warning
is given for deposits of subsoil C containing a layer at least 10 m thick of soft
clays/silts with a high plasticity index (PI > 40) and high water content. Such
soils have very low v_s, low internal damping, and an abnormally extended
range of linear behavior and can therefore produce anomalous seismic site
amplification and soil–structure interaction effects (typically, this is the case
for the lake deposits in Mexico City [O3,S17,S18]). Figure 2.9 is shown in
terms of acceleration ratio. Response for a given site is found by multiplying
ordinates by the peak effective ground accelerations. In many standard cases,
soil–structure interaction does not significantly affect member ductility levels,
since the influence of additional flexibility (Section 5.3.1(a)) is compensated
by period shift and increased damping [C22], but specific studies are required
when the local soil does not match any of the descriptions given or, more
generally, when the importance of the bridge to be designed justifies a specific
study. In this case a model of the local soil has to be used to correct the
ground motion expected at the bedrock. The simplest case is represented by
a series of perfect (ideal) horizontal layers of known thickness and properties,
overlaying bedrock. For each layer it is then simple to derive analytically a
transfer function, where a number of natural periods of vibration are character-

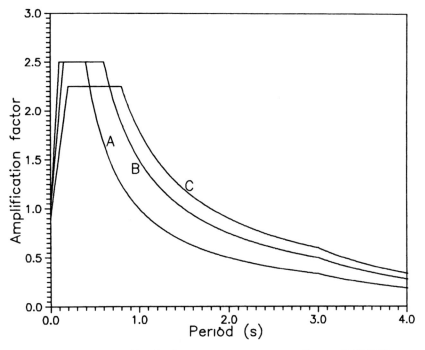

FIG. 2.9 Soil-specific elastic response spectra according to EC8 [E5].

ized by the appropriate frequency and damping. For example, the first theoretical frequency of vibration (f_1) of a layer with thickness t and shear wave velocity v_s is equal to

$$f_1 = \frac{v_s}{4t} \tag{2.8}$$

while damping and amplitude will also depend on the angle of incidence of the shear wave.

A global model of the local soil can then be obtained with a simple physical model where each layer is represented by its stiffness, damping, and mass, as shown in Fig. 2.10. In such models the properties of each layer could account for the possible presence of water, for variation in the pore-water pressure, and for the nonlinear response of soil (depending on the maximum amplitude of the shear wave). Typical variations of the apparent shear modulus and of the damping ratio as a function of the shear distortion are shown in Fig. 2.11. Several computer programs are available to simulate the uniaxial response of soil (e.g., [S15,M15]).

In more complex, yet common cases, the soil cannot be described in terms of horizontal layers, and a two- or three-dimensional model is necessary. A classical finite-element mesh can then be used to model the soil above the

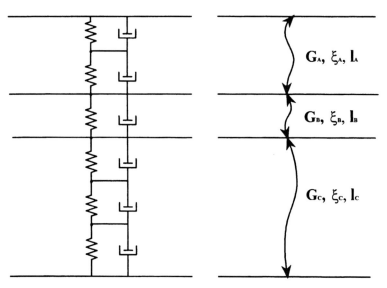

FIG. 2.10   Physical model for a horizontally layered soil.

bedrock. Also in this case, specifically developed computer programs are available, such as that described in [I9], which carries out analysis in the time domain, and in [L3], which operates in the frequency domain. It must be remembered that the usefulness of a refined representation of the local soil conditions is often impaired by uncertainties in determination of the expected bedrock ground motion. When a detailed evaluation of the local soil effects is important, for example when possible nonsynchronous input motions have to be considered (see Section 2.5), several simulations with different input and parametric variations of the soil properties should always be run.

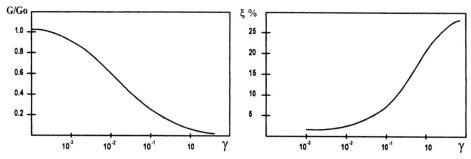

FIG. 2.11   Example of variation of shear modulus and damping ratio for a layer of soil as a function of the shear distorsion amplitude.

2.3.3 Dislocations

It was mentioned in Section 2.2.7 that permanent ground displacements as large as 2.5 m (8 ft) have been measured after seismic events involving liquefac-tion or soil failure. Where appropriate, this kind of possible permanent deformations has to be accounted for in the prediction of the soil response and in the design of the foundation system. In the case of bridges there is another source of possibly large permanent relative displacement between abutments or distant piers, since bridges not infrequently cross active faults, with obvious potential dislocations between the two sides in case of earthquake. A recent example was offered by the Akashi suspension bridge, under construction during the 1995 Kobe earthquake, whose epicenter was in the strait crossed by the bridge. The distance between the two towers of the bridge was 1990 m before and 1990.8 m after the earthquake, due to a relative movement of 1.3 m on a plane inclined to the bridge axis. The design of the deck, still to be built, will be modified appropriately. An upper bound of the dislocation potential to be considered for a bridge crossing a fault system can be obtained multiplying the time from the last earthquake plus the expected life of the bridge by the average rate of relative displacement along the fault system, which can vary from a few to more than a hundred millimeters per year ($\frac{1}{8}$ to 4 in.), as pointed out in Section 2.2.1.

2.3.4 Slumping

It is well known that earthquakes can induce landslides of various kinds. For bridge design, possible rotational slumping and block sliding of the portion of soil where abutments or piers are founded, or possible lateral spreading of approach embankments, are of specific interest (see Fig. 2.12). Earthquake-resistant design for slope stability is often based on experience and sound judgment rather than on detailed numerical simulation. The complexities and uncertainties about loading conditions and soil response associated with earthquake problems often limit or inpair the usefulness of sophisticated analysis for practical design of bridges. Therefore, at sites where it is felt that slope stability is a potential problem, the bridge will be designed for ground movement, with long structural approach slabs and long deck seats. Nevertheless, sophisticated methods and computer programs capable of simulating the soil dynamic response under pulsating inertia forces have been developed and implemented. Their use is generally limited to very special cases and to basic rather than design studies. In practical design a quantitative evaluation of the degree of instability can be obtained using pseudostatic procedures, which assume that the earthquake effects can be replaced by a static horizontal inertial force equal to a fraction of the weight of the potentially sliding soil. A pseudostatic slope stability analysis involves the assumption of a potential

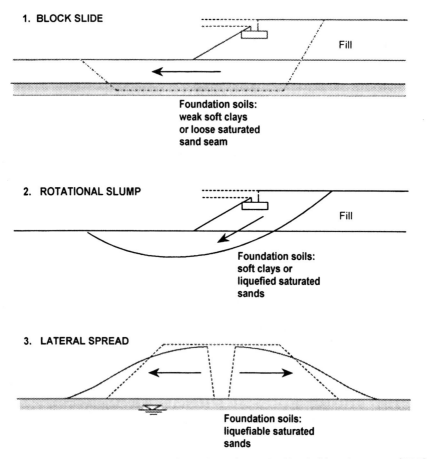

FIG. 2.12 Possible failure mechanisms of portions of soil at bridge abutments [C13].

sliding surface and evaluation of the driving and resisting forces acting on the portion of soil so defined. The factor of safety of the slope is defined as the minimum ratio between resisting and driving force, which will depend on the failure surface selected. A number of trials will generally be needed to identify the worst situation.

The simplest method of calculating driving and resisting forces requires subdivision of the potentially sliding soil portion (again, several trials may be needed to select a critical subdivision) into a number of vertical slices. The driving and resisting forces will then be computed for each slice, neglecting the interslice side forces [C13]. The procedure ("normal method of slices") can be summarized in the following steps:

- A circular failure surface is selected and divided into 10 to 15 vertical slices.
- The total weight of each slice and the pore pressure (if the water table is above the failure surface of the slice) at the circular failure surface are computed.
- The static horizontal force equivalent to the earthquake is computed, assuming an appropriate factor k, as subsequently discussed, multiplying the weight of the slice:

$$V_E = kW \tag{2.9}$$

- The resisting force for each slice is computed using the appropriate shear strength criterion, and the resisting moment is computed according to the following equation (see Fig. 2.13):

$$M_R = [cl + \tan \phi(W \cos \alpha - ul - kW \sin \alpha)]R_c \tag{2.10}$$

where c is the soil cohesion, ϕ the friction angle of soil, W the static weight of the slice, u the pore pressure, α the angle between the tangent to the failure surface and the horizontal, l the slice length, and R_c the radius of the slip circle (as defined in Fig. 2.13).

- The driving moment is computed according to the relationship

$$M_D = (W \sin \alpha + kW \cos \alpha)R_c \tag{2.11}$$

- The factor of safety is computed as the ratio of the sum of the resisting moments over the sum of the driving moments:

$$\mathrm{FS}_E = \frac{\Sigma \, [cl + \tan \phi(W \cos \alpha - ul - kW \sin \alpha)]R_c}{\Sigma \, (W \sin \alpha R_c + kW \cos \alpha)R_c} \tag{2.12}$$

Note that the earthquake effect results in an increase in the driving moment and a decrease in the resisting moment with respect to the case of static equilibrium of the slope. Therefore, the static factor of safety of the slope (FS_s) results from expression (2.12) simply by deleting the seismic terms ($kW \sin \alpha$ and $kW \cos \alpha$).

The simple methods described above can be used for preliminary evaluation of potential problems of slope stability, but more accurate (yet still pseudostatic) methods should be used for final checks. The choice of the method to be used will depend on the nature of the soil, on the data available on the soil properties, and on the geometry of the problem. Several computer programs are also available. A comprehensive discussion of slope stability analysis is presented in [C14]. The key problem is in all cases an appropriate determination of the seismic coefficient (k). An upper limit is offered by the design peak ground acceleration, which would be

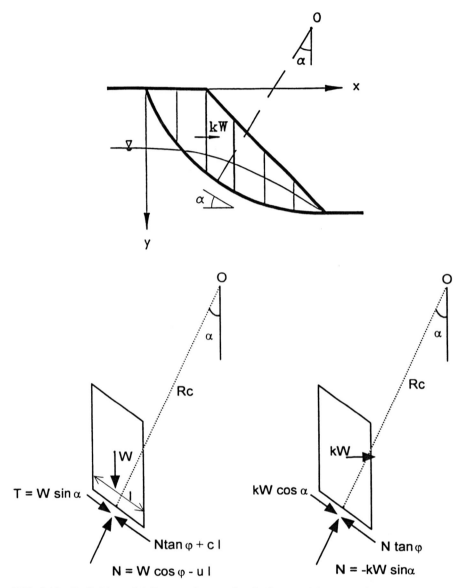

FIG. 2.13 Definition of failure surface and soil slices and forces acting on a typical slice according to the normal method of slices [C13].

an appropriate value if a rigid-body response of the embankment with uniform acceleration throughout is assumed. Empirical values commonly used in the past varied between 0.1 and 0.15. Although some reduction below the peak ground acceleration is reasonable to account for the duration of driving force needed to produce significant displacement on the failure surface, these values appear too low in regions of strong seismicity, where peak ground accelerations may exceed 0.6g.

For design or assessment purposes it will be convenient to compute a yield coefficient (k_y), corresponding to $FS_E = 1$. This factor is a measure of the safety of the slope and is obviously related directly to the value of the static safety factor. A design peak ground acceleration lower than k_y implies insignificant ground deformations during the earthquake. If this is not the case, the expected permanent deformation could be estimated considering a sliding block analogy [N3]: As shown in Fig. 2.14, every time the ground acceleration exceeds the yield coefficient value, some permanent displacement will be accumulated. The total permanent displacement will therefore depend on the time-history characteristics: namely, on the integral of the acceleration time history (i.e., a velocity) exceeding the yield coefficient value.

On the base of extensive parametric studies, the following equation has been proposed to estimate the permanent ground displacement [R3]:

$$\Delta = 0.087 \frac{v_g^2}{Ag} \left[\frac{k_y}{A}\right]^{-4} \tag{2.13}$$

where v_g is the peak ground velocity of the earthquake, A a seismic coefficient (i.e., the peak ground acceleration given as a fraction of the acceleration of gravity), and g the acceleration of gravity (9.81 m/s^2 = 386 in./s^2). Indications of the ratio between peak ground velocity and peak ground acceleration have been given in Section 2.2.7 as a function of the nature of the soil. An average value often used for preliminary evaluation is

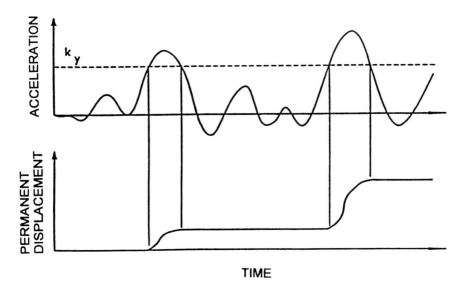

FIG. 2.14 Accumulation of permanent displacement in a sliding block of soil [N3].

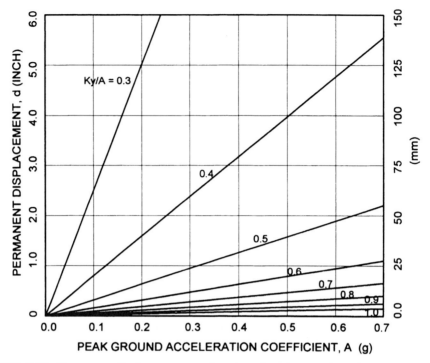

FIG. 2.15 Permanent displacement as a function of peak ground acceleration coefficient (A) and of yield coefficient (k_y), according to Eq. (2.15).

$$v_g = \begin{cases} 0.75A & \text{(distance in meters)} \\ 30A & \text{(distance in inches)} \end{cases} \tag{2.14}$$

Substituting Eq. (2.14) into Eq. (2.13), the following approximate relationship is obtained:

$$\Delta = \begin{cases} 5\dfrac{A^5}{k_y^4} & \text{mm} \\[2mm] 0.2\dfrac{A^5}{k_y^4} & \text{in.} \end{cases} \tag{2.15}$$

Equation (2.15) can be used for preliminary design of embankments and retaining walls whenever permanent displacement is accepted for the design earthquake. In this case the acceptable displacement will be decided and the equation will be solved for k_y. It is evident from Fig. 2.15 that the expected displacement will be limited to less than 50 mm (2 in.) for peak ground acceleration as large as 0.6g provided that $k_y > 0.5A$. This criterion has been adopted in (A2). Obviously, appropriate allowance has to be made for the allowed displacement. It has been shown [W7] that Eq. (2.15) can be regarded as a 90% confidence level design criterion.

When designing retaining walls, the Mononobe–Okabe pseudostatic approach is normally used for preliminary design. The main assumptions of the model are the following:

- The retaining wall can move enough to mobilize the soil strength or active pressure conditions.
- The backfill is cohesionless, with friction angle ϕ, and dry.

The active force that results from the theory (P_{AE}, see Fig. 2.16) is given by the following expression:

$$P_{AE} = 0.5\gamma H^2 (1 - k_v)K_{AE} \tag{2.16}$$

where γ is the unit weight of the soil, H is given in Fig. 2.16, k_v is the seismic acceleration coefficient in the vertical direction, and K_{AE} is the dynamic active earth pressure coefficient. K_{AE} depends on several factors discussed in detail in the literature. The following simple approximation has been suggested [W7]:

$$K_{AE} = K_A + 0.75k_h \tag{2.17}$$

where K_A is the static active earth pressure coefficient, k_h the seismic acceleration coefficient in the horizontal direction, and k_v is neglected; k_h can be assumed as equal to $0.5A$ if some displacement is accepted, as discussed above, or can be assumed equal to A if the displacement potential has to be limited. For example, this could be the case when displacement of the retaining wall

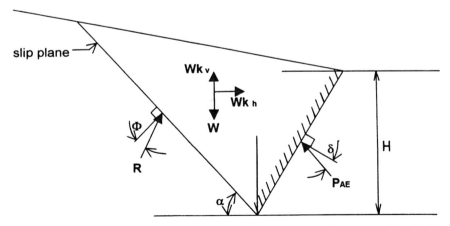

FIG. 2.16 Driving and resisting seismic wedges according to the Mononobe–Okabe pressure theory.

may result in settlement of the approach fill. It should always be checked that the mode of failure will be sliding rather than overturning, to assure a progressive (ductile) failure. It has to be recognized that the generous factors of safety often used for static design generally limit the permanent displacement resulting under earthquake loadings. This may not be case when there is a potential for liquefaction, as discussed in the next section.

2.3.5 Liquefaction

Flow slides associated with liquefaction of saturated sand soils have been a major source of catastrophic failure of bridge abutments in past earthquakes, as discussed briefly in Section 1.2.1. The most common situation that may induce significant ground displacement is associated with the liquefaction of underlying cohesionless soil deposits. It is well known that liquefaction is characterized by the formation of boils and mud spouts at the ground surface, by seepage of water through ground cracks, and by development of quicksand-like conditions over large areas. The basic mechanism of soil liquefaction is also well known, and therefore the general conditions that could lead to soil liquefaction are understood. Saturated sand deposits, subjected to ground vibration, tend to compact and decrease in volume: "If drainage is unable to occur, the tendency to decrease in volume results in an increase in pore-water pressure, and if the pore-water pressure builds up to the point at which it is equal to the overburden pressure, the effective stress becomes zero, the sand loses its strength completely, and it develops a liquified state" [S14].

Liquefaction has been reported to occur essentially in late Holocene saturated deposits of loose sands or silts, according to the gradation curves given in Fig. 2.17, with SPT blow counts generally not greater than 15. The extent of ground displacement depends on peak ground acceleration, duration of ground shaking, thickness of liquefied soil layers, and ground surface topography. The fundamental data required to assess the liquefaction potential and to evaluate the consequent lateral ground displacement are therefore the following:

- Geologic and topographic information
- Grain size distribution curves, soil density, and water table level
- Data from in situ and laboratory tests: namely, SPT blow counts and cyclic shear tests
- Distance to the seismic energy source and expected magnitude of the earthquake (i.e., expected peak ground acceleration and duration)

An assessment of the liquefaction potential can be performed evaluating the average cyclic shear stress τ_1 required to cause liquefaction in N cycles and the average cyclic stress τ_d induced by the expected earthquake in N cycles. The factor of safety against liquefaction can then be defined as

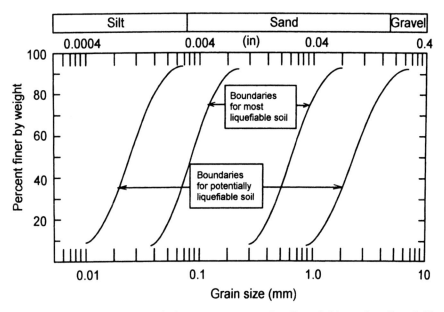

FIG. 2.17 Limits in the gradation curves separating liquefiable and unliquefiable soils [C13].

$$FS_1 = \frac{\tau_1}{\tau_d} \qquad (2.18)$$

τ_d can be estimated approximately from the following expression [S14]:

$$\tau_d = 0.65\gamma H A r_d \qquad (2.19)$$

where γ is the soil unit weight, H the depth of the location under consideration, A the seismic coefficient (i.e., the peak ground acceleration divided by the acceleration gravity), and r_d a stress reduction coefficient which depends on the deformability of the soil column above the location under consideration and therefore depends on the soil profile. The range of possible values for r_d for various soil profiles is shown in Fig. 2.18. In the upper 40 ft (12 m) the average values can be used, with errors generally limited to less than 5%. The number of cycles to be considered depends on the duration of ground shaking and therefore on the magnitude of the design earthquake. The following numbers of cycles have been suggested [S14]: magnitude M6, 5 cycles; M6.75, 10 cycles; M7.5, 15 cycles; and M8.5, 26 cycles.

The average shear stress required to induce liquefaction (τ_1) can be obtained from cyclic simple shear tests or from cyclic triaxial compression tests. In both cases extreme care should be used to obtain good-quality "undisturbed samples" and some judgment will be required to interpret the test data. A

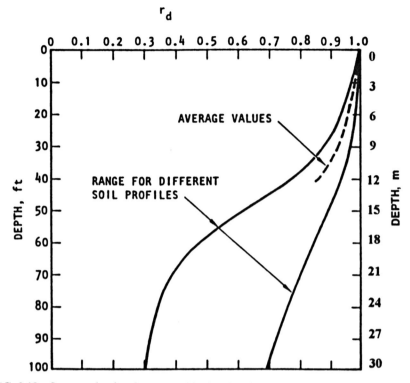

FIG. 2.18   Stress reduction factor used in Eq. (2.19) as a function of soil profile [S14].

convenient parameter to express the effective equivalent seismic action on a sand element to evaluate its liquefaction potential is represented by the ratio of average cyclic shear stress to the initial vertical effective stress (σ_0') acting on the sand before the cyclic stresses were applied. Dividing both terms by σ_0', Eq. (2.19) becomes

$$\frac{\tau_d}{\sigma_0'} = 0.65 A r_d \frac{\sigma_0}{\sigma_0'} \tag{2.20}$$

where σ_0 is the total vertical stress at the depth considered, equal to γH, while to compute σ_0' the effect of the presence of the water table is added. This parameter (τ_d/σ_0') therefore takes into account the earthquake intensity and reflects the influence of the depth of the soil element, the soil relative density, and the depth of the water table.

It has also been noted that the main factors affecting the cyclic loading characteristics of a sand affect the penetration resistance in the same general way. For this reason the penetration resistance has long been used as an index of liquefaction. The standard penetration test (SPT) has provided most of the field test data available for liquefaction studies and for this reason is considered to be a more reliable indicator than other tests. To assess a liquefaction potential the standard penetration resistance, expressed by the number of blows N, is corrected to eliminate the influence of the confining pressure, according to the following relationship [S14]:

$$N_1 = C_N N \tag{2.21}$$

where C_N is a function of the effective overburden pressure at the depth where the SPT was conducted and can be read off from Fig. 2.19. The modified penetration resistance N_1 can be used directly to evaluate the liquefaction potential, using Fig. 2.20 to obtain the cyclic stress ratio required to induce liquefaction (τ_l/σ_0') and comparing the result with the result of Eq. (2.20).

If a potential flow failure has been predicted to occur, it may be of interest to estimate the lateral ground displacement associated with soil liquefaction.

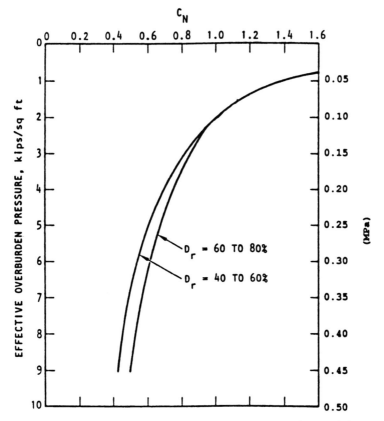

FIG. 2.19 Correction factor to eliminate the effect of the effective confining pressure from the result of a standard penetration test [S14].

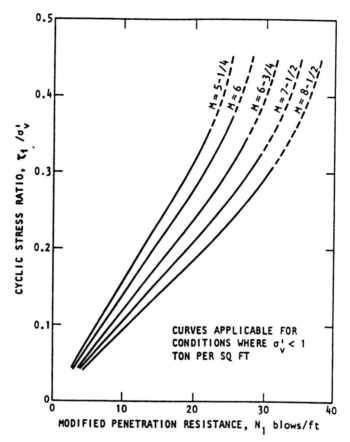

FIG. 2.20   Cyclic stress ratio required to induce liquefaction as a function of modified penetration resistance [S14].

The following empirical expressions [B6] have been derived for this purpose, for a free face and a ground slope, respectively, according to Fig. 2.21:

$$\log(D_H + 0.01) = \begin{cases} -16.366 + 1.178M - 0.927 \log R - 0.013R + 0.657 \log W \\ +0.348 \log T_{15} + 4.527 \log(100 - F_{15}) - 0.922D50_{15} \text{ (free face)} \qquad (2.22) \\ -15.787 + 1.178M - 0.927 \log R - 0.013R + 0.429 \log S \\ + 0.348 \log T_{15} + 4.527 \log(100 - F_{15}) - 0.922D50_{15} \text{ (ground slope)} (2.23) \end{cases}$$

where D_H is the horizontal displacement (m), M the earthquake magnitude, R the horizontal distance from the source (km), W the free face ratio (%, i.e., 100 H/L; see Fig. 2.21), S the ground slope (%), T_{15} the thickness of the saturated sands with blow count $N_{60} < 15$ (m), F_{15} the average fine content in T_{15} with particle size < 0.075 mm, and $D50_{15}$ the average grain size (mm). It is doubtful that these empirical equations would be of practical use for

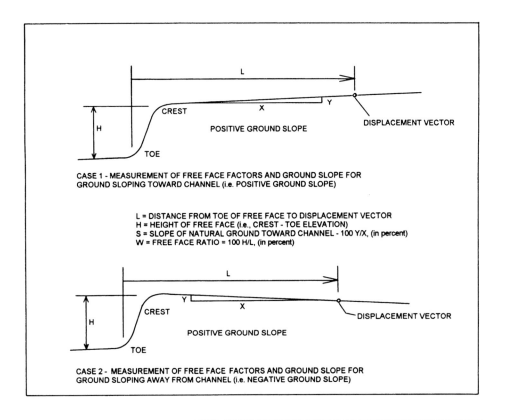

L = DISTANCE FROM TOE OF FREE FACE TO DISPLACEMENT VECTOR
H = HEIGHT OF FREE FACE (i.e., CREST - TOE ELEVATION)
S = SLOPE OF NATURAL GROUND TOWARD CHANNEL - 100 Y/X, (in percent)
W = FREE FACE RATIO = 100 H/L, (in percent)

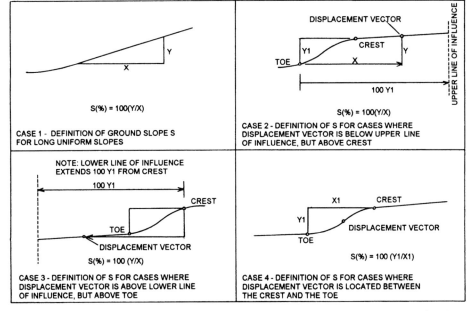

FIG. 2.21 Definition of free face and ground slope factors to be used in Eqs. (2.22) and (2.23) [B6].

93

design, but they could be of some help in the assessment of an existing dangerous situation. They also have the merit of showing the relative importance of several parameters. Their applicability is questionable in epicentral areas (with distance from the epicenter variable from 0.25 km for $M = 6.5$ to 10 km for $M = 8$).

2.4 DESIGN GROUND MOTION

2.4.1 Elastic Response Spectra

The most common way to describe a seismic event is given by its acceleration response spectrum, which condenses information about amplitude and frequency content of the ground motion. On the other hand, any information about duration or number of cycles is not represented. The response spectrum is defined as the maximum response of a single-degree-of-freedom system with damping to dynamic motion or forces, and therefore it depends on the characteristics of the system and on the nature of the ground motion. In principle, the concept is not restricted to linear elastic behavior or to seismic excitation. If a linear elastic response and a constant viscous damping are assumed, the response spectrum becomes a function of the dynamic input and of the period of vibration of the system (i.e., it depends only on the dynamic input amplitude and frequency content).

The quantities commonly studied in terms of response spectra are displacements, velocities, and accelerations, which can be expressed by absolute values (taken with respect to the ground condition before an earthquake) or relative values (taken with respect to the ground during the earthquake). In seismic design absolute acceleration, and relative displacement and velocity are of interest and commonly plotted in response spectra. Instead of true velocity and acceleration, however, slightly different values are commonly used, for reasons of convenience explained below.

The solution of the equation of motion of a linear elastic single-degree-of-freedom system (see Section 4.3) can be written as follows, neglecting the difference between damped and undamped periods of vibration:

$$d(t) = \frac{1}{\omega} \int a_g(\tau) e^{-\xi\omega(t-\tau)} \sin \omega(t - \tau) \, d\tau \qquad (2.24)$$

where $d(t)$ is the system (relative) displacement, ω the angular frequency, $a_g(\tau)$ the ground acceleration, ξ the damping ratio, and t the time. Taking the derivative of Eq. (2.24), the relative velocity can be obtained:

$$v(t) = \int (a_g(\tau) e^{-\xi\omega(t-\tau)} \cos \omega(t - \tau) + \tan^{-1} \frac{\xi}{(1 - \xi^2)^{1/2}} \, d\tau \qquad (2.25)$$

The maximum values of the integrals contained in Eqs. (2.24) and (2.25) are similar, since the second term of Eq. (2.25) is negligible for the case of small damping. It is then possible to write

$$\max(v(t)) \simeq \omega \max(d(t)) \tag{2.26}$$

The equation of motion can also be rearranged to write

$$a(t) + a_g(t) = -\omega^2 d(t) - 2\omega\xi v(t) \tag{2.27}$$

The left-hand side of Eq. (2.27) is the absolute acceleration of the system. The last term of the right-hand side is again small for small damping values; therefore,

$$\max(a(t) + a_g(t)) \simeq \max(\omega^2 d(t)) \tag{2.28}$$

The quantities ωd and $\omega^2 d$ are defined as pseudovelocity (PSV) and pseudoacceleration (PSA), and their maxima values are often plotted in the design response spectra currently used instead of using true velocity and acceleration. The interelations between d, PSV, and PSA enable tripartite response spectra to be plotted on one logarithmic graph, containing information on all three quantities. The practical use of such plots for real accelerograms is limited because of difficulty in extracting values with any reasonable accuracy, their merit being rather in the potential for quick visual comparison of different earthquake motion contents.

As discussed in Sections 2.2.4, 2.2.7, 2.2.10, and 2.3.2, the response spectra of a ground motion at a given site depends on the energy released at the source, the distance from the epicenter, the geography of the site, and the local soil conditions. For example, in Fig. 2.7, different PSA response spectra produced by the same earthquake were compared to show effects of the local soil condition in amplifying and filtering the ground motion. PSA response spectra always start from the ground acceleration at zero period, increase to values in the range of two to three times the ground acceleration at periods of vibration that can vary between 0.1 and 1.5 s, and then decrease, tending to very small values for long periods of vibration.

When important and special local soil effects are not present, PSA spectra start decreasing at values not larger than 0.8 s. PSA response spectra for the three Kobe records of Fig. 2.5 are shown in Fig. 2.22 for comparison. Very different spectral shapes are apparent. Response spectra for structure displacement start from zero at the zero period of vibration, increase almost linearly to a maximum, which is often in the range 3 to 5 s, and then decrease to the peak ground displacement value. Displacement response spectra for the three Kobe records of Fig. 2.5 are shown in Fig. 2.23 for comparison. The very large displacement of 1.5 m for the soil failure case of the Higashi-Kobe-Ohashi Bridge record should be noted.

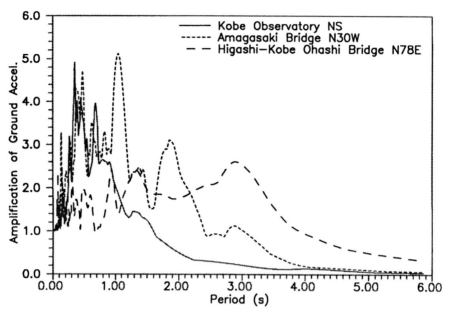

FIG. 2.22 Comparison of acceleration response spectra ($\xi = 2\%$) for 1995 Kobe earthquake records shown in Fig. 2.5.

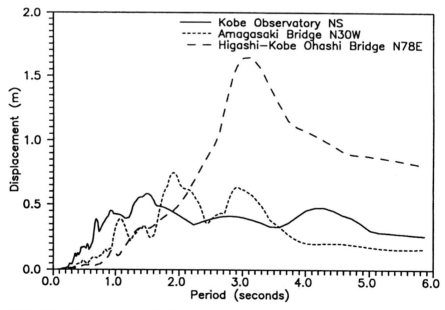

FIG. 2.23 Comparison of displacement response spectra ($\xi = 2\%$) for 1995 Kobe earthquake records shown in Fig. 2.5.

Acceleration and displacement response ordinates are greatly affected by the level of equivalent viscous damping, as is discussed further in Section 4.3.1(c), and illustrated in Figs. 2.24 and 2.25, respectively. The influence is greatest at peak spectral response, and least at very short and very long periods. As discussed further in Sections 4.4.3 and 6.3.2, hysteretic energy dissipation can be converted to equivalent viscous damping. Typical equivalent viscous damping levels for ductile systems are in the range 15 to 30% of critical damping. Variable-damping displacement spectra are the starting point for displacement-based design (Section 5.3.1(c)) using the substitute structure approach (Section 4.4.3).

2.4.2 Inelastic Response Spectra

If the system responds to the dynamic excitation with nonlinear behavior, the initial period of vibration and elastic equivalent viscous damping of the system are not sufficient to obtain the maximum response, which will depend on the actual shape of the force–displacement curve of the system [I7]. To reduce the problem to manageable proportions, it is common practice to assume a linear elastic–perfectly plastic response, as being equivalent to the actual response of the system. The significant parameters of the system will then be its initial stiffness (K), its mass (M), its yield strength (R_y), and its displacement

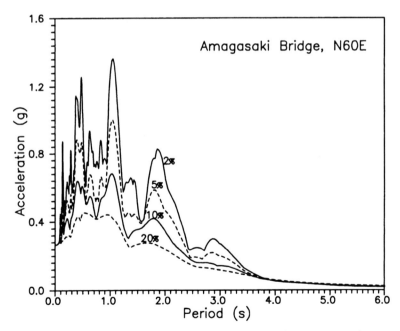

FIG. 2.24 Example of influence of damping on acceleration response. (Amagasaki Bridge record, 1995 Kobe earthquake).

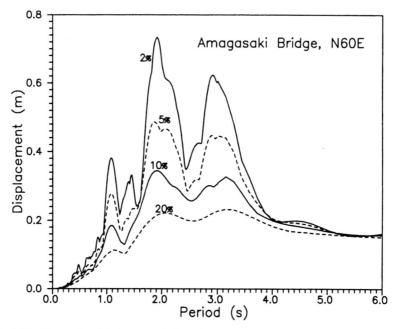

FIG. 2.25 Example of influence of damping on displacement response (Amagasaki Bridge record, 1995 Kobe earthquake).

capacity (Δ_u; see Fig. 2.26). Mass and stiffness can be condensed in a single parameter, the initial period of vibration (T). The ratio between displacement capacity and yield displacement (Δ_y) is defined as displacement ductility capacity (μ_u); the ratio between the displacement required to respond to a given motion (Δ_D) and yield displacement (Δ_y) is defined as displacement ductility demand (μ_D).

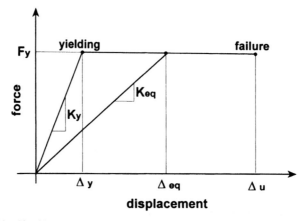

FIG. 2.26 Significant parameters of a linear elastic–perfectly plastic system.

For a given input motion, it is then possible to perform a series of numerical analyses modifying the initial stiffness (i.e., the initial period of vibration) and the yield strength (or displacement) and then construct displacement ductility demand response spectra. Plots of this kind will show a number of different curves characterized by the same yield strength, that is, by the same ratio between yield force and system mass, multiplied by peak ground acceleration ($C_y = \dfrac{R_y}{Ma_g}$; see Fig. 2.27). If the ductility demand is used as a parameter, standard acceleration (force) response spectra can be produced. From this kind of spectrum the yield strength required to remain within a prescribed limit of available ductility for a given input motion and initial period of vibration can be read directly. It has to be noted (see Fig. 2.28) that the ductility demand can be regarded as a measure of the energy dissipated by hysteresis, and it plays a role similar to that of the viscous damping (although, strictly speaking, the dissipated energy depends on the entire history, not only on the peak value; for this reason, other definitions of ductility have been proposed to account for the entire displacement history).

A defect of inelastic response spectra is their typical assumption of a specific hysteretic response characteristic (e.g., elastoplastic). If the characteristics of

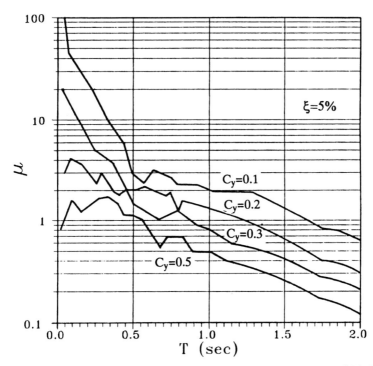

FIG. 2.27 Ductility response spectra for elastic–perfectly plastic systems (Friuli earthquake 1976, PGA = 0.35g).

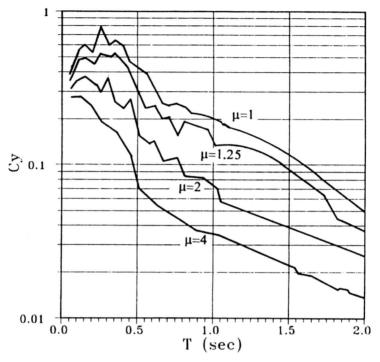

FIG. 2.28 Acceleration response spectra for elastic–perfectly plastic systems (yield strength response spectra) as a function of the displacement ductility demand (Friuli earthquake, 1976, PGA = 0.35g).

the response differ considerably from those assumed, the applicability of the inelastic response spectra may be impaired. In this context it should be noted that the lateral force–displacement response of bridges with column plastic hinges typically results in less than 50% of the inelastic energy dissipation of elastoplastic response. This is discussed further in Section 4.3.

2.4.3 Risk–Importance–Return Period Relationships

It has been pointed out in Section 2.2.3 that small earthquakes occur more frequently than do large ones. They can, nevertheless, generate peak ground accelerations of similar magnitude but over smaller areas and with shorter duration ground shaking. The quantification of seismic risk at a given site thus involves assessing the probability of occurrence of ground shaking of given characteristics as a result of the combined effects of frequent moderate earthquakes occurring close to the site and infrequent larger earthquakes occurring at greater distances. A site-specific seismic risk, generally expressed in terms of annual probability of exceedance of a given level of peak ground acceleration, can be calculated on the basis of many different models. The two basic sources of such models are historical seismicity data and geologic data. For example, the simple model expressed by Eq. (2.5) is based on geologic data, while a recurrence model based on the historical seismicity of Valparaiso (Table 2.1) could be an example of a model based on historical seismicity data. Unfortunately, none of the models proposed can be considered totally reliable, and only in particular cases will historical and geologic data produce similar values of risk. It is therefore highly recommended to apply numerous models and review their results critically before accepting a given probability of recurrence of a certain ground acceleration. A comprehensive discussion of the evaluation of seismic risk can be found in [I11].

The intensity of ground motion will clearly depend also on the level of structural response contemplated, or, in other words, on the limit state (L.S.) assumed as a basis of design. In bridge design it is common to accept three possible limit states, as discussed in Section 1.3.6. The acceptable risk for each L.S. will depend on the importance of the particular bridge: For a very important bridge the requirement for a serviceability L.S. could become so demanding that the other two would be covered automatically, requiring different concepts for capacity design and ductility provisions [C11].

Values of annual probability (p) which might be considered appropriate for ordinary bridges are:

$$p = \begin{cases} \dfrac{1}{100} \text{ to } \dfrac{1}{50} & \text{(serviceability L.S.)} \\[2ex] \dfrac{1}{500} & \text{(damage control L.S.)} \\[2ex] \dfrac{1}{5000} & \text{(survival L.S.)} \end{cases}$$

If bridges of equal importance are considered, the relative relevance of the various limit states will depend on the ductility associated with each and the seismicity of the region: In general, again, only one L.S. will govern the design. There is growing evidence that for events that have very long return periods, appropriate for the survival L.S., the peak ground acceleration (PGA) is rather independent of seismicity or proximity to a major fault, representing the maximum credible earthquake. For example, large regions of the east coast and midwest of the United States are thought to be susceptible to maximum levels of ground shaking similar to those for the much more seismically active west coast, when return periods are measured in thousands of years. However, for short return periods, expected PGAs in moderate- to high-seismicity regions may be one order of magnitude greater than for low-seismicity regions. These trends are included in Fig. 2.29, which plots PGA against annual probability of exceedance for three levels of site seismicity. If a typical multispan reinforced concrete bridge is considered and it is assumed, for example, that $\mu = 1.5$ for the serviceability L.S., $\mu = 4$ for the damage control L.S., and

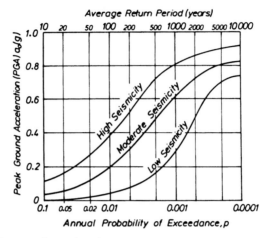

FIG. 2.29 Peak ground acceleration versus annual probability of exceedance for various levels of seismicity [P4].

$\mu = 7$ for the survival L.S., it is possible to compare risk and resistance for the three levels of seismicity.

The discussion above implies that the expected peak ground acceleration can be defined on the basis of limit state and seismicity data for the region. As discussed, the local geographical and soil conditions will then suggest appropriate correction. To complete the description of the design seismic action, information on its frequency content will always be required as well as information on the expected ground motion duration in special cases.

2.4.4 Design Response Spectra

An appropriate generalization of the concept of response spectra leads to the definition of design response spectra, which have to represent a whole class of possible events at a site rather than a specific earthquake. Therefore, in principle the ordinates of a design response spectrum must have a uniform probability of exceedance over all periods. The elastic response spectra shown in Fig. 2.9 offer an example of uniform risk spectra. Although it would be unlikely that any real earthquake would show such long plateau with uniform acceleration, the envelope of the spectra of several possible events could do so.

It has been common in the past to define acceleration response spectra with three regions, as follows:

- For periods smaller than 0.1 to 0.2 s, the acceleration increases linearly with the period.
- For periods of 0.1–0.2 s to 0.4–0.8 s, the acceleration is constant.
- For longer periods the acceleration decreases proportionally to $1/T$.

This implies that the displacement should, correspondingly, vary proportionally to T^3, T^2, and T (i.e., eventually increasing indefinitely in linear fashion with the period of vibration). This obvious inadequacy has been recognized and partially corrected in EC8 [E5], where the displacement is assumed constant for periods larger than 3 s. The resulting displacement spectra are shown in Fig. 2.30. The maximum peak ground displacement, as given in EC8, is also indicated, showing that it can be considerably smaller than the constant displacement obtained from the response at 3-s periods.

Acceleration and displacement response spectra can conveniently be condensed in a single diagram, as shown in Fig. 2.31, plotting acceleration on the ordinates, displacements on the abscissa, and defining lines of constant period of vibration radiating from the origin. Figure 2.31 compares uniform risk spectra as defined in EC8 with real events spectra scaled to the same peak ground acceleration.

As discussed in Section 2.4.3, in most cases structures are designed to resist earthquakes with significant penetration into the nonlinear range. Elastic response spectra are therefore of modest interest for design unless appropriate corrections are made to account for the nonlinear structural response. The problem can be discussed first with reference to an elastic–perfectly plastic system and then to a more realistic response.

For a system of indefinitely long period, the relative structure displacement has to be equal and opposite to the ground displacement, since the mass can be considered immobile with respect to the absolute frame of reference,

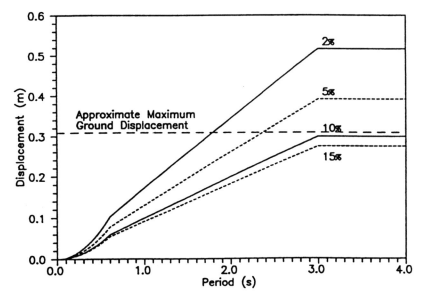

FIG. 2.30 Displacement elastic response spectra as defined in EC8, as a function of soil condition and viscous damping (soil type B, ground acceleration 0.35g).

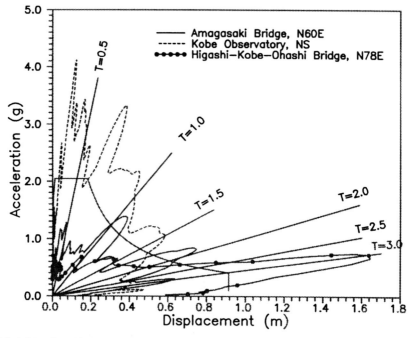

FIG. 2.31 Combined acceleration and displacement elastic response spectra: comparison of real and uniform risk (EC8 for 0.8g PGA) spectra.

regardless of whether the response is elastic or inelastic. A relative displacement conservation region can therefore be defined for which the force-reduction factor can be considered equal to the available ductility [see Fig. 2.32(a)]. In this case it is possible to write

$$F_y = \frac{F_e}{\mu} \qquad (2.29)$$

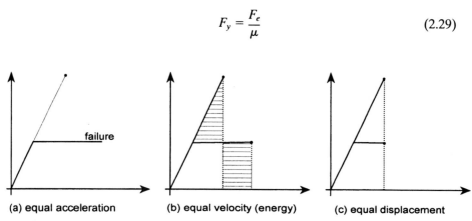

FIG. 2.32 Qualitative response of (a) short-, (b) intermediate-, and (c) long-period elastic–perfectly plastic single-degree-of-freedom systems.

It is found that this approach is also reasonably appropriate for shorter-period structures, provided that the elastic fundamental period of the structure is at least about 1.5 times the period at peak spectral response. The displacement, however, is now period-dependent.

For infinitely short period structures the opposite is true: Mass and ground are moving together with no relative displacement. Therefore, in this case acceleration is conserved regardless of the nonlinear response of the system. In this case ductility does not result in any benefit in the form of reduction of the force to be resisted by the structure. This can be mitigated only slightly by a small period elongation, which may contribute to a reduction of the effective peak acceleration. The implication with reference to Fig. 2.32(a) is

$$F_y = F_e \tag{2.30}$$

For intermediate cases, conservation of energy, or velocity, has been proposed as a means to allow computation of the reduction of the applied force [see Fig. 2.32(b)]. It can be shown that if the single pulses of the accelerogram have a duration (t_d) significantly shorter than the basic period of vibration of the system, an energy conservation principle leads to the following relationship in terms of force reduction:

$$\frac{F_e}{F_y} = (2\mu - 1)^{1/2} \tag{2.31}$$

This provides some advantage in terms of a reduction in the strength to be provided to the system, but with a corresponding increment in the expected displacement and therefore in the potential damage [see Fig. 2.32(b)]. In the case of relatively long pulses (say, $t_d > 4T$) it can be shown that conservation of energy leads to a smaller benefit in terms of force reduction, as follows:

$$\frac{F_e}{F_y} = 2 \left(1 - \tfrac{1}{2} \mu\right) \tag{2.32}$$

In the case of long pulses the maximum reduction in the required strength is then $\tfrac{1}{2}$ if a theoretically infinite ductility is available.

On the basis of earlier discussions, a number of methods to obtain nonlinear response spectra from elastic spectra and available ductility have been elaborated (e.g., [N2]). Real structures do not, however, respond with elastic–perfectly plastic hysteresis loops and the uncertainties on the response and on the effective period of vibration to be considered tend to vanify the elegance of the equations presented above. For this reason very few codes or design recommendations have adopted force-reduction values that depend on natural period, although the qualitative behavior described above has been accepted for many years.

As an example, in Fig. 2.33 the design acceleration spectra recommended in EC8 are shown. The force-reduction factor (q in EC8) is recommended to be taken equal to 3.5 in case of reinforced concrete bridges designed for a ductile response. It can be observed that bridges are often long-period structures and fall into the displacement conservation range. The force-reduction factor is therefore similar to the ductility value without any influence of the shape of the hysteresis loops. In Chapters 4 to 7 we note that alternative design and analysis methods may be used where the basic design parameters are displacements, or deformation, rather than forces. In this case displacement rather than acceleration response spectra will be used, and the problem of realistic versus elastoplastic hysteretic energy dissipation resolved through use of the substitute structure methodology (Section 4.4.3).

2.4.5   Generation of Artificial Ground Motion

There are cases where the simulation of the response using a scaled elastic response spectrum is not felt appropriate, for example, because of the importance of the bridge or because of a particular feature of its behavior (e.g., when isolation devices are used, as discussed in Chapter 6). A time-history integration, as discussed in Section 4.5.4, is in this case the only alternative, with a need for appropriate nonlinear models on the structural side, and for

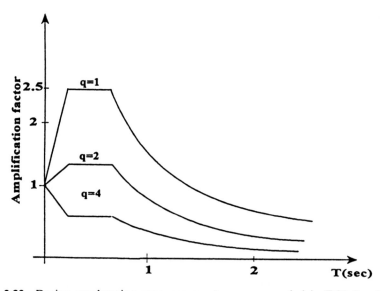

FIG. 2.33 Design acceleration response spectra recommended in EC8 for ductile reinforced concrete bridges (q is the force reduction factor).

appropriate ground motion time history on the action side. Real or artificially generated accelerograms could be used.

The availability of real strong-motion records is still limited because of the relative youth of strong-motion instrumentation with respect to the return period of large earthquakes. Also, each of the available records is significantly affected by effective source mechanism, travel path geology, and local soil conditions, therefore being a unique product that is hardly considerable as representative of a class of earthquake motions that could affect a structure located at a given site. In contrast, artificial earthquake records can be generated prescribing frequency content, acceleration amplitude, shaking duration, and time modulating function, considered to be appropriate to the site.

The methods to generate artificial accelerograms can be classified into three fundamental categories as follows:

1. Methods based on the composition of many harmonic waves with different amplitude and phase angles
2. Methods based on the use of response spectra, generally obtaining the ground motion record filtering a white noise with a single-degree-of-freedom linear system
3. Methods based on the summation of various pulses randomly distributed along the time axis to compose a new wave

It is beyond the scope of this book to enter into more detail as to the generation techniques; computer programs able to generate accelerograms that closely match a given spectrum, duration, and modulating function are readily available [V3] and easy to use. However, the following potential problems have to be kept in mind:

• The number of strong pulses in a generated accelerogram can be much higher than in a real earthquake, because of the tendency of saturating the strong-motion duration at the maximum acceleration level. The accumulated damage can therefore be overemphasized.
• The effective periods of most building structures rarely exceed 2 or 3 s, while this is often not the case for bridges, particularly after some damage has taken place. For very large periods of vibration the response spectra may be ill-defined, since the available knowledge is rather incomplete and uncertain. For large periods of vibration the input power spectral distribution is also concentrated on fewer period lengths, resulting in less uniform results. This problem affects generated as well as real accelerograms, which at large periods often depend more on the filtering process than on the actual ground motion. In all cases the use of a relatively large number of generated accelerograms thus has to be recommended.

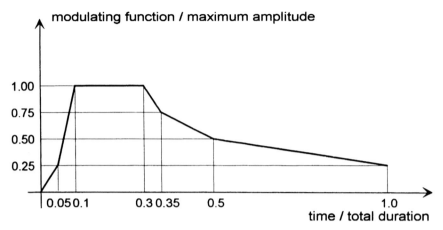

FIG. 2.34 Standard modulating function for artificially generated ground motions [E9].

The time modulating function to be adopted should in principle depend on the seismotectonic characteristics of the region. A possible shape appropriate for usual conditions is shown in Fig. 2.34.

2.5 EFFECTS OF NONSYNCHRONOUS INPUT MOTION

2.5.1 Synchronous and Nonsynchronous Input to Piers

It has been discussed in Sections 2.2.1 and 2.2.8 how an earthquake is generated in a certain part of the earth crust and then propagates by means of waves with different properties and velocities. It is therefore obvious to observe that the earthquake motion does not reach different points of the earth bedrock at the same time. Local soil conditions and geographical features may also play a fundamental role in determining different soil motions at different points of the earth surface. These effects are particularly relevant for bridges, since significantly different input motions can affect the same structure: foundations of different piers can be located at distances that may allow relative delays in the ground motion, the local soil condition for piers located in valleys and peaks can differ quite significantly and the loss of coherency of the wave is enhanced by the irregularity of the soil.

When it is felt that the nonsynchronism of the input motion could be important, the greatest difficulty is the definition of some reasonable input at the different points as a function of the general characteristics of the expected ground motion and of the three sources of nonsynchronism: time delay, loss of coherency, and local soil filtering and amplification. It has been shown [M13] that the loss of coherency has limited effects on the response of the structure, and local soil effects are impossible to generalize. In many cases, however, wave propagation may have significant influence on response, particularly when the wave velocity is slow (i.e., the soil is soft).

2.5.2 Linear Multimodal Approach

The equation of motion of a multidegree-of-freedom structure with a linear response of the matrix form discussed further in Section 4.3.2, that is,

$$M\ddot{x} + C\dot{x} + Kx = F \tag{2.33}$$

can be ordered separating the m free (f) and n restrained ((r), in the case of a bridge, the pier and abutments bases, or the end nodes of the foundation system model, or the end nodes of the soil model) degrees of freedom, obtaining the following equation:

$$\begin{bmatrix} M_{ff} M_{fr} \\ M_{rf} M_{rr} \end{bmatrix} \begin{Bmatrix} \ddot{x}_f \\ \ddot{x}_r \end{Bmatrix} + \begin{bmatrix} C_{ff} C_{fr} \\ C_{rf} C_{rr} \end{bmatrix} \begin{Bmatrix} \dot{x}_f \\ \dot{x}_r \end{Bmatrix} + \begin{bmatrix} K_{ff} K_{fr} \\ K_{rf} K_{rr} \end{bmatrix} \begin{Bmatrix} x_f \\ x_r \end{Bmatrix} = \begin{Bmatrix} 0 \\ F_r \end{Bmatrix} \tag{2.34}$$

The free displacement vector can then be further subdivided into a pseudostatic part (x_s), produced by the static differential displacements at the restrained joints, and a dynamic part (x_d), produced by the dynamic response of the structure:

$$x_f = x_s + x_d \tag{2.35}$$

The pseudostatic displacement vector can be calculated immediately from Eq. (2.34), excluding the dynamic terms (inertial and viscous), as

$$x_s = -K_{ff}^{-1} K_{fr} x_r \tag{2.36}$$

Equations (2.35) and (2.36) can finally be substituted into Eq. (2.34), obtaining the differential equation to be solved to compute the dynamic part of the structure displacement vector:

$$M_{ff}\ddot{x}_d + C_{ff}\dot{x}_d + K_{ff}x_d = (M_{ff}K_{ff}^{-1} K_{fr} - M_{fr})\ddot{x}_r \tag{2.37}$$

A standard modal analysis and mode superposition can be used to solve Eq. (2.37).

It is apparent that a synchronous input motion does not produce any pseudostatic displacement. However, for nonsynchronous input the pseudostatic effects could dominate the design of a stiff structure. It has been shown [Z3] that for low-coherency input ground motions, the total response is essentially equal to the pseudostatic response.

2.5.3   Nonlinear Approach

When the nonlinear response of the structure is explicitly taken into consideration, a modal analysis cannot be performed, but Eqs. (2.36) and (2.37) still apply. A direct numerical integration of Eq. (2.37) will therefore be performed to obtain the dynamic part of the structure displacement vector. A nonlinear analysis of a bridge subjected to nonsynchronous input motion does not present more difficulties than a synchronous analysis, provided that an appropriate ground displacement time history is used (this is actually the real difficulty, as discussed in Section 2.5.5).

Very few nonlinear analyses of bridges subjected to nonsynchronous input have been performed. From the limited results available, it appears that the nonsynchronism results in a kind of redistribution effect, with more uniform ductility demand in the piers than in the corresponding synchronous case [M13]. Until more extended results are available, it seems reasonable to apply the conventional assumption of rigid-body motion, checking the effects of the pseudostatic displacement resulting from an estimation of the nonsynchronous input. The relative differential displacements of the pier bases are considered the most important and dangerous effect of an asynchronous input motion. Out-of-phase accelerations at the pier bases may also result in the effect of exciting otherwise uninfluential higher modes involving flexural/torsional effects on the deck, on the piers, and on the connections between the two.

It is important to note that the problem of pseudostatic displacements at the pier foundations could be governed by surface waves rather than by shear waves, because of their long wavelength, which in soft soil can become comparable to the length of the bridge. Rayleigh waves are characterized by an elliptical motion in a vertical plane, which could cause horizontal and vertical movements as well as tilting of the piers. Love waves propagate in a horizontal plane, resulting in a component of motion toward and away from each other regardless of the direction of approach. Due to the long wavelength, the accelerations associated with even large displacements are usually quite low.

2.5.4   Preliminary Design Approach

If it is assumed that the actual displacement history remains the same over the dimensions of the structure, the relative displacement of the foundations of two piers is given by the difference between two points in the displacement history, separated by the time needed by the waves to travel the distance between the two piers. The ground displacements induced by the earthquake motion will arrive first at the foundation closer to the epicentral area and will reach the other foundations after a time interval (Δt) which is proportional to the distance between the foundations (L) and inversely proportional to the wave velocity (v_s). The angle θ between direction of approach and the axis of the bridge also has to be taken into account, since depending on it there can be no significant delay (when the seismic waves are traveling in a direction perpendicular to the bridge, $\theta = 0$) or only the transversal response could be modified ($\theta = 90°$). With the symbols of Fig. 2.35, the time delay can be expressed as

$$\Delta t = \frac{L \cos \theta}{v_s} \qquad (2.38)$$

Considering that the velocity of S waves can be as low as 100 m/s in local soil above the bedrock, that the distance between piers of ordinary bridges can reach 300 m, and maximizing $\cos \theta$ (i.e., $\theta = \pi/2$ rad), it can be observed that in extreme conditions (a suspension bridge on deep soft soil) a maximum delay of the ground displacement between two piers of a few seconds might be expected. This assumes, however, that the wave propagation is purely dependent on the properties of the upper layers. In most cases the path to adjacent piers will involve wave propagation (horizontally) in the underlying rock with vertical propagation up through the softer upper layers. In such cases the time interval between arrival at adjacent piers will be more dependent on the velocity of S waves in the underlying rock. More reasonable and common conditions could thus lead to delays below 0.1 s, which can still produce significant differential displacement at the pier bases.

The maximum relative displacement (d_m) will occur when its derivative with respect to time is greatest. Since the derivative of displacement with respect to time is velocity, it is reasonable to expect that the maximum relative

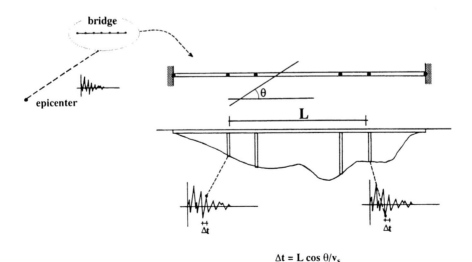

$$\Delta t = L \cos \theta / v_s$$

FIG. 2.35 Nonsynchronous input motion as resulting from earthquake source, bridge geometry, and wave path.

displacement will take place at the peak velocity (v_m) of the earthquake record and is given by

$$d_m = v_m \sin \theta \, \Delta t \tag{2.39}$$

in the longitudinal direction ($\sin \theta$ allow to obtain the component of the velocity parallel to the axis of the structure), and

$$d_m = v_m \cos \theta \, \Delta t \tag{2.40}$$

in the transversal direction.

Substituting Eq. (2.38) into Eq. (2.39) yields

$$d_m = v_m \sin \theta \cos \theta \, \frac{L}{v_s} \tag{2.41}$$

Now maximizing ($\sin \theta \cos \theta$), that is, assuming that $\theta = \pi/4$ rad, the maximum longitudinal differential displacement can be evaluated as

$$\Delta_m = v_m \frac{L}{2v_s} \tag{2.42}$$

Similar considerations applied to transverse response indicate that maximum relative transverse displacement occurs with $\theta = 0$, and is equal to twice Δ_m given by Eq. (2.42). This means that longitudinal differential displacement of 10 to 30 mm between adjacent piers can be expected in rather ordinary conditions, and much larger displacements can occur for deep soft soil (i.e., low wave velocities) and large spans between piers.

In the discussion above we have deliberately considered only the delay due to the wave velocity, neglecting the effects of loss of coherency and effects of local soil at the base of different piers. EC8-2 [E9] suggest that in case of piers or abutments supported on significantly different soil conditions, an averaged response spectrum $[R_a(T)]$ could be adopted, according to the following relationship:

$$R_a(T) = \Sigma \frac{r_i}{\Sigma_j r_j} R_i(T) \tag{2.43}$$

where r_i is the reaction force on the base of pier i when the deck is subjected to a unit displacement, and $R_i(T)$ is the site-dependent response spectrum appropriate to the soil conditions at the foundation of pier i.

It is also more simply suggested that an averaged response spectrum could be obtained considering for each period the largest value of the site-dependent response spectra corresponding to the different soil conditions at the foundations of the bridge. A more consistent approach recently has been suggested [D2] based on the derivation of a multiple-support response spectrum (MSRS rule) as an extension of the well-known CQC rule [W5]. This method, which may become popular in the near future, accounts for the cross-correlations occurring between the support motions and the modes of the structure.

As a consequence of the above-mentioned difficulties in adequately representing the influence of variable ground conditions along the bridge length, together with uncertainties related to lack of coherence and synchronism of input to piers of long bridges, the recommendation is made in Chapter 4 that analyses of very long bridges be simplified by considering the bridge to be subdivided into simpler frames separated by movement joints, each of which is considered to respond essentially independently of the remainder of the bridge. Local soil conditions will then modify the input only to the relevant section of bridge.

2.5.5 Generation of Nonsynchronous Ground Motion Time Histories

It has been recommended [E9] to consider the spatial variability of the ground motion when the bridge is longer than 200 m and geological discontinuities or marked topographical features are present or when the bridge is longer than 600 m. As already pointed out, the spatial variability of motion can be affected by the delay needed by the wavetrain to reach the base of different piers because of the finite velocity of the seismic waves and by the loss of coherency due to reflection and refraction of the waves and to a filtering effect of the local soil. The loss of coherency is strongly frequency dependent, with more significant effects at higher frequencies. For frequencies lower than 1 Hz, the coherence is close to 1; it starts to decrease significantly for frequencies higher than 5 Hz. Examples of coherency factor according to [O1] and [L4] are shown in Fig. 2.36 for a specifically generated event and distances varying between 50 and 150 m. If only low-frequency (long-period) regions of the response spectrum are of interest, depending on the properties of the bridge to be studied, the loss of coherency can be neglected and only the time delay taken into consideration. Vice versa, for long distances and rather stiff structures, totally uncorrelated ground motions with appropriate frequency content (response spectrum) may be generated and used. In general, the approaches discussed in Section 2.4.5 can be used to generate artificial accelerograms with prescribed response spectra and correlation functions of the type presented in [O1] and [L4].

The displacement time history needed for the analysis (and in most cases more important than the acceleration time history for its effect on the structure) can be generated with a double integration of the acceleration time history. Unfortunately, this double-integration process is sensitive to inaccuracy errors, often resulting in a drift from the time axis without any physical meaning. A baseline correction should then be applied to the accelerogram. A common procedure [J3] consists of the following steps:

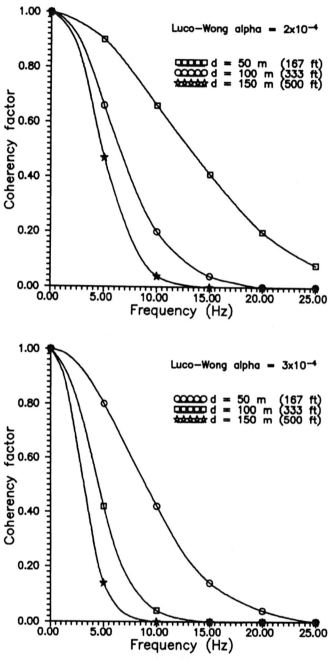

FIG. 2.36 Coherency factor as a function of frequency and distance for a specific event, according to the approaches presented in [L4].

- A second-order polynomial is added to the uncorrected accelerogram:

$$a(t) = a_{unc}(t) - c_0 - c_1 t - c_2 t^2 \qquad (2.44)$$

- The corrected velocity time history is obtained integrating Eq. (2.44):

$$v(t) = \int_0^t a_{unc}(\tau) \, d\tau - c_0 t - \frac{c_1 t^2}{2} - \frac{c_2 t^3}{3} \qquad (2.45)$$

- To minimize the drift, the integral of the velocity squared is minimized, taking the partial derivatives with respect to each one of the constants c_i and setting them equal to zero:

$$\int_0^D v(t) \, \frac{\partial v(t)}{\partial c_i} \, dt = 0 \qquad (2.46)$$

A less rigorous, yet efficient procedure consists of the following, simpler steps, based on the assumption that final displacement is zero, and errors accumulate linearly with time:

- The uncorrected acceleration time history is integrated to obtain an uncorrected displacement time history.
- The displacement time history is corrected adding a linear function passing through the origin and the last point of the displacement time history.
- The corrected acceleration time history is obtained taking the second derivative of the corrected displacement time history.

3

CONCEPTUAL DESIGN

3.1 INTRODUCTION

As discussed in some detail in the following section, structural options for seismic resistance are subject to a number of nonseismic constraints. Before discussing these, it is perhaps pertinent to list the "ideal" structural characteristics for seismic resistance that would apply in the absence of any constraints.

- The bridge should be straight: Curved bridges complicate response.
- The deck should be continuous, with as few movement joints as possible: Simply supported spans are prone to unseating, as are movement joints.
- Foundation material should be rock or firm alluvium: Soft soils amplify structural displacements, and may be prone to slumping or liquefaction.
- Pier heights should be constant along the bridge: Nonuniform heights result in stiffness variation and attraction of damage to stiffer piers.
- Pier stiffness and strength should be the same in all directions: Different stiffnesses and strengths in (say) longitudinal and transverse directions imply inefficiency of design.
- Span lengths should be kept short: Long spans result in high axial forces on columns, with potential for reduced ductility.
- Plastic hinges should develop in columns rather than in cap beams or superstructures and should be accessible for inspection and repair after an earthquake.
- Abutments and piers should be oriented perpendicular to the bridge axis: Skew supports tend to cause rotational response, with increased displacements.

In addition to the foregoing criteria for optimum seismic performance, it is important that the structural systems adopted should be chosen in recognition of the importance of seismic response. Adopting a structural form based solely on gravity-load considerations and then adding seismic resistance by the addition of extra reinforcement is unlikely to be the best solution.

3.2 CONSTRAINTS

3.2.1 Functional Constraints

The structural system adopted for lateral force resistance to seismic actions should be the most suitable possible, given constraints imposed by functional requirements and geographical considerations. Although the structural form will be dominated by these constraints, seismic considerations should be included at the earliest stage and may constitute the critical difference between two or more alternative designs with otherwise equal merits. Functional constraints are imposed by the design traffic density, which will dictate the number of traffic lanes, and hence the width of the bridge, and the alignment, which is controlled by the location, elevation, and direction of traffic at the two ends of the bridge, and by the design traffic speed, which dictates maximum curvature. These aspects are discussed briefly below from the viewpoint of seismic design.

(a)   Alignment. Seismic and functional requirements will often be in conflict relating to alignment. Generally speaking, the ideal bridge structure from a seismic viewpoint is the simplest and most regular. Thus a straight bridge with uniform span lengths and pier heights is preferred. Frequently, this is not possible, particularly for on- and off-ramps of highway interchanges, where curved bridges with variable span lengths and piers of considerable variation in height are common, as shown, for example, in Fig. 3.1(a). The consequence of the structural irregularity is to create uneven demands on the individual piers, with shorter piers resisting a disproportionately higher level of inertia force than taller piers. If the shorter piers are strengthened by increased longitudinal reinforcement ratios, they are further stiffened, thus attracting a still higher proportion of the seismic force. It may thus be impossible to provide a seismic resisting system that satisfies the desirable aim of having all piers reaching their design strength at the same level of excitation. At the other extreme, if piers of different heights are given the same section size and reinforcement, the shorter piers may be subjected to greatly increased ductility demand. The optimum structural solution of reducing the section size of the shorter columns to even out the lateral stiffnesses will normally not be acceptable because of aesthetic considerations. A better solution will be obtained by attempting to regularize column strength and stiffness by placing the shorter columns in structural sleeves, as suggested in Fig. 3.2, with an

(a)

(b)

(c)   (d)

FIG. 3.1 Constraints due to highway geometry. (*a*) Curved intersecting off-ramps; (*b*) Eccentric columns; (*c*) Outrigger cap beam; (*d*) Common columns (two bridges).

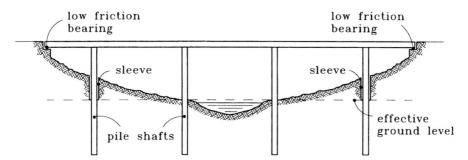

FIG. 3.2   Sleeving of short piers to regularize pier stiffness.

adequate gap between the sleeve and column so that the displacement of the column under seismic response is not impeded. In this way, fixity of each column can be provided at essentially the same level. This is particularly effective with the pile shaft designs discussed in Section 3.3.6. To avoid debris falling into the gap between the sleeve and column, a cover plate is normally attached to the column over the top of the sleeve, free to slide over the sleeve. Sleeving to protect columns in unstable sloping ground is also discussed later, with reference to Fig. 3.6.

Functional constraints on locations of piers are common in multilevel interchanges. Frequently, the optimum column location is not possible, because of traffic lanes on lower levels of the interchange or on surface roads. The result can be a high degree of irregularity in span lengths, which is undesirable from gravity load as well as seismic force considerations, and/or a need to offset the support system from the bridge centerline by the use of eccentric columns or outrigger bent caps, as shown in Fig. 3.1(b) and (c). This situation, although sometimes unavoidable, can result in undesirable seismic response and should be avoided where possible. In such cases it will be tempting to place columns as close as possible to the edge of the constraint. Care is needed where the constraint is the edge of a lower-level bridge. Sufficient clearance must be provided between the column and edge of the bridge to avoid pounding during out-of-phase response of the upper and lower levels of bridging. Damage due to pounding was discussed in Section 1.2.1(c) with reference to an example where a 6-in. (150-mm) separation was insufficient to avoid pounding damage.

Where bridges cross at an angle, it may be tempting to use one common column for the upper and lower bridges on one side, as shown in Fig. 3.1(d). This should be avoided. The lateral stiffness of this column for the upper-level bridge is likely to be many times higher than that of adjacent piers or of the support provided on the opposite side of the bridge, thus attracting extra seismic force or creating excessive ductility demands on the short upper-level column. Actual response of the common column is likely to be strongly influenced by the relative displacement responses of the upper and lower bridges, which may at any given instant be in the same or different directions. The relative directions of response will greatly affect the ductility demand on this critical column. If such a support condition cannot be avoided, it must be designed extremely conservatively, not by providing extra flexural strength, which is likely to be counterproductive, but by the use of increased levels of transverse reinforcement, as discussed in Section 5.3.2(*b*), to increase ductility capacity and shear strength.

The seismic response of curved bridges is greatly influenced by the geometry. Considering the example in Fig. 3.3, subjected to transverse inertial response, the curved plan results in seismic axial forces in the columns, even though the bents have only one column each. These axial forces increase the lateral resistance by an amount $\Delta_F = Cx/h$. The axial forces may create problems at the internal movement joints and must be resisted by torsion of the superstructure, which in turn induces moments at the column tops, of reversed sign to that at the base.

Where the bridge alignment requires a bifurcation of adjacent traffic lanes, as shown in Fig. 3.4, special problems can result. Under transverse response, the two legs, A and B, may respond out of phase. If the connection between these legs is monolithic, high superstructure forces may be developed in the horizontal plane at the crotch of the bifurcation, *D*. It may be difficult to provide adequate strength to resist these large forces. A better structural solution may be to separate the legs of the bifurcation and support by movement joints independently on the body of the bridge, as indicated in Fig. 3.4.

(b) **Width.** Clearly, there is little that can be done to influence the width of the bridge based purely on seismic considerations. The only choice that may be made relevant to seismic response is whether to construct a single bridge for both traffic directions or to support the two directions on individual bridges. In fact, even this choice is more likely to be made based on road alignment requirements than structural aspects. Future possible structural modifications should be considered. It is not uncommon for bridge widening to be required after some years of service, because of increased and unanticipated traffic flow. The increased structural mass creates increased lateral inertia forces.

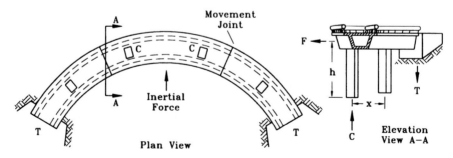

FIG. 3.3 Transverse response of a curved bridge.

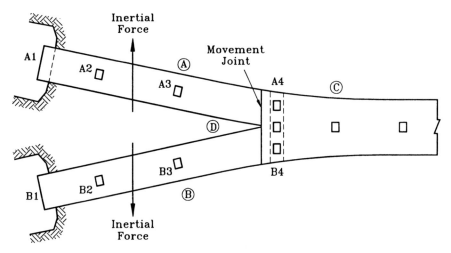

FIG. 3.4 Superstructure bifurcation.

Modification of the substructure can be extremely expensive in comparison with making provision for increased seismic mass in the original substructure design.

3.2.2 Geographical Constraints

The terrain crossed by the bridge will exert the greatest influence on such aspects as bridge length, number of spans, and location of piers. In difficult terrain, or deep-water crossings, long spans and few internal piers will normally be chosen. As a consequence of the need to accommodate thermal movements, the connection between the piers and superstructure may involve some form of sliding bearing. Also, to reduce the number of movement joints, which are always potential locations of continual maintenance, sliding bearings will often be selected in bridges with large numbers of shorter spans. Distances between movement joints up to 1000 m (3300 ft) are now not uncommon. As a consequence of the provision of bearings, there will be no transfer of seismic moments between the piers and superstructure under longitudinal response, simplifying superstructure design but increasing pier seismic moments compared with a monolithic design.

Long-span bridges may develop significant transverse moments in the superstructure as a result of in-plane inertia response. Vertical response accelerations may also be significant under seismic attack, particularly for prestressed decks. Since prestressing acts to counterbalance dead loads, vertical response creates an imbalance that may cause problems for the direction of vertical response corresponding to reduced gravity. If the induced moment imbalance is sufficient to cause cracking, there may be insufficient strength in the section to sustain the applied moments due to the combination of gravity loads, prestress, and seismic forces as a result of the prestress tendon location. This is illustrated in Fig. 3.5, where under seismic response, cracking initiating from the soffit may develop under upward seismic response. The tendon, being located near the top of the section, is poorly located to resist the total moment after cracking occurs. Although it is difficult to imagine failure resulting from this effect because of high superstructure ductility capacity, and we are unaware of any examples of serious damage resulting from this effect, it would be prudent during superstructure design to anticipate this effect and to guard against potential damage by lowering the cable profile over the supports and increasing the level of prestress if necessary.

With long-span bridges or long bridges of many spans, the probability of the input excitation being coherent and synchronous at all supports is greatly diminished, as discussed in Section 2.5.1. Although this may not be particularly significant for transverse response, which is most affected by the local lateral inertia forces, it would appear that resonant response in the longitudinal direction could be destroyed by out-of-phase components attempting to drive different parts of the bridge in different directions. True resonant response in the longitudinal direction may be possible only if the excitation is a shear wave propagating exactly transverse to the bridge axis through perfectly uniform soils, affecting a straight and uniform bridge. Such conditions cannot be expected in practice. Aspects relating to out-of-phase excitation of bridge piers are discussed in Section 2.5.1.

Bridges often cross unstable ground, and it may be necessary to locate columns on steep slopes which have the potential for failure under seismic excitation. In such cases special protection measures may be needed to sustain

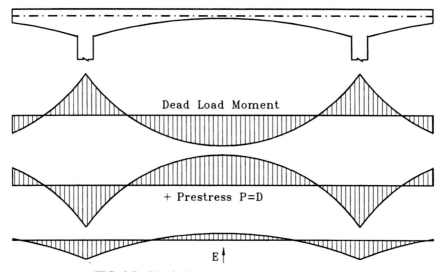

FIG. 3.5 Vertical response of long-span bridges.

the support of the column, without designing for a significant increase in the depth to fixity of the piles resulting from slope failure, which might severely change the seismic response characteristics. On the other hand, slope failure may place unacceptable levels of lateral force on the bridge columns, particularly when the bridge traverses the slope, rather than having an alignment down the slope as would be the case in a valley crossing, where additional restraint may be provided by other piers on more stable foundations. A method of providing additional protection is shown in Fig. 3.6, where the column is protected by a large hollow caisson of sectional shape such that failing ground can move past it without significant restraint. The column is founded some depth below the natural surface, in stable ground, so that the lateral stiffness is maintained at the expected value during and after the earthquake.

Where bridges are required to cross liquefiable ground, the first choice should be avoidance. If possible, bridge routing should be changed to avoid the site, since our knowledge of the extent of loss of support, of potential lateral spreading, and of soil forces on substructures during ground liquefaction is rudimentary. If the site cannot be changed, ground improvement techniques to minimize liquefaction potential should be considered. Attempts should be made to place piers in the least susceptible locations. This will require more extensive geotechnical investigations than that needed for firm ground sites.

There are two schools of thought about the most appropriate structural configuration for crossings of liquefiable ground. One holds that simply supported spans linked together over piers provides the greatest potential for

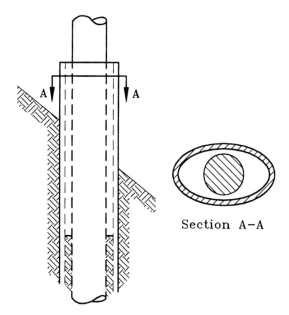

Section A–A

FIG. 3.6 Protective hollow caisson for unstable slope.

accommodating gross displacements occurring during liquefaction, without collapse. In rebuttal, it should be pointed out that this approach has not been particularly successful in past earthquakes (see, e.g., Fig. 1.5). The alternative approach is to ensure full continuity of superstructure and piers, and support of piers on piles driven through the liquefiable layers to firm support. The structural redundancy and the assurity of firm pile tip bearing capacity reduce the susceptibility to failure caused by soil deformation, particularly if the superstructure is monolithically connected to abutments, which are anchored back to supports on firm ground by slack cabling, as shown in Fig. 3.7.

3.2.3 Superstructure Categories

The critical structural elements in the seismic lateral resistance system are generally the bents and substructures. It is, however, relevant to review briefly superstructure section types commonly used in bridge construction, to identify when they are typically used, and what, if any, special seismic considerations must be made.

(a) Concrete Sections. Table 3.1 shows a number of common section shapes and indicates the typical span ranges appropriate for each section. Brief notes

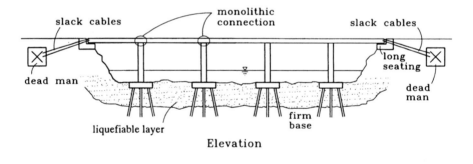

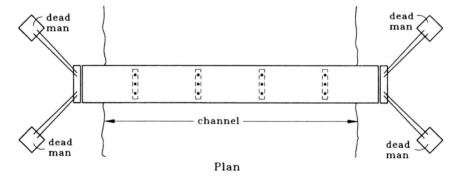

FIG. 3.7 Monolithic structure design for liquefiable sites.

TABLE 3.1   Section Shapes for Concrete Superstructures

TYPE	EXAMPLE	CONSTRUCTION	SPAN RANGE
Solid Slab	300-700mm 900 mm	In Situ	5-15m (15-50 ft)
Voided Slab	450mm 1000mm	Precast	6-15m (20-50ft)
Inverted-Tee	120-150 450-600mm	Precast T In Situ Slab	12-24m (40-80ft)
I-Beam	1200-1800 mm	Precast I In Situ Slab	12-35m (40-120ft)
Double-Tee	≈1500 mm	In Situ	25-40m (80-130ft)
Single-Spine Box Girder	1000-6000 mm	Precast or In Situ	30-200m (100-650ft)
Multi-Cell Box Girder	1000-3000mm	Precast or In Situ	30-100m (100-330ft)
Twin-Spine Box Girder	1000-6000mm	Precast or In Situ	30-200m (100-650ft)
Rectangular Box	2000-6000 mm	Precast or In Situ	30-150m (100-300ft)

on the sections are included below. It is common to divide bridge spans into three categories: short, medium, and long, although the terms have different meanings to different people. In this book, spans of 0 to 30 m (0 to 100 ft) will be considered short, 30 to 60 m (100 to 200 ft) will be considered medium, and spans greater than 60 m (200 ft) will be considered long.

1. *Solid Slab.* For small bridges, the structurally efficient sections shown toward the bottom of Table 3.1 are inappropriate, as the reduction in weight is more than offset by increased construction costs. Solid slabs are still common for road and rail bridges with spans below 15 m (50 ft).

2. *Voided Slab.* Various designs of voided precast slab units are used in different countries for road bridges. The example shown in Table 3.1 has two circular voids and is suitable for spans up to 15 m (50 ft). The units are typically keyed together and lightly prestressed transversely in situ to facilitate load sharing. An advantage of these units is that they form the entire structural cross section. An entire span can be placed in very little time using a mobile crane, thus minimizing disruption when the span crosses existing roading. Continuity over internal supports can be obtained for live load by overlapping mild-steel loops protruding from the end of each present slab, with the joint being filled with in situ concrete, as shown in Fig. 3.8.

3. *Inverted T.* The inverted T section is one of a number of precast section shapes that have been developed for use in the short-span range. Typically, the units are at close centers, enabling the in situ deck to be cast on permanent forms spanning between the inverted T units. Shear connection between T beam and in situ slab is normally by beam stirrups which protrude from the top of the beam and are bent horizontally in the slab for anchorage or by spirals cast half in, half out of the beam.

4. *I Beams.* The most common bridge section in the short-span range consists of precast I beams with an in situ slab. Standardized precast I beams, such as the AASHTO-PCI [A3] standard girders are available in many countries. The I beams are typically spaced farther apart than with inverted T's. For the upper end of the span range 24 to 35 m (75 to 120 ft) the units may be pretensioned for their own weight and post-tensioned in situ for the additional weight of the slab. Diaphragms are provided at supports and frequently at a number of other locations along the span to improve load sharing between the beams.

5. *Double T.* This section shape has recently won some favor with contractors as an alternative to the box girder for in situ construction of bridges at the lower end of the medium-span range. Although not as efficient structurally as the box girder, the reduced complexity of construction resulting from avoid-

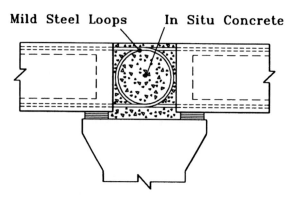

Mild Steel Loops   In Situ Concrete

FIG. 3.8   Continuity of precast spans for live load.

ing the enclosed cell generally more than compensates for the increased dead load. Changes in required section width can be accommodated more easily than with box sections. The double-T section is not suitable for high roadway horizontal curvatures because of poor torsional characteristics.

6. *Box Girders*. Table 3.1 includes four examples of box girders, the first three applicable to road bridges and the last to rail bridges. The sections have the advantage of producing high stiffness and strength for minimum weight. As the span length and thus significance of dead load increases, this becomes a dominant consideration. Box sections also have very high torsional rigidity, which assists in lateral load distribution and makes the sections particularly suited to bridges curved in the horizontal plane.

The choice among single-spine, multispine, and multicell section depends on a number of factors, including section width, section depth, and method of transverse reinforcement. Multicell sections enable reduced deck slab thicknesses to be used but generally have increased total web width compared with a single-spine box and are much more complex to construct. The slight advantage thus tends to decrease as section depth increases, and consequently, multicell boxes are not often used for spans greater than 100 m (328 ft).

Multispine boxes may be used for very wide bridges. Frequently, the spines will be separately constructed and joined by a link slab. Where two spines are used, carrying opposite directions of traffic flow, the link will generally be nonstructural.

For medium-span bridges, a prismatic span will generally be appropriate. For longer spans, nonprismatic sections are common, with increased section depth toward the supports. Web and soffit slab thickness are frequently increased in the support region, but deck-slab thickness will normally remain constant over the span, as it is dictated by transverse bending under live load.

Although diaphragms are often used over supports to assist in transferring load from webs into the pier system, the high torsional rigidity of box sections makes it unnecessary to include diaphragms at regular spacing along the span. In fact, internal diaphragms may have detrimental effects due to increased structural dead weight and restraint of slab shrinkage in the transverse direction, resulting in deck-slab cracking.

It is only recently that prestressed concrete has been used for major rail bridges. However, the hollow rectangular box section is well suited to rail bridging and has been used successfully for spans in excess of 120 m (393 ft). Box girder sections are particularly appropriate for seismic design where monolithic column/superstructure details are adopted, as discussed subsequently. They are also generally adopted for continuous concrete superstructures bearing-supported on bents.

Movement joints in concrete superstructures deserve special attention in seismic regions. These are provided primarily to accommodate longitudinal movements resulting from superstructure temperature fluctuations and from creep and shrinkage of the superstructure. As noted in Section 1.2.1(*a*), span unseating may occur at these locations if adequate movement capacity is not available to accommodate the possible seismic displacements and if a longitudinal restraint system, as discussed briefly in Section 1.4.3(*a*), is not provided. Design approaches for seismic design of movement joints are included in Section 5.8.

The movement joint must also be capable of transferring transverse shear between adjacent sections of the superstructure during seismic attack if this shear transfer is relied upon in the seismic design. Since adjacent superstructure sections may tend to respond out of phase in the transverse direction, transverse shears of significant magnitude may be needed to restrain the joint transversely against relative movement. In such cases it is becoming increasingly common to provide shear keys in the diaphragms provided in the adjacent superstructure sections at the movement joint, as shown in Fig. 3.9. Some transverse clearance between the teeth of such shear keys must be provided to avoid binding under normal operations and to allow adequate seismic lateral rotation between adjacent superstructure sections, which will occur when they respond out of phase. It should also be recognized that high lateral shear force transfer across the movement joint may be accompanied by frictional forces, inducing additional, perhaps unexpected longitudinal force transfer.

An alternative to the provision of positive shear keys as shown in Fig. 3.9 is to allow relative transverse displacement to occur. This may often be considered appropriate when the bridge is long, and rather regular, and hence transverse response of adjacent frames of superstructure is expected to be in phase. It must be recognized, however, that nonsynchronous ground excitation, discussed in Section 2.5.1, may cause the adjacent frames of superstructure to respond out of phase, even if the frames are identical. Thus provision for the maximum possible relative displacement, with the adjacent sections responding out of phase, must be made. Because of uncertainties in the extent

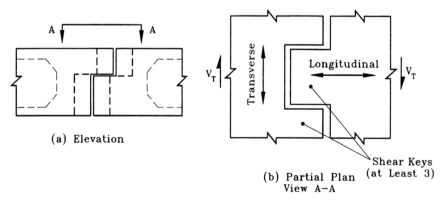

FIG. 3.9 Movement joint shear keys for transverse response.

of frictional shear transfer across the movement joint during seismic response, each longitudinal frame separated by a movement joint should be capable of responding within the required performance criteria, without reliance on support from the adjacent frame.

Although this discussion on movement joints has been directed toward in-span joint locations, it applies equally to superstructure–abutment movement joints. Abutments are discussed further in Section 3.2.2.

(b) Steel Superstructure Sections. For short spans, available stock sizes of steel I-beam sections are often combined with an in situ concrete deck, with shear connectors welded to the top flange of the I beams and embedded in the deck concrete, providing the shear connection. Diaphragms will generally be provided at the ends of the spans by welding I-beam sections between the longitudinal girders or by cross-bracing members. Although these transverse diaphragms may not be necessary for pure gravity load considerations, they are of great importance for seismic response, since the transverse bending strength of the longitudinal I-beam webs is generally insufficient to resist moments induced by the concrete deck slab inertia forces under transverse seismic response. Failure under lateral forces can still occur when the bracing does not extend to the bottom of the webs of the longitudinal girders, as occurred in the 1995 Kobe earthquake with the Hanshin expressway, shown in Fig. 1.29.

For longer-span steel bridges, specially fabricated steel plate girders are often used. These can be designed to be continuous over internal supports with movement joints provided at positions of low gravity-load moment for equivalent continuous structures. The girders are often made of nonprismatic section, to optimize design efficiency, with increased section depth and extra flange plates provided in the support regions. Web stiffeners may be provided to reduce the tendency for web buckling in regions of high shear. In these structures the transfer of deck lateral inertia forces into the piers generally requires more elaborate bracing systems than the provision of a simple dia-phragm above the supports. All elements of the bracing system, including shear transfer between the deck slab and the girders, must be checked carefully to ensure that the lateral load path is capable of resisting seismic lateral forces corresponding to the extreme lateral strength (see Section 1.3.4). That is, the lateral load path must be designed based on capacity design principles as outlined in Section 1.3.3.

Design of movement joints follows the same principles outlined in the preceding section for concrete superstructures. Particular care is needed to ensure that the design assumptions about lateral force transfer transverse to the movement joint are realistic, provide the best option for seismic perfor-mance, and can in fact be developed in practice. Poorly conceived design details can result in buckling of elements in this critical region of force transfer.

Steel box-girder sections have been used in recent years in regions of high seismicity, where medium to long spans are required. These may be combined with an in situ concrete deck slab to form a composite moment-resisting section to carry both dead and live loads, or may form the deck element for cable-stayed or suspension bridges, where the superstructure stiffness is provided primarily for live-load transfer. In either case, the deck and bracing system must be capable of transferring the transverse seismic forces back to the supports.

Steel bridges are inherently lighter than concrete bridges and thus should generally result in lower seismic inertia forces than for equivalent concrete superstructure alternatives. As a consequence, the design of the substructure system should require smaller members, with cost savings. This saving will often be offset by increased superstructure costs for the steel options, and additional complexity, and reduced structural efficiency in the connection between superstructure and substructure.

3.3 SEISMIC DESIGN ALTERNATIVES

3.3.1 Superstructure–Pier Connection

There are two main categories of superstructure–pier connection to be considered at the conceptual design stage. The superstructure may be constructed with a moment-resisting connection, as shown in Fig. 3.10(*a*), or with a bearing support, as shown in Fig. 3.10(*b*). The choice between the two options has a great significance to seismic performance and design force levels. Advantages and disadvantages of the options are discussed in the following.

(*a*) *Moment-Resisting Connection.* This will normally be practical only for concrete superstructures, either prestressed or conventionally reinforced, supported on concrete substructures. It is at least theoretically possible for a steel

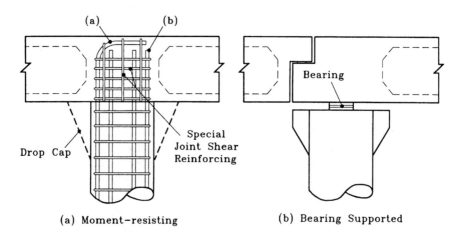

(a) Moment–resisting   (b) Bearing Supported

FIG. 3.10 Superstructure–column connection.

superstructure to be constructed with a moment-resisting connection to a steel substructure, but we are unaware of any examples in seismic regions.

ADVANTAGES. The monolithic sub- and superstructure connection detail of Fig. 3.10(a) is most appropriate for comparatively slender piers or short bridges. The moment-resisting capacity of the connection creates the potential for additional redundancy in the lateral force resisting path, particularly for longitudinal response. Assuming moment fixity at the column base, the potential plastic hinge at the top of the column creates an additional location for energy dissipation during intense seismic attack, compared with the bearing supported alternative. Under longitudinal response, this will place the columns of the bent into double bending, increasing the longitudinal shear resistance for a given section size and reinforcement content for the columns.

If the bent consists of more than one column in the transverse direction, the columns will be in double bending in both longitudinal and transverse response, if moment-resisting connections are provided at the column bases, thus resulting in equal stiffness in the longitudinal and transverse direction. This creates the optimum condition for seismic design where there is no preferential direction for seismic response and thus circular columns may be used.

The fixed-top connection also allows the designer to consider the option of pinned connections at the column base for multicolumn bents. This option is discussed in more detail in Section 3.3.5. A further advantage of the monolithic connection is that it is insensitive to levels of seismic displacement, except insofar as larger displacements may affect the strength of the connection and the rotational capacity of the column–top plastic hinge.

DISADVANTAGES. As a result of the moment connection between the column and the superstructure, seismic moments will be induced in the superstructure under longitudinal response. These will add to, or subtract from, existing gravity load moments and may create critical design conditions for the superstructure. The critical design case will generally be when the seismic moments oppose and exceed the gravity-load moments at the column face, as shown in Fig. 3.11. Special longitudinal reinforcement will need to be placed in the soffit to carry this moment. When the superstructure is comparatively wide and is supported on a single-column bent, the width of superstructure effective in resisting the longitudinal seismic moments may be very much less than the full section width, aggravating the problems. These aspects of longitudinal response are considered in some detail in Section 5.7.

The monolithic column–superstructure connection also places special demands on the connection. Column longitudinal reinforcement of large diameter may need to be anchored in comparatively shallow cap beams. Ideally, this should be effected by bending the reinforcement over the joint as shown by bar (a) in Fig. 3.10(a), although because of the ensuing congestion above the joint, designers prefer to terminate the column bars by straight-bar exten-

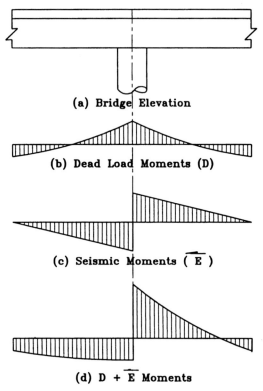

FIG. 3.11 Longitudinal superstructure moments for moment-resisting column–superstructure connection.

sions into the cap beam. It will often be found that such anchorage violates standard code provisions for required length. It is shown in Section 5.5.3 that less conservative anchorage lengths can often be justified by analyses based on an examination of the mechanics of force transfer within the joint and by confirmatory experiments on large-scale units.

Even if the required development length can be provided by straight bar extensions of the column reinforcement, there are additional problems. If a gap is left between the top of the column bars and the superstructure top reinforcement to simplify construction, the mechanism for transferring the tension force from the column tension reinforcement to the superstructure top tension reinforcement is discontinuous and of doubtful integrity. Also, the joint region is subjected to high levels of joint shear force, generally requiring special vertical and horizontal joint shear reinforcing, as shown in Fig. 3.10(a). Aspects of joint shear reinforcement design are considered in some detail in Section 5.4.

When the superstructure is supported by single-column bents, response in the transverse direction will be by simple cantilever action, while longitudinal response will involve double bending of the column. As discussed in the preliminary example on capacity design principles in Section 1.3.3, this can result in a very inefficient substructure unless the column section is made rectangular, with the longer section dimension in the transverse direction. Column section considerations are presented in Section 3.3.4. The imbalance in forces in transverse and longitudinal directions also applies to the foundation design.

With monolithic pier superstructure designs, longitudinal superstructure movements resulting from temperature changes and creep and shrinkage effects must be accommodated by flexing of the columns about the transverse axis. Since the displacements imposed on the columns as a result of these effects are proportional to the distance between superstructure movement joints, it may be necessary to place these at closer intervals than with a continuous superstructure supported on sliding bearings, particularly when the columns are short and stiff.

Finally, the monolithic pier–superstructure design concept is clearly appropriate only for continuous superstructure designs. Where simple spans are more economical, as a result, perhaps, of the use of precast standard beam sections, this concept is inappropriate.

(b) *Bearing-Supported Superstructures.* Bearings between sub- and superstructures may be designed to allow superstructure rotation only, or may also allow relative translation in one or more direction. The most common type of rotating bearing is probably the pot bearing, while translation may be provided by elastomeric bearing pads, by polytetrofluoroethylene (PTFE)/ stainless steel sliders, or by rocker bearings.

Lateral forces transferred through the bearings depend on the bearing design and on whether transverse displacement is restrained by shear keys. Rockers provide almost no longitudinal force capacity and can be unstable if longitudinal relative displacements exceed the design capacity of the rocker. PTFE/stainless steel sliders, which have very low friction coefficients at the low displacement rates corresponding to thermal movements, have significantly higher friction values, typically around 10%, under seismic displacement rates. The actual values depend on the axial bearing stress and the velocity of relative displacement (see Section 6.2.1(c)). The force–deformation relationship can be characterized by Coulomb damping.

Elastomeric bearings provide a resisting force that is proportional to the displacement, and the stiffness can be adjusted over a wide range by choice of bearing dimensions and rubber thickness. Details on bearings are available from manufacturers and from the proceedings of international conferences on bridge bearings [I3,I4]. Advantages and disadvantages of bearing-supported superstructures are discussed briefly below.

ADVANTAGES. Perhaps the greatest advantage with bearing support is that the superstructure is not subjected to seismic moments transferred through the column. Because of this, forms of superstructure section can be chosen that are inappropriate for moment-resisting connections. These include simply supported spans and beam-and-slab construction.

The separation of superstructure and bents by flexible bearings results in a lengthening of the natural period, from a value of T_M, corresponding to monolithic construction, to T_B, corresponding to bearing support. As shown in Fig. 3.12, this can result in very significant reduction in elastic response acceleration levels from point A to B when the initial period is low and when the response spectrum shows a rapid reduction with increasing period, as might be applicable for a rock site. Longer continuous sections of superstructure are possible between movement joints when bearings capable of allowing relative longitudinal displacement are used.

When the support system consists of single-column bents, response in both longitudinal and transverse directions consists of basic vertical cantilever behavior. Consequently, simple circular sections may be used, to result in lateral force–resisting characteristics that are independent of the direction of response. This simplifies design and results in efficient column designs.

The use of bearings provides the designer with greater choice as to how and where seismic forces are to be resisted. Problems with attracting excessive force to short stiff piers can be solved by placing bearings between the columns and superstructure. When elastomeric bearings are utilized, it is possible to compensate for different stiffnesses of different piers by adjusting the bearing stiffness at the top of the piers. Thus flexible piers might be provided with stiff bearings, and vice versa.

DISADVANTAGES. With bearing-supported superstructures, the design is more sensitive to seismic displacement than with monolithic connections. Peak re-

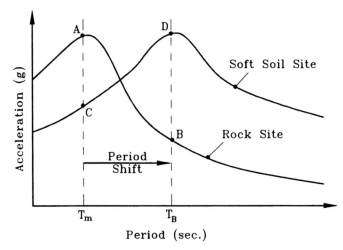

FIG. 3.12 Significance of period shift from bearing support of superstructure with reference to acceleration response spectrum.

sponse displacements are also likely to be significantly larger than with monolithic designs because of the lower stiffness built into the structure. Multicolumn bents act as vertical cantilevers longitudinally but are placed in double bending under transverse response, with inefficiencies similar to those noted previously for single columns with moment-resisting connections to the superstructure. The option of providing pinned column-base connections to footings to reduce foundation seismic design forces is not available for this class of structure.

Where the site acceleration response spectrum rises to a peak at a comparatively long period, as might be the case for a soft soil site, the period shift associated with bearing support of the superstructure may result in increased acceleration response, as shown in Fig. 3.12, where response levels of C and D correspond to monolithic and bearing-supported designs, respectively.

If a ductile column design is adopted, design displacement ductility levels must be restricted to avoid excessive member ductility demand. As suggested in Fig. 3.13, where shear keys have not been added to restrict bearing transverse displacement, bearing deformation Δ_B may constitute much more than 50% of total center-of-mass displacement at yield, Δ_y. Since the column shear force remains essentially constant as plastic displacement Δ_p develops, all of Δ_p results from plastic rotation in the column plastic hinge. The *column* displacement ductility factor can be expressed as

$$\mu_{\Delta c} = 1 + \frac{\Delta_p}{\Delta_s}$$
$$= 1 + (\mu_\Delta - 1)\frac{\Delta_y}{\Delta_s} \tag{3.1}$$

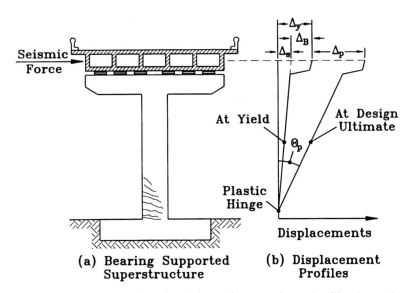

(a) Bearing Supported Superstructure

(b) Displacement Profiles

FIG. 3.13 Influence of bearing deformation on column ductility demand.

where μ_Δ is the structure displacement ductility factor ($\mu_\Delta = 1 + \Delta_p/\Delta_y$) and Δ_s is the portion of Δ_y resulting from deformation of the column alone. Thus if a design value of $\mu_\Delta = 5$ were chosen, with $\Delta_s = 0.3\Delta_y$ and $\Delta_B = 0.7\Delta_y$, Eq. (3.1) would indicate a column displacement ductility demand of $\mu_{\Delta c} = 14.3$, nearly three times that for the structure as a whole.

Clearly, the effects described above can be greatly reduced for transverse response by the use of shear keys restraining bearing transverse displacement but will always exist for longitudinal response. In such cases, ductile column designs may be inappropriate, with reliance being placed on elastic response and/or force transfer back to abutments. A further option, involving special energy dissipation devices, is discussed below.

Under extreme seismic events, much larger than the design level, the additional articulation resulting from separation of the superstructure and substructure may make the bridge more susceptible to damage and collapse, as a result of vertical response accelerations exceeding $1.0g$ and bearing displacements exceeding the physical capacity of the bearings.

Superstructures supported on sliding bearings are likely to develop large residual displacements, particularly under long-duration seismic attack, as a result of the lack of restoring-force characteristics of sliders and the inevitable preference for relative displacement in one direction. This effect can be reduced by combining sliders with elastomeric bearing pads or other elastic devices with restoring-force characteristics.

(c) Isolated Superstructures. With bearing-supported superstructures, the option exists of incorporating special seismic bearings and energy-dissipating devices to reduce resonant buildup of displacement. Such systems include elastomeric bearings fabricated from special rubber with high damping characteristics, or standard elastomeric bearings incorporating lead plugs which dissipate energy in shear. Mechanical devices based on flexural, torsional, or shear yield of steel have also been used, as have combinations of friction sliders and elastomeric bearings. A full treatment of this category of support, including appropriate design information, is given in Chapter 6.

ADVANTAGES. The incorporation of energy-dissipating devices will generally result in significant reduction of seismic displacements compared with bearing-supported designs that do not incorporate energy dissipation. Typical reductions of 20 to 40% can be expected. If the lateral force–deformation characteristics of the bearings and isolators are known with sufficient accuracy, the maximum lateral forces developed in the bents and foundations can be determined. A capacity design approach following the procedures outlined in Section 1.3.3 and developed in more detail in Section 5.3.4 may then be adopted to ensure that elements of the substructure remain in the elastic range under the design-level earthquake. As a consequence, no damage should occur. It will be noted that this approach can also be applied to bearing-supported designs without dissipaters. In this case design forces may be even lower than with dissipaters, although design displacements will inevitably be greater.

DISADVANTAGES. As noted in relation to Fig. 3.12, increased acceleration response may develop when bearing-supported designs, with or without dissipaters, are used at soft soil sites. Long-period structures will rarely experience significantly improved response from seismic isolation.

If the bridge is subjected to significantly higher response levels than anticipated in the design, lateral design forces will increase, possibly resulting in the development of plastic hinges in regions of the substructure that have not been designed for ductility. Even small structure displacement ductility levels may result in unacceptably large plastic rotations, especially if the bents are short and stiff, as discussed previously in relation to Fig. 3.13. Consequently, it is advisable to incorporate ductile detailing at critical potential hinge locations even if a capacity design process has been adopted to ensure elastic response of the substructure. Because of the criticality of the isolating and dissipating devices to satisfactory seismic response, provision must be made for regular inspection and maintenance and for replacement after a major seismic event.

Depending on the design characteristics of the isolators, the bridge seismic response may be sensitive to variations in axial force on the bearings, due to seismic overturning effects or vertical accelerations. There is some analytical and field evidence to suggest that designs with elastometric bearings may amplify short-period vertical response, causing unloading of bearings, and hence loss of effectiveness, or tearing of rubber laminates if the bearing is positively connected to both bent and superstructure and vertical tensions are developed.

3.3.2 Superstructure–Abutment Connection

Because of the importance of soil–structure interaction effects to the seismic response of abutments, their performance is less readily predicted than is that of internal bents of bridges. Unfortunately, this often results in a rather cursory consideration from a seismic resistance viewpoint. The fact that abutment failure rarely results in catastrophic collapse of the bridge is also instrumental in prevalence of this attitude. However, as will be seen from the examples of Section 1.2.2, the consequences of substandard abutment design can be extremely severe.

There are various possible details for connection between superstructure and abutment, with the appropriate choice depending on ground conditions, bridge size, expected displacements, and choice of seismic resistance approach for the bridge at whole. Figure 3.14 shows some examples.

(a) Monolithic Connection. Two alternative details for monolithic abutment–superstructure connections are shown in Fig. 3.14(a) and (b), respec-

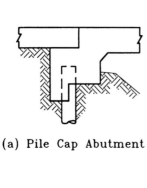

(a) Pile Cap Abutment

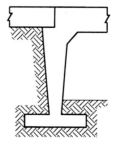

(b)   Rigid Frame Abutment

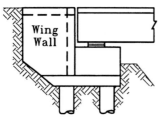

(c)   Seat-Type Abutment

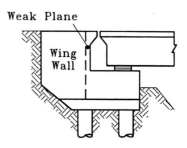

(d) Knock-off Back Wall Detail

FIG. 3.14   Abutment types for longitudinal response.

tively. In the first, gravity loads are carried by piles while longitudinal seismic forces are transferred by passive soil pressure against the backwall and by lateral resistance of the piles. As will be readily appreciated, the degree of fixity provided by this detail can be difficult to determine and will be different for directions of longitudinal seismic force driving the abutment into or away from the soil. In the latter case, resistance will be low. The detail shown in Fig. 3.14(*b*) is more dependable because of the degree of fixity afforded by the footing, which could also be pile supported.

Monolithic connection details between abutment and superstructure are appropriate for one- and two-span bridges and occasionally for somewhat larger bridges. Because of the relative stiffness of the abutments compared with any internal bents, it is common to assume that all seismic resistance is provided by the abutments. Thus columns of internal bents can be designed for gravity effects alone, although potential hinge regions at tops and bottoms of columns should be detailed for ductility to ensure adequate displacement capacity. By locking-in the bridge to the approach foundation material, high reliance is placed on the integrity of approach embankments, which thus merit special consideration. It should be noted that this class of bridge structure has performed very well in recent California earthquakes, despite recorded response acceleration levels as high as 0.6*g*. Normally, since deformations of the bridge relative to ground will be small, it is sufficient to assume that the bridge will respond as a rigid unit, at the effective peak ground acceleration. Elements of the abutment system must be able to resist the soil pressures resulting from this level of response.

(b) Bearing-Supported Superstructure. The alternative to a monolithic connection is to provide some form of bearing support to the superstructure, via a seat-type abutment, as shown in Fig. 3.14(c) and (d). In the first case, a small gap, suitable for temperature, creep, and shrinkage movements is provided between the backwall and the superstructure. Under longitudinal seismic response, high stiffness and resistance are provided in the closing direction after the initial gap has been closed. In the opposite direction, the stiffness and resistance depend primarily on the bearing characteristics. Abutment design is thus governed by forces generated in the closing direction. However, it may be difficult to design the backwall to resist these forces, and it has been common practice to accept structural damage to the backwall as a result of impact and passive soil pressure.

An alternative detail, shown in Fig. 3.13(d), minimizes the damage and results in more dependable longitudinal response characteristics. The top of the backwall is provided with a weak plane that fails under strong seismic attack. This provides a large gap between the backwall and the superstructure, as shown, allowing the design longitudinal displacements to develop unimpeded in either sense of direction. After a major earthquake, damage is easily inspected and repaired.

(c) Isolated Superstructure. The detail of Fig. 3.14(d) is also suitable for incorporation of isolating bearings, in general with the provision of special energy-dissipating elements. It should be noted that if significant force is to be transmitted through the bearings to the abutment, this force will be largely resisted by the piles alone when the bridge moves longitudinally away from the abutment.

(d) Transverse Response. The design for transverse response deserves special consideration, as it may not be possible to develop such large levels of resistance as afforded by passive soil pressure under longitudinal response. A typical seat-type abutment is shown in Fig. 3.15(a), where transverse resistance is provided primarily by the use of wing walls. Under the direction of acceleration shown, only the wing wall at the right side will provide resistance, and strength will be limited by the moment and possibly by the shear capacity of the corner detail under opening moment. Additional resistance will be provided by any piles supporting the abutment seat, and it may be better to rely on these for transverse resistance. If ductile detailing is incorporated in the piles (see Section 5.6.5), this support system may be designed for forces lower than those corresponding to elastic response.

The lateral resistance of the abutment can also be increased by the use of counterforts behind the backwall, as shown in Fig. 3.15(b). These will have

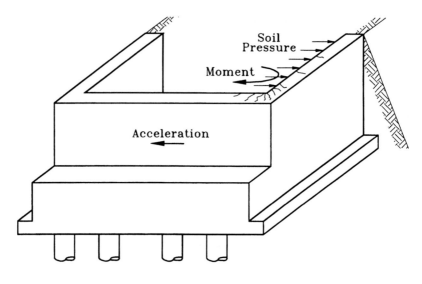

(a)   Resistance by Wing Wall

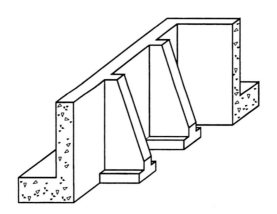

(b)   Back Wall with Counterforts

FIG. 3.15   Abutment resistance to transverse response.

the dual function of strengthening the backwall for longitudinal resistance and acting as lateral bending elements under transverse response, thus adding to the resistance provided by the wing walls and abutments.

Shear keys are often placed between superstructure and abutment to facilitate shear force transfer in the transverse direction when bearing-supported details are adopted. Although this may be suitable for short bridges of one or two spans, in longer structures it can result in high force transfer to the abutments as a consequence of high stiffness relative to that of bents. Damage to the shear keys in moderate earthquakes can thus be expected, and it is common to design the transverse lateral force–resisting system on the basis of two separate scenarios: one with displacements constrained at the abutments, the other with unconstrained displacements, corresponding to conditions after shear key failure.

To some extent this rather undesirable approach is the consequence of uncertainties in maximum feasible shear key forces. However, as shown in Section 5.8.2, these forces can be predicted, even with ductile response of columns, with reasonable certainty using capacity design procedures. It is thus felt that if shear keys are adopted, they should be designed to remain serviceable under the design-level earthquake.

An alternative approach to limiting abutment damage, implemented in the design of some bridges replacing those that failed in the 1994 Northridge earthquake, has been to separate the end diaphragm of the bridge from an approach structure by a movement joint and to support it on cast-in-drilled-hole (CIDH) piles to provide a measure of flexibility to the abutments, both longitudinally and transversely. If the CIDH piles are sleeved for some distance below ground level, as discussed in relation to Fig. 3.2 for short stiff bents, the flexibility of the abutment structure can be tuned to that of the internal bents, regularizing response and avoiding damage concentration in one location.

(e)   Approach Settlement. As discussed in Section 1.2.2, slumping of material behind abutments is common in earthquakes. Although the design approach should be to avoid this by appropriate geotechnical measures, the uncertainty of the science involved indicates that further measures are appropriate. One of the most effective is the use of a settlement slab, which rests on the top of the abutment backwall and the approach fill material, but is separated from the side wing walls by a small gap. A length of 3 to 5 m (10 to 16 ft) is common. If the approach fill material fails, the settlement slab settles at the end away from the backwall, providing a ramp up which emergency services vehicles may have immediate access after an earthquake. Settlement slabs have functioned very effectively in recent earthquakes (e.g., E4).

3.3.3   Bent Configurations

The advantages of single- and multicolumn bents have been partially discussed from a seismic viewpoint in Section 3.3.1. A summary is included in the following with reference to Fig. 3.16.

(a)   Single-Column Bents

ADVANTAGES

- If the superstructure is bearing supported, response characteristics can be made equal transversely and longitudinally, optimizing the seismic design. However, if shear keys are used to restrain transverse displace-

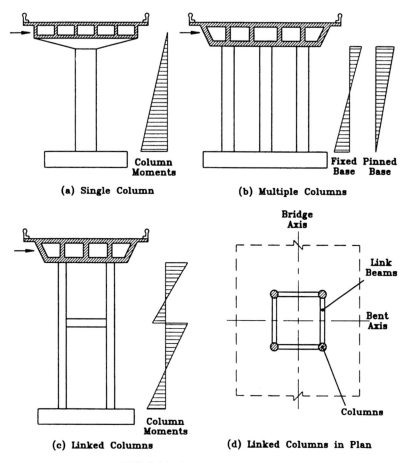

FIG. 3.16 Bent configurations.

ment, transverse and longitudinal period, and hence seismic force, will
be different.

- Since there will be only one plastic hinge location, the behavior is easy
  to determine, with a high degree of accuracy.
- With a monolithic column–superstructure connection, structure and column ductility requirements are identical.

DISADVANTAGES

- With a monolithic column–superstructure connection, seismic design moments will be larger transversely than longitudinally, and longitudinal
  shear demand will be considerable, as illustrated in the capacity demand
  example of Section 1.3.3.

- Moments induced in the column will be higher than for a multicolumn design, as is apparent from examination of Fig. 3.16.
- Displacements of the superstructure will be higher than for a multicolumn design.
- The column must be fixed at the base, resulting in high moments being transferred to the foundation structure.

(b) Multicolumn Bents

ADVANTAGES

- With a monolithic column–superstructure detail, response characteristics in the longitudinal and transverse direction may be made equal.
- If the column bases are fixed to the foundation structure, moments induced in the columns will be significantly less than for the single-column option, even allowing for obvious reduction by the number of columns chosen.
- Foundation design forces can be reduced if the option of pinning the column bases is chosen, but at the expense of increasing column moments.
- As a result of increased redundancy of the system, less reliance is placed on satisfactory performance of a single critical plastic hinge.
- Displacements will be reduced compared with the single-column option.
- A better distribution of column-top moment into the superstructure is possible for monolithic connection details compared with single-column designs.

DISADVANTAGES

- If the superstructure is bearing supported, seismic design moments will be greater longitudinally than transversely.
- Plastic hinges at the critical regions will not develop simultaneously, as a consequence of axial load variations and the influence of cap beam stiffness on end fixity. Thus ductility demand at critical hinges is more difficult to determine.
- Column ductility demands will exceed those of the overall structure as a consequence of cap beam flexibility. This and the preceding point are discussed in more detail in Section 5.3.1(b).
- With monolithic column/superstructure connection, the cap beam may be subjected to very high seismic moments and shears under transverse response.

(c) Linked-Column Bents. For very tall bents, displacements and column design moments can be reduced by connecting columns by one or more link

beams, as shown in Fig. 3.16(*c*). This detail has sometimes been used with a four-column design, with the columns arranged to form a square in plan, as shown in Fig. 3.16(*d*), to provide equal stiffening and strengthening in both transverse and longitudinal directions. With this design it would be normal to provide a bearing-supported superstructure detail, from a platform connecting the column tops.

3.3.4 Column–Pier Section Alternatives

Figure 3.17 illustrates a number of common alternatives for solid column sections, appropriate for seismic resistance.

(a) Circular Sections. Sections A–A and B–B represent the common choice of columns with a circular distribution of longitudinal reinforcement contained within transverse spirals or hoops. The external surface may be circular, octagonal, or some other faceted shape. These sections are efficient, economical, and simple to construct. The continuous curve of the transverse reinforcement provides a continuous confining pressure to the concrete and an inward restraint against buckling to each longitudinal bar. Flexural strength, shear strength, and ductility capacity are independent of the direction of seismic response. Where moment demands are equal in orthogonal directions, the circular section will always be the preferred shape for seismic resistance.

For architectural reasons and to provide better support of the cap beam under eccentric live load, it is common to flare the columns in the upper region, as shown in section C–C, Fig. 3.17, where the column has been flared only in the transverse direction. Although this is the most common choice, columns may also be flared longitudinally. In section C–C the column longitudinal reinforcement has been kept in the original circular configuration, with additional smaller-diameter bars placed in the flare region with additional transverse hoops. The purpose of this reinforcement is to confine the large cover regions adequately external to the main reinforcement. An alternative approach would be to flare the main reinforcement to follow the changing section shape. This results in construction difficulties, with complex bending requirements for the large-diameter longitudinal bars and continuous variation in hoop dimensions with height. It also results in inadequate confinement of concrete and longitudinal bars close to the longitudinal axis. This may be critical where ductility is required at a column-top plastic hinge under longitudinal response. Additional hoop reinforcement may also be required to resist the radially outward component of tension in the curved longitudinal bars.

With the detail shown in section C–C, there will be little enhancement of longitudinal moment capacity over the height of the flare, and the critical section will generally still be at the cap beam soffit. The small gradient of moment strength over the flare will in fact be beneficial, as it will tend to extend the effective length of the plastic hinge, thus reducing peak curvatures

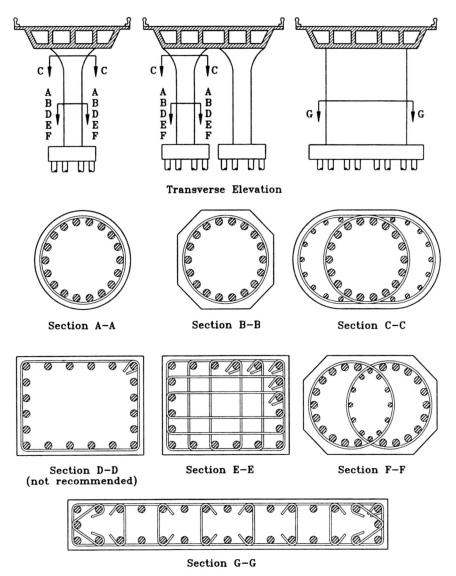

FIG. 3.17 Solid-section column options.

and delaying the onset of cover spalling. For single-column bents, the flare is thus desirable.

When multicolumn bents have flared columns using the detail of section C–C, substantial flexural strength enhancement will occur under transverse response as a consequence of the increased lever arm from the center of the tension reinforcement to the extreme compression fiber. It may thus be difficult to define the column-top plastic hinge location accurately. Unless the effects of flexural strength enhancement are considered fully and carefully, column shear forces corresponding to formation of a lateral plastic mechanism may be severely underestimated, leading to a propensity for shear failure. A detail structurally separating the flare from the cap beam by a layer of expansion joint material has sometimes been adopted.

(b)   Rectangular Columns. Sections D–D to F–F of Fig. 3.17 show alternatives for solid rectangular sections. The option shown in section D–D has only peripheral hoop reinforcement. This does not provide effective confinement to the core concrete or effective lateral restraint to longitudinal bars except those at or immediately adjacent to the corners. As a consequence, this detail should not be used where ductile response is required. Some restricted ductility, the amount dependent on axial load level, longitudinal reinforcement ratio, and column aspect ratio, will be available but should not be relied upon in design.

In building columns it is common to use rectangular columns for seismic resistance [P4]. To provide adequate confinement to concrete and antibuckling restraint for longitudinal reinforcement, overlapping rectangular and octagonal hoops are provided. This is possible because it is uncommon to use more than 16 longitudinal bars in the section. Bridge columns, being typically larger in section than building columns generally require large numbers of longitudinal bars. To provide adequate restraint in these cases may require impractical combinations of transverse hoops, as shown in section E–E. This section has 20 longitudinal bars, which is probably the maximum that could be confined by rectangular hoops, thus limiting its applicability to comparatively small column sizes.

For large rectangular columns, the detail using longitudinal reinforcement contained within intersecting spirals, as shown in Fig. 3.17, section F–F, should be considered. Large chamfers are provided at the corners to avoid excessive cover with consequent spalling problems. The spirals must overlap by a sufficient amount to ensure fully composite action for shear under transverse response. Details are given in Sections 5.3.2(c)(iv) and 5.3.4(b). Square column sections with spirally confined longitudinal reinforcement should be faceted to a fully octagonal section, as in section B–B.

(c)   Rectangular Piers. When longitudinal seismic response is resisted entirely by abutments, or where large variations in bent heights make it uneconomical to build significant longitudinal resistance into a tall bent, an elongated rectangular pier section, as shown in Fig. 3.17, section G–G, may be adopted. This has high strength and stiffness in the transverse direction, acting essentially as a structural wall, but low stiffness in the longitudinal direction, thus attracting little longitudinal shear. The end regions of the wall will need special confinement if ductility is required under transverse response. Under longitudinal response it will generally be impractical to fully confine the long faces of the section, and only occasional cross links will be provided. This will normally be adequate, since longitudinal displacements will be limited by the

stiffer seismic-resisting parts of the bridge (e.g., abutments or shorter bents). Also, pier sections typically have low axial load and longitudinal reinforcement ratios and have considerable ductility capacity even if not confined [P15].

(d) Hollow-Section Columns. With large, long-span bridges, bent section sizes may need to be very large. If the bents are tall, it may be desirable to use a hollow section for the columns. This has the advantage of maximizing structural efficiency in terms of the strength/mass and stiffness/mass ratios and reducing the mass contribution of the column to seismic response (see Section 4.3.1(a)). Tall heavy columns may also develop significant seismic moments as a result of self-weight inertial response.

The hollow circular section of Fig. 3.18(*a*) might appear to be the optimum choice. Care is needed, however, with the two-layer hoop reinforcement shown. Tensions induced in the inner hoop as a result of confinement or shear resistance have a radially inward component which must be resisted by radial links wrapped around the inner hoop, not just around the longitudinal bars. If compression strains at the inside surface approach the crushing strain, the section may implode [Z1]. Note again that the inner hoop has a negative confining effect, acting against the radial links. Because of these effects, the tube must be sufficiently thick to ensure that compression strains at the inside surface are kept to acceptable levels (less than 0.004) even after spalling of the outer cover concrete.

The rectangular section of Fig. 3.18(*b*) is less susceptible to these effects, because of the greater width of the effective compression zone and since there is no deleterious radial component of confining force. However, as with the circular section, effective confinement of the section requires large numbers of transverse links or hoops (and construction is time-consuming and relatively expensive).

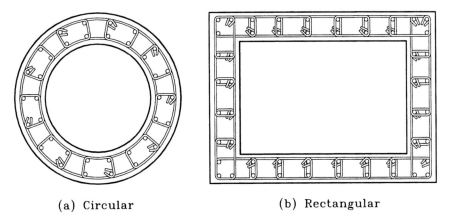

(a) Circular   (b) Rectangular

FIG. 3.18 Hollow-section options for tall columns.

3.3.5 Bent–Foundation Connection

The possibility of providing a pinned-base connection to columns of a multicolumn bent with monolithic column–superstructure connections was discussed
previously. The advantage is that foundation forces are reduced in comparison
with those for a fixed column base condition, at the expense of higher moments
at the top of the column (see Fig. 3.16). The foundation forces for the two
options are examined in more detail in Fig. 3.19, with reference to a footing
supported by foundation cylinders. For the pinned-base condition, foundation
forces result from the column axial load, which may include a seismic component, and from the shear force transferred through the pin. Axial forces in
the foundation cylinders will be nearly equal, allowing design economies. The
footing is subjected to comparatively low bending moment and shear force,
as shown in Fig. 3.19(a).

Transference of the column base moment in Fig. 3.19(b) results in much
higher foundation forces. For the direction of seismic forces shown, significant
uplift forces may develop in the foundation cylinders at the left side. As a

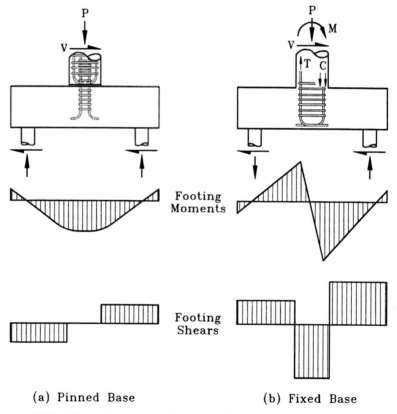

(a) Pinned Base   (b) Fixed Base

FIG. 3.19 Influence of column-base fixity on foundation forces.

consequence, the compressed cylinders at the right side of Fig. 3.19(*b*) are subjected to axial forces more than twice those of the cylinders in the pinned-base condition, for the same column axial load. Footing moments and shears are greatly increased as a consequence. Figure 3.19(*b*) shows the shape of the moments and shear distributions in the footing, treating the column tension and compression stress resultants T and C as forces applied to the footing. Positive moments will be much higher than for the pinned-base condition, and negative moments will also develop if the left cylinders are subjected to tension, as shown. This will require additional top reinforcement above the nominal levels that would be placed for the pinned-base condition. Shear forces will be much higher than for the pinned-base case, with highest values occurring under the column between the tension and compression stress resultants. This "joint" shear force has typically been ignored in all but the most recent bridge designs, but has been shown in experiments to be capable of causing shear failure to footings (see Section 5.6.4) when special joint shear reinforcement has not been provided.

3.3.6 Foundation System

The choice of foundation system will have a very substantial influence on seismic response of the bridge as a whole and on the distribution of column and foundation forces. Figure 3.20 illustrates a number of alternatives.

(a) Spread Footings. With firm ground or rock conditions, the most economical solution will be a spread footing. As shown in Fig. 3.20(*a*), a plastic hinge can be forced into the column base provided that the footing dimensions are such that resistance provided by gravity forces exceeds the overturning moment at the base of the footing. Thus

$$(P + W_f)x > M + Vh_f \tag{3.2}$$

where *M*, *V*, and *P* are the moment, shear, and axial force transmitted from the column to the footing, W_f is the footing weight, and *x* is the distance from the centroid of soil pressure resistance on the footing base. Column forces should be based on overstrength estimates of column capacity. As suggested in Fig. 3.20(*a*), it is not necessary for the entire footing to be in compression at maximum load.

Shear will be taken out by friction on the base of the footing and end bearing on the vertical face of the footing. The latter will be dependable only if the footing is cast against a cut face of natural foundation material rather than having material backfilled against the surface. To improve shear transfer it may be advisable to provide shear keys to positively engage the foundation material at the base of the footing, as shown by the dashed lines in Fig. 3.20(*a*).

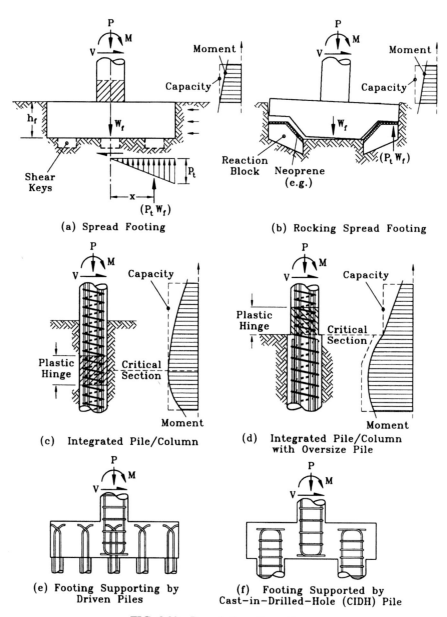

FIG. 3.20 Foundation alternatives.

(b)   Rocking Spread Footings. An alternative approach is suggested in Fig. 3.20(*b*), also applicable for strong ground conditions, where the footing is deliberately proportioned such that the inequality of Eq. (3.2) is not satisfied. In this case the footing will tend to rock at a lower level of lateral response than that corresponding to development of a column base plastic hinge, as indicated in the moment profile to the right of Fig. 3.20(*b*). This will protect

the column from damage during an earthquake, at the expense of significantly increased lateral response displacements at the center of mass. To prevent plastic deformation to the foundation material under the toe during rocking, which could reduce the effective bearing width of the footing, it may be advisable to provide reaction blocks as shown, which will also assist in providing horizontal shear transfer. Rocking response is discussed further in Section 6.4.

(c) Integral Pile-Shaft/Columns. Two alternative concepts for integral pile shaft–column designs are shown in Fig. 3.20. Integral pile shaft–columns are increasingly popular because of the comparative economy of constructing large cast-in-drilled-hole [CIDH] piles compared with driven piles with pile cap footings. They have the added advantage that problems associated with the critical column–footing connection discussed in relation to Fig. 3.19 are avoided. In Fig. 3.20(c) there is no physical distinction between the pile shaft and the column except for a construction joint, normally located at or close to ground level. The maximum moment forms at a depth of typically 1.5 to 2.5 pile diameters. The resulting plastic hinge is comparatively long as a consequence of the gradual variation in moment close to the critical section [see Fig. 3.20(c)]. Spalling of cover concrete will thus not be expected until large plastic rotations have developed. A disadvantage of the detail is that the extent of damage to the plastic hinge region will not be apparent after an earthquake without excavation of foundation material. As shown in Section 5.3.1(b), elastic displacements at the center of mass resulting from flexibility of the soil–pile system may be high, resulting in high ratios of hinge rotational ductility demand to structure displacement ductility demand.

With the alternative detail of Fig. 3.20(d), the pile moment capacity is increased above that of the column in accordance with capacity design principles to ensure that hinging occurs at the base of the column, where damage can be inspected after an earthquake. Because of the greater pile shaft diameter, elastic soil–pile displacements will be less than for the detail in Fig. 3.20(c), resulting in a more advantageous relationship between hinge and structure ductility and lower overall displacements under seismic response. The advantages will be counteracted to some extent by the shorter plastic hinge length and hence by earlier spalling and lower plastic rotational capacity. There is also obviously a cost penalty with this detail.

(d) Pile-Supported Footings. Conventional pile-supported foundation options are illustrated in Fig. 3.20(e) and (f). A variety of driven pile types could be used for the foundation of Fig. 3.20(e), including steel H piles, concrete-filled steel shell (pipe) piles, and prestressed or reinforced concrete piles, with the choice depending primarily on foundation material capacity. In all cases positive connection to the footing should be provided by reinforcement or strand anchored as high as possible in the footing, to assist with shear transfer mechanisms, which can be critical in the footing.

As mentioned earlier, current economics tend to favor cast-in-drilled-hole [CIDH] piles, and it is becoming increasingly common to design footings supported by a comparatively small number of large-diameter piles [750 to 1200 mm (30 to 48 in.) diameter], as indicated in Fig. 3.20(*f*). Full moment-resisting connections between the piles and the footing need special consideration under seismic action.

For most pile-supported footings, the design philosphy will be to force plastic hinging to occur at the base of the column if a moment-resisting column–footing connection is provided, and to keep the piles elastic during seismic response by appropriate capacity design measures. Since the magnitude of pile moments is strongly influenced by the soil stiffness, possible variations in the value of the latter must be investigated when determining column design moments. In certain circumstances it may not be practical to force plastic hinging into the columns or piers of the bent, and limited plastic hinging of the piles can be permitted. This is considered in Section 5.6.5.

Although lateral displacements of the pile cap footing will normally be quite small compared with structure displacements, rotation of the footing caused by axial force variations in the piles can contribute significantly to seismic displacements at the center of lateral force and should always be considered at an early stage of design.

3.4 DEEP-VALLEY-CROSSING CONSIDERATIONS

Deep valley crossings require special consideration for both seismic and non-seismic loads. Figure 3.21 shows three options for a given site. In the first [Fig. 3.21(*a*)], a nonprismatic section bridge constructed by segmental construction from an anchored land span at each end is considered. The end supports are provided by tension members anchored to deep piles ending in belled ends to resist the uplift forces. Typically, the tension piles will have small diameters, functioning primarily to protect prestressing tendons providing the vertical resistance. As a consequence, they will have little seismic resistance, which will be provided solely by the internal compression columns. These columns must thus be designed very conservatively. Consideration must be given to the possibility of geographical amplification of the ground motion, causing the ground on opposite sides of the valley not only to have enhanced seismic input, but possibly to respond asynchronously. As a consequence, the compression columns should be reasonably flexible so that unanticipated out-of-phase ground displacements can be accommodated without excessive damage. Extra confinement of potential hinge regions should also be provided. Vertical seismic response may cause greatly increased tension forces in the end tension piles and needs careful consideration.

The option in Fig. 3.21(*b*), where the superstructure is supported on tall piers, will be less susceptible to asynchronous ground motion but may sustain excessive displacement levels. Column stiffness can be increased by the use

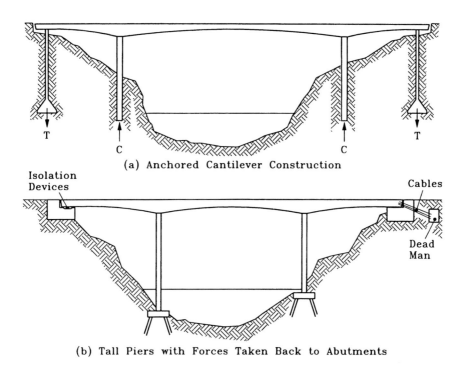

(a) Anchored Cantilever Construction

(b) Tall Piers with Forces Taken Back to Abutments

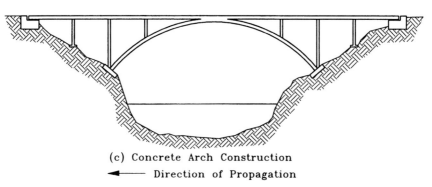

(c) Concrete Arch Construction

◄──────── Direction of Propagation

FIG. 3.21 Structure options for deep valley crossings.

of linked column details such as those described in Fig. 3.16(c) and (d). A variation of this design would be to provide transverse stiffness and strength by linked columns but to rely on the abutments to provide the longitudinal resistance. Longitudinal forces could be taken out by isolation devices, as shown at the left abutments, or by tendons anchoring the end diaphragm back to a "dead man," as shown at the right abutment. The advantage of this detail over conventional restrainer design linking the end diaphragms to the abutment is that the system can be made more flexible, reducing forces in the cables, and the abutment can be protected from the high reaction forces from the restrainers, which can be difficult to accommodate.

Concrete arches [Fig. 3.21(c)], which were common for deep valley crossings from the 1880s to the 1930s, have recently enjoyed renewed popularity as a consequence of new construction techniques involving segmental arch construction supported during construction by cable stays from temporary pylons. An arch design is inherently nonductile, and thus seismic force levels should be based on elastic response levels. Again, special consideration will need to be given to potential problems associated with asynchronous ground motions at springings of the arch caused by geographical amplification or by traveling-wave effects (see Section 2.5).

3.5 LONG-SPAN-BRIDGE CONSIDERATIONS

Major bridges with spans exceeding 200 m (650 ft) in seismic zones will always be subject to special seismic studies and as such, need little general consideration here. Superstructure types may include nonprismatic segmental concrete construction, steel trusses, steel or concrete tied arches, cable-stayed bridges, or suspension bridges. Because of the importance, cost, and general lack of alternative routes, such crossings should always be designed very conservatively. As a consequence, it is common to require elastic or nearly elastic response to a design earthquake of very low probability of occurrence.

Considerations of asynchronous support motion, including the possibility of significant relative displacements of adjacent piers due to traveling-wave effects may be more important than resonant response considerations, particularly when the superstructure form is stiff, such as truss, arch, or segmental construction on soft ground. These effects will be emphasized when the pier heights are small or the pier construction is very rigid. The more flexible superstructure forms, such as suspension or cable-stayed bridges, will generally have ample displacement capacity for seismic response. Fundamental periods of vibration will be long, and response is likely to be at the peak ground displacement level. It will often be prudent to assume that this occurs 180° out of phase at adjacent piers when these are separated by considerable distances, as will often be the case. Vertical response should also receive special consideration.

In the 1995 Kobe earthquake, a number of long-span structures, including tied arches, suspension bridges, and cable-stayed bridges on the Wangan expressway, which is located largely on reclaimed land in Osaka Bay, were subjected to strong ground motion. In general, the long-span structures fared very well. Problems were related primarily to differential ground movements at adjacent piers (see Fig. 1.7) of stiff structural forms with simple supports.

4

MODELING AND ANALYSIS

4.1 INTRODUCTION: SEISMIC BRIDGE ASSESSMENT AND DESIGN TOOLS

The quantification of seismic bridge response in terms of overall structural displacements, member forces, and local deformations is accomplished with the help of mathematical models and analysis techniques. The most common modeling and analysis tools for quantitative seismic bridge response assessment are discussed in this chapter in terms of their usefulness, applicability, and limitations, to provide the reader, wherever possible, with general or specific guidelines for the appropriate model development and analysis execution.

The modeling and seismic analysis of bridge structures have seen a major evolution over recent decades linked directly to the rapid development of digital computing. Both static and dynamic analysis of bridge systems experienced major breakthroughs when finite-element techniques were developed in the mid-1950s [T5] and in the 1970s, when the first authoritative texts on the dynamics of structures and earthquake engineering were written [C10, N1]. Within a very short time, the limitations of hand calculations and iterative solution strategies were overcome by continuously improving hardware and software tools, making it possible to model complete multiframe bridge systems and analyze their dynamic time-history response to incoherent input ground motions along the length of the bridge, considering both material and geometric nonlinear effects.

However, this advance in computational capabilities has not been fully reflected in improved seismic design of new, or the vulnerability assessment and retrofit of existing, bridge structures. To the contrary, more innovative bridge systems were designed and new construction techniques were developed prior to and without the use of digital computers, and the advanced modeling and analysis tools currently available sometimes seem to obscure rather than aid the process of designing new or retrofitting existing bridge structures subjected to earthquake loads. This unfortunate development can be attributed to the fact that due to their complexity and specialization, these advanced models and analyses have developed an existence of their own and are no longer seen just as a tool to aid one of the many necessary steps within the overall design process. Rather, global models and complex analyses are frequently used as the driver for the overall seismic bridge assessment or design effort.

In a perfect world, with deterministic earthquake force input and a well-defined model in terms of known boundary conditions, material properties, and stiffness and damping characteristics, an analysis-driven design seems at least plausible. In reality, the unknown nature of the seismic event, uncertainties in material properties, and unknown boundary conditions, among other imponderables, do not support such an approach but suggest instead that a design process which deals iteratively with all these uncertainties rather than deterministic mathematical models and analyses needs to be the driver. Thus, similar to concept selection and detailing, modeling and analysis must be an integrated part of the overall seismic bridge assessment and design process, providing the necessary tools to quantify seismic demands and capacities. To select the most appropriate model and type of analysis to quantify specific design issues is an art in itself and requires in-depth understanding of (1) the overall seismic bridge design process, (2) the dynamic response of bridge structures under earthquake loads, (3) the consequences of inaccuracies in modeling assumptions, and (4) the available modeling and analysis techniques, with all their limitations and pitfalls. Thus, similar to conceptual design and detailing, experience is required to select the appropriate modeling and analysis tools. In this chapter we can only provide an overview of some of these tools and guidelines toward their useful application.

Furthermore, despite the availability of advanced models and analysis tools, there still exist many seismic bridge response characteristics where limited physical understanding of the actual response mechanism precludes or invalidates the application of complex mathematical models, since basic input parameters are not readily available. Problem areas where seismic bridge response modeling and analysis still needs further understanding, development, and improvement are, among others, (1) soil–structure interaction at abutments and piers, (2) movement joint characterization, (3) fully cyclic (hysteretic) load–deformation characteristics and damping, (4) shear and bending interaction with increasing ductilities, (5) deformations in joints and connection regions, and (6) dynamic interaction of bridge sections with different response characteristics and/or nonsynchronous earthquake excitations.

In the following, an attempt is made to provide some guidance to seismic bridge modeling and analysis in general and to address some of the critical issues outlined above. The purpose of modeling and analysis tools for seismic bridge design and assessment is evaluated first, followed by a limited review of some of the fundamentals in the dynamic response of bridges. Models and modeling assumptions for various seismic bridge response aspects are discussed and different levels of analyses are evaluated based on their usefulness within the bridge seismic design or assessment process. Applications of some of the modeling and analysis tools are presented throughout the chapter, with idealized examples designed to demonstrate specific models or techniques, and finally, by means of a comprehensive seismic bridge response assessment example at the end of this chapter.

4.2 MODELING AND ANALYSIS OBJECTIVES

The principal objective of modeling and analysis tools is the quantification of the seismic response of bridges in terms of structural displacements and member forces and deformations. This quantification is required for both the seismic design of new bridges and the seismic assessment of existing bridges. For the sizing and detailing of new bridges, modeling and analysis tools are used primarily to determine the seismic demand in the form of required member forces and deformations, whereas for the seismic vulnerability assessment of existing bridges the emphasis is on the available deformation and strength capacity quantification. The overview of the seismic bridge analysis process depicted in Fig. 4.1 shows the two approaches schematically.

Design models developed to quantify the *seismic demand* are often based on approximate member dimensions from a preliminary design, utilizing estimated effective section properties and nominal or design material characteristics. To capture the seismic demand, models representing the entire or global structural system are developed and various analysis techniques, mostly linear elastic, provide quantification of member forces for equivalent static or dynamic earthquake load input. Based on these member forces, dimensions and detailing can, if necessary, be iteratively refined. In direct support of the capacity design philosophy and principles outlined in Section 1.3, these analyses will be used primarily to determine (1) the flexural strength characteristics of critical plastic hinge regions, and (2) the required strengths of other members or sections to be protected by capacity design considerations discussed further in Section 5.3.3.

The seismic vulnerability assessment of existing bridges is typically aimed at a quantification of *available capacities* based on known dimensions and design details, effective section properties, and probable, or where possible measured, current material properties. The approach consists of a detailed characterization of the most probable force–deformation behavior of individual bridge components which are subsequently combined in pushover frame

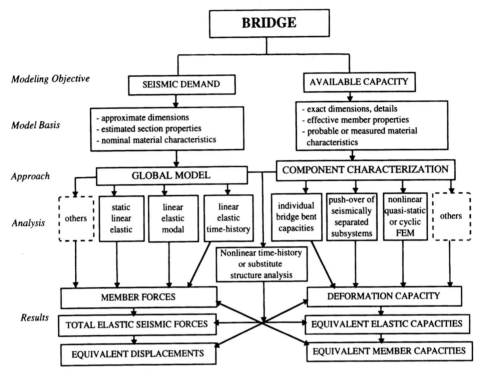

FIG. 4.1 Seismic bridge analysis process.

or complete nonlinear finite-element analyses to obtain deformation capacity estimates of seismically separate subsystems of the bridge structure.

The final step consists of a comparison of demand and capacity results, which requires a determination of equivalent displacements or elastic forces obtained from a corresponding design or assessment model. Only a nonlinear time-history analysis or a substitute structure analysis, as outlined in Fig. 4.1 and discussed further in Section 4.5, combines the nonlinear component characterization and a simulated seismic excitation for a direct seismic bridge response determination. The various modeling and analysis tools available for the quantification of seismic demands and capacities are discussed in the following sections.

Within the seismic bridge analysis process, the model is the tool that facilitates the mathematical formulation of the geometry and behavior characteristics of the prototype bridge structure. The formulation of a mathematical model to describe the geometric domain of a prototype structure is referred to as *discretization* since discrete mathematical elements and their connections and interactions are used to describe the prototype behavior. Various levels of discretization are possible within the mathematical model development, ranging from phenomenological lumped-parameter models and structural component models to detailed finite-element models, as outlined in Fig. 4.2.

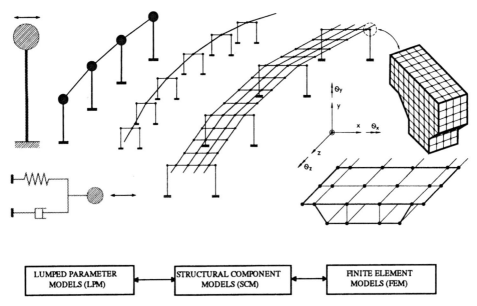

FIG. 4.2 Levels of modeling for seismic bridge analysis.

Lumped-parameter models (LPMs), in which bridge characteristics such as mass, stiffness, and damping are conveniently lumped or concentrated at discrete locations, are simple in their mathematical formulation but require significant knowledge and experience to formulate equivalent force–deformation reelationships of the few idealized elements in order to represent prototype bridge behavior. *Structural component models* (SCMs) are based on idealized structural subsystems that are connected to resemble the general geometry of the bridge prototype, and phenomonological response characterization is provided in the form of member end force–deformation relationships for each structural component or subsystem. Finally, *finite-element models* (FEMs) discretize the actual geometric domain of the bridge structure with a large number of small elements with performance characteristics derived directly from the constituent structural materials. The geometric discretization effort increases significantly from the lumped-parameter models to the structural component models and on to the finite-element models, and can be quantified by the number of defined unknown response quantities. Since most bride analysis models are displacement based, these unknown response quantities are typically expressed in the form of independent deformations at the model joints or nodes and are referred to as *degrees of freedom* (DOFs), where movement of a to-be-determined magnitude is possible. On the other hand, the modeling effort in terms of individual member characterization can be automated to a large degree in FEMs but requires significant definition and engineering judgment for SCMs and LPMs. The modeling effort also increases, when instead of linear elastic behavior, nonlinear monotonic or nonlinear-cyclic response models need to be developed. Simultaneously, the computational effort increases significantly when instead of a linear elastic static analysis, nonlinear and/or time-history analyses are to be performed. The combined discretization and computational effort is the reason why frequently the more complex models or discretizations are developed for simpler analysis techniques, while simple discretizations are utilized in the more complex analysis procedures.

The correct choice of *modeling and analysis* tools depends on (1) the desired response quantity for which the bridge analysis is performed and the level of accuracy needed, (2) the design or assessment limit state for which the bridge model is analysed, (3) the importance of the bridge structure and to which degree better results can be obtained through the use of more complex tools, and (4) the qualifications and experience of the designer and to what degree more detailed results can be utilized in the seismic design or vulnerability assessment process. It is important to reemphasize that both modeling and analysis are viewed as tools which should be used selectively within the bridge seismic design process to quantify the seismic response of the bridge structure. Both sets of seismic design tools are discussed in more detail in subsequent sections of this chapter.

4.3 FUNDAMENTALS OF SEISMIC BRIDGE BEHAVIOR: STRUCTURAL DYNAMICS

Application of the appropriate modeling and analysis tool to the bridge seismic response problem requires a general understanding of some of the key principles of structural dynamics. Those principles that relate directly to simplified bridge design and assessment are reviewed in the following: detailed closed-form or numerical solution procedures for the governing differential equations are left to comprehensive texts on structural dynamics and earthquake engineering [C10,N1]. Examples developed in this section to demonstrate the use of some of these basic dynamic structural analysis principles represent significantly simplified bridge models to keep the focus on the analytical tool rather than on an accurate bridge prototype representation. Refined bridge models and analysis techniques representing and capturing more closely the most probable seismic bridge response are discussed in Section 4.4.

4.3.1 Dynamic Behavior of Bridges: Equation of Motion

The dynamic excitation and response of a bridge subjected to earthquake ground motion in the form of ground acceleration $\ddot{u}_g(t)$ can best be explained by means of a single-degree-of-freedom (SDOF) model of a bridge structure (Fig. 4.3). Such a simplified SDOF model of a single-column bent viaduct under transverse earthquake ground motion input can provide an approximation to

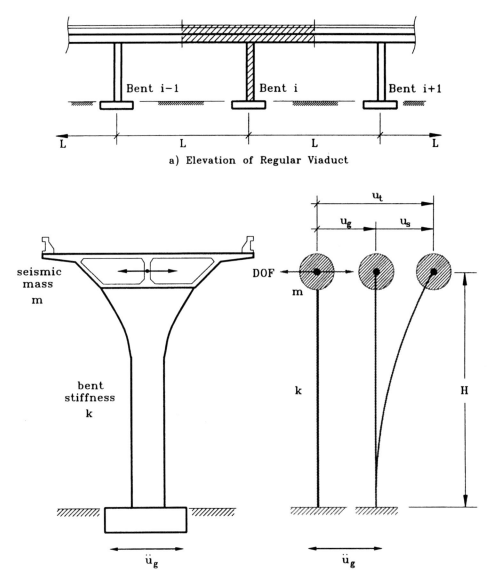

a) Elevation of Regular Viaduct

b) Section   c) Model and Deformations

FIG. 4.3 Transverse dynamic bridge response model.

the true seismic response of the bridge prototype as long as the bridge is straight and consists of a large number of equal spans and piers of equal height or stiffness as shown in Fig. 4.3(a), with all piers exposed to the same coherent earthquake ground acceleration $\ddot{u}_g(t)$ perpendicular to the bridge axis, and with a superstructure that can be assumed to move as a rigid body. For these simplified assumptions, the response of all bents will be the same as the response of the overall bridge and can be represented by the single column shown in Fig. 4.3(a) and (b) with tributary mass from the two adjacent half spans of the superstructure.

In this simplified bridge model, the seismic mass m is assumed to be lumped at the top of the single-column bent at a height H above ground level which represents the distance to the centroid of superstructure mass. The cantilever stiffness of the assumed massless bridge pier can be expressed by k, which is the force required to produce a unit displacement at the center of mass relative to the column base. Furthermore, if damping of the bridge system can be expressed in the form of viscous damping, the characteristic damping force required to resist a unit velocity at the point of mass can be expressed as c. To describe the dynamic response of this simplified bridge model requires a distinction between the structural displacement u_s of the cantilever pier and the total displacement u_t in the form

$$u_t = u_s + u_g \tag{4.1}$$

at the centroid of mass at all times, where u_g is the ground displacement relative to an absolute frame of reference.

Equilibrium of all forces acting on the system at this single displacement degree of freedom requires that (1) the inertia force $f_i(t) = m\ddot{u}_t(t)$, which resists the total acceleration of the seismic mass m, (2) the viscous damping force $f_d(t) = c\dot{u}_s(t)$, which resists the velocity of the mass m through damping in the pier expressed in the form of equivalent viscous damping, and (3) the restoring force $f_s(t) = ku_s(t)$, which resists the structural deformation u_s in the bridge pier by means of internally stored strain energy, can be combined in a single equilibrium equation at the unknown displacement degree of freedom (DOF) in the form

$$m(\ddot{u}_g + \ddot{u}_s) + c\dot{u}_s + ku_s = 0 \tag{4.2}$$

or

$$m\ddot{u}_s + c\dot{u}_s + ku_s = -m\ddot{u}_g \tag{4.3}$$

which represents the equation of motion of the single-degree-of-freedom bridge model in Fig. 4.3(c) under transverse earthquake ground accelerations.

The same SDOF formulation for the single-column bent model in Fig. 4.3 can be used even when significant mass contributions from the pier itself no longer justify the assumption of a massless column. A generalized SDOF system [C10] denoted by *, can be formulated for a single generalized displacement DOF $u^*(t)$ as long as the general shape of the column deformation during earthquake response can be defined as

$$u(x,t) = \psi(x)Z(t) \tag{4.4}$$

where $\psi(x)$ denotes the known deformation shape of vibration and $Z(t)$ represents the variation with time.

A brief discussion of the three structural system parameters of mass, stiffness, and damping in Eq. (4.3) as they relate to dynamic bridge response follows.

(a) Mass. The mass of a bridge system, which contributes to the bridge seismic response in the form of inertia forces, can be characterized by the weight W_s of the moving portion of the bridge divided by the gravitational constant g:

$$m = \frac{W_s}{g} = \frac{W_s}{9.807 \text{ m/s}^2} = \frac{W_s}{32.2 \text{ ft/s}^2} \tag{4.5}$$

Two types of inertia forces can contribute to the bridge dynamic response, translational and rotational inertia forces. When inertia forces are referenced to the mass centroid of a member, these two inertia force components can be expressed as

$$\left.\begin{array}{l} f_i^T(t) = m\ddot{u}(t) \\ f_i^R(t) = j\ddot{\theta}(t) \end{array}\right\} \tag{4.6}$$

where m represents the translational mass and j the rotational mass moment of inertia subjected to translational and rotational accelerations $\ddot{u}(t)$ and $\ddot{\theta}(t)$, respectively.

For bridge systems most of the mass or seismic weight is typically contributed by the bridge superstructure and can often be expressed as a distributed mass $\overline{m}$ along the length of the bridge. Since the in-plane axial and flexural stiffness of a bridge superstructure are generally large compared to the lateral stiffness of the supporting bents, a rigid-body dynamic system for the superstructure with reference to the mass centroid can be a reasonable first assumption. For the example of a four-bent bridge frame in Fig. 4.4 with in-plane rigid superstructure of mass $\overline{m}(x) = \overline{m} = $ const. as depicted in Fig. 4.4(a) and (b), the lumped mass can be expressed as

$$m = \overline{m}L \tag{4.7}$$

for translational motion of the mass centroid, and as

$$j = \int_{-L/2}^{L/2} \overline{m}(x)x^2 \, dx = \frac{\overline{m}L^3}{12} = \frac{mL^2}{12} \tag{4.8}$$

for rotation about the vertical axis through the center of mass. When the

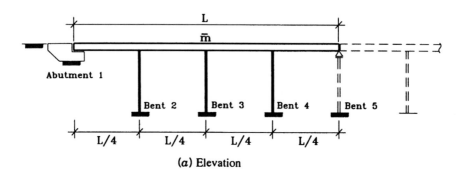

(a) Elevation

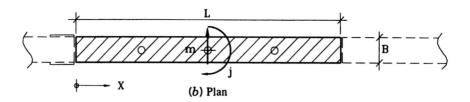

(b) Plan

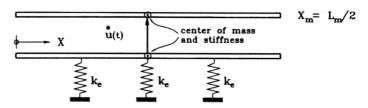

(c) Plan model with free abutment condition

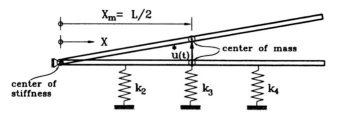

(d) Plan model with hinged abutment condition

FIG. 4.4 Multibent bridge frame.

frame length L is not significantly larger than the superstructure width B (i.e., for wide multilane viaducts), the rotational mass moment of inertia can be expressed as

$$j = m \frac{L^2 + B^2}{12} \qquad (4.9)$$

which for $B \ll L$ takes on the form of Eq. (4.8).

Rotational inertia effects about the bridge longitudinal axis may be more significant than those about the vertical axis. Lateral displacement of the center of mass of a single-column bent such as that shown in Fig. 4.3(b) will be accompanied by vertical displacements of the edges of the superstructure of a magnitude that may be comparable with that of the lateral displacement, if the superstructure width B is similar to, or larger than, the pier height H. The system cannot then be adequately represented by a SDOF model, since the rotational inertia of the superstructure will modify the fundamental period, and induce moments of significant magnitude at the center of mass. These moments may be of the same or reversed sign to the base moment at any instant, as a consequence of the two-mode response of the structure. Failure to model this may result in underestimation of the column shear force in the capacity design process of Section 5.3.3. In these cases, the rotational mass moment of inertia to be used in the two-mode model is given by Eq. (4.9) with $L = 0$. Alternatively, the bent can be modelled by a two-mass representation, where the inertia mass m is represented by two masses of 0.5 m located at the radius of gyration $r = B/\sqrt{12}$ on either side of the column, on the cap beam centerline.

The simplest case of mass model used in many preliminary bridge design and assessment analyses assumes that the entire bridge mass is concentrated in the superstructure and that the mass of the bridge pier is negligible. For the single-column bent with monolithic superstructure connection, shown in Fig. 4.5(a), the entire superstructure mass m can thus be lumped at the superstructure centroid without any contribution from the assumed massless pier. However, if the mass of the bridge pier is large, a tributary mass from the bridge column of clear height H_c can be added to the superstructure mass at height H. The amount of tributary column mass to be added to the generalized displacement degree of freedom $u^*(t)$ at the superstructure centroid can be determined as the mass addition that results in the same dynamic response characteristics as the system that takes the distributed nature of column mass into account.

Since, as outlined in Section 1.3, the bridge columns are expected to form flexural plastic hinges under the design earthquake, and since subsequent to the plastic hinge formation most of the bridge displacement can be attributed to inelastic rotation in these plastic column hinges, a linear deformation model between hinge locations for the bridge column response, as shown in Fig. 4.6, is a reasonable assumption. With the deformed column shape $\psi(x)$ expressed as

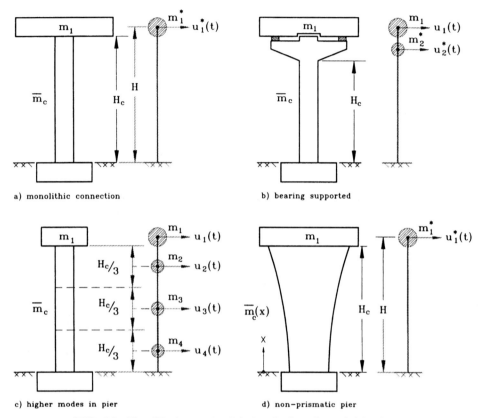

FIG. 4.5 Simplified mass models for single-column bridge bents.

a) monolithic connection

b) bearing supported

c) higher modes in pier

d) non-prismatic pier

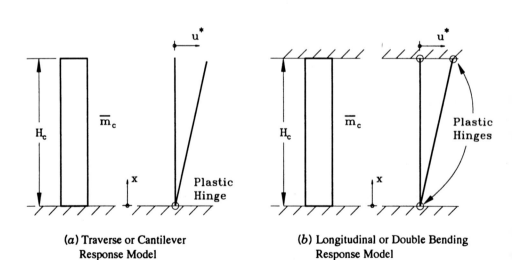

(a) Traverse or Cantilever
Response Model

(b) Longitudinal or Double Bending
Response Model

FIG. 4.6 Idealized inelastic column response models.

$$\psi(x) = \frac{1}{H_c} x \qquad (4.10)$$

and an assumed uniformly distributed mass $\overline{m}_c$ along the column height, the generalized mass m^* which characterizes contributions from the distributed column mass to the generalized displacement DOF u^* can be expressed for the mass components m and j from Eqs. (4.7) and (4.8) at the column centroid $(x_c = H_c/2)$ as

$$m^* = \overline{m}_c[\psi(x_c)]^2 + j[\psi'(x_c)]^2 \qquad (4.11)$$

resulting in a generalized mass contribution from the column of

$$m^* = \overline{m}_c H_c \left(\frac{1}{4}\right) + \frac{\overline{m}_c H_c^3}{12} \frac{1}{H_c^2} = \frac{\overline{m}_c H_c}{3} \qquad (4.12)$$

Since the clear column height H_c in typical bridges is significantly larger than the superstructure depth D, the general approximation of $H_c \simeq H$ in Fig. 4.5 can be made, which allows the generalized mass m^* with contribution of tributary column mass from Eq. (4.12) and superstructure mass m in Fig. 4.5(a) to be expressed as

$$m^* = m_1 + \frac{\overline{m}_c H_c}{3} \qquad (4.13)$$

The decision as to when it is important to include the column mass can be made based on the expected change in dynamic response characteristics, which are shown in Section 4.3.2 to be proportional to the square root of the mass. Therefore, a column with total mass equal to one-half the mass of the tributary superstructure contributes only $\sqrt{1 + 0.5/3} = 1.08$, or an 8% change in dynamic response characteristics if the generalized mass formulation of Eq. (4.13) is employed for a generalized SDOF model as shown in Fig. 4.5(a). However, since maximum member elastic forces are approximately proportional to the effective mass, rather than its square root, we recommend that the formulation of Eq. (4.13) be adopted when $\overline{m}_c H_c \geqslant 0.10 \, m_1$.

For bridges where the superstructure is supported on bearings, as shown in Fig. 4.5(b), the increased displacements in the bearings typically require a two-mass model, mass m_1 for the superstructure and mass m_2 for the cap beam and column, and a multidegree-of-freedom (MDOF) representation of unknown displacement coordinates. Mass m_2 can again be derived as generalized mass m_2^* from Eq. (4.13) as sum of the cap-beam mass and one-third of the tributary mass from the prismatic column portion.

For tall slender columns where higher modes can contribute significantly to the dynamic response of the pier, a multiple lumped-mass model, as shown in Fig. 4.5(c), can be employed. To capture second- or third-mode effects, at least two or three discrete mass locations and associated DOFs need to be modeled. Typically, only translational contributions in the form of Eq. (4.7) are considered since the rotational contributions are small, as can be seen from Eq. (4.12). Finally, nonprismatic columns with flared tops, as shown in Fig. 4.5(d), can have significant contributions to the generalized mass m^*.

The tributary mass or generalized mass at the column top can be expressed based on the linear deformation models in Fig. 4.6 and a known column mass distribution $\overline{m}_c(x)$ as

$$m^* = \int_0^{H_c} \overline{m}_c(x)\psi^2(x)\, dx \tag{4.14}$$

which for the linear displacement shape function of Eq. (4.10) with the generalized displacement DOF at height H of $\psi(x) = x/H$ results in

$$m^* = \int_0^{H_c} \overline{m}_c(x)\left(\frac{x}{H}\right)^2 dx = \frac{1}{H^2} \int_0^{H_c} \overline{m}_c(x)x^2\, dx \tag{4.15}$$

The generalized mass m^* in Fig. 4.5(d) can then be expressed as

$$m_1^* = m_1 + \frac{1}{H^2} \int_0^{H_c} \overline{m}_c(x)x^2\, dx \tag{4.16}$$

For a prismatic column with $\overline{m}(x) = \overline{m} = $ const. and the previously stated simplifying assumption of $H = H_c$, Eq. (4.15) results in the tributary mass derived in Eq. (4.12) for prismatic bridge columns. Thus Eq. (4.15) represents the basic form for the derivation of generalized mass for the column displacement DOF at height H and a linear deformation mode.

(b) Stiffness. The restoring force term $f_s(t) = ku_s(t)$ in the general equation of motion (4.3) depends on the stiffness k of the bridge system. The translational stiffness for slender bridge piers can be expressed as

$$k = \alpha \frac{EI_e}{H_e^3} \tag{4.17}$$

where E is the modulus of elasticity, I_e the effective moment of inertia of the cross section, H_e the effective column height, and the coefficient α represents the boundary conditions. These various parameters affecting the stiffness k are discussed below.

For the bridge frame in Fig. 4.4, lateral stiffness is provided by abutment 1 and bridge bents 2 to 4. For the idealized rotational and translational unconstrained abutment case in Fig. 4.4(c), the bridge piers in the transverse directions deform as cantilevers in the deformation mode, depicted in Fig. 4.7(a), and the stiffness for each pier can be expressed as

$$k_e^T = \frac{\alpha EI_e}{H_e^3} = \frac{3EI_e}{H_e^3} \tag{4.18}$$

Similarly, in the longitudinal bridge direction, assuming a stiff or rigid superstructure, the stiffness term for a column in double curvature bending with both ends fully constrained against rotation as shown in Fig. 4.7(b) can be expressed as

$$k_e^L = \frac{\alpha EI_e}{H_e^3} = \frac{12EI_e}{H_e^3} \tag{4.19}$$

and the total stiffness of the bridge frame in Fig. 4.4(c) with rigid superstructure and the same bent stiffness k_e for all three bents as

$$k^T = 3k_e^T = \frac{9EI_e}{H_e^3} \quad \text{and} \quad k^L = 3k_e^L = \frac{36EI_e}{H_e^3} \tag{4.20}$$

for transverse and longitudinal response, respectively.

   In reality, limited superstructure and foundation flexibility will result in a coefficient α smaller than indicated for the ideal boundary conditions in Fig. 4.7 and can be determined by standard direct stiffness calculations [G3] as

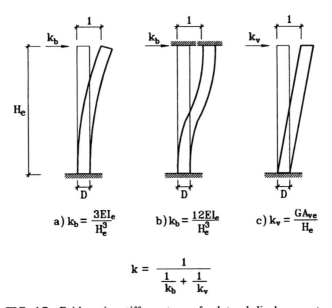

$$a)\, k_b = \frac{3EI_e}{H_e^3} \qquad b)\, k_b = \frac{12EI_e}{H_e^3} \qquad c)\, k_v = \frac{GA_{ve}}{H_e}$$

$$k = \frac{1}{\dfrac{1}{k_b} + \dfrac{1}{k_v}}$$

FIG. 4.7 Bridge pier stiffness terms for lateral displacements.

long as the boundary flexibilities are known. Also, rotational member end stiffnesses are readily available in any textbook on structural analysis, in cases where rotational DOFs are considered in the analysis.

For squat bridge piers, where the clear column height H_c is no longer significantly larger then the column depth D, shear deformations can become significant in comparison with the flexural deformations. The shear deformation for a unit load, or the shear flexibility, in Fig. 4.7(c) can be expressed as

$$f_v = \frac{H_e}{A_{ve}G} \tag{4.21}$$

where A_{ve} represents the effective shear area (to be discussed later) and G the shear modulus of the pier cross section. As a general rule of thumb, shear deformations can become significant when the shear span M/V of the pier is less than three times the pier depth D, or

$$\frac{M}{V} \leq 3D \tag{4.22}$$

where M and V represent the maximum moment M and corresponding shear force V in the bridge pier, respectively. The shear flexibility f_v can be combined with the flexural or bending flexibility

$$f_b = \frac{1}{k_b} = \frac{H_e^3}{\alpha EI_e} \tag{4.23}$$

to form the combined stiffness

$$k = \frac{1}{f_b + f_v} = \frac{1}{H_e^3/\alpha EI_e + H_e/A_{ve}G} \tag{4.24}$$

Since bridge columns are, as outlined in Chapter 1, expected to respond under the design earthquake inelastically, effective member properties H_e, I_e, and A_{ve}, which reflect the extent of concrete cracking and reinforcement yielding, should be used in the modeling and analytical member characterization to obtain realistic seismic response quantification.

From Eq. (4.17) it can be seen that rather than gross-section-based moments of inertia I_g and shear areas A_v, effective properties I_e and A_{ve} are specified, and for the height H in the stiffness formulation an effective height H_e is used. Constitutive parameters E and G are assumed to be constant for most types of bridge analyses and can be determined for concrete piers from the nominal concrete compression strength f_c' based on standard ACI or CEB procedures. The modulus of elasticity E can be determined as outlined in

Section 5.1, Eqs. (5.1) and (5.2), and the shear modulus G with a Poisson ratio v for concrete between 0.15 and 0.2 for homogeneous material assumptions as

$$G = \frac{E}{2(1 + v)} \tag{4.25}$$

While, strictly speaking, even E and G vary depending on loading or unloading and on the orthogonal strain states, these variations are not accounted for in most bridge analyses except in very detailed nonlinear finite-element analyses, discussed in Section 4.5.

The effective column height H_e is different from the clear column height H_c for longitudinal or double bending bridge response, or from the cantilever height to the mass centroid for transverse response, since based on the design philosophy developed in Chapter 1, plastic column hinges are expected to form at one or both ends of the column member. With increasing levels of ductility in these members and hinges, yield penetration of the column reinforcement into the adjacent footing or cap beam occurs, which provides additional flexibility to these yield penetration regions. This added flexibility can conveniently be expressed by an increase of effective column height to H_e for SDOF formulations or with additional nodes and members for MDOF models. For plastic hinges directly adjacent to a concrete cap or footing, the amount of yield penetration into the joint can be estimated as outlined in Chapter 5 from Eq. (5.39) as

$$L_{pj} = \begin{cases} 0.15 f_y d_{bl} & (f_y \text{ in ksi}) \\ 0.022 f_y d_{bl} & (f_y \text{ in MPa}) \end{cases} \tag{4.26}$$

This yield penetration length L_{pj} can be added to the clear column height H_c as outlined in Fig. 4.8 for longitudinal and transverse response, respectively, to form the effective column height H_e. In addition, footing springs, as shown in Fig. 4.8, modeling the effects of soil deformations (discussed in Section 4.4.2) should be employed to obtain the correct stiffness characteristics for bridge bent models.

To reflect the cracked state of a concrete bridge column in the seismic response analysis, an effective or cracked-section moment of inertia I_e should be employed. The effective stiffness EI_e does not reflect only the effect of cracking but also the state of the bridge column determined at first theoretical yield of the reinforcement and can be determined from sectional moment–curvature analyses as

$$EI_e = \frac{M_{yi}}{\Phi_{yi}} \tag{4.27}$$

where M_{yi} and Φ_{yi} represent the ideal yield moment and curvature for a bilinear moment–curvature approximation as discussed further in Section

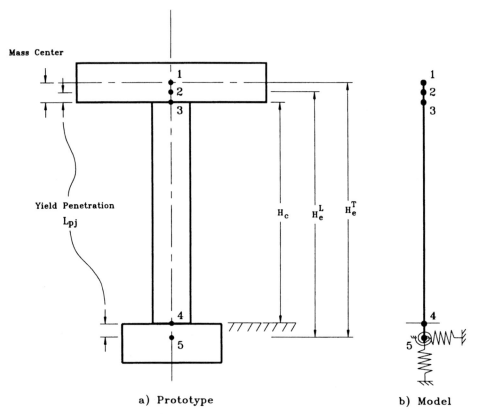

a) Prototype   b) Model

FIG. 4.8   Effective column height.

5.3.2. The effective stiffness I_e depends on the axial load ratio $P_{axial}/(A_g f_c')$ and the longitudinal reinforcement ratio A_{st}/A_g, where A_g and A_{st} represent the gross concrete area and the total longitudinal reinforcement areas, respectively. These effective stiffnesses are represented in graphical form in Fig. 4.9 for typical circular and square column cross sections and show that for typical column reinforcement ratios between 1 and 3% and axial load ratios between 10 and 30%, a reduction in effective section moment of inertia to between 35 and 60% of the gross section moment of inertia I_g is not uncommon. A similar reduction in effective stiffness applies to other concrete bridge members, such as cap beams and superstructure girders, and appropriate values for I_e can also be determined from Eq. (4.27) or Fig. 4.9 for corresponding axial load levels and reinforcement ratios.

Finally, an effective shear stiffness GA_{ve} rather than the shear stiffness based on the shear area A_v should be employed to reflect the increased shear deformations in flexurally cracked concrete members. Again, a dependency similar to the effective flexural stiffness EI_e on the axial load and the reinforcement ratio can be expected prior to significant shear distress, which

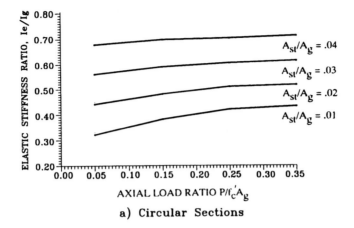

a) Circular Sections

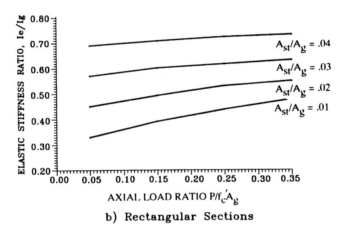

b) Rectangular Sections

FIG. 4.9 Effective stiffness of cracked reinforced concrete sections [N4].

should not occur in capacity design–protected bridge members. Due to lack of specific research data, it can be assumed that the effective stiffness reduction in shear can be considered proportional to the effective stiffness reduction in flexure:

$$GA_{ve} = GA_v \frac{EI_e}{EI_g} \quad \text{or} \quad A_{ve} = A_v \frac{I_e}{I_g} \tag{4.28}$$

until more experimental and analytical research data become available.

(c) Damping. The third term in the general equation of motion is a viscous or velocity proportional damping force $f_d(t) = c\dot{u}_s(t)$, which is used primarily

for mathematical or numerical convenience and stability rather than for the phenomenological modeling of actual bridge damping characteristics. Viscous damping is physically correct only for an oil-filled dashpot and is difficult to rationalize for other forms of damping actually encountered in bridge or other structural systems. More common damping types in bridges are (1) Coulomb damping, (2) radiation damping, and (3) hysteretic damping. Coulomb or friction damping occurs primarily in bridge superstructure bearings and movement joints and is independent of velocity or displacement. To a lesser degree, friction damping can occur in cracks of reinforced concrete structures. Radiation damping in bridges occurs due to soil structure interaction (SSI) and energy dissipated by waves radiating out into the half-space of soil surrounding the bridge footings. The most common and physically most obvious form of damping or energy dissipation in bridge structures is in the form of hysteresis of the force–deformation response.

To conform to the simple mathematical form of the equation of motion, other forms of damping, in particular, hysteretic damping, encountered in bridge systems are conveniently expressed in the form of an equivalent viscous damping coefficient c_{eq}. The equivalent viscous damping coefficient c_{eq} is commonly expressed by the equivalent damping ratio ξ_{eq} and the critical damping coefficient c_{cr}, which is the smallest amount of damping for which no oscillation occurs in free dynamic response [C10]:

$$c_{eq} = \xi_{eq} c_{cr} \tag{4.29}$$

The hysteretic damping or energy loss per cycle, represented by the area A_h in Fig. 4.10 for one complete idealized load–displacement hysteresis loop, can then be converted for the same displacement amplitude to an equivalent viscous damping ratio:

$$\xi_{eq} = \frac{A_h}{2\pi V_m \Delta_m} = \frac{A_h}{4\pi A_e} \tag{4.30}$$

where V_m and Δ_m represent the average peak force and displacement values [C10]. The area A_e represents the elastic strain energy stored in an equivalent linear elastic system under static conditions with effective stiffness

$$k_{eff} = \frac{V_m}{\Delta_m} \tag{4.31}$$

The equivalent viscous damping coefficient can then be obtained from Eq. (4.29). To assess the magnitude of equivalent viscous damping from hysteretic damping, typical force–displacement hysteresis loop shapes for various bridge members are depicted in Fig. 4.11.

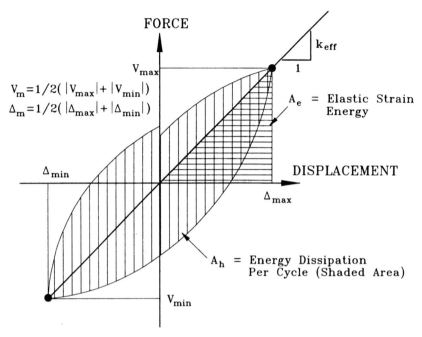

FIG. 4.10 Hysteretic energy dissipation and effective stiffness for cyclic response.

It is obvious from Fig. 4.10 that the maximum equivalent viscous damping ratio which can be obtained with Eq. (4.30) is $\xi_{eq} = 2/\pi = 0.64$ for a system that cycles with rigid–perfectly plastic force–deformation characteristics. This rigid–perfectly plastic loop shape is not very realistic for typical local inelastic mechanisms in bridges that are depicted in Fig. 4.11 for one representative cycle. Even frequently used elastic–plastic response idealizations such as those shown in Fig. 4.11(*a*) apply in very few cases. Only friction slider bearings as shown in Fig. 4.11(*c*) can approach this value. Beam hinges with no or low axial load levels can also exhibit significant hysteretic energy absorption, as shown schematically by the large loops in Fig. 4.11(*b*), and can result in equivalent viscous damping ratios of 30% or higher. High axial loads on a bridge member such as columns or prestressed cap beams result in pinched hysteretic loop shapes, as shown in Fig. 4.11(*d*), resulting in reduced equivalent viscous damping between 10 and 25%. The rocking response of a bridge pier as depicted in Fig. 4.11(*e*) is essentially nonlinear elastic without noticeable hysteresis and thus very little equivalent hysteretic damping. However, in the case of foundation rocking, additional energy is dissipated in the form of radiation damping in the surrounding soil, as outlined in Section 6.4.

For bridge members with nonsymmetric response characteristics, as shown in Fig. 4.11(*f*), the displacement-dependent equivalent viscous damping coefficient approach no longer applies strictly since displacements in the two directions are of different magnitude, but an averaged procedure, as suggested

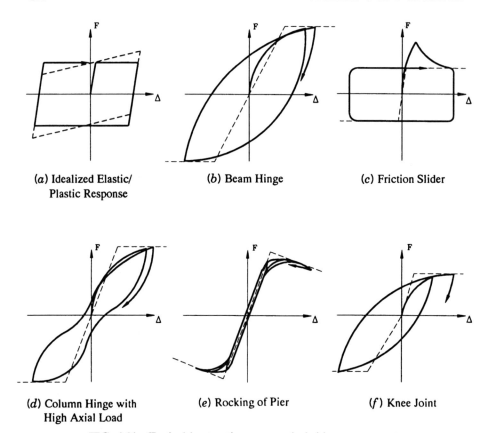

(a) Idealized Elastic/   (b) Beam Hinge   (c) Friction Slider
    Plastic Response

(d) Column Hinge with   (e) Rocking of Pier   (f) Knee Joint
    High Axial Load

FIG. 4.11   Typical hysteretic response in bridge components.

in Fig. 4.10, can still be employed. The same difficulty exists with any dynamic analysis where equivalent viscous damping is employed since the damping values are derived for the maximum amplitudes and are subsequently applied in the analytical model also to all smaller-amplitude cycles.

From the discussions above it is obvious that determination of a correct equivalent viscous damping coefficient for analytical bridge models is difficult at best. Thus empirical damping values are frequently employed to reflect the sum of all possible damping contributions, as well as the fact that most of the cyclic dynamic response in an earthquake is expected to occur at smaller-than-expected maximum displacement levels.

Thus, for steel structures, damping values between 2 and 5% of critical damping are commonly assumed, while for concrete structures a range from 2 to 7% is used to reflect the most representative dynamic response range. In light of these uncertainties, the commonly assumed 5% viscous damping coefficient in structural dynamic analysis can hardly be argued with. Only in cases where (1) soil–structure interaction plays an important role, (2) special energy absorption devices are employed, and (3) high hysteretic energy dissipation is relied upon should higher damping coefficients be employed. It should also be noted that most analytical models will be based on initial elastic stiffness, as discussed in the preceding section, and the damping adopted should represent the elastic phase of response. Where inelastic time-history analysis is used, the hysteretic energy dissipation will be directly modelled by the force-displacement hysteresis rules adopted in the analysis. Only with the substitute structure analysis procedure, discussed further in Section 4.5.2(*b*), where the effective stiffness represents that at maximum displacement, rather than at yield, should the effective damping be increased to include the effects of hysteretic damping.

4.3.2 Bridge Dynamic Response Characteristics

Independent of the specific dynamic input, each bridge system is represented within the elastic range by dynamic response modes typically referred to as the *natural modes of vibration,* characterized by independent mode shapes Φ_i with corresponding periods of vibration T_i. The number of characteristic mode shapes and vibration periods of a bridge model depend on the selected number of dynamic degrees of freedom defined during the analytical model discretization.

While the prototype bridge features an infinite number of vibration modes, bridge analysis models feature a selected finite number of DOFs and associated modes of vibration. However, the governing dynamic response of a bridge can typically be captured by the contribution of a limited number of vibration modes. The fundamental or lowest mode of vibration can often provide a good indication of the dynamic response of a bridge, making single-degree-of-freedom models, which approximate the fundamental dynamic response of the prototype bridge, invaluable design tools.

(a) Single-Degree-of-Freedom Characteristics. The fundamental or first mode of vibration characteristics can be found for simple systems from a single-degree-of-freedom (SDOF) model like the one shown in Fig. 4.3 once the lumped mass and stiffness characteristics are known. As long as damping is significantly less than critical damping, which is the case for essentially elastic response, damping has very little influence on the dynamic response characteristics and is typically ignored.

For a single-degree-of-freedom bridge model with lumped mass m and effective stiffness k, the undamped free vibration can be expressed from Eq. (4.3) as

$$m\ddot{u}(t) + ku(t) = 0 \qquad (4.32)$$

Assuming that the displacement $u(t)$ with time follows a harmonic motion, as shown in Fig. 4.12, of the form

$$u(t) = A \sin(\omega t - \alpha) \tag{4.33}$$

where ω is the circular natural frequency, α a phase shift for the sinusoidal response, and A a scaling factor that determines the amplitude of the harmonic motion, Eq. (4.33) and its second time derivative can be substituted into Eq. (4.32), resulting in the characteristic equation

$$ku - \omega^2 mu = (k - \omega^2 m)u = 0 \tag{4.34}$$

For arbitrary displacements u, Eq. (4.34) can be satisfied when

$$|k - \omega^2 m| = 0 \tag{4.35}$$

which occurs only for a specific circular frequency ω or the eigenvalue of Eq. (4.35). The eigenvalue solution of this scalar equation represents the circular frequency ω at which Eq. (4.35) is satisfied as

$$\omega = \sqrt{\frac{k}{m}} \tag{4.36}$$

and from the undamped natural circular frequency ω in Eq. (4.36) the cyclic natural frequency f and natural period of vibration T for a SDOF bridge model can be found as

$$f = \frac{\omega}{2\pi} \tag{4.37}$$

$$T = \frac{1}{f} = \frac{2\pi}{\omega} \tag{4.38}$$

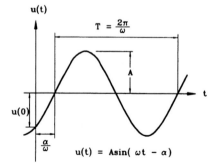

FIG. 4.12   Free undamped harmonic response of SDOF system.

Combining Eqs. (4.36) and (4.38), the fundamental or natural undamped period of vibration can be expressed as

$$T = 2\pi \sqrt{\frac{m}{k}} \tag{4.39}$$

The corresponding natural mode shape consists of a displacement at the designated SDOF (i.e., the transverse motion of the lumped mass in Fig. 4.3), of arbitrary sign and magnitude.

In bridge systems where high damping expressed in the form of an equivalent viscous damping ratio ξ_{eq} is present, the damped circular frequency ω_d can be expressed as

$$\omega_d = \omega \sqrt{1 - \xi_{eq}^2} \tag{4.40}$$

and the corresponding damped natural period of vibration of a SDOF system as

$$T_d = 2\pi \sqrt{\frac{m}{k(1 - \xi_{eq}^2)}} \tag{4.41}$$

From Eqs. (4.40) and (4.41) the small influence of the damping ratio on the dynamic response characteristics is evident. For example, an equivalent viscous damping ratio of 10% only increases the natural period of vibration by 0.5%, and even a 50% damping ratio only results in a 15% natural period elongation. Since typically maximum expected equivalent viscous damping ratios in bridges are less than 15%, the influence of damping on the dynamic response characteristics can be neglected.

As discussed above, m and k in Eq. (4.39) represent the effective seismic mass W_s/g and the effective stiffness k_{eff} with reference to the single displacement degree of freedom at the mass centroid. This single-degree-of-freedom concept to estimate the fundamental dynamic bridge response characteristics can also be applied to structures with distributed parameters in the form of a generalized single-degree-of-freedom system as outlined by Eq. (4.4).

Expressions for a generalized single-degree-of-freedom response of a distributed parameter systems can be found in [C10,N1], and only the special case of lumped stiffness and mass characteristics along the length x of a bridge structure are discussed below since discrete springs (modeling the stiffness of individual bents) and discrete masses (modeling lumped superstructure inertia) can readily be identified for typical bridge systems. In the case of discrete translational and rotational masses m_i and j_i, the principle of virtual work can be applied to a generalized single-degree-of-freedom system in the form of Eq. (4.4) such that a new generalized mass m^* can be derived as

$$m^* = \sum_i m_i \psi_{(xi)}^2 + \sum_i j_i \psi_{(xi)}'^2 \qquad (4.42)$$

and a new generalized stiffness k^* as

$$k^* = \sum_i k_i \psi_{(xi)}^2 \qquad (4.43)$$

The analytical relationships for the dynamic response characterization of a SDOF model are illustrated in the following with the example of the four-span bridge bent in Fig. 4.4.

In this example, realistic bridge conditions, such as abutment stiffness and flexibility, as well as movement joint constraints with the adjacent frame, are replaced by idealized free or fixed boundary conditions to demonstrate and emphasize the analytical procedures. More realistic prototype characteristics are discussed in Section 4.4.2.

Example 4.1. For the example of the multicolumn bent in Fig. 4.4 with an in-plane rigid bridge deck, the generalized mass and stiffness parameters can be found as follows, based on the idealized boundary conditions.

Case 1: Abutment 1 Free to Move. With idealized free boundary conditions of the three-column bent as shown in Fig. 4.4(c) and the assumption of the same effective stiffness k_e in all three bents, the generalized parameters for the transverse translation of the mass centroid at $x_1 = L/2$ can be expressed with

$$\psi(x) = 1 = \text{const.} \qquad (4.44)$$

as

$$\begin{aligned} m^* &= m_1 \psi_1^2 + j_1 \psi_1'^2 \\ &= \overline{m}L + 0 = \overline{m}L \end{aligned} \qquad (4.45)$$

and

$$k^* = \sum_i k_i \cdot 1 = 3k_e \qquad (4.46)$$

with the effective stiffness k_e for each bent derived from Eq. (4.18). The fundamental period of vibration can then be found from Eq. (4.39) as

$$T = 2\pi \sqrt{\frac{m^*}{k^*}} = 2\pi \sqrt{\frac{\overline{m}L}{3k_e}} \qquad (4.47)$$

Case 2: Abutment 1 Pinned. For the case of an idealized pinned in-plane abutment boundary condition as shown in Fig. 4.4(d), an infinite transverse stiffness at the abutment is added to the system and the generalized displaced shape can be expressed as

$$\psi(x) = \frac{1}{x^*} x \tag{4.48}$$

where $x^* = L/2$ represents the distance from the center of stiffness to the generalized coordinate or displacement DOF $u^*(t)$. Lumped-mass properties m and j can be found at $x = L/2$ from Eqs. (4.7) and (4.8), respectively. The generalized mass can now be found as

$$m^* = m\psi^2 + j\psi'^2 = \overline{m}L \left(\frac{L/2}{x^*}\right)^2 + \frac{\overline{m}L^3}{12} \left(\frac{1}{x^*}\right)^2 = mL + \frac{m}{12}\frac{L^3}{(L/2)^2} = \frac{4}{3}\overline{m}L \tag{4.49}$$

and the generalized stiffness as

$$k^* = \sum_i k_i \psi^2 = \sum k_i \left(\frac{x_i}{x^*}\right)^2 = \frac{7}{2} k_e \tag{4.50}$$

resulting in a natural period of vibration of

$$T = 2\pi \sqrt{\frac{m^*}{k^*}} = 2\pi \sqrt{\frac{8\,\overline{m}L}{21\,k_e}} \tag{4.51}$$

(b)   Multidegree-of-Freedom Characteristics. For multidegree-of-freedom (MDOF) bridge models, Eq. (4.32) still describes the undamped free vibration response, but now **k** and **m** are no longer scalars but matrices containing stiffness and mass coefficients, which correspond to the vector of chosen displacement degrees of freedom **u** and their interaction. Denoting vector and matrix quantities by bold type, the general undamped free vibration of a MDOF system can be expressed as

$$\mathbf{m}\ddot{u}(t) + \mathbf{k}u(t) = 0 \tag{4.52}$$

The stiffness matrix **k** can be obtained from standard static displacement-based analysis models [G3] and features coupling between DOFs in the form of off-diagonal terms, whereas the mass matrix **m,** due to the negligible effect of mass coupling, can best be expressed in the form of tributary lumped masses to the corresponding displacement degree of freedom, resulting in a diagonal or uncoupled mass matrix.

The characteristic dynamic equation for the n-MDOF bridge model can now be expressed as

$$(\mathbf{k} - \omega^2 \mathbf{m})\mathbf{u} = 0 \tag{4.53}$$

which for arbitrary displacements requires that

$$|\mathbf{k} - \omega^2\mathbf{m}| = 0 \tag{4.54}$$

and n roots or solutions ω_n^2 can be found that satisfy Eq. (4.54), representing the n natural circular frequencies of motion of the bridge model.

Once the vector ω with characteristic or modal frequencies is obtained, the corresponding mode shapes or natural modes of vibration can be obtained by substituting the individual modal frequencies into Eq. (4.53), and arbitrarily prescribing the magnitude of one of the DOFs for the harmonic motion, since with time, any magnitude of displacements between zero and the maximum values is possible.

Example 4.2. The basic concepts of the dynamic response characterization of a MDOF bridge model is explained by the example of Fig. 4.13, which represents an idealized bridge bent for a double-deck viaduct and the corresponding simplified two-degree-of-freedom analysis model. In this model, for illustration purposes, axial and shear deformation of members are neglected and the bridge superstructures are assumed to be rigid flexural elements. More realistic bridge modeling assumptions for multicolumn bents are provided in Section 4.4.2.

From Eq. (4.19) the flexural stiffness for each upper column can be determined as $12EI/H^3$ and from Eq. (4.18) the individual lower column translational stiffness is $3EI/H^3$, resulting in a stiffness matrix $\mathbf{k}$ of

$$\mathbf{k} = \begin{bmatrix} \dfrac{24EI}{H^3} & \dfrac{-24EI}{H^3} \\ \dfrac{-24EI}{H^3} & \dfrac{30EI}{H^3} \end{bmatrix} = \dfrac{6EI}{H^3}\begin{bmatrix} 4 & -4 \\ -4 & 5 \end{bmatrix} \tag{4.55}$$

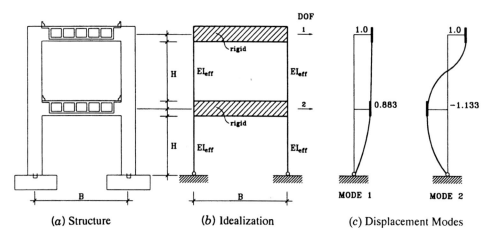

(a) Structure   (b) Idealization   (c) Displacement Modes

FIG. 4.13 Double-deck viaduct bridge bent.

The mass matrix can be expressed with the cap beam length B and the assumed distributed constant mass $\overline{m} = m/B$ as

$$\mathbf{m} = \begin{bmatrix} \overline{m}B & 0 \\ 0 & \overline{m}B \end{bmatrix} = \overline{m}B \begin{bmatrix} 1 & 0 \\ 0 & 1 \end{bmatrix} = m \begin{bmatrix} 1 & 0 \\ 0 & 1 \end{bmatrix} \tag{4.56}$$

and the characteristic Eq. (4.54) requires that the determinant of

$$\left| \frac{6EI}{H^3} \begin{bmatrix} 4 & -4 \\ -4 & 5 \end{bmatrix} - \omega^2 m \begin{bmatrix} 1 & 0 \\ 0 & 1 \end{bmatrix} \right| = 0$$

Substituting $A = \omega^2 mH^3/6EI$ results in

$$\left| \frac{6EI}{H^3} \begin{bmatrix} 4 - A & -4 \\ -4 & 5 - A \end{bmatrix} \right| = 0$$

and $A^2 - 9A + 4 = 0$ with roots $A_1 = 0.469$ and $A_2 = 8.531$.

The circular natural frequencies can now be obtained as

$$\omega_i = \sqrt{\frac{6EIA_i}{mH^3}}$$

and the mode shapes ϕ_i can be determined from Eq. (4.53) assuming that $u_1 = 1$ as

$$\phi_1 = \begin{Bmatrix} 1.0 \\ 0.883 \end{Bmatrix} \quad \text{and} \quad \phi_2 = \begin{Bmatrix} 1.0 \\ -1.133 \end{Bmatrix}$$

These mode shapes are depicted schematically in Fig. 4.13(c) and show that in mode 1 the two lumped masses move in phase, whereas the sign change in mode 2 constitutes out-of-phase motion of the two mass points.

4.3.3 Elastic Seismic Response of Bridges: Maximum Response Values

The general forced vibration of a structure in the case of an earthquake was expressed by Eq. (4.3) with the mass times the acceleration as the forcing function. For a discrete displacement-based analytical bridge model, Eq. (4.3) can be written in matrix form as

$$\mathbf{m}\ddot{\mathbf{u}}(t) + \mathbf{c}\dot{\mathbf{u}}(t) + \mathbf{k}\mathbf{u}(t) = \mathbf{p}_{\text{eff}}(t) \tag{4.57}$$

with the forcing function $\mathbf{p}_{\text{eff}}(t)$ expressed as

$$\mathbf{p}_{\text{eff}}(t) = -\mathbf{m}\mathbf{r}\ddot{u}_g(t) \tag{4.58}$$

The vector **r** represents the displacement of each of the DOFs for a unit ground displacement u_g and determines which mass DOFs are directly excited by the earthquake ground motion. For only translational DOFs in the direction of earthquake ground motion, the influence coefficient vector **r** takes on the form of a unit vector {**1**}. Equations (4.57) and (4.58) apply equally to lumped SDOF, generalized SDOF, or MDOF systems with the number of simultaneous equilibrium conditions or the problem size to be solved represented by the number of selected displacement DOFs in the analytical bridge model.

The dynamic bridge response analysis can be carried out in various forms: through (1) a rarely used frequency-domain analysis, (2) a direct integration of the coupled equilibrium equations in the time domain, or (3) a transformation to normal or modal coordinates and solution and superposition of uncoupled or orthogonal modal response. All three techniques lead to a complete time-history displacement response of the bridge model, and details of the various solution strategies can be found in the technical literature on structural dynamics [C10,N1]. In general, since both the frequency-domain and modal time-domain analyses rely on the principle of superposition, these techniques are only applicable to linear or linearized bridge systems, whereas the direct integration can also be employed for nonlinear time-history analyses. However, linearized iterative analysis techniques based on load-dependent Ritz vectors [W8] rather than natural made shapes can be used to solve for nonlinear time-history response. Since earthquake ground motions tend to excite the lowest modes of vibration more than the higher modes, good approximations of the earthquake response of a bridge can be obtained from only a few modes, making the modal superposition analysis a powerful tool for bridge systems, particularly when large numbers of DOFs are involved.

The seismic assessment or design of bridges is generally based on extreme or maximum dynamic response quantities and does not necessarily require a complete time-history response. Although these maximum response quantities can be obtained by scanning time-history response records, it is typically sufficient and more convenient to determine maximum modal response quantities by means of response spectra, as discussed in Section 2.2, once the dynamic response characteristics in the form of natural periods of vibration and mode shapes have been determined.

In the combination of maximum response values for individual modes of vibration, two issues need to be considered: (1) that each of the modes of vibration has a different earthquake participation factor, and (2) that the maximum response in each mode does not occur at the same time and in the same direction during the earthquake duration. For a bridge model with mass matrix **m**, normalized mode shapes Φ_i, and ground motion influence coefficient vector **r**, the participation of each mode can be obtained as the modal participation coefficient

$$p_i = \frac{\Phi_i^T \mathbf{m} \mathbf{r}}{\Phi_i^T \mathbf{m} \Phi_i} \qquad (4.59)$$

The modal participation coefficient p_i for mode i is a function of the ith mode shape, the mass distribution of the structure, and the direction of earthquake excitation. The modal participation coefficient provides a measure as to how strongly a given mode participates in the dynamic response. Since mode shapes Φ_i can be normalized in different ways, the absolute magnitude of the participation factor has no meaning, only its relative magnitude with respect to the other participating modes. The maximum spectral response for mode i can now be found as

$$\mathbf{R}_i = p_i \cdot \Phi_i \cdot S_i \tag{4.60}$$

where S_i represents the ordinate of the employed response spectrum at the natural period T_i.

A direct combination of all the response maxima for the individual modes can be obtained as

$$\mathbf{R} = \sum_i |\mathbf{R}_i| = \sum_i |p_i| \cdot |\Phi_i| \cdot S_i \tag{4.61}$$

Clearly, Eq. (4.61) overestimates the actual maximum response since sign and time differences of maximum modal contributions are not considered. A more reasonable estimate of the probable response maxima can be obtained from the square root of the sum of the squares (SRSS) of the modal maxima in the form

$$\mathbf{R} = \sqrt{\mathbf{R}_1^2 + \mathbf{R}_2^2 \cdots} = \sqrt{\sum_i \mathbf{R}_i^2} \tag{4.62}$$

For cases where modal response characteristics are not well separated in time (i.e., $T_n \approx T_m$), the SRSS combination technique has been shown to lead to erroneous results [C10]. In these cases a complete quadratic combination (CQC) of modal response maxima is recommended [W5] in the form

$$\mathbf{R} = \sqrt{\sum_i \sum_j \mathbf{R}_i \cdot \rho_{ij} \cdot \mathbf{R}_j} \tag{4.63}$$

where ρ_{ij} represents the cross-modal coefficients which depend on the modal damping ratio ξ_i and ξ_j and the modal period ratio $t = T_i/T_j$ as

$$\rho_{ij} = \frac{8\sqrt{\xi_i \xi_j}\,(\xi_i + t\xi_j)t^{3/2}}{(1 - t^2)^2 + 4\xi_i\xi_j t(1 + t^2) + 4(\xi_i^2 + \xi_j^2)t^2} \tag{4.64}$$

Other proposed combination techniques that follow the cross-correlated format of Eq. (4.63) differ from the CQC approach in defining the derivation of correlation coefficients ξ_{ij} [C10] but are not widely used in seismic bridge analysis.

In cases where the bridge analysis considers multisupport excitations in the form of effects of wave passage, incoherence of support motions, and spatially varying local site conditions, the effects of correlation between the support motions and the dynamic modes of the bridge can be accounted for using the multiple support response spectrum (MSRS) method [D2].

Example 4.3. To demonstrate the spectral earthquake response analysis of a bridge structure, consider again the simple idealized 2-DOF model of the double-deck viaduct bent in Fig. 4.13. Assuming dimensions $B = 50$ ft (15 m) and $H = 15$ ft (4.5 m), a weight per area of bridge deck of 0.3 kips/ft^2 (14.3 kN/m^2) which includes the bent self-weight for each roadway level, and a regular bridge structure with 80-ft spans between bents, the mass m at each DOF is $m = 80 \times 50 \times 0.3/32.2 = 37.3$ kips-s^2/ft (544 tonnes). With an effective stiffness $EI_e = 6 \times 10^6$ kip-ft^2 2.5 $\times$ 10^6 kNm^2) based on $0.5I_g$, the stiffness and mass matrices of Eqs. (4.55) and (4.56) take the form

$$[K] = \begin{bmatrix} 42{,}670 & -42{,}670 \\ -42{,}670 & 53{,}330 \end{bmatrix} \text{(kips/ft)} \quad \text{and} \quad [M] = \begin{bmatrix} 37.3 & 0 \\ 0 & 37.3 \end{bmatrix} \text{(kips-s}^2\text{/ft)}$$

resulting in the natural periods of vibration

$$\{T\} = \begin{Bmatrix} 0.543 \\ 1.27 \end{Bmatrix} \text{(s)}$$

with mode shapes

$$[\phi] = \begin{bmatrix} 1.0 & 1.0 \\ 0.883 & 1.133 \end{bmatrix}$$

as shown in Fig. 4.13(c).

With a ground motion influence coefficient vector

$$\{r\} = \begin{Bmatrix} 1.0 \\ 1.0 \end{Bmatrix}$$

the modal participation factors p_i can be obtained from Eq. (4.59) as

$$p_1 = \frac{\begin{Bmatrix} 1.0 \\ 0.883 \end{Bmatrix}^T \begin{bmatrix} 37.3 & 0 \\ 0 & 37.3 \end{bmatrix} \begin{Bmatrix} 1.0 \\ 1.0 \end{Bmatrix}}{\begin{Bmatrix} 1.0 \\ 0.883 \end{Bmatrix}^T \begin{bmatrix} 37.3 & 0 \\ 0 & 37.3 \end{bmatrix} \begin{Bmatrix} 1.0 \\ 0.883 \end{Bmatrix}} = \frac{70.236}{66.382} = 1.058$$

$$p_2 = \frac{\begin{Bmatrix} 1.0 \\ -1.133 \end{Bmatrix}^T \begin{bmatrix} 37.3 & 0 \\ 0 & 37.3 \end{bmatrix} \begin{Bmatrix} 1.0 \\ 1.0 \end{Bmatrix}}{\begin{Bmatrix} 1.0 \\ -1.133 \end{Bmatrix}^T \begin{bmatrix} 37.3 & 0 \\ 0 & 37.3 \end{bmatrix} \begin{Bmatrix} 1.0 \\ -1.133 \end{Bmatrix}} = \frac{-4.961}{85.182} = -0.058$$

which indicates predominantly first-mode participation. Assuming that both modal response periods fall within the horizontal portion of a 5% damped acceleration response design spectrum with maximum ordinate $S_a = 1.5g$ as shown in Fig. 4.14,

$$\{S_a\} = \begin{Bmatrix} 1.5 \\ 1.5 \end{Bmatrix} 32.2 \ (\text{ft/s}^2) = \begin{Bmatrix} 48.3 \\ 48.3 \end{Bmatrix} (\text{ft/s}^2) \left(\begin{Bmatrix} 14.7 \\ 14.7 \end{Bmatrix} (\text{m/s}^2) \right)$$

and the maximum modal horizontal displacement response can be found as

$$u_{1,\max} = p_1 \{\phi_1\} \frac{S_{a1}}{\omega_1^2} = 1.058 \begin{Bmatrix} 1.0 \\ 0.883 \end{Bmatrix} \frac{48.3}{11.581^2} = \begin{Bmatrix} 0.381 \\ 0.336 \end{Bmatrix} (\text{ft}) = \begin{Bmatrix} 4.57 \\ 4.04 \end{Bmatrix} (\text{in.}) \left(\begin{Bmatrix} 116 \\ 103 \end{Bmatrix} (\text{mm}) \right)$$

$$u_{2,\max} = p_2 \{\phi_2\} \frac{S_{a2}}{\omega_2^2} = -0.058 \begin{Bmatrix} 1.0 \\ 1.133 \end{Bmatrix} \frac{48.3}{49.392^2} = \begin{Bmatrix} 0.0011 \\ 0.0013 \end{Bmatrix} (\text{ft}) = \begin{Bmatrix} -0.014 \\ 0.016 \end{Bmatrix} (\text{in.}) \left(\begin{Bmatrix} -0.36 \\ 0.41 \end{Bmatrix} (\text{mm}) \right)$$

Combining the modal response maxima, direct combination according to Eq. (4.61) results in

$$\{u\} = \begin{Bmatrix} 4.584 \\ 4.056 \end{Bmatrix} (\text{in.}) \left(\begin{Bmatrix} 116.4 \\ 103.0 \end{Bmatrix} (\text{mm}) \right)$$

Using SRSS or Eq. (4.62), the combination of modal maxima results in

$$\{u\} = \begin{Bmatrix} 4.57 \\ 4.04 \end{Bmatrix} (\text{in.}) \left(\begin{Bmatrix} 116.1 \\ 102.6 \end{Bmatrix} (\text{mm}) \right)$$

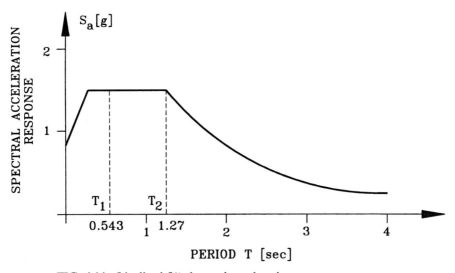

FIG. 4.14 Idealized 5% damped acceleration response spectrum.

and using CQC from Eq. (4.63) with $t = T_1/T_2 = 4.28$, $\xi = 0.05$ for both modes, and the cross-modal coefficients from Eq. (4.64)

$$[\rho] = \begin{bmatrix} 1.0 & 0.003 \\ 0.003 & 1.0 \end{bmatrix} \quad \text{results in} \quad \{u\} = \begin{Bmatrix} 4.57 \\ 4.04 \end{Bmatrix} \text{(in.)} \left(\begin{Bmatrix} 116.1 \\ 102.6 \end{Bmatrix} \text{(mm)} \right)$$

which shows that for well-spaced modes, both CQC and SRSS combination techniques produce the same results, which is also reflected in the small magnitude of cross-correlation coefficients or off-diagonal terms ρ_{ij}. In general, when the frequencies of two modes differ by more than 20%, the effects of cross-modal correlation are insignificant and CQC and SRSS procedures provide the same results. Note that had the response been quantified by mass acceleration rather than displacement, differences between the modal combination rules would have been more apparent.

Finally, a bridge structure can be exposed to seismic excitations in two perpendicular directions x and y simultaneously, and any time-dependent response quantity $z(t)$ of interest can be expressed as an upper bound by

$$\mathbf{z}(t) = \mathbf{z}^x(t) + \mathbf{z}^y(t) \qquad (4.65)$$

where $\mathbf{z}^x$ and $\mathbf{z}^y$ are contributions to the same response quantity produced by horizontal earthquake ground motions in the x and y directions, respectively. Since maximum response quantities $\mathbf{R}^x$ and $\mathbf{R}^y$ are not likely to occur at the same time, and since the input in the two perpendicular directions will have very little cross-correlation, the cross-correlation of maximum response quantities $\mathbf{R}^x$ and $\mathbf{R}^y$ will also be low, and SRSS can be used to combine maxima as

$$\mathbf{R} = \sqrt{(\mathbf{R}^x)^2 + (\mathbf{R}^y)^2} \qquad (4.66)$$

which produces results [C10] within 5% of those obtained with the commonly used 30% rule of

$$\mathbf{R} = |\mathbf{R}^x| + 0.3|\mathbf{R}^y| \qquad (4.67)$$

where absolute maxima of the same response quantity are combined for 100% of earthquake input in one direction and 30% of earthquake input in the perpendicular direction. For simultaneous horizontal and vertical ground motions, the cross-correlation is expected to be even smaller due to time separation in the maximum intensity of ground shaking between vertical and horizontal motion. Thus either no combination or SRSS combination seems appropriate.

All of the combination techniques for maximum bridge response values outlined above apply strictly only to structures with linear elastic response. The usefulness of response maxima from individual modes and their subsequent combinations are questionable for inelastically responding bridge systems since response characteristics constantly change and first-mode response will become more dominant with increasing inelastic deformations.

4.4 MODELING OF BRIDGE STRUCTURES

The basic modeling objective in seismic bridge analysis is to provide the simplest mathematical formulation of the true bridge behavior which satisfies a particular assessment or design requirement for a quantitative response determination. Assuming that appropriate analytical tools (Section 4.5) exist to provide the numerical quantification, the model has to capture the physical and mechanical interactions of earthquake input and structure response.

4.4.1 General Modeling Issues

The objective of the analytical model is to describe the geometric domain, the seismic mass, the connection and boundary conditions, and the loading of the prototype as closely as possible to facilitate the engineering interpretation of numerical response quantities. To accomplish this, bridge superstructure and bent geometry are described in the model by a spatial relationship similar to the one given by the prototype. Individual elements simulating structural parts or complete bridge components are connected at nodes and the nodal displacements are used as unknowns or DOFs in the analysis process. For each element, member end forces and deformation relationships are defined by:

1. *Kinematics:* relating the nodal displacements to internal deformations and strains
2. *Constitutives:* relating internal strains to stresses
3. *Statics:* relating internal stresses to nodal loads, for any type of linear or nonlinear bridge behavior

The modeling effort consists of defining these relationships by "first principle" mechanics or in phenomenological form.

In addition to the modeling of the geometric domain and the member load–deformation characterization, the correct seismic weight or mass associated with each DOF also needs to be determined when, as usual, inertia forces contribute to the bridge response. Furthermore, the connections between individual bridge frames, the abutments and the foundation are complex and typically not very well known in the prototype, and simplifying modeling assumptions need to be made.

With the earthquake loading to the bridge generated by ground motion, the soil structure interface and loading history need to be characterized. In some cases, particularly when soft soils, massive foundations, and/or liquefaction potential are present, soil structure interaction (SSI) should be modeled through description of appropriate impedance functions, soil springs, or actual soil modeling in the form of a continuous half-space or a portion thereof [L3].

Particularly for longer multispan and multiframe bridges, questions arise concerning the necessity and validity of modeling the entire bridge in a global model. For the seismic demand assessment, a global model that can capture the effects of complex geometries such as curves in plan and elevation, effects of highly skew supports, contributions of ramp structures, as well as the interaction between frames, seems appropriate. However, it needs to be realized that with increasing bridge length, the uncertainties in spatially varying ground motion input or errors in the typically assumed uniformity or coherence of input at all supports increase, and thus the response quantities obtained are not necessarily better or more reliable. Furthermore, the sheer volume of input and output data management for a global model of a long bridge structure can in some cases hinder rather than improve the basic understanding of the important seismic response characteristics.

The best description of the geometric domain of the prototype bridge would comprise a model of every single member or structural element in its correct spatial and physical relationship to other members, but this is not always practical, as can be seen from the example of the San Francisco Oakland Bay bridge in Fig. 4.15(a), which shows a top view of a portion of the East Bay bridge system, and Fig. 4.15(b), which shows one of the connection points in the cantilever truss portion for eight laced and riveted members. Since as can be seen in Fig. 4.15(a), every truss member is built up from many individual structural elements, and the connection point of the laced truss members with riveted gusset plates consists again of a multitude of individual elements and a complex spatial arrangement, as shown in Fig. 4.15(b), a detailed discretization or analytical description of each of these structural elements and the

(a)   (b)

FIG. 4.15 East Bay spans of the San Francisco-Oakland Bay bridge. (a) Top view of parts of the East Bay structure; (b) Laced member and connection detail.

characterization or modeling of their individual force–deformation and connection characteristics is prohibitive for the total bridge system. The need to separate the total bridge system into manageable subsystems, frames, bents, and individual structural elements for modeling purposes is quite obvious, and the various modeling approaches are discussed in the following.

(a) Structural Systems. The total structural bridge system consists of the superstructure and the substructures. The superstructure, particularly for larger bridges, is separated into sections by expansion joints which allow thermal- and time-dependent expansion or contraction of the bridge to occur without introducing large stresses or strains into individual bridge members or by articulated construction joints provided as part of a particular bridge and erection system. Joints where relative deformations between parts of the superstructure can occur are important for seismic response since the structural discontinuities at these movement joints allow individual bridge sections to respond with different characteristics and with complex interaction to the earthquake input.

Together with their respective substructures or supporting bents consisting of the piers or columns and the foundations, these separated superstructure sections, referred to as *frames,* play a major role in the earthquake response quantification, due to their individual dynamic response characteristics. As outlined in Section 4.3, the individual frame response characteristics are controlled by the mass, contributed primarily by the superstructure and by the stiffness of the individual bents. The superstructure geometry and stiffness control how the individual bents are coupled in their dynamic response. To reflect the importance and differences of these individual subsystems in terms of the analytical modeling for seismic bridge response quantification, a distinction is made among (1) global models, (2) frame models, and (3) bent models.

(i) Global Bridge Models. Global models of the entire bridge structure have a limited usefulness except for cases where (1) the bridge is short and consists only of a single frame, (2) the expected response is in the essentially elastic range, and (3) a quantifiable basis for ground motion input variations along the length of the bridge and movement joint characterization can be established. Global bridge models are used predominantly for seismic demand quantifications in the form of linear elastic modal response spectrum analyses (discussed further in Section 4.5.3), to determine elastic displacements and equivalent elastic member forces. Since the real value of global analytical models is primarily in the response characterization of the total bridge system, they should be used primarily at the end of the seismic bridge design or assessment process to verify parameter quantifications from individual frame and bent models in the framework of overall system response, which is not considered in the other models. In models where limited inelastic response is expected, substructuring of the linear elastic portions of the bridges based on mathematically rigorous condensation techniques [C10,G3] can significantly reduce the

computational effort. In particular, bridge systems with irregular geometry, such as curved bridges, skew bridges, intersections, and separations of bridges, as well as ramp and distribution structures, can exhibit dynamic response characteristics which are not necessarily obvious and which may not be captured in a separate analysis of a subsystem. However, the implications of ground motion characterization at different piers and movement joint modeling should always be considered and, where necessary, evaluated through sensitivity studies on the assumed parameters.

(ii)   Frame Models. Individual frame models of bridge sections between movement joints provide a powerful tool for seismic bridge response quantification since the dynamic response characteristics of an individual or stand-alone frame can be assessed with reasonable accuracy, and the length of a particular frame as well as the variability of soil conditions over this limited length are such that coherency of ground motion input to the individual bents can be a reasonable assumption. Due to the limited geometric domain of an individual frame, detailed frame models, based on the inelastic response characteristics of individual bents, can be developed which can provide realistic quantifications of both the seismic demand and capacity.

The modeling and analysis of individual frames as stand-alone frames without consideration of constraints by adjacent frames or abutments provides not only for simple analytical models, but also in most cases critical or upper-bound response values for the individual frame since interaction with other parts of the bridge system with different dynamic response quantities has the tendency to spoil or reduce harmonic excitation and with it, the dynamic resonance phenomenon. A frame-by-frame assessment in the form of stand-alone frame models and analyses of all the individual frames will provide the designer not only with these upper-bound response values, but also with a better understanding of the individual response characteristics of sections of the bridge, an understanding that can be difficult to extract from a global response analysis.

Interaction with adjacent frames can be considered even with frame-by-frame models in the form of springs or boundary frames, which are typically formulated with linear elastic characteristics. Another commonly used approach is the modeling of groups of frames where three to five frames are modeled simultaneously with appropriate assumptions for movement joint characteristics [discussed further in Section 4.4.2(f)], and only the resulting response quantities for the inner frame or frames are considered as representative design or assessment data.

(iii)   Bent Models. The development of realistic frame models requires a detailed characterization of individual bents, since generally the stiffness for the frame models is contributed directly by the bents. Models of individual bents are therefore used primarily to develop effective bent stiffness characteristics and deformation limit states. Individual bent models should include

foundation flexibility effects (Section 4.4.2(d)) and can be combined in frame models by means of superstructure elements, impedance relationships in the form of stiffness transfer matrices, or simply by geometric constraints. The high in-plane stiffness of most bridge superstructures allows, as a first approximation in many cases, rigid-body movement assumptions for the superstructure (see Fig. 4.4), which greatly simplifies the combination of individual bent models.

(b) Individual Structural Members. Bent, frame, or global models utilize individual members to describe the physical behavior characteristics of elements between nodal points or joints defined in the mathematical discretization of the bridge prototype. Elements used can be classified by their geometry and their principal structural action as defined by structural mechanics. The three groups of structural members or elements used in bridge models are (1) line elements, (2) plates and shells, and (3) solid elements. Without detailed reference, these different elements were introduced in Fig. 4.2: line elements in the form of springs, dashpots, beam, or column elements, used primarily in lumped parameters or structural component models, plate-and-shell elements or three-dimensional solid elements in the finite-element category.

Elements in a structural model are connected to the nodes defined in the structural discretization and are compatible at these nodes with the defined unknown modal response quantities, typically nodal deformations in the form of displacement DOFs, as outlined in Fig. 4.16 for various element types. Line

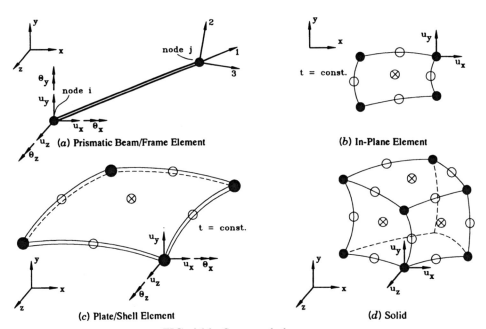

FIG. 4.16   Structural elements.

elements are represented in Fig. 4.16 by a beam element, shown in Fig. 4.16(*a*) with six DOFs or unknown member end deformations at each joint. Two-dimensional elements are represented by an in-plane element in Fig. 4.16(*b*) and in a flat shell element in Fig. 4.16(*c*) with four- to nine-node discretization. While the four-corner-node formulation is simple, a large number of elements or fine discretization with a large number of DOFs is typically necessary to overcome the limited flexibility in these lower-order or four-node elements. The addition of midside and center nodes provides added flexibility to model the most common deformation and strain states. As shown in Fig. 4.16(*b*), in-plane elements have only two DOFs per node and are used in plane-stress or plane-strain problems, where constant stress or strain states through the thickness of a structural member can be assumed. The in-plane element in Fig. 4.16(*b*) is typically an integral part of the plate element in Fig. 4.16(*c*), representing the in-plane or membrane stress-strain state. As shown in Fig. 4.16(*c*), only five nodal DOFs are typically present in plate and flat shell elements. Finally, three-dimensional solid elements, shown schematically in Fig. 4.16(*d*), feature three displacement DOFs per node and similar to the planar elements, can have either a lower-order formulation with corner nodes only or a higher-order formulation with midside and center nodes. Detailed derivations and characterizations of these various elements are beyond the scope of this book and can be found in the general structural analysis and finite-element literature [H5,C19].

(i)   Line Elements.   Line elements are characterized by their one-dimensional geometric representation even though their behavior can be fully three-dimensional and they can be arbitrarily oriented and fully utilized in three-dimensional space. Line elements can be one-directional in nature in the form of springs, dashpots, or truss members with simple one-directional force–deformation relationships or can take on the form of bending elements such as beams and columns connected at their member ends to form bents, frames, or superstructure grillages. Between the nodal points at the member ends, a variety of bending element formulations are possible, depending on the performance characterization. For linear elastic response, standard prismatic beam elements with cubic displacement variation along the element length are most commonly used. For inelastic member characterization, member end force–deformation relationships can be defined through inelastic hinge elements or with special nonlinear beam elements which can incorporate stiffness degradation and/or fully hysteretic response characterization through modifications of the element stiffness matrix.

The local stiffness degradation can be captured with sectional moment–curvature analyses which are based on layered or filamented discretizations of the member cross section, as shown in Fig. 4.17(*a*) for different rectangular or circular reinforced concrete sections. A layered approach is sufficient for one-directional bending about a known axis and simple cross-section geometry, whereas the filamented discretizations are used when the loading directions

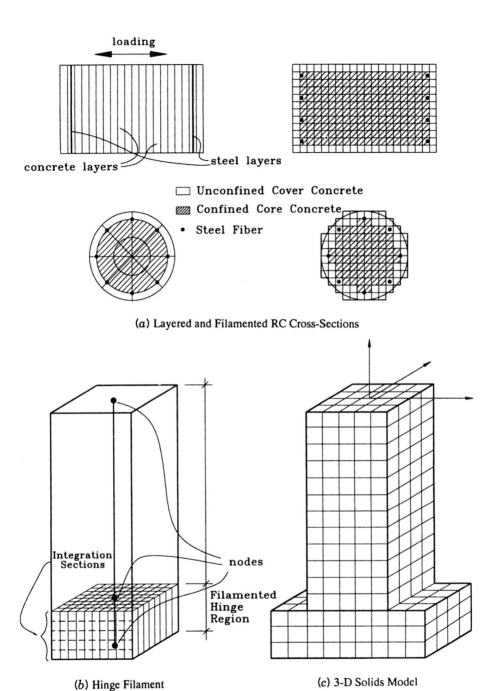

loading

concrete layers — steel layers

☐ Unconfined Cover Concrete
▨ Confined Core Concrete
• Steel Fiber

(a) Layered and Filamented RC Cross-Sections

Integration Sections

nodes

Filamented Hinge Region

(b) Hinge Filament   (c) 3-D Solids Model

FIG. 4.17   Special bridge pier elements.

can change and the cross section is of arbitrary geometry, including hollow-core members. Individual layers or filaments are treated as uniaxial members with uniaxial constitutive characteristics which can be modified for different levels of lateral confinement. The centerline element representation between nodes utilizes the Bernoulli hypotheses for sections perpendicular to the element axis. While standard stiffness-based filamented or layered elements will need an empirical relationship for an effective plastic hinge length [see Section 5.3.2, Eq. (5.39)] to define moment–rotation relationships, force- or flexibility-based filamented members or hinge/end regions [P25], as outlined in Fig. 4.17(*b*), can directly capture the extent of the inelastic region and the full nonlinear response characteristics of a flexural plastic hinge. All of the line element models discussed above have been shown to be capable of characterizing the flexural response of bridge columns and beams but fall short in the characterization of shear deformations and shear failure modes. In general, the inelastic interaction between flexure and shear represents an area where significant improvements in the analytical modeling are still possible and necessary. Until these developments are completed, the phenomenological response models for flexure and shear interaction with increasing ductilities developed in Section 5.3.4 can be employed to limit or assess potentially nonductile response.

(ii) Plates and Shells. Plate or shell elements in bridge models are used primarily to determine local stress levels in cellular bridge superstructures or in cellular piers and caissons. Rarely are nonlinear layered plate elements employed for seismic response assessment. Nonlinear plane-stress elements have been used to model local joint or connection regions and to assess cracking and yield propagation. These models are commonly based on smeared rotating orthotropic properties [D3,C4] but have limited application to seismic bridge analysis even on the capacity determination side, due to lack of comprehensive characterization of effects of bond slip, confinement, and anchorage of reinforcement for different geometries in terms of concrete cover, confinement by adjacent members, cover concrete spalling, and increasing local ductilities.

(iii) Solids. Three-dimensional solids models are used only in very limited cases in the form of linear elastic models, where quantifications of principal stress states in joint regions or regions of complex geometry are required, as indicated in Fig. 4.17(*c*), or as nonlinear smeared continuum models for detailed localized failure investigations. However, all the concerns and limitations expressed above for in-plane models in the nonlinear domain apply and need to be considered in a three-dimensional nonlinear analytical model with solid elements.

4.4.2 Modeling of Bridge Components

This section deals with specific modeling issues for bridge components, to provide basic guidance for the geometric modeling and member characterization of typical bridge parts.

(a) Superstructure. Since every bridge features a superstructure consisting
of the bridge deck and a support system for the bridge deck spanning piers
or bents, the modeling of the bridge superstructure is a key component in the
analysis process. As outlined in Section 1.3, under seismic force input the
bridge superstructure is expected to remain essentially elastic, limiting nonlinear modeling considerations to joints between superstructure segments, connections with supporting bents, and to the assessment analyses of older bridges,
where the superstructure is not protected against inelastic action by capacity
design principles.

   In modeling the bridge superstructure, issues of (1) geometry and effective
member characterization, (2) support and connection definition, and (3) permanent load effects and mass participation need to be addressed. Most bridge
structures, by definition, bridge or span long distances with their superstructure
and can therefore be considered as linear structures, where the span length
L between bents is larger than the width B or depth D of the superstructure
(Fig. 4.18). For the seismic analysis of a bridge it is typically not necessary to
model the full three-dimensional domain of the superstructure with finite
elements in the form of three-dimensional solids or plates; rather, significantly
simpler models suffice, provided that they represent effective stiffness characteristics and mass distribution.

   In many cases, the bridge superstructure, due to its in-plane rigidity, can
be assumed to move as a rigid body under seismic loads, and the entire
modeling problem is reduced to the stiffness modeling of the bents with

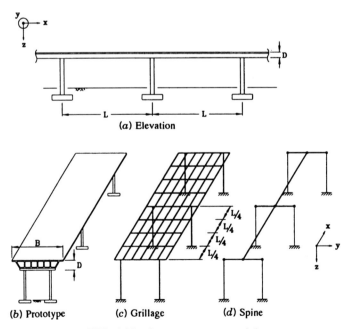

(a) Elevation

(b) Prototype   (c) Grillage   (d) Spine

FIG. 4.18   Superstructure models.

geometric constraints simulating the rigid superstructure. Where vertical superstructure flexibility reduces the fixity of the top boundary conditions for columns and piers, a reduction in the stiffness coefficient α in Eq. (4.9) can be determined.

In cases where the superstructure cannot be considered rigid (i.e., long and narrow bridges, interchange connectors, etc.), the superstructure can be modeled as a grillage of beam elements as shown in Fig. 4.18(c) or a spine with beam elements following the center of gravity of the cross section along the length of the bridge, as shown in Fig. 4.18(d). Equivalent member properties for the spine or beam elements need to be derived which represent the overall effective superstructure stiffness. For reinforced concrete superstructures (e.g., multicell box girders), superstructure bending under longitudinal seismic loads can be expected to cause or enhance already existing cracking from gravity loads and live loads, and effective or cracked stiffness properties should be assigned for I_{yy}, the moment of inertia of the entire cross section about the transverse or y axis as defined in Figs. 4.18(d) and 4.19, based on the effective or cracked stiffness defined by Eq. (4.27). While detailed cracked-section stiffness analyses can be performed for each girder or the complete superstructure as shown in Fig. 4.19, it is often sufficient to calculate the gross-section stiffness I_g and reduce it to $I_e = 0.5I_g$ for reinforced concrete and assume no reduction for essentially uncracked prestressed concrete superstructures, resulting in an effective flexural stiffness of

$$EI_e = \begin{cases} 0.5EI_g & \text{(reinforced concrete)} \\ 1.0EI_g & \text{(prestressed concrete)} \end{cases} \qquad (4.68)$$

The torsional rigidity J for grillage or spine elements can also be determined from standard mechanics principles and can be considered as fully effective as long as the torsional cracking moment is not exceeded, at which point the torsional stiffness significantly reduces.

The torsional stiffness of composite girder and deck bridges is very small due to the open cross section and does not need adjustment for cracking. The

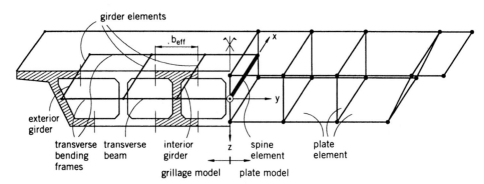

FIG. 4.19   Equivalent superstructure member definitions.

torsional model characteristics for a cellular superstructure can be found from the parameters and relationships outlined in [C5] and depicted in Fig. 4.20. Torque T and twist θ are related for isotropic beams or shafts as

$$T = GJ \cdot \theta \qquad (4.69)$$

where G is the shear modulus and J the torsional moment of inertia. For thin-walled hollow sections the torsional moment of inertia J can be found as

$$J = \frac{4A_0^2 t}{p_0} \qquad (4.70)$$

with A_0 and p_0 represent the area and perimeter of the shear flow in the tubular section of wall thickness t. These parameters are also defined in Fig.

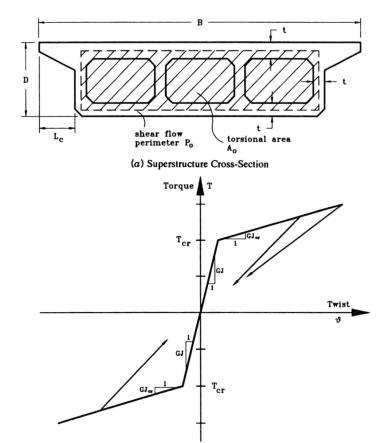

(a) Superstructure Cross-Section

(b) Idealized Torque-Twist Characteristics

FIG. 4.20 Superstructure torsion.

4.20, where a three-cell box girder superstructure with idealized constant deck, soffit, and web thickness t is shown. For bridges with different and varying thickness t_i, an averaged t_{av} can provide a close approximation.

Based on the same assumption for principal tension cracking p_t^{cr} and degradation p_t^{deg} outlined in Section 5.4, and associated nominal stress thresholds of $p_t^{cr} = 3.5\sqrt{f_c}$ psi ($p_t^{cr} = 0.29\sqrt{f_c}$ MPa) and $p_t^{deg} = 5\sqrt{f_c}$ psi ($p_t^{deg} = 0.42\sqrt{f_c}$ MPa), the nominal torsion limit states can be approximately defined based on relationships developed in [C5] as

$$T = \frac{A_0}{p_0} \gamma \sqrt{f_c} \sqrt{1 + \frac{f_{pc}}{\gamma\sqrt{f_c}}} \tag{4.71}$$

where f_c is the concrete compression strength, f_{pc} the prestress level in the cross section defined as the total prestress force P divided by the gross concrete area of the cross section A_g, and γ a principal tensile stress coefficient in the general relationship

$$t_p = \gamma \sqrt{f_c} \tag{4.72}$$

The cracking torque can now be obtained from Eq. (4.71) for $\gamma_{cr} = 3.5$ for stresses expressed in psi and $\gamma_{cr} = 0.29$ for stresses expressed in MPa. Similarly, the ultimate torque can be obtained by substituting the nominal principal tension stress state at which degradation of the concrete aggregate interlock mechanism is expected as $\gamma_u = 5.0$ for stresses in psi and $\gamma_u = 0.42$ for stresses in MPa substituted into Eq. (4.71).

The cracked-section torsional moment of inertia can be obtained [C5] as

$$J_{cr} = \frac{4\overline{A}^2 E_s}{\overline{p} E_c} \sqrt{\frac{A_t}{s} \frac{(A_e + A_p)}{\overline{p}}} \tag{4.73}$$

where E_s and E_c are the moduli for steel and concrete, $\overline{A}$ and $\overline{p}$ can be taken as $0.85A_0$ and $0.9p_0$ defined in Fig. 4.20, A_t and s are the transverse reinforcement area of a stirrup and the stirrup spacing, and A_e and A_p the total longitudinal mild and prestressed reinforcement areas.

As a general rule of thumb, without the more detailed torsional stiffness calculations, the torsional stiffness for superstructure spine analyses can be taken as

$$GJ_e = \begin{cases} 1.0GJ & \text{(uncracked)} \\ 0.05GJ & \text{(cracked)} \end{cases}$$

However, in most bridge superstructures, except for highly curved in-plane arrangements, the torsion levels in the earthquake case will be significantly below the cracking torque limit state and no torsional stiffness reduction needs to be considered.

For superstructures carrying wide roadways, the spine model may produce erroneous results, particularly when combinations of earthquake forces with gravity loads and live loads need to be investigated. In a highly skewed bridge, as shown in Fig. 4.21 for the example of the geometry of the Mission/Gothic undercrossing on California State Route 118, which collapsed during the January 17, 1994, Northridge earthquake [P5], a simple spine model with rigid cross links at the bents as shown for one of the two parallel bridge structures in Fig. 4.21(c) cannot capture even gravity-load distributions to individual columns and abutment bearings, since loads in a spine model are typically applied along the spine axis only.

A two-dimensional grillage model for the superstructure, as shown in Figs. 4.18(b) and 4.21(c), can capture these effects as long as sufficient transverse distribution beams between girders are provided and the applied loads are distributed to all nodes over the bridge deck area. The longitudinal girder placement in the grillage model follows the prototype girder lines, and proper-

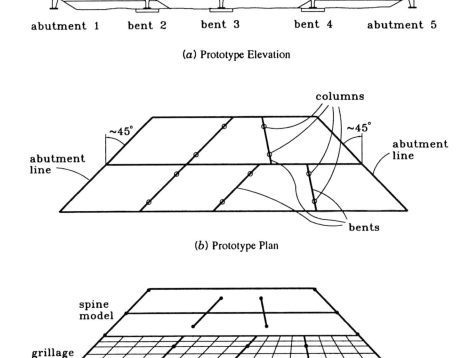

(a) Prototype Elevation

(b) Prototype Plan

(c) Schematic Spine and Grillage Models

FIG. 4.21 Superstructure modeling of mission/gothic undercrossing [P5].

ties can be derived in conjunction with Eq. (4.68) for individual girders and tributary flange widths, as outlined in Fig. 4.19. Torsional stiffness characteristics derived from Eqs. (4.70) and (4.73) for the total superstructure are distributed equally to the number of longitudinal girder elements. The placement and characterization of the transverse distribution beams is more arbitrary. As a general rule, transverse distribution beams should be applied at least at the center and quarter points of each span, and their properties should be equivalent to the transverse bending behavior of the superstructure.

For superstructures with individual girders connected only by a deck, the transverse stiffness is small and can be captured by the gross bending stiffness of the deck slab in the transverse direction and an effective width as defined below for cellular superstructures. For cellular superstructures such as thin-walled multicell box girders, the shear deformations in the transverse direction can be significant [H2] and the equivalent bending stiffness for transverse grillage beams can be approximated by the simple addition of the gross-section moduli of the deck and soffit slabs as

$$EI_{\text{trans}} = E(I_{\text{slab}} + I_{\text{soffit}}) = \frac{Eb_{\text{eff}}}{12}(h_{\text{slab}}^3 + h_{\text{soffit}}^3) \qquad (4.74)$$

where h_{slab} and h_{soffit} represent the deck slab and soffit slab thickness and b_{eff} is the tributary width, or half the distance to two adjacent transverse beams.

Increased transverse stiffness can be obtained at locations where transverse diaphragms are present in the superstructure. Transverse diaphragms can be modeled with transverse grillage beam elements with characteristics derived from T or I sections and an effective flange width of eight times the slab thickness on either side of the diaphragm. Only in rare cases where detailed local stress-level quantification in the superstructure is required will a full three-dimensional model with plate elements as outlined in Fig. 4.19 be necessary. These three-dimensional plate models for the superstructure are more important for truck and lane load distribution effects than in seismic response simulations.

Finally, long-span steel bridges frequently feature very articulated super-structures in the form of riveted or bolted trusses in which each member can be laced and comprised of numerous elements, as shown in Fig. 4.15. Superstructure modeling for the seismic demand assessment of these bridges requires the development of superelements by (1) substructuring and static condensation, or (2) by extensive calibration of equivalent span elements with detailed finite-element models. The superelements connect to the displacement DOFs at the top of the piers or bents and represent force–deformation characteristics for the entire span at the span ends. In addition, mass distributions along the superelement and over the element cross section need to be determined to model representative dynamic response characteristics in the translational and torsional modes of vibration. Calibration of the superelement is required not only for static loads and deformations but also for dynamic response characteristics.

(b)   Single-Column Bents. In the seismic response analysis of bridges, the bents or supports are the critical structural elements that provide gravity-load and earthquake force transfer to the ground and ground motion input to the bridge superstructure. Since current seismic design philosophy generally places any inelastic seismic demand in the supporting columns as discussed in Section 1.3, correct analytical modeling of the columns is of primary importance.

Mass and stiffness characterizations of single-column bents were discussed in detail in Section 4.3.1; the actual structural discretization will be discussed here. For a generic single-column bent as shown in Fig. 4.22(*a*), different discretizations need to be considered, depending on the bent geometry and the expected seismic response. For essentially elastic response and a prismatic column, as depicted in Fig. 4.22(*b*), a single-column element connected at nodes 2 and 3 at the superstructure soffit and the top of the footing is sufficient to model the seismic response as long as generalized mass distributions in accordance with Fig. 4.5(*a*) and effective properties in accordance with Eq. (4.24) are considered. When significant inelastic action is expected in the form of plastic hinges at node 2 or 3, strain penetration into the adjacent members as discussed in Eq. (4.26) will provide added flexibility, which can be modeled with additional beam elements between nodes 1 and 2 or 3 and 4 of length equal to the expected strain penetration length. The same effective or cracked member stiffness as determined for the column section should be used for penetration elements into the adjacent joints. Footing flexibilities in the form of equivalent spring stiffnesses for all six footing DOFs should be determined based on available geotechnical information.

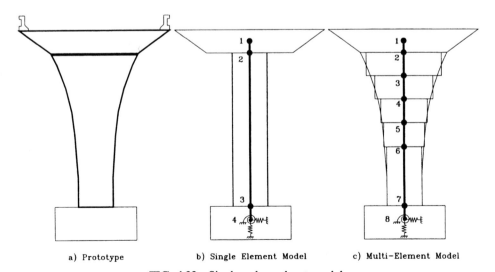

a) Prototype   b) Single Element Model   c) Multi−Element Model

FIG. 4.22   Single-column-bent models.

For tapered or flared single-column bents or bents where column mass effects may be important, the discretization or placement of nodes and elements for the analytical model follows either the mass distribution requirements outlined in Section 4.3.1(a) or, since typically only prismatic beam/column elements are available, a stepped discretization as outlined in Fig. 4.22, which approximates the actual geometric domain. Continued refinements of the column discretization are typically not necessary and can be shown to be negligible by simple comparison of the resulting dynamic response characteristics.

Nonlinear response characteristics and associated single-column-bent models follow the general guidelines given in Section 4.4.1 and are based on moment–curvature analyses for the critical column section, taking axial load levels as well as confinement effects into account. The lateral displacement response of bridge piers during a seismic event is typically small compared to the column dimensions and in most cases allows nonlinear geometric effects to be neglected. Exceptions are very tall or slender columns, for which the combined influence of gravity load and displacement can be significant even though the displacement may be small compared to the overall bridge dimensions. In these cases, rather than complete nonlinear geometric behavior, only second-order or P–Δ effects need to be considered, by evaluating equilibrium in the deformed configuration. One way of accounting for P–Δ effects in the analytical modeling is by means of reduced capacity for lateral loads, as schematically shown in Fig. 4.23, and the yield and ultimate deformation limit states in Fig. 4.23(b) and (c), respectively, together with bending moment contributions from the lateral earthquake force F and the vertical load P due to gravity. P–Δ analyses techniques are discussed further in Section 4.5. While the total moment displacement relationship in Fig. 4.23(d) can still be idealized as a ductile elastoplastic bilinear relationship, the actual lateral force contribution to the total moment reduces with increasing P–Δ moment by an amount ΔM which increases with increasing displacement ductility. This effect can be

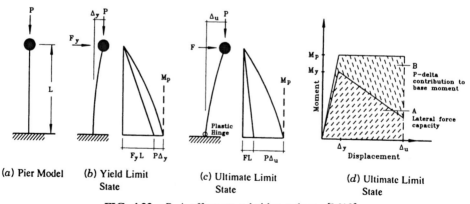

FIG. 4.23 P–Δ effect on a bridge column [M10].

interpreted as an effective lateral capacity reduction for increasing ductilities and modeled by means of a single-column-bent characteristic which follows line A in Fig. 4.23(d), that is, a characteristic with a negative postyield stiffness. This type of modeling of P-Δ effects can be employed in lieu of iterative solution strategies that establish equilibrium in the deformed configuration. P-Δ effects should be considered either by the appropriate analysis technique or by the above-outlined capacity reduction in the element characterization once the P-Δ moment at the displacement limit state under consideration reaches magnitudes defined by Eq. (5.162).

(c)   Multicolumn Bents.   In multicolumn bents the framing action or coupling between columns contributes to the seismic response in terms of stiffness, capacity, and axial load levels in the various frame members during cyclic earthquake loading. In the analytical model, all of these effects can be incorporated in a planar frame model along the bent axis, consisting of beam and column elements with effective member properties as discussed in Section 4.3.1 and, where needed, special elements for joints and boundary conditions as illustrated in Fig. 4.24. For equivalent linear elastic analysis models, typically a single-column element from the top of the footing to the bottom of the cap beam suffices, extended by link elements into the footing or cap. Effective or

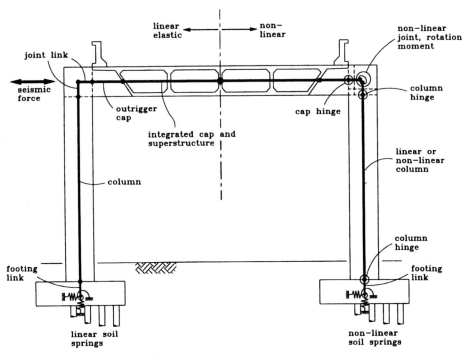

FIG. 4.24   Multicolumn-bent model.

cracked member stiffness properties for EI_e and GA_{ve} as outlined in Section 4.3.1(b) should be employed, and because of the framing action, EA_e, an effective axial stiffness, should also be included. Similar to the effective shear stiffness, the effective axial stiffness can be reduced from the gross-section stiffness in proportion to the effective flexural stiffness as

$$EA_e = EA_g \frac{EI_e}{EI_g} \tag{4.75}$$

to reflect the influence of axial load, flexural cracking, and reinforcement ratios.

Bent caps are modeled with beam elements connected to nodes at the column face and extended into the joint with link elements. The face nodes at column and cap face extensions are typically placed at these locations to capture critical response quantities in terms of maximum bending moments, and the link extensions to the joint center provide the geometric connectivity and compatibility. Link extensions into the joint are modeled using gross column or cap beam stiffness properties as long as no significant inelastic action is expected at the adjacent cap or column face. Where flexural hinges with high ductility demands occur at the column ends, the link extensions may be modeled with cracked or effective column properties to reflect yield penetration into the joint region and increased flexibilities from joint shear distress. Use of rigid end blocks to represent these regions is inappropriate in all cases and results in overestimated stiffness values. Different ratios of top and bottom reinforcement in the cap beam may result in separate elements with different stiffness along the length of the cap beam, to model positive and negative bending characteristics.

The correct axial force level for the member characterization of a multicolumn bent depends directly on the seismic force input at the top of the bent due to the coupling characteristics of the bent cap. Thus member capacities and stiffness characteristics should be updated and adjusted continuously based on the axial force state from gravity loads and seismic force input. Although very few analytical tools such as SC-Push3D [S8] exist that actually adjust member stiffness and capacities based on the axial load state, it is typically sufficient to determine the axial loads in a two-dimensional multicolumn-bent model from the theoretical collapse limit state. An initial estimate of the column moments at the lateral strength of the bent is made and the column axial forces are formed from statics. These axial forces are used to determine the appropriate member stiffness for a detailed plastic collapse analysis. Some iteration may be required since the lateral strength of the bent may be affected by the column axial force levels.

Outrigger cap beams, as shown in Fig. 4.24, can also carry significant torsional forces under longitudinal seismic action in their uncracked state. However, as discussed in Section 4.4.2(a), as soon as torsional cracking occurs, the effective torsional stiffness and with it the torsional forces are reduced significantly. Torsional stiffness and capacity of outrigger cap beams can be evaluated with Eqs. (4.69) to (4.73) derived for a cellular superstructures, as long as the area A_0 enclosed by the shear flow centerline with perimeter p_0 are determined from the centerline dimensions of the cap beam stirrups. To be effective in the cracked state, cap beam reinforcement needs to be detailed for torsion in the form of closed stirrups and equal amounts of longitudinal cap reinforcement distributed around the cap beam perimeter. Reduced amounts of longitudinal cap beam reinforcement are permissible [C5] as long as appropriate levels of cap beam prestressing are provided.

In bridge bents where the cap beam is integrated into the superstructure, as shown in Fig. 4.24, an effective superstructure width should be taken into account for effective cap beam stiffness and capacity characterization. Although experimental evidence points to very large effective superstructure width contributions with increasing crack pattern development in the superstructure, a separate characterization for design or assessment models is proposed since the actual effective width is difficult to determine.

For design models, the effective superstructure width should be limited to ensure sufficient reinforcement placement in close proximity along the cap to avoid inelastic superstructure action from plastic column moment input. For T- or I-girder superstructures, only the deck slab will contribute to the flexural stiffness of the cap element, whereas for box girders, both deck and soffit slabs participate. For design models, the ACI guidelines [A5] for tributary width to stiffness and capacity determination are recommended to ensure the capacity protection against inelastic action in the superstructure. Thus, unless more detailed effective width analyses are performed, effective slab width contributions of

$$b_{\text{eff}} \le 8t_{\text{slab}} \quad \text{or} \quad \frac{L_{\text{cap}}}{4} \tag{4.76}$$

on either side of the cap beam should be used. Due to the thinness of typical deck and soffit slabs, other limitations tied to the distance between beams usually do not apply.

For assessment models, experimentally observed larger effective widths can be approximated by assuming a 45° spread of tributary force flow, as indicated in Fig. 4.25. The effective width at exterior columns can be approximated with reference to the inside column face, as shown in Fig. 4.25(b), based on a 45° influence spread as

$$b_{\text{eff}} = B_c + 2(D_c + L_o) \quad \text{(exterior columns)} \tag{4.77}$$

and at an interior column or between interior columns as

$$b_{\text{eff}} = B_c + L_c \quad \text{(interior)} \tag{4.78}$$

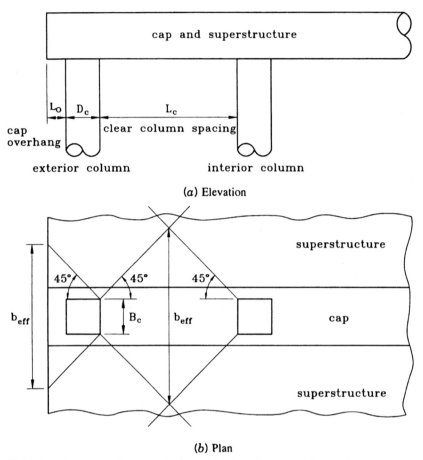

FIG. 4.25 Effective cap beam width in multicolumn bents with integral superstructure.

where L_c represents the clear distance between adjacent columns, and L_o is the overhang (see Fig. 4.25(b)).

Models for nonlinear analysis techniques discussed in Section 4.5 can include special nonlinear hinge elements in flexural plastic hinge zones or special nonlinear spring elements in joint regions and at the foundation, as depicted in Fig. 4.24 to represent appropriate nonlinear response characteristics. Joints between the cap beams and columns in multicolumn bridge bents have received significant attention since the 1989 Loma Prieta earthquake, where the collapse of a 1-km-long section of the Cypress viaduct on Interstate 880 in Oakland, California, was attributed at least partially to joint failures in the cap–column connections at the lower-deck level [H3], as shown in Fig. 4.26(a).

Developing appropriate load–deformation characteristics for these joint regions in the form of special joint elements is difficult at best. Strut and tie models [I10] for these joint regions have been shown to be very effective as

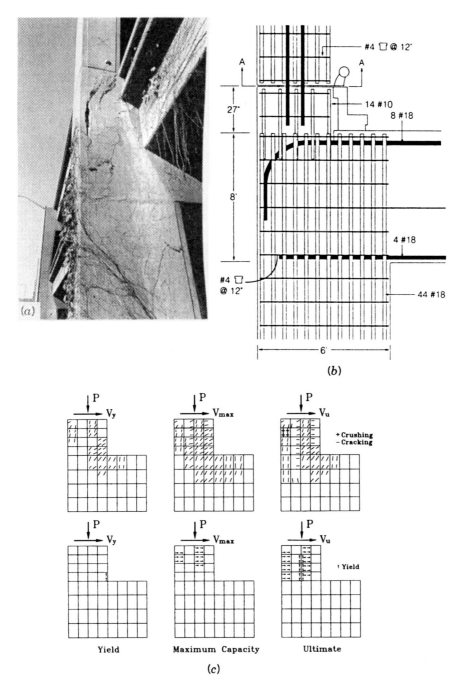

FIG. 4.26 Collapse of the Cypress viaduct and associated joint models. (*a*) Joint-damage; (*b*) Pedestal and joint reinforcement; (*c*) Crack and yield pattern.

design tools to determine force capacities but are limited in determining deformation limit states, whereas two- or three-dimensional nonlinear finite-element analyses can be useful to capture local distress patterns in terms of the extent of cracking and yielding as well as limited deformations, but they are still severely limited in characterizing local debonding of reinforcing bars, the effect of spalling of cover concrete, and confinement effects from adjacent members.

An example of the crack and yield pattern development analysis of a typical lower-deck beam–column joint at the Cypress viaduct [S19] is depicted in Fig. 4.26 with results from a two-dimensional nonlinear in-plane analysis using the computer code PCYCO [K6]. The finite-element model utilized four-node in-plane elements that were modified with additional incompatible displacement modes to improve the bending behavior. Reinforcement in PCYCO is overlaid and may be modeled as either a smeared layer of steel uniformly distributed over the element domain or as discrete bars. The latter option was chosen to model the cap beam top and bottom reinforcement shown in Fig. 4.26(b), consisting of No. 18 (D57 mm) bars. The constitutive behavior of the elements is based on a biaxial orthogonally anistropic model [D3] and on modified compression field theory [V2]. The model can be classified as a tensile-strength smeared-rotating-crack model where discrete crack locations can be estimated from strain integration perpendicular to the principal tensile stress direction. Progressive crack and yield patterns derived with this model for the beam–column connection in Fig. 4.26(a) and (b) for increasing horizontal upper column shear forces V are depicted in Fig. 4.26(c) for the yield, maximum capacity, and crushing limit states of the pedestal and joint region. As can be seen from Fig. 4.26(c), cracking and yielding start vertically and horizontally in the pedestal portion, and cracking continues into the cap with maximum tensile strains following the bend of the cap-beam top reinforcement, as indicated by the schematic crack lines. Wide-open flexural cracks can be located even with a smeared-crack model in regions where reinforcement yield is exceeded, (i.e., along vertical lines in the pedestal region). Finally, crushing occurs on the compression side of the pedestal. The analytically obtained distress patterns in Fig. 4.26(c) correlate with the distress pattern shown in Fig. 4.26(a) in portions of the Cypress viaduct that did not collapse. Limitations of these nonlinear finite-element models are discussed in Section 4.5.2.

(d) *Foundations.* Foundations of bridge piers are the structural elements at or below ground level that support the pier and provide vertical, lateral, and rotational resistance to gravity loads and seismic forces. The way in which this resistance is developed depends on (1) the type and geometry of the foundation, (2) the characteristics of the surrounding soil, and (3) the interaction between soil and structure and is discussed extensively in the geotechnical literature [L3,L6].

The three most common footing types for bridge piers and abutments are (1) spread footings for stiff soil sites, (2) pile-supported cap footings for soft

soil sites or soil layers with liquefaction potential, and (3) cast-in-drilled-hole
(CIDH) pile shafts, which can be drilled without casing in stable soils and
with casing in less competent and water-saturated soils.

(i)   Spread and Pile Footings.   Spread or pile footings are typically considered
to be rigid bodies that allow support conditions to be modeled at a single
point with boundary springs at the bottom of the column or pier model at
the end of the effective length extension link into the footing, as shown
schematically in Fig. 4.27. For a two-dimensional column, bent, or frame
model, only a vertical, a translational, and a rotational boundary spring need
to be defined, whereas in a three-dimensional model, six springs, one for each
possible DOF at the column base, are required.

In spread footings the soil resistance is provided in the vertical direction
by direct bearing pressure, in the horizontal direction by passive soil pressure
in front of the footing and friction along the footing base and sides, and in
the rotational direction by the soil overburden on top of the footing and

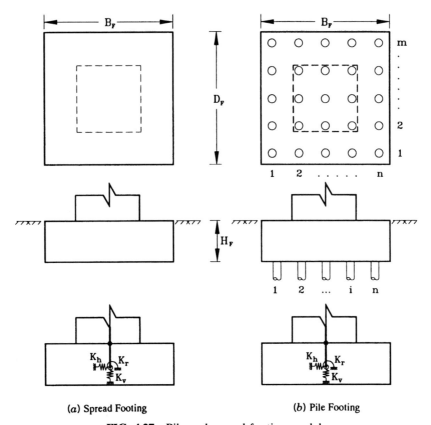

(a) Spread Footing   (b) Pile Footing

FIG. 4.27   Pile and spread-footing models.

gravity-load effects. Since spread footings are typically built on stiff and competent soils, stiff springs or fixed boundary conditions are assumed for the translational springs, and rotation is considered only when uplift and rocking of the entire footing, as discussed in Section 6.4, can occur. In cases where a soil reaction coefficient k_s (force/length^3) is available, the vertical and rotational foundation springs shown in Fig. 4.27(a) can be derived by integration over the linear soil resistance profile for a unit displacement and rotation as

$$K_v = B_F D_F k_s \qquad\qquad (4.79)$$

$$K_r = \frac{1}{12} B_F^3 D_F k_s \qquad\qquad (4.80)$$

respectively.

For pile footings there is evidence that the rotational stiffness is of greater significance than lateral stiffness on overall bridge response [L6]. The rotational stiffness and capacity of pile footings are largely related to pile axial stiffness (including tip penetration effects) and the pile axial capacities in compression and uplift. Assuming the axial stiffness of each pile to be k_p, the total vertical stiffness and rotational stiffness for a pile-supported footing can be found as

$$K_v = mnk_p \qquad\qquad (4.81)$$

$$K_r = m \sum_{i=1}^{n} k_p x_i^2 \qquad\qquad (4.82)$$

respectively, for m equal rows of n piles parallel to the loading direction [see Fig. 4.27(b)]. The lateral stiffness is controlled by the bending stiffness of the piles, their connection with the pile cap, and the soil stiffness. Winkler spring models along the depth of a pile model can be used to evaluate the lateral p–y curve for each pile, as shown in the following section for CIDH pile shafts.

For typical pile footings with less than 20 piles at 3-diameter center-to-center spacing or more, pile group effects can be ignored [L6]. However, for close-spaced piles and for large pile groups at major river or bay crossings, which can consist of several hundred piles, pile group or shadowing effects can become significant and should be evaluated by a qualified geotechnical engineer.

Pile capacities in bending can be established as outlined in Section 5.6 based on reinforcement, axial load, and confinement levels. The axial capacity of piles is typically specified as a nominal compression or bearing capacity. Under dynamic short-term loads, the ultimate compression capacity of piles is often assumed to be four times the nominal pile capacity, whereas the ultimate tensile capacity is often assumed to be one-half the nominal capacity.

However, the pile-to-pile cap connection must be checked to ensure that the assumed ultimate pile tension capacity can be transferred.

(ii)   Pile Shafts. Where soil conditions permit, the use of CIDH pile shafts has become a cost-effective and widely used bridge bent foundation, as discussed in Section 3.3.6. Either the column is continued as a pile shaft extension with the same diameter into the ground and the possibility of an inground plastic hinge exists, as is the case in Fig. 4.28(*a*), or an oversized shaft is cast typically to ground level and a column with reduced cross section is continued above ground, providing a well-defined location for the bottom plastic hinge.

In either case, the flexibility of the pile shaft and the surrounding soil should be modeled in a seismic demand analysis. Where geotechnical data on soil conditions in the form of Young's modulus E_s (force/length^2) exist, a Winkler reaction spring system for lateral support by the surrounding soil can be assumed and individual soil spring stiffnesses along the pile length can be

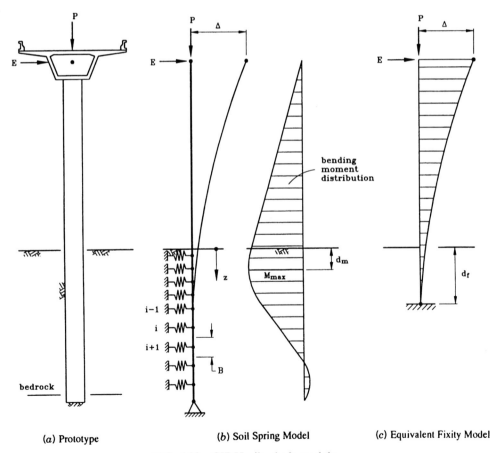

(a) Prototype   (b) Soil Spring Model   (c) Equivalent Fixity Model

FIG. 4.28   CIDH pile shaft models.

determined (Fig. 4.28), based on the following considerations. Assume that a contact pressure p at the soil–pile interface can be expressed as a function of the soil deformation Δ_s:

$$p = k_s \Delta_s \quad \text{(force/length}^2\text{)} \tag{4.83}$$

where k_s (force/length^3) represents a soil reaction coefficient. Then a Winkler soil reaction modulus or spring constant k along the length of the pile with diameter D can be determined as

$$k = Dk_s \quad \text{(force/length}^2\text{)} \tag{4.84}$$

and is often referred to as a reaction modulus. For cohesionless soils and normally consolidated clays, a linear increase of k with depth z measured from the ground surface can be a reasonable assumption, and the Winkler soil reaction modulus k can be expressed as a function of depth z as

$$k(z) = k^*z \tag{4.85}$$

where k^* (force/length^3) represents the depth-independent subgrade reaction modulus.

Discrete soil spring stiffnesses K_i at depth z_i, as shown in Fig. 4.28(b), can now be determined for a given tributary length B_i of pile shaft:

$$K_i = k^*z_iB_i = k^*z_i \cdot \frac{|z_{i+1} - z_{i-1}|}{2} \tag{4.86}$$

Values for k or k^* can be obtained from the geotechnical literature in relationship to Young's modulus E_s or subgrade modulus E_s^* for the soil, which in turn can be found (although with significant variability) from standard penetration tests, shear wave velocity measurements, or direct bearing tests carried out by an experienced geotechnical engineer.

The technical literature is divided about the dependency of Eq. (4.86) on the pile diameter. While Eq. (4.86) does not explicitly depict D, it is contained in k^* through Eqs. (4.84) and (4.85), and Eq. (4.86) could be expressed as

$$K_i = k_s^*z_iB_iD \tag{4.87}$$

as a function of the pile diameter D and the depth-independent subgrade soil reaction coefficient k_s^*. While earlier theoretical considerations [T9] implied that the product of subgrade reaction coefficient and pile diameter $k_s^*D = k^*$ is constant, which makes Eq. (4.86) diameter independent, more recently experimental data [C18,L5] indicate that only the subgrade reaction coefficient k_s is independent of the pile diameter and that a subgrade reaction modulus

k_2 for a pile with diameter D_2 can be obtained from a known coefficient k_1 and pile shaft diameter D_1 as

$$k_2 = k_1 \frac{D_2}{D_1} \qquad (4.88)$$

A comprehensive discussion of these issues is provided elsewhere [P23]. As shown in Fig. 4.28(*b*), an analytical model can provide reasonable approximations to the pile boundary conditions but does not represent dynamic soil–structure interaction (SSI) since no soil inertia effects, soil wave radiation effects, or viscous effects of soils movement around the pile shaft and modifications of these characteristics by the pile stiffness or the density of pile groups are considered.

An alternative approach to modeling soil flexibility effects on pile shaft systems with discrete soil springs is the effective fixity approach [R4]. In this approach the equivalent depth to fixity d_f, depicted in Fig. 4.28(*c*), is determined as the depth that produces in a fixed-base column with soil removed above the fixed base the same top-of-the-column lateral displacements under the ideal seismic lateral force E_i applied at the column top. Both the axial column load P and any top-of-the-column moments M (not shown in the cantilever example of Fig. 4.28) at force level E_i need to be considered.

Depth-to-fixity charts, similar to the one shown in Fig. 4.29 for cohesionless sand, can be established in nondimensional form based on the subgrade reac-

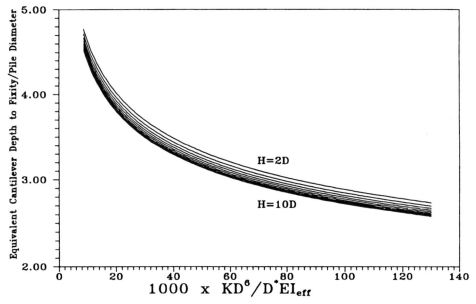

FIG. 4.29 Depth to fixity for an equivalent cantilever versus nondimensional system stiffness.

tion modulus k, the pile diameter D, the effective pile stiffness EI_{eff}, and a reference pile diameter D^* taken as $D^* = 1.83$ m (6 ft) for the relationships developed in Fig. 4.29. Similar curves have been established by Caltrans [C12] for sand and clay soil conditions based on blow counts from standard penetration tests. Although this equivalent depth-to-fixity modeling approach provides good stiffness estimates for seismic bridge analyses, it is important to note that the maximum in-ground moment does not occur at the equivalent point of fixity, but rather, closer to the ground surface, as is apparent in Fig. 4.28, which amplifies the column shear forces due to the higher moment gradient or shorter shear span compared to the equivalent depth-to-fixity model. Thus for nonlinear analyses the discrete soil spring model approach of Fig. 4.28(b) is recommended, with at least four discrete soil springs above and below the theoretical point of fixity. If predetermined plastic hinge locations are used in the form of inelastic elements or joints, their location should be defined based on the guidelines for plastic hinges in pile shafts provided in Section 5.3.1(b).

The load–deformation characteristics of a CIDH pile can be refined further by assuming nonlinear soil behavior. Simplified bilinear elastic–plastic behavior with the plastic hardening portion taken as one-fourth of the initial slope or reaction modulus k is a reasonable first assumption in considering nonlinear effects. Transition from the elastic to the plastic state occurs at a lateral pile deflection Δ_y independent of depth z, resulting in the load–deformation characteristics shown in Fig. 4.30(a). For typical cohesionless soil, this yield deformation Δ_y occurs at approximately 25 mm or 1 in. Unloading from the plastic range can be assumed to occur parallel to the initial stiffness k down to zero load and subsequent gapping of the soil, resulting in the hysteretic load–deflection characteristics for each soil spring depicted in Fig. 4.30(b).

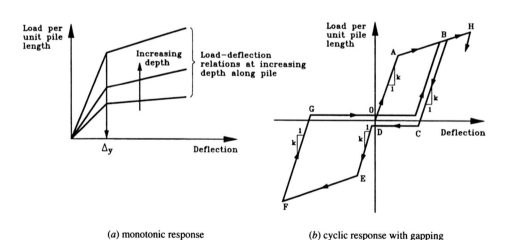

(a) monotonic response   (b) cyclic response with gapping

FIG. 4.30 Pile shaft/soil deformation characteristics (a) monotonic response; (b) cyclic response with gapping.

Where dynamic inelastic time-history analyses [discussed further in Section 4.5.4(b)] are considered, the computer code should be capable of modeling this type of hysteretic response.

(e) Abutments. Modeling assumptions made for abutment stiffnesses and capacities as well as damping of the surrounding soil mass can have a significant effect on the analytical dynamic response characteristics, particularly in shorter bridge structures. Three abutment types (Section 3.3.2) need to be considered: (1) monolithic or diaphragm abutments, (2) seat abutments, and (3) CIDH-shaft-controlled abutments. The first two types represent traditional bridge abutments with the back wall and superstructures built either monolithically or separated by joints and bearings; the third abutment type reflects recent Caltrans developments that attempt to control inelastic abutment displacements through large-diameter CIDH piles by means of predictable inelastic load–deformation characteristics in well-confined CIDH pile shafts built integral with or in close proximity to the abutment.

All three abutment types have in common that they are (1) massive structures; (2) mobilize and interact with large soil masses; (3) based on their geometry, exhibit significantly higher stiffness values than do other bridge bents and thus attract proportionally higher seismic forces; and (4) feature some or all of the following highly nonlinear elements and behavior characteristics: breakaway shear keys, expansion joint restrainers, sacrificial wing and back walls, and a potential for inelastic pile action.

New bridge designs with ductile bents or retrofitted bridges in which previous weak links in the seismic chain of response have been eliminated have the potential to tax abutment capacities in future seismic events more than the damage levels to abutments observed in past earthquakes. However, very limited research data exist to date in support of abutment modeling and design, so that only directions for improvements to current abutment modeling practice can be provided.

Abutment capacity and stiffness characteristics are often based on empirical relationships. Caltrans, for example, utilizes an abutment capacity based on a maximum effective soil pressure of 5 kips/ft^2 (239 kPa), amplified by about 50% to 7.7 kips/ft^2 (368 kPa) for dynamic or earthquake loads, capacity levels recently verified by large-scale abutment tests [M16]. For the projected abutment area A_{eff} in the loading direction, the nominal dynamic abutment capacity can then be determined as

$$F_{\text{abut}} = 7.7 \ (\text{kips/ft}^2) \cdot A_{\text{eff}} \quad [368 \ (\text{kPa}) \cdot A_{\text{eff}}] \qquad (4.89)$$

For protection against possible higher forces, the back wall in the longitudinal and the shear keys in the transverse direction may be designed to break off in the form of sacrificial elements, which will protect the abutment–pile system against significant inelastic action.

Prior to engagement of abutment shear keys or the back wall in seat abutments, and subsequent to failure of these components, only the pile stiffness k_p from the abutment piles is considered, neglecting any friction. In the push direction, upon closure of the seating gap, the backwall soil stiffness k_s and pile stiffness k_p are additive (Fig. 4.31) until the soil capacity is exceeded, at which point the pile stiffness k_p alone controls the force–deformation behavior. In the pull direction, only the pile stiffness contributes to the abutment resistance characteristics. Since in common linear elastic spectral analyses for bridge structures only one linear spring stiffness for abutment movement in a given direction can be employed, an iterative solution strategy similar to the substitute structure analysis explained in detail in Section 4.5.2(b) (iii) is frequently employed [L6] to find compatible abutment displacement and resistance values. Thus, in spectral design analysis models, abutment stiffness values are iteratively adjusted in successive linear elastic analyses based on assumed displacements and corresponding effective stiffness values, a procedure outlined schematically in Fig. 4.31, until a stable displacement response is reached. Due to this iterative procedure, the assumed initial abutment stiffness is not very important. Current Caltrans guidelines set the initial abutment stiffness estimates at

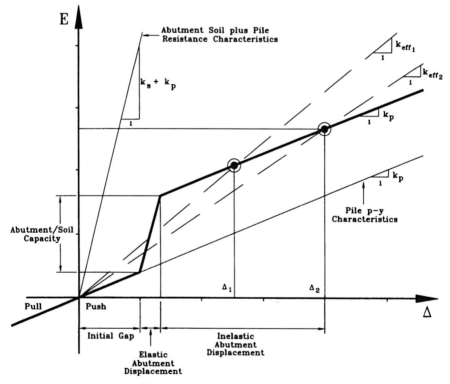

FIG. 4.31 Abutment force–deformation characteristics.

$$k_a = k_s + k_p = \begin{cases} [200 \text{ (kips/in./ft)} \cdot B] + [40 \text{ (kips/in.)} \cdot n_p] \\ [115 \text{ (kN/mm/m)} \cdot B] + [7 \text{ (kN/mm)} \cdot n_p] \end{cases} \quad (4.90)$$

where B represents the effective abutment width and n_p is the number of piles. In the transverse direction the effective width B is taken as the length of the wing walls multiplied by a factor of $\frac{8}{9}$ to account for differences in participation of both wing walls [C12]. However, unless the secant stiffness or substitute structure approach is extended to the entire bridge systems analysis including equivalent bent characteristics, resulting displacement values may significantly underestimate actual abutment displacements.

The initial stiffness from Eq. (4.90) is very high and small displacements exceed the available abutment capacities, requiring both the above-outlined sacrificial elements and iterative analysis procedure as well as subsequent design measures to accommodate the displacement levels calculated. For the other abutment types, the analysis approach outlined invariably results in inelastic actions in the abutment substructures and displacement levels need to be designed appropriately.

Although the iterative abutment stiffness approach described above has some merits for design models, in the actual deformation capacity assessment of existing designs or structures such an analysis approach is not very satisfactory. Rather, probable stiffness data in the form of soil springs for the abutment should be derived from geotechnical investigations. The two extreme cases of a demand assessment in terms of fully fixed and completely released abutment conditions can also be very useful in providing additional insight into the bridge seismic response expected.

Recent large-scale experimental tests on abutment characteristics at the University of California–Davis [M17] resulted in the proposed force–deformation relationship presented in Fig. 4.32, which can provide some guidance in establishing more realistic abutment stiffness estimates once the abutment backwall has engaged fully in the push direction. As mentioned above and outlined in Fig. 4.31, prior to the initial seating gap closure and in the tension direction the abutment stiffness is controlled by pile deformations.

(f) Movement Joints and Restrainers. Long bridges are divided into frames by movement or expansion joints to compensate for deformations from initial shortening due to prestressing, time-dependent effects such as creep and shrinkage, and environmental effects such as temperature deformations. Seismically, these movement joints have the tendency to allow separate frames to develop their own characteristic dynamic response and to modify this individual dynamic response through complex interaction through restraining devices installed in these movement joints, as discussed in Section 5.8.3. In addition to these restraining devices, friction during movement, and banging or pounding upon closure of these joints during a seismic event need to

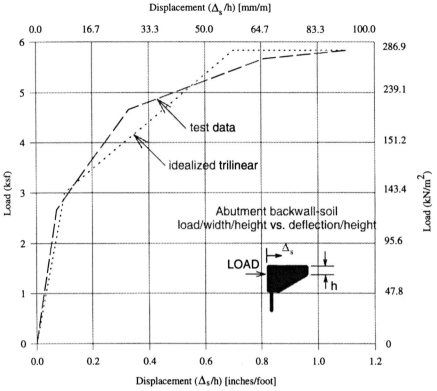

FIG. 4.32 Proposed characteristics and experimental envelope for abutment back wall load–deformation [M16].

be considered. This dynamic response modification can manifest itself in a reduction of dynamic response levels due to increased damping from friction in the joints, spoiled harmonic response from restrainer engagement and pounding, and by a possible longer seismic force-transfer path and increased tributary seismic force to stiff elements of the bridge due to the connectivity or coupling between frames. Very few quantified performance data from field or laboratory tests for the analytical modeling of the dynamic movement joint response characteristics exist, and until such data become available, guidelines for movement joint modeling can only be provided in general terms.

Movement joints typically allow deformations in the form of translation in the longitudinal bridge direction and flexural rotation about the movement joint axis but restrict translations perpendicular to the bridge axis by means of shear keys. Vertical shear transfer is provided through bearing seats and vertical restrainers. However, movement joints cannot be viewed only as longitudinal bridge separations in a seismic event since transverse seismic deformation input can open and close movement joints to various degrees depending on the geometry of the bridge structure, as depicted in Fig. 4.33. Frames with bents of unbalanced transverse stiffness, and in particular stiff abutments, can cause rotations of the bridge superstructure in the plane of the bridge deck even for straight bridges and transverse seismic deformations, as shown in Fig. 4.33(a), resulting in movement joint opening and pounding. The same is true for curved bridges under transverse movements, as depicted in Fig. 4.33(b), for seismic displacements in arbitrary directions, which will provide uneven movement joint displacements, particularly in joints parallel to the earthquake direction. Finally, as shown in Fig. 4.33(c), bridges with skew movement joints will see the opening of the joint in the same direction, independent of the direction of transverse seismic motion, since the engagement of the obtuse corner will always result in a force couple between inertia and restoring forces with the same rotational orientation, as discussed in Section 1.2.1.

The nonuniform opening and closing of movement joints makes it essential that movement joints be modeled with their exact geometry. This can also be accomplished in a spine-type bridge model through rigidly connected link elements along the movement joint, as depicted in Fig. 4.33(c). Support bearings, cable restrainers, pipe extenders (all elements discussed in Section 8.5.1), and shear keys can then be placed in their correct geometric position and spatial relationships.

Stiffness characteristics of cable restrainers and pipe extenders in the longitudinal bridge direction can be obtained from direct tension characteristics once the unit has engaged based on restrainer area A_r, restrainer modulus E_r, and restrainer length L_r. For joint closure a very high, numerically difficult to model infinite stiffness can be assumed at the contact location, and during the slack period of the restrainer, typically no stiffness contribution or only sliding friction and associated damping are present. While the extreme stiffness values of restrainers are easily quantified in the form of spring stiffnesses as

$$K_s = \frac{E_r A_r}{L_r} \quad \text{(tension)}$$

$$K_s = 0 \qquad \text{(slack)} \qquad\qquad (4.91)$$

$$K_s \to \infty \qquad \text{(compression)}$$

the stiffness changes during cyclic loading necessitate the need for nonlinear analysis.

Since linear elastic modal analysis tools are used most commonly for bridge analysis, the California Department of Transportation [C12] has adopted the dual analysis procedure of tension and compression models, where in the tension model all joints are provided with linear elastic restrainer springs $K_s = E_r A_r / L_r$ and in the compression model all joints are rigidly connected in the bridge axial direction, but free to rotate about the vertical axis. Maximum

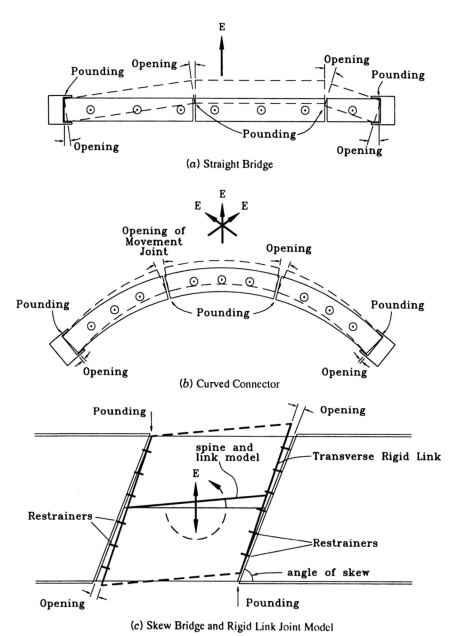

(a) Straight Bridge

(b) Curved Connector

(c) Skew Bridge and Rigid Link Joint Model

FIG. 4.33 Opening and closure of movement joints under seismic loads.

response quantities from either model are used subsequently for seismic design or assessment purposes. Although this dual analysis model approach results in reasonable estimates of maximum member forces and displacements, it cannot be used to predict relative displacements at the movement joints since the maximum displacements of adjacent frames calculated occur at different times during an earthquake. Relative-movement joint displacement values obtained using combination techniques for maximum response values from elastic modal analyses as outlined in Section 4.3.2 significantly overestimate displacement values obtained from nonlinear time-history analyses [Y1], discussed further in Section 5.8.

For nonlinear dynamic time-history analyses, joint element models (Fig. 4.34) with nonlinear stiffness characteristics, gapping, and Coulomb friction damping have been developed [R1,Y1]. The expansion joint model in Fig. 4.34 consists of three separate components: (1) a restrainer model, (2) a bearing/sliding friction model, and (3) a shear key or Coulomb friction model. The restrainer model simulates the influence of longitudinal restraining devices [discussed further in Section 8.5.1(a)] on the longitudinal opening of the joint. In compression, the model is almost perfectly rigid once the initial gap has closed, whereas in the tension following slack elimination, the restrainer stiffness outlined by Eq. (4.91) is activated.

Depending on the restraining device, yield and hysteretic behavior can result, as shown in Fig. 4.34(c). The shear key will engage in the transverse direction once the initial transverse gap has closed and will provide a rigid constraint until shear key failure occurs. Subsequent to shear key failure, hysteretic force–deformation characteristics in the transverse direction based on the axial force P transferred through the shear key and a friction coefficient $\mu \geq 1$ for Coulomb friction can be employed. As a minimum even in the absence of normal forces P on the shear key, the shear friction force provided by the shear key dowel reinforcement can be activated and quantified based on standard shear friction principles. Finally, the primary shear or bearing force transfer occurs through bearing pads in the movement joint, as indicated in Fig. 4.34(a), and the force–displacement characterization is in the form of sliding friction, as outlined in Fig. 4.34(c), with friction coefficients between 2 and 20%, depending on the slider or bearing. All of the foregoing effects can be modeled with the discrete spring and dashpot assembly shown in Fig. 4.34(b), together with vertical spring characterizations resulting from vertical bearing stiffness and restraining devices. The correct parameter quantification for such a complex joint model is quite difficult and requires extensive sensitivity studies to qualify resulting response assessments.

4.4.3 Substitute Structure Models

With recognition of the importance of displacements rather than forces for the survival, damage-control, and serviceability of a bridge following a major

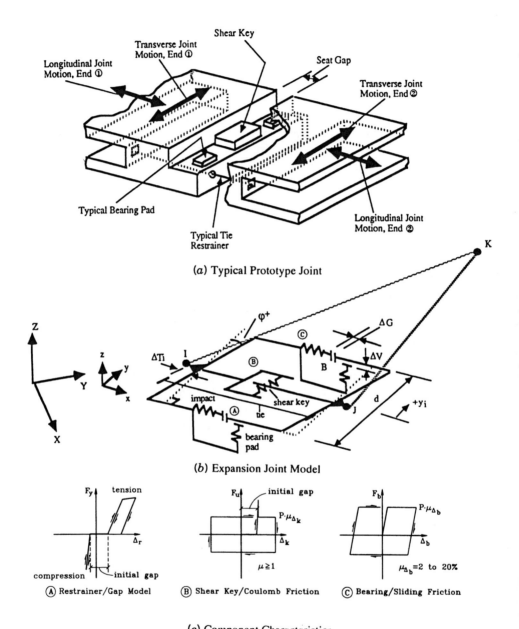

(a) Typical Prototype Joint

(b) Expansion Joint Model

(A) Restrainer/Gap Model   (B) Shear Key/Coulomb Friction   (C) Bearing/Sliding Friction

(c) Component Characteristics

FIG. 4.34   Three-dimensional expansion joint element [R7].

earthquake, the substitute structure analysis approach [G2,S16] [discussed further in Section 4.5.2(b)] has received renewed attention with respect to a completely displacement-based design approach for seismic bridge response [K3,C3].

The fundamental concept of the substitute structure analysis approach is outlined in Fig. 4.35, where the entire inelastic prototype bridge is described by an idealized linear elastic system or substitute structure which describes the fundamental dynamic force–deformation behavior of an inelastic bridge model with equivalent linear elastic stiffness and damping properties. While the mass of the substitute structure model is kept constant, both effective systems stiffness and effective systems damping ratio are adjusted such that the displacement of the inelastic prototype response is the same as the displacement of the substitute structure model.

The two parameters needed for a substitute structure model characterization, the effective stiffness k_{eff} and the effective damping ξ_{eff}, are depicted schematically in Fig. 4.36 as a function of the structural displacement ductility μ_Δ to show their variational tendencies by means of examples discussed in the literature [G2,K8]. For idealized elastic–perfectly plastic systems response, the effective stiffness decreases geometrically with increasing ductility, as outlined in Fig. 4.36. Depending on the dominant inelastic mechanism and the corresponding hysteretic load–displacement response, various effective damping ratio relationships as a function of ductility can be derived as shown schematically for two cases in Fig. 4.35, with effective damping ratios ranging from below 2% at ductility 1 to over 20% at ductility 6, for the example of a Takeda column hinge mechanism as shown in Fig. 4.44(d), with unloading slope coefficient $\alpha = 0.5$ and an axial load ratio of 15% of the nominal gross

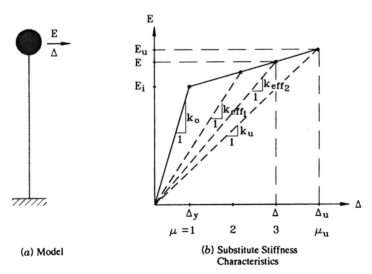

(a) Model   (b) Substitute Stiffness
                                        Characteristics

FIG. 4.35 Substitute structure model.

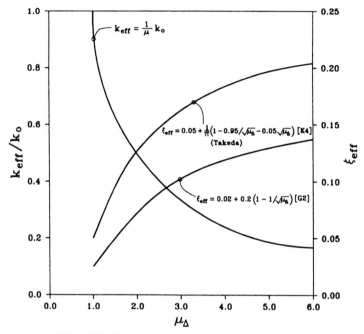

FIG. 4.36 Substitute structure model parameters.

section concretè capacity. For beam hinges with low axial load levels, the maximum value for effective damping ratio can significantly exceed the 20% level, as outlined in Section 4.3.1(c), with an upper limit of 64% for perfectly rigid–plastic response mechanisms. Since the substitute structure analysis relies on a linear elastic solution to the general equation of motion (4.3) with viscous damping only, the principles outlined in Section 4.3.1(c) for finding equivalent viscous damping coefficients from the hysteretic response of the prototype structure apply to the determination of the effective damping ratio in the substitute structure approach. Relevant analysis procedures for evaluating the substitute structure response for displacement-based bridge design are discussed in Sections 4.5.2(b)(iii), and 5.3.1(c).

4.5 METHODS OF ANALYSIS

Analysis tools for the seismic response quantification of bridges range from simple linear elastic static analyses in the form of hand calculations to three-dimensional dynamic nonlinear response history analyses for global bridge systems with large numbers of DOFs ($>10^6$ in some cases). In the following, an attempt is made to categorize the various analysis tools based on the desired quantification of expected bridge performance and to provide limited guidance to the appropriateness and application of the various analytical tools in the general seismic bridge design and assessment process.

4.5.1 Types of Analysis Tools

The analysis tools provide the numerical mathematical process to extract response quantifications from the models described in Section 4.4. Based on the expected response of the individual members, the form in which the seismic load is simulated, and the consideration of geometric effects, the various types of analyses can be classified into tools that can provide solutions for (1) linear $\langle L \rangle$ or nonlinear $\langle N \rangle$ materials responses, (2) static $\langle S \rangle$, dynamic $\langle D \rangle$, or spectral response $\langle R \rangle$ simulation, and (3) geometric effects such as P–Δ $\langle P \rangle$ or full nonlinear or large deformations geometry $\langle G \rangle$.

Since the analysis objective is the seismic response evaluation of bridges in the form of capacity and demand quantification, as outlined in Fig. 4.1, useful analytical tools based on the foregoing characteristics can best be separated by the way in which the seismic force input is provided. A schematic overview of the most commonly used bridge analysis tools based on seismic force simulation is given in Table 4.1, together with an indication of their primary application in terms of demand and/or capacity quantification, as well as comments concerning the general analysis tool characteristics. As can be seen from Table 4.1, the analysis type $\langle G \rangle$, or full geometric nonlinear effects, was omitted since very few bridge structures exhibit deformations during design- or service-level earthquakes, which are large compared to the dimensions of the bridge structure and thus do not require a full geometric nonlinear analysis. In most cases, where combinations of deformations and axial loads in towers, slender bridge piers, or tall members result in significant contributions to member forces, their effect on the overall deformation state of the structure is still small enough that only equilibrium considerations in the deformed state (i.e., P–Δ effects), not complete nonlinear geometric effects, need to be considered. Furthermore, linear elastic dynamic time-history analyses were omitted from Table 4.1 since there is little merit in knowing the complete response time history of a linear elastic bridge model for a specific earthquake input, particularly when maximum response quantities can conveniently be obtained from a much simpler response spectrum analysis.

Table 4.1 also shows that while most analysis tools provide some form of seismic demand quantification, only models that include nonlinear material characteristics can provide actual capacity quantifications. Brief discussions of some of the key features of the most commonly used analytical tools summarized in Table 4.1 are provided in the following sections.

4.5.2 Static or Quasistatic Analysis Tools

In many cases of seismic analysis it is most convenient to apply the seismic actions in the form of an equivalent static force to the bridge model, particu-

TABLE 4.1 Seismic Bridge Analysis Tools

Load Simulation	Analysis Type[a]	Objective[b] Demand	Objective[b] Capacity	Comments
Static ⟨S⟩	⟨L⟩	X		Linear elastic frame or finite-element analysis with equivalent lateral loads to determine member forces and equivalent displacements
	⟨N⟩	x	X	Pushover or nonlinear analysis to determine available deformation capacities, equivalent elastic member forces, or plastic deformation demands
	⟨L, P⟩	X		Linear elastic demand analysis for tall bridges or slender piers or members where deformations and axial loads can change member demands
	⟨N, P⟩	x	X	Nonlinear pushover or cyclic analysis for tall bridges or slender piers with reduced capacities reflecting P–Δ effects
Response spectrum ⟨R⟩	⟨L⟩	X		Seismic demand analysis based on maximum modal response and appropriate modal combination techniques; member demands reflecting inelastic structural response can be obtained from reduced or "nonlinear response spectra" but still based on linear modal analyses
Dynamic ⟨D⟩	⟨N⟩	X	X	Nonlinear time-history analysis to determine the complete seismic response with nonlinear member characteristics for specific earthquake records
	⟨N, P⟩	X	X	Nonlinear time-history analysis for tall bridges or bridges with slender members and piers, including P–Δ effects

[a] L, linear materials, N, nonlinear materials; P, P–Δ effects.
[b] X, primary tool; x, secondary tool.

larly when seismic force distributions or likely deformation modes can be estimated. The magnitude of the equivalent static force is either specified as absolute acceleration coefficient a_s or obtained from the design- or assessment-level earthquake in the form of an expected peak ground acceleration which then needs to be amplified by an appropriate spectral acceleration response factor for specific site or soil characteristics, as shown schematically in Fig. 4.37, based on the expected fundamental period assuming 5% equivalent viscous damping, described in Section 4.3.1(c). The acceleration coefficient a_s is then multiplied by the total seismic weight W_s to determine the earthquake force

$$E = W_s a_s \tag{4.92}$$

lumped at the centroid of seismic mass, or distributed proportional to the expected fundamental mode shape. This force and distribution pattern is then applied to the structure either as static monotonic force or in the form of a quasi-static stepwise monotonic or cyclic force.

(a) Solution Strategies. Depending on the ability of the analytical model to capture linear, nonlinear, unloading, or cyclic response, different solution strategies need to be adopted.

(i) Monotonic Loading. Monotonic forces are either applied in one step for linear ⟨L⟩ or incremental for nonlinear ⟨N⟩ analyses. For nonlinear analyses,

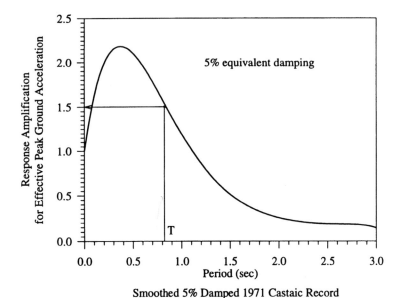

Smoothed 5% Damped 1971 Castaic Record

FIG. 4.37 Normalized spectral acceleration response amplification shapes.

analytical tools are required which allow an iterative solution strategy to balance internal and external forces for a given deformation level. The most common nonlinear solution strategies are

1. Step-by-step solutions with a tangent stiffness update at each step
2. Step-by-step solutions with equilibrium correction at the beginning of each new load step
3. Newton–Raphson iterations, which update the tangent stiffness and correct for any unbalanced loads iteratively during each load step
4. Constant stiffness iterations, which use a constant stiffness rather than an updated tangent stiffness for unbalanced load correction during each load step

While the step-by-step solution strategies require the load to be applied in small increments, with the correct force–deformation path better approximated with a larger number of smaller steps, the iteration analysis procedures are independent of step size as long as the applied load does not exceed the available capacity, and iterations are carried out until a specified accuracy or tolerance is achieved. For highly nonlinear bridge systems the Newton–Raphson solution strategies typically provide faster convergence with less computational effort, whereas the constant stiffness iteration strategies are numerically more efficient for systems that are not highly nonlinear and are more robust for very ductile systems with large inelastic deformations at small variations of load (i.e., regions where the bridge structure features a close to zero or even negative tangent stiffness).

(ii) Cyclic Loading. Quasistatic cyclic analyses can also be performed as long as analytical tools are available which can characterize unloading and reloading of the bridge model. These tools are based on iterative solution strategies similar to those presented above with increased complexity to follow any arbitrary cyclic response path. For seismic bridge design or assessment, fully cyclic analyses are rarely used due to uncertainties in the cyclic component characteristics, lack of reliable cyclic damage accumulation models, and cyclic load patterns which may not realistically simulate the actual seismic input.

Areas where cyclic analyses are performed are on the research side to predict or diagnose large or full-scale bridge component tests or to develop cyclic force–deformation characteristics with detailed finite-element analyses for structural components or bridge elements to be used in other seismic bridge analysis models. A conceptual example of structural component characterization is depicted in Fig. 4.38, where an individual diagonal bracing member in a pier of the Richmond–San Rafael bridge in California was modeled with a detailed finite-element model and subjected to cyclically increasing axial member end forces until local and overall buckling, as well as fracture in the built-up laced and riveted member occurred [A10]. The individual steel

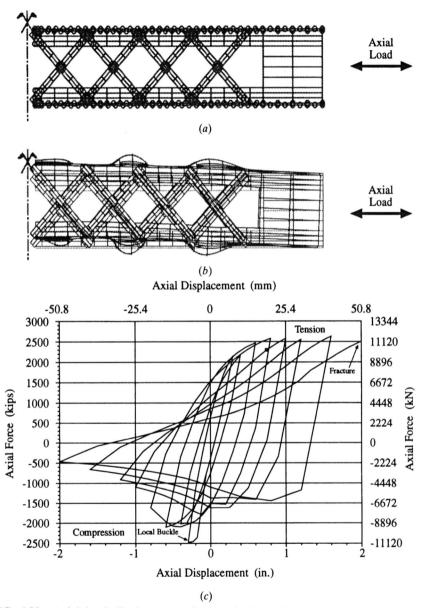

(a)

(b)

(c)

FIG. 4.38 Axial load–displacement characterization of typical longitudinal tower brace in the Richmond–San Rafael bridge [A10] (a) Topview of laced and riveted member discretization; (b) Buckeled shape of laced brace; (c) Axial load-displacement behavior.

channels and lattice members that comprise the laced and riveted brace were modeled with plate elements as depicted in Fig. 4.38(*a*), with nonlinear materials characteristics and nonlinear geometric effects to trace local and global instabilities. The cyclic force history was applied in displacement control and the resulting cyclic force–deformation behavior, shown in Fig. 4.38(*b*), was generated as a characteristic structural component axial force–deformation relationship which can subsequently be used in other analysis models where diagonal braces of the same geometry, boundary conditions, and materials are modeled by individual truss or beam elements with the derived force–deformation model.

(b)   Analysis Tools. Within the quasistatic analysis domain, the most commonly used analysis tools for the quantification of seismic bridge response are (1) linear elastic analyses, (2) pushover analyses, and (3) substitute structure analyses. All three types of analyses can still be performed by hand calculations at least for individual parts of the bridge structure, such as columns, bents, or frames.

(i)   Linear Elastic Analyses. Linear elastic quasistatic analyses are used as seismic bridge analysis tools to (1) predict or define stiffness characteristics of columns, bents, or frames modeled with effective linear elastic properties as defined in Section 4.3.1(*b*), (2) determine deformation and force response in the linear elastic response range for an equivalent quasistatic seismic load input, or (3) determine structural displacements for the inelastic response range under the static simulated seismic load assuming that equal displacement principles are applicable or that a realistic equivalent response estimate based or equal energy principles with the help of Eq. (5.22) can be derived. Analytical tools for linear elastic bridge analyses consist either of hand calculations utilizing simple beam, frame, or truss models, or of standard structural analysis programs ranging from basic two-dimensional frame codes to general-purpose three-dimensional finite-element packages with element libraries which include elements such as those depicted in Fig. 4.16 and provide amenities such as graphic pre- and postprocessors, automated mesh generation, and shape, frequency, or weight optimization routines.

(ii)   Collapse Mechanism Analyses. One of the most powerful equivalent seismic load analysis tools is the *event scaling procedure,* used primarily to determine the sequence of inelastic actions, the formation of local mechanisms, and the formation of a global collapse mode. Since this "collapse" analysis is typically performed for lateral seismic forces, these analyses are often referred to as *pushover analyses,* since for a given seismic lateral load distribution pattern, the pushover or lateral force failure mode is determined through the stepwise formation of local mechanism or plastic hinges. The pushover analysis utilizes the same seismic force distribution as discussed above with an arbitrary or unit magnitude, and all forces and deformations are scaled until the first specified event in the form of a nonlinear or inelastic action occurs. Change

of member stiffness due to cracking, formation of a flexural hinge, yielding of a soil spring, and so on, can constitute events. At each event the structural model is physically altered to reflect the occurrence of the event in the form of a changed member stiffness or the introduction of a hinge mechanism. In this way the pushover analysis consists of a sequence of linear elastic analyses with a stepwise changing structural system and can be performed with hand calculations or any linear elastic structural analysis program.

Requirements for a pushover or event scaling analysis are (1) a linear elastic structural model, (2) initial or conditioning loads (i.e., gravity loads), and (3) the characterization of all important nonlinear actions or events, typically in the form of bi- or trilinear force–deformation relationships. Since some of these events, such as the flexural hinge capacity, depend on changing axial force levels, critical levels of axial load are either predetermined based on the expected collapse mode or iteratively adjusted in each event increment. Clearly, such an iterative capacity adjustment is feasible only in an automated process. Special-purpose programs such as SC-PUSH3D [S8] iterate for axial load effects, update the structural stiffness at each time step, and track inelastic deformations.

The results of a pushover analysis are ultimate deformation capacities of bents or frames, as well as inelastic deformation demands on local mechanisms, related to predetermined capacities of sections or members found using the models developed in Chapters 5 and 7. Since effective stiffness characteristics of bridge subsystems such as bents and frames can also be readily obtained from pushover analysis, pushover models can also help in spectral seismic demand assessment. Recent recognition of these capabilities of pushover analyses has led to the development of special pushover computer codes for bridge structures [S8], even though the same concept of event scaling has been available for many years in limited forms in generic nonlinear analysis codes such as the DRAIN family of computer programs [P25].

Example 4.4. The basic concept of a pushover analysis is demonstrated on the example of a two-column bridge bent with geometry and properties, as shown in Fig. 4.39(a) and (b). For simplicity it is assumed that the posttensioned cap beam acts as a rigid flexural and axial link, that the seismic weight including column and cap contribution is $W_s = 1000$ kips (4448 kN), and that the center of seismic mass is at $\overline{H} = 17$ ft (5.2 m). Furthermore, the column height from the footing hinge to the bottom of the cap beam is assumed to be 12 ft (3.7 m).

The collapse mode for this bent has been reached when flexural plastic hinges form at the column tops at locations B and C. The initial axial load at locations B and C, respectively, is assumed to be $W_s/2 = 500$ kips (2224 kN). With this axial load, given column geometry, and reinforcement ratios, a moment–curvature analysis is performed, resulting in a bilinear elastoplastic approximation as shown in Fig. 4.39(b) with $M_{yi} = 1544$ ft-kips (2095 kN · m). The lateral earthquake force E corresponding to the collapse mechanism capacity can now be determined, assuming plastic hinges at the column tops with capacities M_{yi} each, as

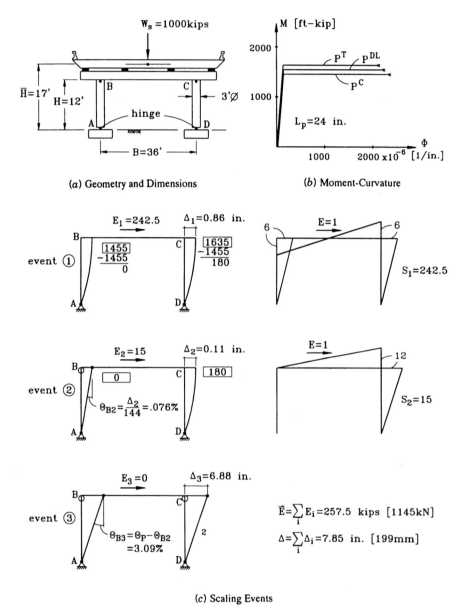

(a) Geometry and Dimensions

(b) Moment-Curvature

(c) Scaling Events

FIG. 4.39 Pushover analysis of a two-column bent.

234

$$E = \frac{2M_{yi}}{H} = \frac{2(1544)}{12} = 257.3 \text{ kips } (1144 \text{ kN})$$

From the bent geometry the change in axial force in each column due to the lateral load E can now be determined as

$$\Delta P = \pm\frac{\overline{H}E}{B} = \pm\frac{17}{36}E = \pm0.472E = \pm122 \text{ kips } (543 \text{ kN})$$

A reanalysis of the moment–curvature relationships, including earthquake axial force contributions, results in moments and curvatures for the various axial force levels listed in Table 4.2 where Φ_{yi} and Φ_u are the equivalent bilinear yield and ultimate curvatures, as discussed in Section 5.3.1(d)(viii). Since no dead-load moments are assumed in the columns, the ideal moment capacities in Fig. 4.39(b) and Table 4.2 are fully available to resist the applied earthquake forces. These seismic moment capacities are listed again in the capacity box at the top of the columns for event 1 in Fig. 4.39(c). The event scaling analysis now consists of the following steps.

Event 1: A unit load $E = 1$ is applied to the top of the bent model, resulting in the moment demand distribution indicated. Event 1 is reached when the demand moment with the smallest scale factor reaches the available capacity, which occurs in the tension column at location B. The scale factor S_1 can now be determined as

$$S_1 = \frac{1455}{6} = 242.5$$

$$E_1 = 1(242.5) = 242.5 \text{ kips } (1101 \text{ kN})$$

$$\Delta_1 = \frac{E_1 H^3}{6EI_{\text{eff}}} = 0.86 \text{ in. } (21.8 \text{ mm})$$

EI_{eff} is taken as the average from tension and compression column in Table 4.2. The remaining flexural capacities are now calculated, a flexural hinge is introduced at location B, and a new linear elastic analysis for event 2 is performed.

TABLE 4.2 Section Properties

Column Type	P (kips)	Φ_{yi} (in^{-1})	Φ_u (in^{-1})	M_{yi} (ft-kips)	EI_{eff} (in^2-kips)
Compression	622	135×10^{-6}	2125×10^{-6}	1635	1.45×10^8
Gravity load	500	134×10^{-6}	2226×10^{-6}	1544	1.39×10^8
Tension	388	132×10^{-6}	2347×10^{-6}	1455	1.32×10^8

Event 2: Again a unit load $E = 1$ is applied, resulting in the moment demand distribution indicated (Fig. 4.39). Event 2 is reached when the demand moment at C reaches the remaining capacity at C at a scale factor S_2 with

$$S_2 = \frac{180}{12} = 15$$

$$E_2 = 1(15) = 15 \text{ kips (67 kN)}$$

$$\Delta_2 = \frac{E_2 H^3}{3EI_{\text{eff}}} = 0.11 \text{ in. (2.8 mm)}$$

With a flexural hinge added at location C, a full collapse mechanism is obtained and the last event is reached when the first plastic hinge reaches the plastic rotation capacity θ_p.

Event 3: From relationships outlined in Section 5.3.2(*b*)(i) the plastic rotation capacity of the two hinges can be determined as $\theta_p = L_p (\phi_u - \phi_y)$, resulting in

$$\theta_p^B = 24(2347 - 132) \times 10^{-6} = 5.32\%$$

$$\theta_p^C = 24(2125 - 135) \times 10^{-6} = 4.78\%$$

respectively. During event 2 the plastic capacity of hinge B was already reduced by $\Delta_2/H = 0.08\%$, which leaves a remaining rotation capacity at B of

$$\theta_{p,\text{rem}}^B = 5.32 - 0.08 = 5.24\%$$

which is still larger than θ_p^C. Event 3 is therefore reached when the displacement Δ_3 causes a rotational demand at C of 4.78% at a displacement of

$$\Delta_3 = 4.78 \frac{1}{100} 144 = 6.88 \text{ in. (175 mm)}$$

The final lateral earthquake force E and the ultimate total displacement of the cap beam can now be determined from the event scaling analysis as

$$E = \sum_i E_i = 242.5 + 15 = 257.5 \text{ kips (1145 kN)}$$

$$\Delta = \sum_i \Delta_i = 0.86 + 0.11 + 6.88 = 7.85 \text{ in. (199 mm)}$$

which represent the available lateral seismic load and displacement capacities of the two-column bridge bent.

(iii) Substitute Structure Analysis. The objective of a substitute structure analysis is to model the seismic response of an inelastic bridge system with an equivalent elastic system of effective systems stiffness and effective systems damping, as outlined in Section 4.4.3. The goal is to determine with a simple linear elastic model in a secant stiffness approach, as outlined in Fig. 4.35, the inelastic response of the bridge to a specific seismic input given in the form of a response spectrum. The substitute structure analysis procedure is illustrated by the following examples.

Example 4.5. Assume that the actual seismic response of the two-column bent in Fig. 4.39 is to be estimated for the Eurocode 8, 0.35g acceleration response spectrum for subsoil class B shown in Fig. 4.40.

Step 1: From moment–curvature analyses of the critical column sections where inelastic action is expected, the effective member stiffness for each member and the complete bent system can be determined, as shown in Fig. 4.39(b), and an ideal yield displacement at the systems capacity level E_i for the structural system can be found. The initial system can be characterized by

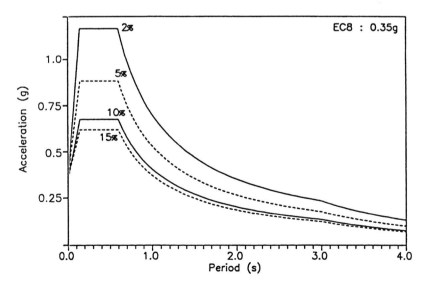

FIG. 4.40 Eurocode 8 elastic acceleration response spectra for maximum ground accelerations of 0.35g on subsoil class B.

$$K_{eff,0} = \frac{6EI_{eff}}{H^3} = \frac{6(1.34 \times 10^8)}{144^3} = 277 \text{ kips/in.} = 3327 \text{ kips/ft (48.5 MN/m)}$$

$$E_i = 257 \text{ kips (1.14 MN)}$$

$$\Delta_{yi} = \frac{E_i}{K_{eff,0}} = \frac{257}{3327} = 0.077 \text{ ft} = 0.93 \text{ in. (24 mm)}$$

$$T_0 = 2\pi \sqrt{\frac{W_s}{gK_{eff}}} = 2\pi \sqrt{\frac{1000}{(32.2)3327}} = 0.62 \text{ s}$$

For an assumed 5% damped system, the response acceleration coefficient $S_a(g)$ is found as $S_a = 0.85g$ and an equivalent elastic earthquake force and displacement can be determined as

$$E_1 = 0.85 \times 1000 = 850 \text{ kips (3.78 MN)}$$

$$\Delta_1 = \frac{E_1}{K_{eff,0}} = \frac{850}{3327} = 0.26 \text{ ft} = 3.1 \text{ in. (78 mm)}$$

Step 2: A new effective structural stiffness can now be determined from the systems capacity E_i and the displacement Δ_1 as

$$K_{eff,1} = \frac{E_i}{\Delta_1} = \frac{257}{3.1} = 82.9 \text{ kips/in.} = 995 \text{ kips/ft (14.5 MN/m)}$$

The displacement level 1 implies a systems ductility of

$$\mu_1 = \frac{\Delta_1}{\Delta_{yi}} = \frac{3.1}{0.93} = 3.33$$

for which an effective damping based on [K8] and the Takeda relationship in Fig. 4.36 of $\xi_{eff,1} = 17.3\%$ can be obtained. With these new effective system properties, the system's response can be found as

$$T_1 = 2\pi \sqrt{\frac{1000}{32.2(995)}} = 1.11 \text{ s}$$

$$S_a = 0.35g \rightarrow E_2 = 350 \text{ kips (156 MN)}$$

$$\Delta_2 = \frac{350}{995} = 0.35 \text{ ft} = 4.22 \text{ in. (107 mm)}$$

Step 3: The effective systems properties are now iteratively adjusted following the procedure established in steps 1 and 2, that is,

$$K_{\text{eff},2} = \frac{257}{4.27} = 60.9 \text{ kips/in.} = 731 \text{ kips/ft} \ (10.7 \text{ MN/m})$$

$$\mu_2 = \frac{4.22}{0.93} = 4.54 \rightarrow \xi_{\text{eff},2} = 22\%$$

$$T_2 = 2\pi \sqrt{\frac{1000}{32.2(731)}} = 1.32 \text{ s} \rightarrow S_a = 0.25g \rightarrow E_3 = 250 \text{ kips} \ (1.1 \text{ MN})$$

Step 4: A further iteration results in

$$K_{\text{eff},3} = \frac{257}{4.1} = 62.7 \text{ kips/in.} = 752 \text{ kips/ft} \ (10.9 \text{ MN/m})$$

$$\mu_3 = \frac{4.1}{0.93} = 4.41 \rightarrow \xi_{\text{eff},3} = 19\%$$

$$T_3 = 2\pi \sqrt{\frac{1000}{32.2(752)}} = 1.27 \text{ s} \rightarrow S_a = 0.257g \rightarrow E_y = E_i = 257 \text{ kips}$$

$$\Delta_4 = \frac{257}{752} = 0.34 \text{ ft} = 4.1 \text{ in.} \ (104 \text{ mm})$$

The effective systems properties for the substitute linear elastic structure which result in the same response as the actual inelastic bridge system for the given earthquake input are

$$K_{\text{eff}} = 752 \text{ kips/ft} \ (10.9 \text{ MN/m})$$
$$\xi_{\text{eff}} = 19\%$$

and result in a maximum systems displacement $\Delta_{\max} = 4.1$ in. (104 mm).

Example 4.6. Since substitute structure analyses are tools in support of a displacement-based design approach, as outlined in Section 5.3.1(c), direct use of displacement response spectra is more appropriate. Assume that the structure evaluated in Example 4.5 is to be analyzed utilizing the displacement response spectra for the same Eurocode 8, 0.35g, subsoil class B, input as shown in Fig. 4.41.

Step 1: The initial systems response is defined by $K_{\text{eff},0} = 3327$ kips/ft (48.5 MN/m), $\Delta_{yi} = 0.93$ in. (24 mm), and $E_i = 257$ kips (1.14 MN). A first displacement estimate can be made assuming an equivalent elastic response level at 1g as

$$\Delta_1 = \frac{W_s}{K_{\text{eff},0}} = \frac{1000}{3327} = 0.3 \text{ ft} = 3.6 \text{ in.} \ (91 \text{ mm})$$

$$\mu_1 = \frac{3.8}{0.93} = 3.9 \rightarrow \xi_1 = 18.3\%$$

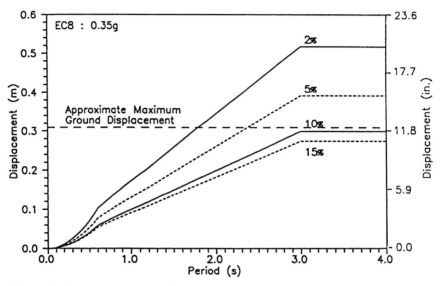

FIG. 4.41 Eurocode 8 elastic displacement response spectra for maximum ground accelerations of 0.35g on subsoil Class B.

With $\Delta_1 = 91$ mm and $\xi_1 = 18.3\%$, the new systems period can be estimated from Fig. 4.41 as $T_1 = 1.25$ s. From Eq. (4.39) a new effective systems stiffness can be determined as

$$K_{\text{eff},1} = \frac{W_s}{g}\frac{4\pi^2}{T^2} = \frac{1000}{32.2}\left(\frac{4\pi^2}{1.25^2}\right) = 795 \text{ kips/ft (11.4 MN/m)}$$

Step 2: With the new effective stiffness $K_{\text{eff},1}$ and the ideal capacity E_i, a new displacement estimate of

$$\Delta_2 = \frac{E_i}{K_{\text{eff},1}} = \frac{257}{785} = 0.331 = 3.9 \text{ in. (99 mm)}$$

can be obtained. Following the procedure outlined in step 1, a new systems response ductility of $\mu_\Delta = 3.9/0.93 = 4.2$ and effective damping coefficient $\xi_2 = 18.8\%$ can be determined from Fig. 4.41 and a new systems response period $T_2 = 1.3$ s is obtained. An updated effective stiffness of

$$K_{\text{eff},2} = \frac{1000}{32.2}\left(\frac{4\pi^2}{1.3^2}\right) = 725 \text{ kips/ft (10.6 MN/m)}$$

can be established, resulting in a new systems displacement estimate of 4.25 in. (108 mm), indicating convergence to the equivalent substitute structure properties and response levels obtained in Example 4.5. The stepwise analysis process for both examples is depicted in Fig. 4.42 and shows that rapid convergence can be obtained for both analyses.

This substitute structure analysis procedure is ideal for cases where the bridge can be modeled as an equivalent SDOF system and the dominant inelastic response can be captured by a single mode of vibration. Where higher modes of vibration contribute to the inelastic response, the principle of superposition used for linear elastic modal analyses and small displacements strictly no longer applies and results become questionable. Developments to apply a substitute structure approach to displacement-based inelastic MDOF models with significant higher-mode contributions are being investigated [C3].

4.5.3 Response Spectrum Analyses

Modal spectral analyses are demand analysis tools to determine maximum response quantities from the spectrum of a given ground motion or from

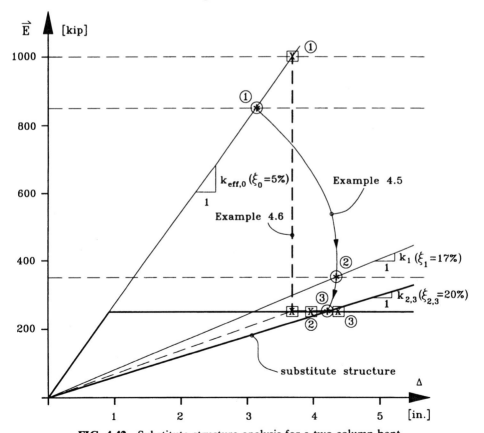

FIG. 4.42 Substitute structure analysis for a two-column bent.

smoothed design spectra. Analysis models used for modal spectral analysis are linear elastic models based on effective stiffness properties and on assumed equivalent viscous damping ratios. With these requirements response spectrum analyses can be performed (1) for bridge systems which are expected to perform essentially in the linear elastic range based on cracked or effective stiffness properties, (2) for inelastic response of bridge systems where the equivalent response is linearized to the initial effective stiffness and subsequently modified by means of equal energy or equal displacement principles as outlined in Section 5.3.1(c), and (3) for substitute structure analyses as outlined in Section 4.5.2.

Results from linear elastic modal spectral analyses are most applicable to displacement response estimates since bridge structures respond typically with a fundamental period T greater than the period T_0 at maximum intensity in the response spectrum (i.e., $T > T_0$), where equal displacement principles apply. Equivalent member forces need to be adjusted based on Eq. (5.22). However, it should be recognized that force adjustments based on Eq. (5.22) are strictly valid only for the members in which the inelastic actions occur and that other members are protected by the inelastic members and mechanisms in the prototype bridge against higher forces determined by the strictly linear elastic modal analysis.

Furthermore, it should be recognized that only maximum modal response values are determined which do not occur at the same time during the earthquake time history. The modal combination techniques outlined in Section 4.3.2(b) account for this effect of linear elastic bridge systems. However, since modal analysis techniques rely on the principle of superposition, these techniques are valid only as long as in addition to the linear elastic response, only small displacements occur. During the inelastic seismic response of bridge structures, displacements can exceed the small displacement as stipulated in small displacement theory, and frequencies and mode shapes can no longer be considered simple harmonics since they depend on the displacement amplitude. With the validity of the principle of superposition in question, any combination technique of modal maximum response quantities needs to be questioned in the inelastic response domain.

Many modal analysis programs provide an effective mass or mass participation factor for each mode. This mass participation factor, which is similar in physical meaning to the modal participation factor of Eq. (4.59) and has the same form as Eq. (4.59) except that the numerator is squared, can be used to determine how many modes should be considered in a given response direction. The sum of the effective masses for all modes in a given response direction must equal the total mass of the bridge. Effective mass participation of 80 to 90% of the total bridge mass in any given response direction can be considered sufficient to capture the dominant dynamic response of the bridge structure.

Some bridge seismic design codes allow or specify the use of inelastic response spectra, which implies that the acceleration response spectrum has been modified to account for ductile member actions that result in reliable structural displacement ductilities. Member forces obtained from these inelastic acceleration response spectra can then be directly considered as actual maximum member demands. Note that the basic analysis tool is still exactly the same as that used for elastic response spectra utilizing dynamic characteristics from a linear bridge model, with the exception that the linear elastic acceleration response spectrum has been modified by division with the force reduction factor Z from Eq. (5.22) and is expressed for different structural displacement ductility levels μ_Δ. Use of an inelastic acceleration response spectrum requires that (1) all elements and joints are detailed to achieve the postulated structural displacement ductility level, and that (2) it applies only to the determination of forces, not to velocities or displacements. Maximum displacements will be found from the yield displacement Δ_{yi} and the ductility level adopted for design as

$$\Delta_m = \mu_\Delta \Delta_{yi} \tag{4.93}$$

where μ_Δ and Z are related by Eq. (5.22).

4.5.4 Time-History Analyses

The final analysis category comprises tools that utilize a particular earthquake ground motion input and provide bridge response quantification for this earthquake input in the form of time histories of the various response quantities. Similar to the static or quasistatic analyses, models with linear elastic materials behavior $\langle L \rangle$, nonlinear cyclic materials characteristics $\langle N \rangle$, and geometric nonlinearities as $P–\Delta$ $\langle P \rangle$ or full nonlinear geometry $\langle G \rangle$ can be analyzed. However, now in addition to two or three spatial dimensions, an additional dimension, the time t, has to be accounted for.

For the earthquake time-history analysis of bridge models, three analysis tools are available: (1) step-by-step integration in the time domain, (2) superposition of normalized modal time histories in the time domain, and (3) evaluation of frequency-dependent response contributions with transformation to and superposition in the time domain. Since little design information can be gained from linear elastic time-history analyses with a specific earthquake ground motion input, methods 2 and 3, which are in their general form limited to the linear elastic domain due to the inherent reliance on the principle of superposition are not discussed further here.

However, stepwise linearized modal time-history analyses have been successfully employed in the nonlinear dynamic response analysis of global bridge models [W8]. For each linearized timestep very efficient mode superposition procedures based on load dependent Ritz vectors rather than in the standard eigenvectors are employed to obtain displacements, velocities, and accelerations for the subsequent time step in which deviations encountered in the restoring force can be applied as corrections to the loadvector or iteratively reduced.

Step-by-step time integration is the most general approach and investigates the dynamic response of the structure to a sequence of individual time-dependent force pulses of length or integration step Δt. The time-dependent force $\mathbf{P}_{eff}(t)$ [see Eq. (4.57)] is divided into n time steps of duration Δt and the bridge response to the impulse force $\mathbf{P}_{eff}(t_i$ to $t_{i+1})$ is evaluated with appropriate initial conditions for nodal displacements, velocities, and accelerations $u(t_i)$, $\dot{u}(t_i)$, and $\ddot{u}(t_i)$, respectively. Time-integration solution strategies range from conditionally stable explicit schemes to unconditionally stable implicit integration schemes, the main difference being the numerical stability of the solution.

Numerical integration schemes for the time domain can have problems with accuracy or period distortion as well as numerical stability when the integration time step Δt is not small enough. As a general rule, numerical stability in conditionally stable explicit time integration schemes, such as the commonly used Newmark explicit method [N1,C10], can be achieved when the time step Δt is selected such that

$$\Delta t \leq \frac{T_n}{\pi} \tag{4.94}$$

where T_n represents the period of the highest significant mode of vibration. Since detailed MDOF models can have many modes with very small values of T_i for the higher ones, time steps based on Eq. (4.94) can become extremely small and prohibitive in terms of computational effort. In these cases higher modes can be eliminated by means of numerical damping as outlined below, or unconditionally implicit analysis tools can be employed. Once numerical stability is ensured, selection of the time integration step Δt is governed only by the desired resolution of the highest mode of vibration to be traced in the dynamic response; for example, $\Delta t = T/10$ results in five data points for half of a response cycle. Detailed guidelines for time step selection, dependent on the response period of the highest mode or numerical damping for higher modes to achieve numerical stability are provided in the technical literature [C10,H4] for the various integration schemes.

For MDOF time-history analysis an additional requirement arises when step-by-step or direct integration of the equation of motion in the form of Eq. (4.45) is to be performed, namely the need for an explicitly defined damping matrix $\mathbf{C}$ and not just modal damping coefficients. While it is very difficult to estimate the magnitude of the damping coefficients c_{ij} and since as discussed earlier the treatment of damping in the form of viscous damping is a mathematical necessity and not a bridge systems property, the use of damping matrices which are either mass or stiffness proportional in the form:

$$\mathbf{C} = a_0\mathbf{M} \quad \text{or} \quad \mathbf{c} = a_1\mathbf{K} \tag{4.95}$$

are convenient to use in MDOF time integration analyses. In physical terms, mass proportional damping varies geometrically with frequency and affects primarily the low-frequency components, whereas stiffness proportional damping has a linear relationship with frequency with virtually no damping for low, and high damping for the high-frequency components (Fig. 4.43). The best results are provided by a combination of the two in the form of Rayleigh damping as

$$\mathbf{C} = a_0\mathbf{M} + a_1\mathbf{K} \tag{4.96}$$

resulting in the relationship between damping ratio ξ_n and frequency ω_n of

$$\xi_n = \frac{a_0}{2\omega_n} + \frac{a_1\omega_n}{2} \tag{4.97}$$

This equation has the shape depicted in Fig. 4.43 and can be adjusted with the two parameters a_0 and a_1 to satisfy at least two damping ratios, ξ_j and ξ_k, associated with two of the modes. Modes j and k, for which damping is specified, are typically the fundamental frequency (i.e., $j = 1$) or the dynamic response mode with the highest mass participation, and one of the higher modes that contributes significantly to the structural response, which results in all modes with frequencies greater than ω_k to be higher damped and with very high frequency modes effectively eliminated.

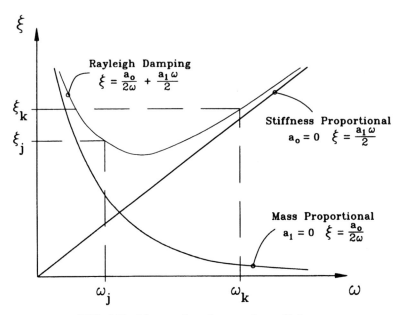

FIG. 4.43   Viscous damping matrix coefficients.

For nonlinear time-history analyses the loading and response are divided into a sequence of short time intervals Δt during which the system is linearized or evaluated as a linear system with the systems characteristics that exist at the beginning of the time interval. At the end of the time step Δt the systems properties are adjusted to reflect the new internal stress and strain states in the form of a new tangent stiffness matrix which is used during the next time step. While similar problems concerning numerical systems stability arise with conditionally stable or explicit time integration schemes as outlined for linear systems, the use of implicit integration schemes is no longer as easily implemented since stiffness information from the subsequent step is required. Again, the technical literature on numerical integration techniques provides various solution strategies for these problem types [H4]. Due to the required complexity in fully cyclic member characterization, nonlinear time-history analyses are typically limited to frame type models in two or three dimensions.

Since nonlinear time-history analyses combine both the demand side of seismic evaluation in the form of earthquake ground motion input and the capacity side in the form of fully cyclic nonlinear member characterization, shown schematically in Fig. 4.1, seismic capacities and demands for the global bridge structure can be evaluated and compared simultaneously. However, such a simultaneous evaluation is meaningful only if (1) local nonlinear response characteristics can describe the expected prototype nonlinearities, (2) boundary conditions between soil and footings and abutments can be defined in the form of realistic boundary conditions, (3) the interaction of bridge components at abutment and in-span expansion joints can be characterized, and (4) appropriate equivalent viscous damping values for the overall bridge response can be defined in addition to the explicitly modeled hysteretic response of individual elements.

The correct choice of nonlinear time-history analysis tool depends on the importance of multidirectional response quantification and on the ability to model the various nonlinear cyclic member characteristics. Programs can be evaluated by the extent to which a complete library of hysteretic member force–deformation models exists, the availability of cyclic capacity degradation rules, and the substructuring capabilities to develop reasonable structural component models to minimize the problem size. Many different hysteresis rules in the form of force–deformation relationships have been developed to represent inelastic bar, beam, or column behavior [C20]. The most common rules are summarized in Fig. 4.44 and consist of (a) simple elastoplastic bilinear patterns with initial stiffness unloading and reloading characteristics used, for example, for soil spring characterization, and (b) gap models in conjunction with (a) to model soil gapping and impact as discussed in Section 4.4.2(d), or restrainer behavior in movement joints as outlined in Section 4.4.2(f). Beam or column flexural hinges with various degrees of pinching and unloading and reloading stiffness can be described with the Takeda models shown in Fig. 4.44(c) and (d).

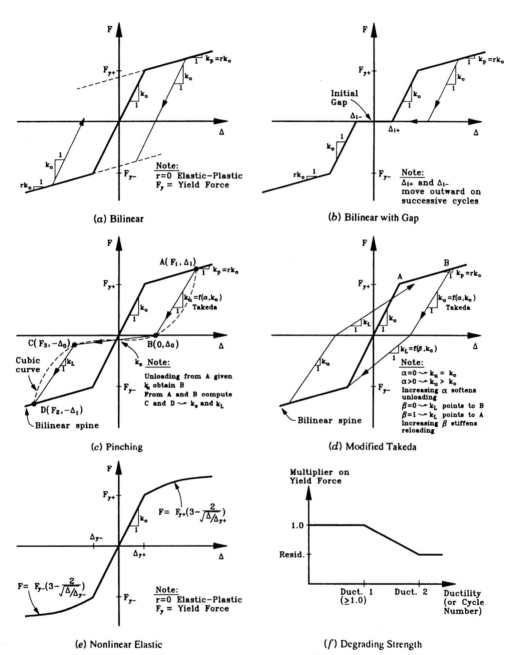

FIG. 4.44 Typical hysteresis and capacity degradation models for nonlinear analysis programs [C17].

Takeda force–deformation hysteresis rules [T3] are characterized by a softened unloading stiffness with increasing ductility of the form

$$k_u = k_0\,\mu^{-\alpha} \tag{4.98}$$

where k_0 represents the effective initial stiffness to the yield point, μ the ductility level defined by the maximum expected deformation divided by the yield deformation, and k_u the resulting unloading stiffness for $\mu > 1$. Calibration of Eq. (4.98) on experimentally obtained hysteresis loops from numerous reinforced concrete bridge component models tested for Caltrans at the University of California–San Diego showed a range of α from 0.25 to 0.7 [M10]. An unloading factor α of 0.25 can be used to describe the behavior of reinforced concrete beam hinge mechanism without axial load [Fig. 4.44(d)] and column flexural hinging with pinching characteristics can be modeled with the pinching relationship in Fig. 4.44(c) and $\alpha = 0.5$ to 0.7 with the higher levels for higher axial loads exceeding 15% of nominal concrete capacity of the gross column section. Rocking response analyses of footings or piers can be described with nonlinear elastic models as shown in Fig. 4.44(e). Since cyclic loading, yield penetration, shear degradation, and P–Δ effects can decrease capacities with increasing ductility levels, rules for force-degradation as shown in Fig. 4.44(f), based on increasing ductility, must be available. Where multi-directional response is required three-dimensional frame programs with appropriate yield and failure surfaces need to be employed. Due to the large amount of data required for the model development, the quantities of response data produced by the analysis, and the inherent complexities of such a nonlinear program, graphics pre- and postprocessors, as well as experienced and qualified analytical engineering support are essential for this type of analysis.

The major limitations to date for nonlinear time-history analyses are (1) the complexity of the programs in general, which limits the recommended applications to experienced designers and analysts; (2) the lack of nonlinear shear behavior and failure models and their interaction with the flexural inelastic action; (3) the lack of nonlinear cyclic models describing common failure phenomena of reinforcement slippage, lap-splice debonding, and bar-bend opening; and (4) the lack of reliable damage accumulation models and their ability to predict or incorporate local effects such as cover concrete spalling and bar buckling, which can initiate critical failure limit states. Since a nonlinear time-history analysis only evaluates the response of the bridge model to a particular earthquake ground motion input, and since intensity, duration, frequency content, and spatial and time variation characteristics of the actual seismic event for a given bridge site are associated with a large degree of uncertainty, such a nonlinear time-history analysis can be used in bridge seismic design or assessment only when the model with all the structural parameter variations is exposed to not just one but a suite of representative ground motion inputs.

A suite of several different input ground motions derived from different potential earthquake sources and modified with variability in local soil conditions can be expected to provide a reasonable basis for design or assessment response quantifications. For example, UBC (Uniform Building Code) guidelines, which recommend the use of the maximum response of three or the average response of seven ground motion inputs as the basis for building design, could be adopted. Alternatively, synthetic ground motion time histories generated to approximate specific spectral response characteristics can be used. Due to all of the above-described variabilities and uncertainties on the ground motion input side, as well as in the structural component and interaction characterization, the time-history analysis tool should be used at the end of the seismic bridge evaluation effort to check or verify specific bridge design or assessment limit states rather than for their sole numerical definition.

4.6 BRIDGE RESPONSE ANALYSIS EXAMPLE: RESPONSE ASSESSMENT OF A LONG REGULAR VIADUCT

To illustrate some of the analysis and modeling issues encountered in seismic bridge response quantification, a detailed analysis example is presented next, incorporating different aspects outlined earlier.

Example 4.7

Problem Statement. A portion of a long viaduct with uniform 40-m (131-ft) spans and identical bents with a clear height of 10 m (32.8 ft), as shown in Fig. 4.45(*a*), is to be evaluated for transverse seismic performance. The viaduct is part of a three-lane freeway with a total superstructure self-weight of 200 kN/m (13.7 kips/ft). The viaduct substructures consist of single-column bents with circular columns 1.83 m (6 ft) in diameter. The columns are continued into the ground to $z = -15$ m (-49 ft) as cast-in-drilled-hole (CIDH) pile shafts with the same dimensions and reinforcement details above and below ground. Column reinforcement consists of 30 D36 mm (No. 11) longitudinal bars and D19 mm (No. 6) spirals with a pitch of 115 mm (4.5 in.). Assumed material properties are $f_y = 414$ MPa (60 ksi) yield for the reinforcement and $f_c' = 24.1$ MPa (3500 psi) nominal concrete strength. The soil is characterized as very dense sand and a depth z linearly varying subgrade reaction coefficient of $k_s^* z = 10z$ (MPa/m) [$k_s^* z = 64z$ (kips/ft^3)]. Ground accelerations of 0.4g are to be considered for the seismic input characterized by the 5% damped Caltrans ARS Spectra [C12] for 80 to 150 ft alluvium, shown in Fig. 4.46, and by two recorded accelerograms from past earthquakes.

Modeling. The transverse seismic excitation and the dynamic response of the uniform viaduct section is assumed to be synchronous along the length of the

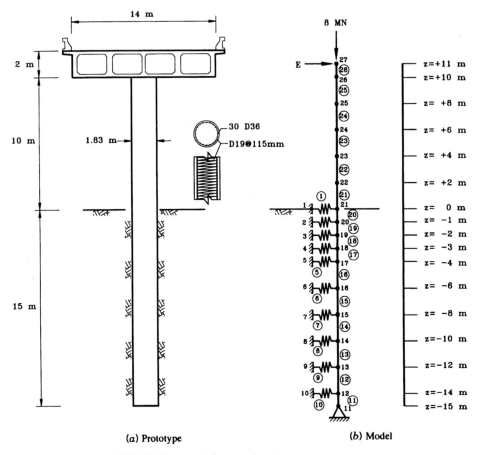

(a) Prototype   (b) Model

FIG. 4.45 Analysis model and spring constants.

viaduct, which allows the response analysis to be performed on a single-column-bent model with tributary seismic weight from the superstructure section between adjacent midspan points or a tributary superstructure length of 40 m (131 ft). The tributary seismic superstructure weight of $W_s = 8$ MN (1800 kips) is assumed to act at superstructure midheight or 11 m (36 ft) above ground level. The single-column-bent model includes soil springs, as shown in Fig. 4.45(b), to account for the influence of soil flexibility.

A moment–curvature analysis for an expected belowground plastic hinge in the CIDH pile shaft section with an axial load of 8.85 MN (2000 kips) (including the column weight) and an idealized bilinear response approximation as shown in Fig. 4.47 result in an effective initial stiffness of

$$EI_{eff} = \frac{M_y}{\phi_y} = \frac{13,700}{0.00265} = 5.17 \times 10^6 \text{ kN-m}^2 \ (12.5 \times 10^6 \text{ kip-ft}^2)$$

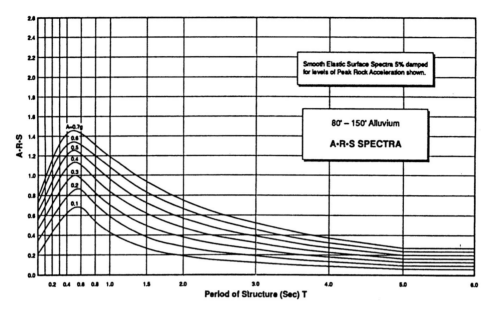

FIG. 4.46 Caltrans design spectra used for Example 47 [C12].

CIDH−Pile Shaft Plastic Hinge Characteristics

1.83 m (6 ft) Dia. shaft, 30 bars 36 mm Dia. (30 No. 11 bars) ρ_s =1.15
19 mm Dia. spiral at 0.115 m (spiral No. 6 at 4.5 in) ρ_s =0.58
P_{axial} = 8850 kN, cc=75 mm (3 in)

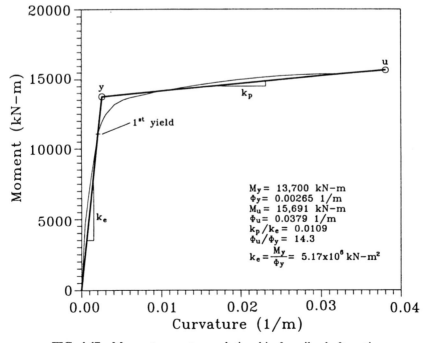

M_y = 13,700 kN−m
Φ_y = 0.00265 1/m
M_u = 15,691 kN−m
Φ_u = 0.0379 1/m
k_p/k_e = 0.0109
Φ_u/Φ_y = 14.3
$k_e = \dfrac{M_y}{\Phi_y}$ = 5.17x10^6 kN−m^2

FIG. 4.47 Moment−curvature relationship for pile shaft section.

From Eq. (4.84) the soil reaction modulus for the 1.83-m (6-ft)-diameter pile can be determined as

$$k = Dk_s = 1.83 \times 10 = 18.3 \text{ MPa/m (384 kips/ft}^3)$$

and from Fig. 4.29 with $H = 6D$ and a soil–pile stiffness parameter of

$$\frac{1000 \, KD^6}{D*EI_{\text{eff}}} = \frac{1,000 \times 18.3 \times 10^3 \times 1.83^6}{1.83 \times 5.17 \times 10^6} = 73$$

an equivalent depth to fixity for a cantilever column of $3.0D$ or approximately -5.5 m (-18.0 ft) can be estimated based on the guidelines developed in Section 4.4.2(d)(ii). Furthermore, based on elastic hinge location estimates, outlined in Section 5.3.2(b)(iii) and Fig. 5.31 with the same soil–pile stiffness parameter of 73, the expected plastic hinge location can be determined as $0.6D$ or -1.0 m (-3.3 ft) below the ground surface for this very dense sand. The pile shaft model discretization is selected such that at least four nodes or soil springs are modeled both above and below the equivalent point of fixity of -5.5 m (-18.0 ft) and that one node is placed close to the expected plastic hinge location at -1 m (-3.3 ft). This discretization is rather coarse and could be refined in realistic examples. To model mass contributions along the free column height, nodes at 2-m intervals are selected, resulting in the discretization shown in Fig. 4.45(b).

Member properties are derived for the column and CIDH pile shaft as effective properties from the moment–curvature analysis in Fig. 4.47 and for the soil springs from Eq. (4.87) as

$$k_i = k_s^* z_i B_i D = 10 z_i B_i (1.83) = 18.3 z_i B_i \text{ (kN/m) } [384 z_i B_i \text{ (kips/ft)}]$$

and resulting spring stiffness values are summarized in Table 4.3. Tributary seismic weights to the model nodes are listed in Table 4.4.

Seismic Response Assessment. The seismic response assessment for the single-column bent defined will consist of (1) a quasistatic response determination using the given acceleration response spectra, and (2) time-history analyses to demonstrate dynamic time-history analysis principles as well as dynamic response characteristics. Both types of analysis address modeling issues of effective stiffness and $P–\Delta$ effects.

1. Quasistatic Response Assessment. For the quasistatic response assessment of the single-column bent under transverse seismic excitation, an equivalent lateral load at the center of mass of the superstructure is assumed to

TABLE 4.3 Soil Spring Stiffness Values

Spring, i	z (m)	B_i (m)	K_i (kN/mm)
1	0.25	0.5	2.3
2	1	1	18.3
3	2	1	36.6
4	3	1	54.9
5	4	1.5	109.8
6	6	2	219.6
7	8	2	292.8
8	10	2	306.0
9	12	2	439.2
10	14	1.5	384.3

simulate the seismic load input. To show the influence of effective stiffness and P–Δ effects, an equivalent lateral earthquake force corresponding to the bilinear yield moment capacity obtained in Fig. 4.47, and an estimated effective cantilever length of $11 + 1 = 12$ m (39 ft) to the expected plastic hinge location, can be assumed as

$$E = \frac{13,700}{12} = 1,150 \text{ kN (260 kips)}$$

TABLE 4.4 Tributary Seismic Weights

Node	Tributary Length (m)	Volume (m³)	Seismic Weight W_s (kN)
27	—	—	8000[a]
26	1	2.63	63.1
25	2	5.26	126.2
24	2	5.26	126.2
23	2	5.26	126.2
22	2	5.26	126.2
21	1.5	3.95	94.7
20	1	2.63	63.1
19	1	2.63	63.1
18	1	2.63	63.1
17	1.5	3.95	94.7
16	2	5.26	126.2
15	2	5.26	126.2
14	2	5.26	126.2
13	2	5.26	126.2
12	1.5	3.95	94.7

[a] Additional seismic weight of superstructure.

Under this ideal yield load the deflected shape and bending moment distribution along the length of the column and in-ground pile shaft are shown in Fig. 4.48 for the three cases of (1) gross stiffness section properties, (2) effective stiffness section properties, and (3) effective stiffness properties and P–Δ effects. While gross-section properties are based on an uncracked concrete cross section of the pile shaft, cracked or effective stiffness properties are based on the initial stiffness of the idealized bilinear moment–curvature relationship shown in Fig. 4.47, which corresponds to 40% of EI_g. The same result can be obtained from Fig. 4.9(a) for a longitudinal reinforcement ratio of 1.1% and an axial load ratio of 14%.

The deflected shape of the CIDH pile shaft in Fig. 4.48(a) shows that the lateral displacement at the superstructure centroid, as expected, more than doubles with the cracked or effective pile shaft stiffness and that P–Δ effects can add another 20% to the effective stiffness-based top displacement. Of particular interest also are the horizontal displacements and rotations at the ground surface, which, as can be seen from Fig. 4.48(a), contribute significantly to the lateral displacement at the center of mass or the level of the bridge superstructure. Since the applied quasistatic lateral force of $E = 1150$ kN (260 kips) corresponds to the lateral load at the ideal yield point in Fig. 4.47, the maximum displacement of over 0.318 m (1.04 ft) without P–Δ, and 0.39m (1.28 ft) with P–Δ, can be considered as the nominal yield displacements or the reference displacement levels for structural systems ductility assessment. Furthermore, the maximum soil displacement of 0.065 m (2.5 in.) at this lateral

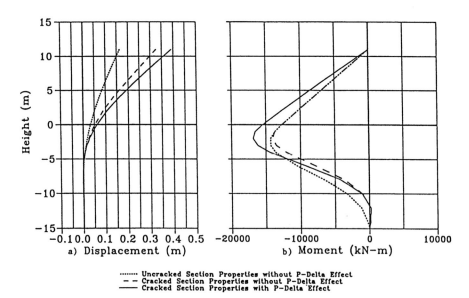

●●●●●●● Uncracked Section Properties without P–Delta Effect
– – Cracked Section Properties without P–Delta Effect
—— Cracked Section Properties with P–Delta Effect

FIG. 4.48 Transverse static analysis of 6-ft (1.93-m)-diameter column–shaft model under $P = 1150$ kN.

force level already exceeds a commonly assumed yield level at which soil stiffness changes, as shown in Fig. 4.30(a), and an at least bilinear load–deformation characterization for the soil springs should be used based on qualified geotechnical input. However, for clarity and simplicity in this assessment example, linear elastic soil spring characteristics and no inelastic soil action with possible gapping were assumed for the subsequent analyses.

The effective stiffness assumptions for the pile-shaft system have a significantly smaller effect on the maximum moment demand in the pile shaft, as can be seen in Fig. 4.48(b). The maximum moment occurs as expected below ground, between 1 and 2 m (3 to 6 ft) below the surface and tends to migrate upward with cracked section assumptions and P–Δ effects (i.e., with a softening structural system). Note that the in-ground location of maximum moment and the theoretical point of fixity of a pile-shaft system do not coincide. Rather the maximum moment occurs closer to the ground surface than the equivalent point of fixity, a phenomenon that needs to be considered when plastic shear forces are evaluated in an equivalent fixed-base model.

From the applied lateral load and the resulting displacement, an effective lateral stiffness for the seismic mass at the superstructure centroid for the case including $P - \Delta$ effects of

$$K_{eff} = \frac{P}{\Delta} = \frac{1150}{0.39} = 2950 \text{ kN/m } (2170 \text{ kips/ft})$$

can be obtained. For an equivalent single-degree-of-freedom response with generalized coordinate at the superstructure centroid, the generalized mass, including one-third of the column mass above the plastic hinge as derived in Eq. (4.12), can be expressed by a generalized seismic weight of

$$W_s^* = 8000 + \tfrac{1}{3} \times 63.1 \times 12 = 8300 \text{ kN } (1850 \text{ kips})$$

The fundamental response period of the structure can now be found from Eq. (4.39) as

$$T = 2\pi \sqrt{\frac{8300}{9.81 \times 2950}} = 3.4 \text{ s}$$

With the 0.4g acceleration response spectrum of Fig. 4.46, the expected lateral seismic demand would be 0.30g or 2490 kN (560 kips). This lateral seismic demand requires a displacement ductility level of $\mu_\Delta = 2.17$. For comparison, the uncracked or gross-section stiffness based dynamic response characteristics assessment would have resulted in

$$T = 2\pi \sqrt{\frac{8300}{9.81 \times 6700}} = 2.2 \text{ s}$$

for fundamental response period.

2. *Time-History Analysis Response Assessment.* For the time-history analysis response assessment, two recorded earthquake ground motions, the 8244 Orion Blvd., S90W, record from the 1971 San Fernando Valley ($M = 6.6$) earthquake and the James Rd. Array Station 5, component 230 Deg., record from the 1979 Imperial Valley earthquake ($M = 6.8$), were utilized. Both acceleration traces and their corresponding 5% damped acceleration response spectra are depicted in Fig. 4.49. Figure 4.49 shows that while the James Rd. record in its original unscaled form produces a spectral response acceleration of close to the 0.35 to 0.5g target value for the range of characteristic response periods of the structure, the Orion Blvd. record was scaled by a factor of 1.82 to generate spectral accelerations in the required range. Even though as discussed in Section 4.5.4, linear time-history response analyses are not very informative since they provide only one maximum seismic demand data point which can typically be much easier obtained using the corresponding response spectrum, linear elastic time-history analyses were performed here for comparative purposes, to demonstrate by comparison the influence of nonlinear structural systems behavior on the dynamic response. All dynamic time-history analyses were performed using the computer code Ruaumoko [C17].

The time-history analyses included equivalent viscous damping of 5% in the form of Rayleigh damping [see Eq. (4.96)], with coefficients a_0 and a_1 determined for the example of the cracked or effective stiffness case with $T_1 = 3.4$ s and $T_3 = 0.13$ s and $\omega_1 = 2\pi/T_1 = 1.85$ rad/s and $\omega_3 = 2\pi/T_3 = 48.3$ rad/s as

$$\begin{Bmatrix} a_0 \\ a_1 \end{Bmatrix} = \frac{2\omega_1\omega_3}{\omega_3^2 - \omega_1^2} \begin{bmatrix} \omega_3 & -\omega_1 \\ -\frac{1}{\omega_3} & \frac{1}{\omega_1} \end{bmatrix} \begin{Bmatrix} \xi_1 \\ \xi_3 \end{Bmatrix} = \frac{178.71}{2319} \begin{bmatrix} 48.3 & -1.85 \\ -0.021 & 0.54 \end{bmatrix} \begin{Bmatrix} 0.05 \\ 0.05 \end{Bmatrix} = \begin{Bmatrix} 0.178 \\ 0.002 \end{Bmatrix}$$

For the initial time-history response analyses the inelastic in-ground hinge was modeled using bilinear stiffness characteristics for the pile-shaft elements as outlined in Fig. 4.44(a) with a postelastic stiffness $k_p = 0.0109k_0$ based on the moment–curvature relationship in Fig. 4.47. The influence of $P–\Delta$ effects was considered separately in a subsequent parameter study to be discussed later.

The resulting top and ground-level displacement traces are shown in Fig. 4.50 for linear elastic and nonlinear time-history analyses, respectively for the two earthquake records in Fig. 4.49. While both linear analyses show oscillations about the original structure position, the nonlinear analyses depict a permanent offset of up to 0.27 m (0.9 ft) in the superstructure following even these moderate-level earthquakes. A summary of maximum displacements

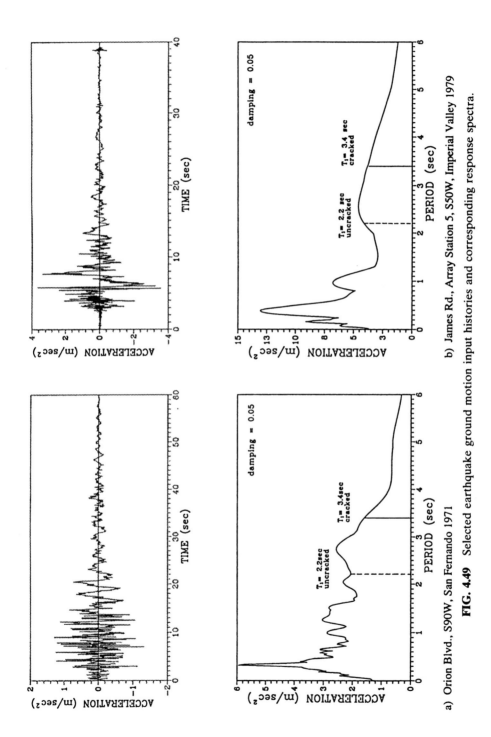

a) Orion Blvd., S90W, San Fernando 1971

b) James Rd., Array Station 5, S50W, Imperial Valley 1979

FIG. 4.49 Selected earthquake ground motion input histories and corresponding response spectra.

257

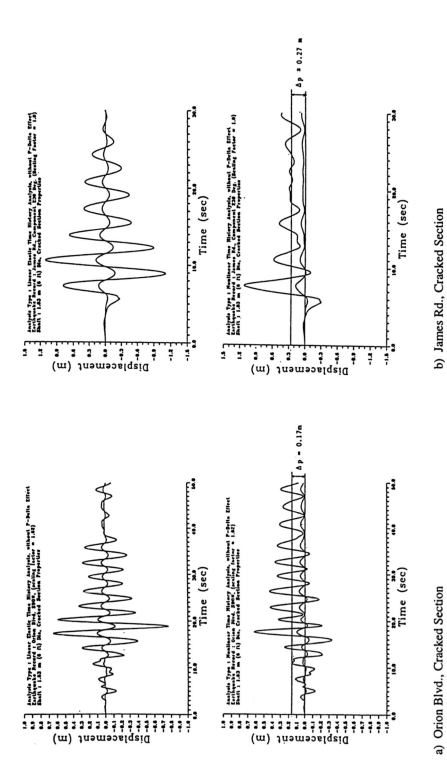

a) Orion Blvd., Cracked Section

b) James Rd., Cracked Section

FIG. 4.50 Displacement response histories at center of mass and ground level.

obtained for the various analyses, including the two quasistatic linear elastic lateral load analyses, is provided in Table 4.5.

A comparison of the top displacements of the cracked section dynamic linear elastic and nonlinear analyses shows the validity of the equal displacement assumption for long-period structures with almost the same maximum deformation levels. Large discrepancies between top displacements from linear and nonlinear analyses are obtained when gross-section properties are used for the linear analyses.

The deformation demands obtained from the nonlinear analyses were evaluated by means of three different ductility coefficients. The first, μ_Δ, is the system displacement ductility defined as Δ_{max}/Δ_y. The second, μ_Δ^* is the member displacement ductility, defined as $1+(\Delta_{max} - \Delta_y)/\Delta_y^*$ where Δ_y^* is the component of yield displacement resulting from structural deformation above the plastic hinge. The third μ_ϕ is the curvature ductility, related to the bilinear yield curvature $\phi_y = M_y/EI$. The derivation is further clarified for the cracked-section dynamic nonlinear response to James Rd. (case 6).

system ductility, μ_Δ:

$$\Delta_y = 0.318\text{m (no } P\text{-}\Delta \text{ effects)}$$
$$\Delta_{max} = 1.125\text{m}$$
$$\mu_\Delta = \frac{1.125}{0.318} = 3.54$$

member displacement ductility, μ_Δ^*
Taking depth to maximum moment $= 1$m, and using the bilinear yield curvature of $\phi_y = 0.00265$ m^{-1},

$$\Delta_y^* = \frac{0.00265 \times 12^2}{3} = 0.127\text{m}$$
$$\mu^* = 1 + (1.125 - 0.318)/0.127 = 7.35$$

TABLE 4.5 Summary of Deformation Analyses[a]

Analysis	Load	Section Properties	$\Delta_{max,top}$ (m)	$\Delta_{max,ground}$ (m)	Max. M (kN·m)	$\Delta_{resid,top}$ (m)	$\mu_\Delta/\mu_\Delta^{*b}$	$\mu_{\phi,hinge}$ (rad)
1 Static-linear	1150 kN	Uncracked	0.164	0.033	14,300	—	—	—
2 Static-linear	1150 kN	Cracked	0.327	0.052	14,100	—	—	—
3 Dynamic-linear	Orion Blvd. × 1.82	Cracked	0.766	0.010	32,970	—	—	—
4 Dynamic-nonlinear	Orion Blvd. × 1.82	Cracked	0.622	0.099	14,240	0.17	1.95/3.4	5.2
5 Dynamic-linear	James Rd.	Cracked	1.112	0.180	47,980	—	—	—
6 Dynamic-nonlinear	James Rd.	Cracked	1.125	0.174	15,370	0.27	3.54/7.35	12.1
7 Dynamic-linear	James Rd.	Uncracked	0.529	0.106	46,380	—	—	—
8 Dynamic-nonlinear	James Rd.	Uncracked	0.978	0.157	15,460	0.15	5.75/13.6	21.9

[a] $M_y = 13,700$ kN·m, $\phi_y = 0.00265$ m^{-1}.
[b] μ_Δ, systems ductility; μ_Δ, fixed-base systems ductility.

curvature ductility factor, μ_ϕ

The plastic rotation at the plastic hinge center is

$$\Theta_p = (\Delta_{max} - \Delta_y)/L = (1.125 - 0.318)/12 = 0.067 \text{ rad.}$$

with a plastic hinge length of $L_p = 1.25D$ (from Fig. 5.30), the plastic curvature is thus

$$\phi_p = 0.067/2.29 = 0.0293 \text{ m}^{-1}$$

and the curvature ductility is thus

$$\mu_\phi = 1 + 0.0293/0.00265 = 12.1$$

For the uncracked section cases, the yield curvature is taken as $\phi_y = M_y/ EI_{gross} = 0.00134 \text{ m}^{-1}$, the yield displacement from Fig. 4.48(a) as $\Delta_y = 0.17\text{m}$, and structural yield displacement as $\Delta_y^* = 0.00134 \times 12\,2/3 = 0.064\text{m}$. It should be noted, however, that the concepts of ductility for the uncracked section analysis are incompatible with the physical realities of the problem.

From Table 4.5 it can be observed that the soil deformations for the CIDH pile shaft have a significant influence on the structural ductility demand, estimated between 2 and 4 for the investigated ground motions and cracked sections. Comparative member ductility demands are more than twice as large. As can be seen from Table 4.5, substantially larger ductility demands were obtained from the uncracked section model. Curvature ductilities in the plastic hinge region are significantly larger than system or member ductility levels, and the plastic hinge ductility demand of $\mu_\phi \simeq 12$ for case 6 indicates that the available plastic hinge curvature ductility of $\mu_\phi \simeq 14$ (see Fig. 4.47) is almost reached.

The variations of results in Table 4.5 show the need to determine the most likely structural response through a series of representative input records, a procedure that should be established particularly for important (essential) bridge structures as standard design practice. Furthermore, the need for a clear definition of inelastic bridge performance in the form of systems or local ductilities, as well as the need for effective or cracked stiffness assumptions in the analytical modeling, have been demonstrated.

Finally, the effect of hysteresis modeling of the in-ground inelastic flexural hinge is investigated below together with the $P-\Delta$ analysis. For the James Rd. earthquake input of Fig. 4.49(b) and the cracked or effective initial stiffness case as derived from the moment–curvature relationship, of Fig. 4.47, bilinear and modified Takeda hysteresis rules were employed with and without $P-\Delta$ analysis. Parameter variations address the effects of postyield stiffness, unloading stiffness, and reloading stiffness assumptions for parameters r, α, and β in Fig. 4.44(a) and (d).

The time-history displacement responses at the top of the structure for the two hysteresis base models ($r = 0$, $\alpha = 0$, $\beta = 0$) are compared in Fig. 4.51. Without P-Δ effects, the response is very similar, with slightly lower response levels for the Takeda analysis, based on the difference in reloading stiffness, which (with $\beta = 0$) points toward the previously obtained maximum displacement level, as depicted in Fig. 4.44, and is thus decreasing with increasing displacements. Both hysteresis models, however, become unstable as soon as P-Δ effects are activated, and inelastic deformation levels are reached after approximately 10 s of earthquake ground motion input as shown in Fig. 4.51. The three hysteresis rule parameters r, α, and β for the modified Takeda model can now be adjusted to investigate the significance to P-Δ response.

Of the three parameters, the postyield stiffness r seems to have the most significant impact on P-Δ analysis stability. Figure 4.52(a) shows for comparison the unstable displacement response with no postyield stiffness, whereas

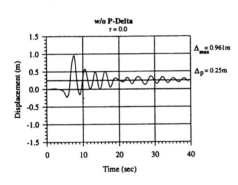
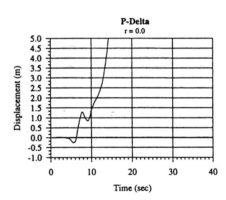

(a) Bilinear Model

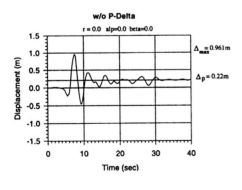
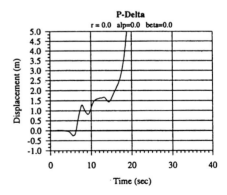

(b) Modified Takeda Model

FIG. 4.51 Effect of different hysteretic response rules and P-Δ.

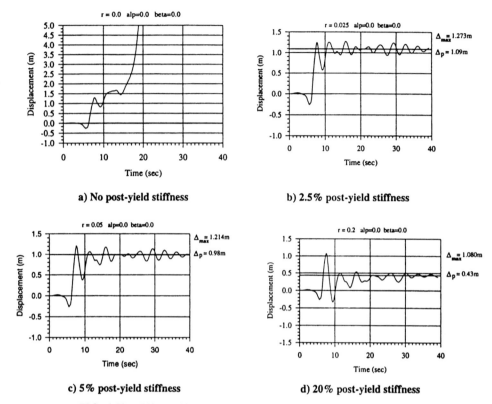

FIG. 4.52   Effect of postyield stiffness on the response with $P-\Delta$.

the 2.5% postyield stiffness shown in Fig. 4.52(*b*) returns stable results but with a very large permanent offset of over 1 m (3.3 ft). This permanent offset decreases with increasing postyield stiffness, as shown in Figs. 4.52(*c*) and (*d*). Postyield stiffness values greater than the 20% of the initial stiffness shown are not very realistic. As defined in Fig. 4.44(*d*) and in Eq. (4.98), adjustment of the unloading parameter α resulted in the displacement response traces shown in Fig. 4.53. Small values of α up to 30% were not effective in stabilizing the dynamic response. An unloading stiffness coefficient $\alpha = 0.3$ returned a stable response but again at a very large permanent offset of over 1 m (3.3 ft), as shown in Fig. 4.53(*b*). This permanent offset decreased at $\alpha = 0.7$ to 0.85 m (2.8 ft). This variation of α from 0.3 to 0.7 again represents the realistic range for flexural hinge mechanisms in bridge members with low and moderate axial load levels, respectively. Finally, Fig. 4.53(*d*) shows that a combination of a small postyield stiffness, $r = 0.05$, and a value of $\alpha = 0.3$ (realistic for beams or columns with low axial loads) stable dynamic response at reduced permanent offset levels of 0.5 m (1.6 ft), can be obtained. The reloading parameter β did not contribute to stabilizing the nonlinear $P-\Delta$ analyses.

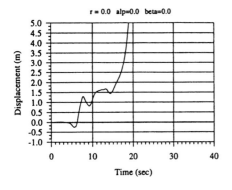

a) No unloading stiffness reduction

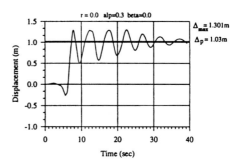

b) Small unloading stiffness reduction

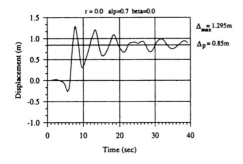

c) Large unloading stiffness reduction

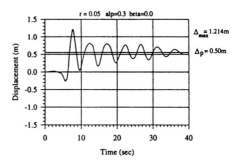

d) Combined unloading stiffness reduction
and small post-yield stiffness

FIG. 4.53   Effect of unloading stiffness reduction on response with P–Δ.

These hysteresis model parameter studies show that for the earthquake and the structural geometry and axial load level, various parameter combinations exist to stabilize the dynamic response. The importance of the hysteresis parameter selection becomes obvious beyond the stability criterion in the widely varying maximum and permanent displacements obtained. Significant engineering judgment and experience are required for the correct model generation, the definition of realistic response parameters, and the correct interpretation of these nonlinear dynamic analysis results for the correct structural response assessment.

The observed nonlinear dynamic response behavior changes significantly for different geometries and axial load ratios and should only be used as indicators of the variability of results based on the different modeling assumptions. Also, only one ground motion input record was used for the shown parameter studies, whereas as discussed in Section 4.5.4, a suite of earthquake input ground motions needs to be employed before quantitative response assessments can be made.

The dynamic analyses, including $P - \Delta$ effects, indicate that this structure is very sensitive to P-Δ moments. This would be anticipated by the design criterion of Eq. 5.162, which, for a peak displacement of 1.12m, would require a yield strength equal to 31% of the weight, for P-Δ effects to be ignored. The actual strength is only 14% of weight, and thus P-Δ criticallity is certain.

5

DESIGN

5.1 INTRODUCTION

The principles of capacity design were introduced in Section 1.3.3 and discussed in relation to conceptual design aspects in Chapter 3. Briefly restated, the principles can be expressed in terms of the following steps:

1. Locations of potential inelastic flexural action (plastic hinges) are identified in the conceptual design phase.
2. From an examination of the deflection characteristics, the relationship between flexural strength and plastic hinge rotation is established. This may be by explicit calculations or by use of general approximations appropriate for the adopted structural system. These relationships enable the system ductility, and hence the required flexural strength of plastic hinges, to be determined.
3. Flexural reinforcement requirements for plastic hinges are then determined.
4. The maximum feasible flexural strength for the plastic hinge regions appropriate for the expected ductility level is calculated. From this, considering realistic variations in live load, section forces (moments, shears, axial forces, and torsion) corresponding to flexural overstrength are calculated.
5. The strength design of members and actions to be protected against inelastic deformation is then carried out.

6. Aspects of detailing required to ensure that assumptions of expected behavior are achieved in practice are then attended to. These include the design of transverse reinforcement for ductility, the location and detailing of anchorage, development and splicing of reinforcement, and the detailing of connections between members.

The relevant design aspects of this process are outlined in some detail in this chapter, following a brief review of material properties relevant to seismic design and retrofit of bridges.

5.2 MATERIAL PROPERTIES FOR SEISMIC RESISTANCE

5.2.1 Unconfined Concrete

Since many books give in-depth consideration to the mechanical properties of concrete, only a brief summary is provided here, with special emphasis on aspects relevant to seismic resistance. For more complete coverage, the reader is referred elsewhere [P3,M2].

(a) Compression Strength f'_c. Compression strength depends on cement content and type, initial water/cement ratio of the mix design, aggregate shape strength and gradation, type and quantity of admixtures, and age of concrete. Strengths in the range 22.5 MPa $\leq f'_c \leq$ 45 MPa (3.25 to 6.5 ksi), are commonly used in seismic design. Higher strengths are not commonly adopted because of increased brittleness.

Compression strength is almost universally defined by the specified 28-day strength. Since actual 28-day strength typically exceeds specified strength by 20 to 25%, on average, and since concrete continues to gain strength with age, the actual strength of the concrete when seismic attack occurs is likely to considerably exceed the specified strength. Tests in California on cores of concrete taken from bridges constructed in the 1950s and 1960s have tested out between 1.5 and 2.7 times the specified strength. This can have considerable importance in seismic assessment of older bridges and should also be recognized in new bridge design. Strength enhancement occurring at seismic strain rates is also significant.

(b) Modulus of Elasticity. The modulus of elasticity, E_c, used for design is generally based on secant measurement under slowly applied compression load to a maximum stress of $0.5f'_c$. Design expressions relate compression modulus of elasticity to compression strength by equations of the form

$$E_c = \begin{cases} 0.043w^{1.5}\sqrt{f'_c} & \text{(MPa)} \\ 33w^{1.5}\sqrt{f'_c} & \text{(psi)} \end{cases} \tag{5.1}$$

for values of concrete unit weight w between 1400 and 2500 kg/m³ (88 to 156 lb/ft³). For normal-weight concrete,

$$E_c = \begin{cases} 4700\sqrt{f'_c} & \text{(MPa)} \\ 57,000\sqrt{f'_c} & \text{(psi)} \end{cases} \tag{5.2}$$

For reasons given above in the discussion of compression strength, the actual modulus of elasticity can be expected to exceed that corresponding to specified 28-day compression strength by 20 to 50%.

(c) Tensile Strength of Concrete. The contribution of the tensile strength of concrete to the dependable strength of members under seismic action must be ignored because of its variable nature and the possible influence of shrinkage- or movement-induced cracking. However, it may be necessary to estimate member tension or flexural behavior at the onset of cracking to ensure in certain cases that the capacity of the reinforced section is not exceeded. For this purpose, the following conservatively high values for tensile strength may be assumed:

$$f'_c = \begin{cases} 0.5\sqrt{f'_c} \text{ MPa} = 6\sqrt{f'_c} \text{ psi} & \text{(concrete in direct tension)} \quad (5.3a) \\ 0.75\sqrt{f'_c} \text{ MPa} = 9\sqrt{f'_c} \text{ psi} & \text{(concrete in flexural tension)} \quad (5.3b) \end{cases}$$

At high strain rates, tension strength may considerably exceed these values. It must be emphasized that although concrete tension strength is ignored in flexural strength calculations, it has a crucial role in the successful resistance to actions induced by shear, bond, and anchorage.

5.2.2 Confined Concrete

(a) Confining Effect of Transverse Reinforcement. In bridge design, ductility will normally be provided by column plastic hinges. The effect of axial compression in these members is to initiate spalling of cover concrete at rather low displacement ductilities. Unless adequate, properly anchored transverse reinforcement is provided to confine the compressed concrete within the core region, and to prevent buckling of the longitudinal compression reinforcement, failure may occur. In conjunction with longitudinal reinforcement, close-spaced transverse reinforcement acts to restrain the lateral expansion of the concrete that accompanies the onset of crushing, maintaining the integrity of the core concrete, and enabling higher compression stresses, and more important, much higher compression strains to be sustained by the compression zone before failure occurs.

Because of their shape, spirals or circular hoops are placed in hoop tension by the expanding concrete and thus provide a continuous confining line load around the circumference, as illustrated in Fig. 5.1(a). The maximum effective lateral pressure f_l that can be induced in the concrete occurs when the spirals or hoops are stressed to their yield strength f_{yh}. Referring to the free body of Fig. 5.1(b), equilibrium requires that

$$f_l = \frac{2f_{yh}A_{sp}}{D's} \tag{5.4}$$

where D' is the diameter of the hoop or spiral, which has a bar area of A_{sp}, and s is the longitudinal spacing of the hoop or spiral.

Square hoops, however, can apply confining reactions only near the corners of the hoops because the pressure of the concrete against the sides of the hoops tends to bend the sides outward [as illustrated by dashed lines in Fig. 5.1(c)]. The confinement provided by square or rectangular hoops can be improved significantly by the use of overlapping hoops or hoops with cross-ties, which results in several legs crossing the section. The better confinement resulting from the presence of a number of transverse bar legs is illustrated in Fig. 5.2(b), (c), and (d). The arching is more efficient since the arches are shallower, and hence more of the concrete area is effectively confined.

The presence of a number of longitudinal bars well distributed around the perimeter of the section, tied across the section, will also aid the confinement of the concrete. The concrete bears against the longitudinal bars and the transverse reinforcement provides the stabilizing reactions to the longitudinal bars [Fig. 5.2(e) and (f)]. However, large numbers of longitudinal bars require complex patterns of transverse hoops and cross ties to provide antibuckling restraint [see Section 5.3.2(c)(ii)]. Since bridge columns are typically large, requiring many longitudinal bars, circular columns are generally preferred to rectangular columns unless the reinforcement in the rectangular column is disposed around intersecting circles, as discussed earlier in relation to Fig. 3.17.

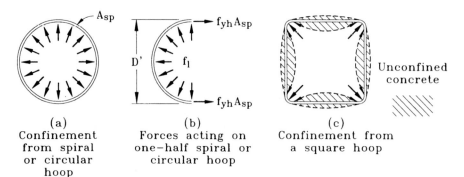

(a) Confinement from spiral or circular hoop

(b) Forces acting on one–half spiral or circular hoop

(c) Confinement from a square hoop

Unconfined concrete

FIG. 5.1   Confinement of concrete by circular and square hoops [P3].

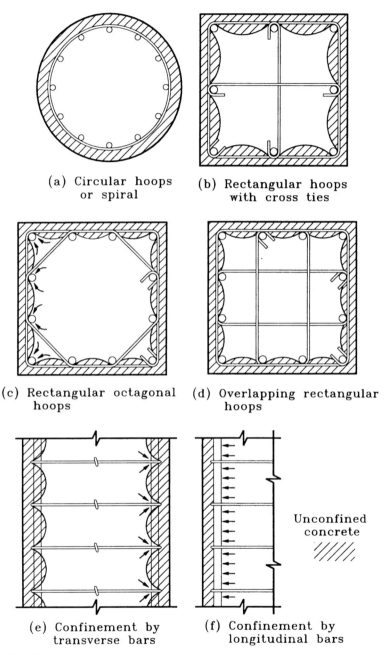

(a) Circular hoops
or spiral

(b) Rectangular hoops
with cross ties

(c) Rectangular octagonal
hoops

(d) Overlapping rectangular
hoops

Unconfined
concrete

(e) Confinement by
transverse bars

(f) Confinement by
longitudinal bars

FIG. 5.2 Confinement of column sections by transverse and longitudinal reinforcement.

Clearly, confinement of the concrete is improved if transverse reinforcement layers are placed relatively close together along the longitudinal axis. There will be some critical spacing of transverse reinforcement layers above which the section midway between the transverse sets will be ineffectively confined, and the averaging implied by Eq. (5.4) will be inappropriate. However, it is generally found that a more significant limitation on longitudinal spacing of transverse reinforcement s is imposed by the need to avoid buckling of longitudinal reinforcement under compression load. This is discussed further in Section 5.3.2(c)(ii).

(b) Compression Stress–Strain Relationships for Confined Concrete. The effect of confinement is to increase the compression strength and ultimate strain of concrete as noted above and illustrated in Fig. 5.3. Many different stress–strain relationships have been developed [B1,K2,M4,S1,V1,E6] for confined concrete. Most of these are applicable to a restricted range of conditions (e.g., circular sections, or rectangular sections). A recent model applicable to all section shapes [M4] and all levels of confinement is defined by the following equations, referring to the nomenclature of Fig. 5.3:

$$f_c = \frac{f'_{cc}xr}{r - 1 + x} \tag{5.5}$$

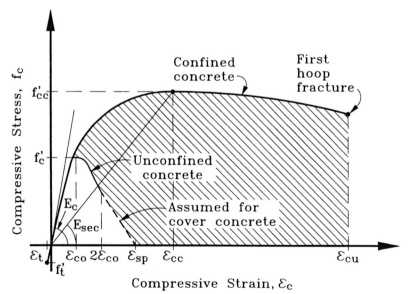

FIG. 5.3 Stress–strain model for concrete in compression [M4].

where

$$f'_{cc} = f'_c \left(2.254 \sqrt{1 + \frac{7.94f'_l}{f'_c}} - \frac{2f'_l}{f'_c} - 1.254 \right) \tag{5.6}$$

$$x = \frac{\varepsilon_c}{\varepsilon_{cc}} \tag{5.7}$$

$$\varepsilon_{cc} = 0.002 \left[1 + 5 \left(\frac{f'_{cc}}{f'_c} - 1 \right) \right] \tag{5.8}$$

$$r = \frac{E_c}{E_c - E_{\text{sec}}} \tag{5.9}$$

$$E_c = \begin{cases} 5000\sqrt{f'_c} & \text{(MPa)} & (5.10a) \\ 60{,}000\sqrt{f'_c} & \text{(psi)} & (5.10b) \end{cases}$$

$$E_{\text{sec}} = \frac{f'_{cc}}{\varepsilon_{cc}} \tag{5.11}$$

In Eqs. (5.5) to (5.11), f'_{cc} and ε_{cc} are the concrete stress and strain at peak stress and f'_l is the effective lateral confining stress. With $f'_l = 0$, Eqs. (5.5) to (5.11) produce an equation appropriate for unconfined concrete.

The effective lateral confining stress, f'_l, is related for circular sections to the average confining stress of Eq. (5.4) by the expression

$$f'_l = K_e f_l \tag{5.12}$$

For rectangular sections, with different transverse reinforcement area ratios ρ_x and ρ_y in the principal directions, different confining stresses are developed, in accordance with the relationships

$$f'_{lx} = K_e \rho_x f_{yh} \tag{5.13a}$$

$$f'_{ly} = K_e \rho_y f_{yh} \tag{5.13b}$$

In Eqs. (5.12) and (5.13), K_e is a confinement effectiveness coefficient, relating the minimum area of the effectively confined core (see Fig. 5.2) to the nominal core area bounded by the centerline of the peripheral hoops [M4]. Typical values of K_e are 0.95 for circular sections, 0.75 for rectangular sections, and 0.6 for rectangular wall sections.

For a rectangular section with unequal effective confining stresses f'_{lx} and f'_{ly} given by Eq. (5.13), the confined strength f'_{cc} may be calculated from the relationship for f'_{cc}/f'_c shown in Fig. 5.4, where $f'_{lx} > f'_{ly}$.

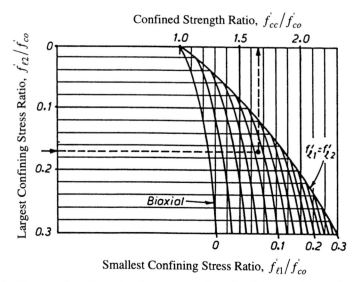

FIG. 5.4 Compression strength enhancement of confined rectangular sections related to orthogonal confining stresses [M4].

The useful limit to compression strain is taken to occur when fracture of transverse confining steel initiates, as shown in Fig. 5.3. This may be estimated by equating the strain-energy capacity of the transverse steel as it is strained to peak stress f_{uh} to the increase in energy absorbed by the concrete, resulting from confinement. This increase in absorbed energy is shown shaded in Fig. 5.3 [M4]. A conservative estimate for ultimate compression strain is given by

$$\varepsilon_{cu} = 0.004 + \frac{1.4\rho_s f_{yh}\varepsilon_{su}}{f'_{cc}} \tag{5.14}$$

where ε_{su} is the steel strain at maximum tensile stress and $\rho_s = 4A_{sp}/D's$ is the volumetric ratio of confining steel. For rectangular sections $\rho_s = \rho_x + \rho_y$. Typical values for ε_{cu} range from 0.012 to 0.05, a 4- to 16-fold increase over the traditionally assumed value for unconfined concrete. Equation (5.14) has been formulated from considerations of confined sections under axial compression. When used to estimate ultimate compression strain of sections subjected to bending, or combined bending and axial compression, Eq. (5.14) tends to be conservative by at least 50%. This conservatism may be thought of as an ultimate strain reduction factor, ensuring an adequate margin of safety to allow for uncertainties in ductility demand.

It should also be noted, however, that Eq. (5.14) has been developed based on behavior in the central regions of axial compressed elements. In many cases, the critical section occurs immediately adjacent to a supporting member (e.g., a footing or cap beam). The stiffness of the supporting member provides additional confinement to the critical region. Under seismic strain rates, concrete exhibits small but significant enhancement to compression strength, tension strength, and modulus of elasticity. These effects may generally be ignored in design and analysis, as they diminish under repeated loading [P4]. The confined-concrete stress–strain model outlined above has been presented in some detail, as it forms the basis of many analyses described later in this chapter and is very suitable for incorporation in computerized moment–curvature analyses.

5.2.3 Reinforcing Steel

(a) Monotonic Characteristics. Ductility of reinforced concrete structural elements relies on the capacity of reinforcing steel to sustain repeated cycles to high levels of plastic strain without significant reduction in stress. As shown in Fig. 5.5 for typical reinforcing steels, behavior is characterized by an initial linear elastic portion of the stress–strain relationship with a modulus of approximately 200 GPa (29,000 ksi), up to the yield stress f_y, followed by a strain plateau of variable length and a subsequent region of strain hardening. Maximum or ultimate stress is reached at about $f_{su} = 1.5f_y$ for typical North American reinforcement, although the ratio decreases for higher-strength steels.

In Europe, *tempcore steel* is rapidly becoming the standard for reinforcing steel. The manufacturing process, involving rapid cooling of the external layers

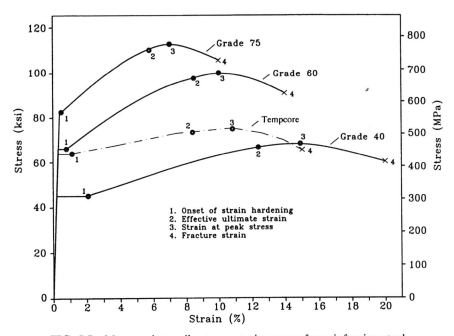

FIG. 5.5 Monotonic tensile stress–strain curves for reinforcing steel.

while the core remains hot, uses the internal heat to temper the outer layers. The process allows a high strength [typically, $f_y \approx 545$ MPa (79 ksi)] to be obtained from a low carbon content (C equivalent $< 0.52\%$), thus ensuring a weldable steel. Tempcore steels have high elongation and can be bent to a tighter radius in hooks and bends than can other reinforcing bars. However, this is obtained at the expense of a low ratio of ultimate to yield strength (typically, $f_{su} = 1.19f_y$). This makes the steel more prone to inelastic buckling because of the reduced strain-hardening modulus and causes a reduction in effective plastic hinge lengths compared with U.S. steels. A consequence of this is the need to use closer-spaced transverse reinforcement in plastic hinge zones [see Eq. (5.50)] and to check steel strains at high ductilities.

After maximum stress is reached, strain softening occurs with deformation concentrating in a local weak spot. In terms of structural response, the strain-softening portion of the curve should be ignored, since it imparts little additional ductility to members where considerable lengths are subjected to effectively constant reinforcement stress. In design and analysis, a reduced effective ultimate tensile strain should be adopted, since there is evidence that under cyclic loading involving sequential tensile and compressive strains, the ultimate tensile strain is less than under the monotonic testing. A simple rule of thumb [D1] is that the effective ultimate tensile strain should be the monotonic tensile strain at peak stress reduced by the maximum expected compression strain under the reversed direction of seismic response. This is illustrated in Fig. 5.6. Alternatively, the simpler requirement that $\varepsilon_s \le 0.75\varepsilon_{su}$ will normally be adequately conservative except for members with high axial compression forces.

It should be noted that the strain-hardening portion of the curve is a desirable attribute that spreads plasticity over a reasonable length of the member, ensuring that tensile strains are not excessive at the design ductility limit. However, excessively rapid increase in stress after development of yield

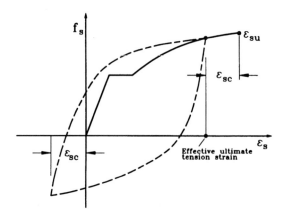

FIG. 5.6 Effective ultimate tensile strain for reinforcing steel.

strain can result in excessive overstrength of plastic hinges, requiring high strengths of capacity protected members and elements (see Section 5.3.3).

A related concern in seismic design of reinforced concrete elements for seismic resistance is the variability of yield strength from the specified or nominal yield strength. In the United States, A706 grade 60 reinforcement, which has the tightest restriction on yield strength range, has a permitted yield of 414 MPa $\leq f_y \leq$ 538 MPa (60 ksi $\leq f_y \leq$ 78 ksi). Thus it is feasible and not uncommon for a member to have reinforcing steel with a yield strength 30% above the specified value. With other grades of reinforcement, particularly where grade 40 (f_y = 275 MPa nominal), which is normally "failed" grade 60 steel whose yield is in the range 380 to 410 MPa (55 to 59 ksi), is specified, the excess over specified strength may be greater. This may result in excessive flexural strength in plastic hinges and the need to make capacity-protected members excessively strong to avoid unanticipated inelastic modes of deformation.

Two approaches can be adopted to reduce the severity of the problem. First, grade 40 reinforcement should not be specified for ductile members unless a tight control on yield strength is assured. Second, consideration should be given to specifying a restricted range of acceptable reinforcement yield strength in the construction contract documents. Reinforcement suppliers normally have mill certificates identifying batches of their stock by yield strength and have little trouble in satisfying a requirement that, say, 450 $\leq f_y \leq$ 510 MPa (65 ksi $\leq f_y \leq$ 74 ksi). The adoption of such a restriction translates into safer and more economical structures, since capacity protection can be assured without the need for such high overstrength factors. Relevant aspects are discussed in more detail in Sections 5.3.1(d) and 5.3.3.

Properties of reinforcing steel specified for seismic design of bridges in Europe [E5] require an ultimate strain of $\varepsilon_{su} \geq 0.09$, a minimum ratio of ultimate to yield stress of $f_u/f_y > 1.2$, and an actual yield strength no more than 20% above the nominal yield strength. This results in less potential overstrength than with steels common in the United States at the expense of less spreading of the plastic hinge region and an increased propensity for buckling as a consequence of the reduced strain-hardening modulus.

(b) Inelastic Cyclic Response. Under cyclic loading the characteristic stress–strain curves of Fig. 5.5 may not form an accurate envelope to the inelastic response. Bauschinger effects result in nonlinear behavior developing at a strain lower than yield stress on unloading from a previous inelastic excursion. Figure 5.7 shows the results of two different types of cyclic testing of reinforcing steel. In Fig. 5.7(a), the cyclic inelastic excursions are predominantly in the tensile strain range, which is typical of beams or columns with low axial compression. For such a response the monotonic stress–strain curve provides a reasonable envelope to the cyclic response in the tension range but not in the compression range.

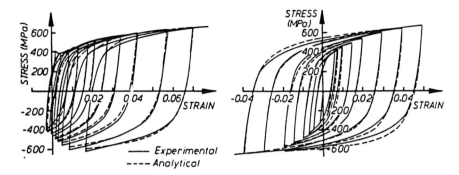

(a) With Unsymmetrical Strain Cycles   (b) With Symmetrical Strain Cycles

FIG. 5.7 Cyclic inelastic stress–strain response of reinforcing steel [f_y = 380 MPa (55 ksi)] [L2].

For columns with high compression stress levels and high reinforcement ratios, reinforcing bars may be subject to strain reversals of almost equal magnitude, implying a neutral-axis position close to the section centroid. As illustrated in Fig. 5.7(b), under cyclic response, the stress level for a given strain increases and can substantially exceed the stress indicated by the monotonic stress–strain curves.

Ductile bridge columns typically have low to moderate axial compression levels and are better characterized by the behavior of Fig. 5.7(a). Hence the monotonic stress–strain curve can reasonably be used to determine peak response. Appropriate equations are discussed in Section 5.3.1(d)(viii). Full description of the stress–strain response under arbitrary rather than cyclic response is complex and is dealt with in a number of specialized research papers [M4,D1]. As with concrete, material properties are influenced by strain rate [M4], with increases of yield strength up to 20% possible on initial loading. However, the effects quickly dissipate under cyclic response, and shake-table testing of reinforced concrete elements does not indicate significant strength enhancement compared with quasistatic test results.

(c)   Temperature and Strain-Aging Effects. Below a certain temperature (typically, about −20°C), the ductility of reinforcing steel is lost and it behaves in a brittle fashion on reaching the yield stress. Care is thus needed when designing structures for ductile response in cold climates. A related effect relevant to warmer climates is the gradual increase, with time, of the threshold temperature between brittle and ductile steel behavior subsequent to plastic straining of reinforcement [P8]. The threshold temperature may rise to as high as +20°C in time. Thus reinforcing steel that has been plastically strained to form a bend or standard hook will eventually exhibit brittle characteristics in the region of the bend. It is thus essential to

ensure, by appropriate detailing, that the steel stress in such regions can never approach the yield stress.

It would appear that structures which have responded inelastically during earthquake response may be subject to the effects of strain aging and could behave in brittle fashion in a subsequent earthquake. This possible effect deserves more research attention. Strain-aging embrittlement of reinforcing steel appears to have been a contributing factor in the fracture of reinforcement in a bridge bent during the 1989 Loma Prieta earthquake, discussed in relation to Fig. 1.27.

5.2.4 Structural Steel

Structural steel used for bridge construction typically has material properties similar to lower grades of reinforcing steel, and the discussion in Section 5.2.3 may be taken to be applicable.

5.2.5 Prestressing Steel

A wide variety of proprietary prestressing systems is available, and it is beyond the scope of this book to review them. Prestressing steels may be

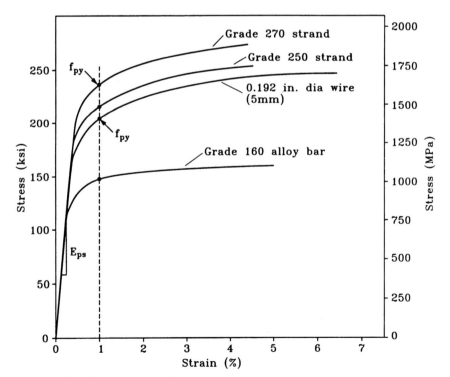

FIG. 5.8 Typical tensile stress–strain curves for prestressing steel.

TABLE 5.1 Mechanical Properties of Prestressing Steels

	Nominal Diameter		Ultimate Tensile Stress		0.02% Proof Stress		Modulus of Elasticity		Ultimate Strain, ε_{su}
	(mm)	in.	MPa	ksi	MPa	ksi	GPa	ksi	
Alloy-steel bar	12–40	0.5–1.58	1100	160	940	136	185	27,000	0.05
Wire	5, 7	0.197–0.276	1175	250	1390	200	200	29,000	0.07
Strand	10–15	0.39–0.60	1725–2075	250–300	1390–1650	200–240	190	27,500	0.06

characterized as cold-drawn or work-hardened steels, provided in the form of bar, strand, or wire. Typical stress–strain curves shown in Fig. 5.8 are characterized by high elastic strain limits, followed by gradual softening without a clearly defined yield plateau and with relatively low ultimate tensile strain. Since the limit of elastic proportionality is not well defined, the onset of strain softening is generally measured by the stress at the intercept of the stress–strain curve with a line parallel to the elastic portion, but offset by 0.02%, and termed the 0.02% proof stress, or by the stress at a specified strain (e.g., 0.75% or 1.0%).

High-tensile-strength alloy steel bars vary from 12 mm (0.5 in.) to 40 mm (1.58 in.) diameter and tend to have lower tensile strength than strand or wire. They frequently have coarse threads rolled on the surface to facilitate anchorage by nuts. Systems using multiwire tendons are generally assembled from either 5 mm (0.197 in.) or 7 mm (0.276 in.) wires of higher strength than bar. Similar strength characteristics are obtained from seven-wire strands, generally of 10 mm (0.39 in.), 12 mm (0.47 in.), or 15 mm (0.60 in.) nominal diameter, but the twisted strand results in lower effective section area and slightly lower effective modulus of elasticity than straight wire.

It is normal to permit maximum steel stress levels during tensioning up to 80% of nominal ultimate tensile stress (UTS), with a maximum stress no greater than 0.7 UTS after transfer. Ultimate tensile strain is significantly lower than for mild steel, and values in the range 5 to 7% are common. Strains of this magnitude can normally be achieved only away from anchorages, since the methods of clamping the prestressing steel in the anchorages tend to create local stress raisers which, while not affecting ultimate strength significantly, reduce the strain at failure. Table 5.1 summarizes typical values of key mechanical parameters for prestressing steels.

5.2.6 Advanced Composite Materials

The term *advanced composite materials* is generally applied to synthetic fiber materials such as fiberglass, carbon fibers, and aramids embedded in some form of polymer matrix (epoxy or ester), developed principally for use in aerospace and defense industry applications but which are now starting to be considered in civil engineering applications. This recent interest in the civil engineering profession for these high-tech materials is partly the result of reducing costs, and new applications where the inherent suitability of the materials outweighs the high material costs, which currently run anywhere between U.S. \$5 and \$30 per pound (10 to 60 ECU/kg). The use of carbon fiber and glass as prestressing materials has been investigated for some years, and a few trial applications have been made, particularly for areas where there is concern about corrosion and stress embrittlement of conventional prestressing steel. Consideration of the use of composite materials as the major structural material for bridges is currently under investigation [S3], and trial bridges have been constructed in China [S4] and Scotland [S5]. In seismic retrofit work, a number of composite materials have been considered as alternatives for steel in jacketing columns for enhanced shear strength, flexural ductility, and lap-splice performance. Details of this promising application are given in Chapter 8.

The mechanical properties of the different materials currently being considered for bridge applications differ widely in terms of ultimate stress and strain as well as elastic moduli. Behavior characteristics are predominately linear all the way to fracture, in the direction parallel to the fibers. The high modulus of elasticity and tensile strength rapidly decrease with increasing deviation angle between fiber orientation and loading direction, and the deformation characteristics become increasingly dominated by the properties of the resin. Tables 5.2 and 5.3 summarize key properties for common composite materials, and Fig. 5.9 shows the influence of fiber orientation on tensile properties for the example of a graphite–epoxy laminate used to produce a filament-wound shell system for new concrete columns without steel reinforcement. The designer should be aware, however, that there are many other material properties that must be considered before using composite materials, such as relaxation under sustained strain, moisture sensitivity, sensitivity to ultraviolet radiation, and temperature effects.

TABLE 5.2 Properties of Advanced Composite Materials

Material	Modulus of Elasticity		Ultimate Tensile Strength, f_u		Ultimate Strain, ε_u	Approximate Cost, 1995	
	GPa	Msi	MPa	ksi	(%)	ECU/kg	\$U.S./lb
Fibers							
Carbon	160–270	25–40	1400–6800	200–1000	1.0–2.5	24–80	12–40
Aramid	62–83	9–12	2800	400	3.6–4.0	16–24	8–12
(Kevlar 29)							
Glass	81	12	3400	500	4.9	2–6	1–3
Polyethylene	117	17	2600	380	3.5	2	1
(Spectra 900)							
Resin							
Epoxy	2.0–4.5	0.3–0.65	27–62	4–9	4–14	2–4	1–2
Vinylester	3.6	0.49	80	12	4	2–3	1–1.5

TABLE 5.3 Realistic Composite Moduli for Bridge Construction or Retrofit (60% Fiber Volume)

	E	
Type	GPa	Msi
Quasi-isotropic		
Hand layup glass/resin	14–21	2–3
Mechanized glass/resin	21–28	3–4
Carbon/resin	42–56	6–8
Hybrid: carbon 20% glass 80%	28–35	4–5
Unidirection, parallel to fiber		
Glass (S or E)	35–42	6–8
Carbon (T300, AS4)	140	18–20

5.3 CAPACITY DESIGN PROCESS

5.3.1 Flexural Strength Requirements for Plastic Hinges

As explained in Section 1.3.2, the required flexural strength and ductility of plastic hinge regions are interrelated. Most seismic design codes treat this interaction implicitly, by specifying force-reduction factors that are applied to elastic force levels calculated from 5% damped elastic acceleration response spectra specified for the region or site. Strength design to these force levels is coupled with prescriptive requirements for transverse reinforcement of

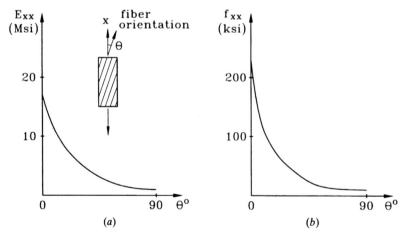

FIG. 5.9 Example of tensile characteristics for a composite material: carbon AS4D/anhydride laminate 55% fiber volume. (*a*) Influence of angle of fiber orientation on modulus of elasticity; (*b*) Influence of angle of fiber orientation on ultimate stress.

plastic hinge regions, which are expected to ensure that the level of ductility implied by the force reduction factor is available. This approach has the merit of simplicity and is appropriate for routine regular structures such as uniform bridges supported on fixed-base cantilever bents. Essentially, this is force-based design since displacements are not calculated at the initial design stages. Aspects of this conventional design approach are considered in the next section.

(a) Force-Based Design (Conventional Design). This standard and well-accepted procedure, where strength is first determined based on specified force-reduction factors, and prescriptive confinement detailing provided to ensure adequate ductility will probably continue to be standard practice for some years, particularly for regular and simple structures. Consequently, elements of the interaction between member and structure ductility are considered in the following to obtain force-reduction factors appropriate for different characteristic structural systems, typical of bridge substructures. An alternative displacement-based design procedure is outlined in Section 5.3.1(c). It is implicit in most design codes that structure ductility and member ductility may be considered interchangeably. That is, the rotational ductility of a plastic hinge is considered equal to the displacement ductility of the structure. It can readily be represented that this is not, in fact, the case. Three different situations that demonstrate this are considered below.

(i) Cap Beam Flexibility in Multicolumn Bents. In multicolumn bents, elastic flexibility of the cap beam contributes to the yield displacement, while all plastic displacement originates in the column hinge regions if the bent is designed according to current design philosophy. This is illustrated in Fig. 5.10 for a twin-column bent with column bases pinned to footings to reduce foundation forces [refer to Section 3.3.5]. If the cap beam is rigid, the yield displacement is $\Delta_y = \Delta_c$, resulting solely from column flexibility. With a plastic

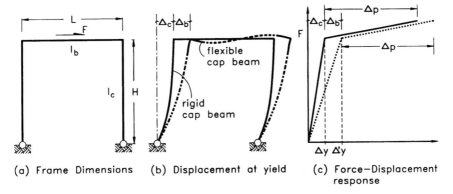

(a) Frame Dimensions (b) Displacement at yield (c) Force–Displacement response

FIG. 5.10 Reduction of displacement ductility capacity caused by cap beam flexibility.

displacement of Δ_p corresponding to the rotational capacity of the column hinges, the structure ductility is

$$\mu_{\Delta r} = \frac{\Delta_y + \Delta_p}{\Delta_y} = 1 + \frac{\Delta_p}{\Delta_c} \tag{5.15}$$

Cap beam flexibility will increase the yield displacement by an amount Δ_b but will not result in additional plastic displacements, since this still is provided solely by column hinge rotation. For bent dimensions $H \times L$, as shown in Fig. 5.10, and cracked-section moments of inertia for beam and column of I_b and I_c, respectively, the yield displacement is increased to

$$\Delta_y' = \Delta_c + \Delta_b = \Delta_c\left(1 + \frac{0.5I_cL}{I_bH}\right)$$

and the structural displacement ductility capacity is reduced to

$$\mu_{\Delta f} = 1 + \frac{\Delta_p}{\Delta_c + \Delta_b} = 1 + \frac{\Delta_p}{\Delta_c(1 + 0.5I_cL/I_bH)}$$

That is,

$$\mu_{\Delta f} = 1 + \frac{\mu_{\Delta r} - 1}{1 + 0.5I_cL/I_bH} \tag{5.16}$$

(ii)   Footing and Bearing Flexibility. A similar effect can occur with single- or multicolumn bents as a result of footing or bearing flexibility, illustrated in Fig. 5.11 for a simple vertical cantilever. Under the lateral force F_y at yield, the overturning moment $M_f = F_y(H + h_f)$ causes rotation θ_f of the footing. For a spread footing this rotation is given by

$$\theta_f = \frac{M_f}{K_f} \tag{5.17}$$

where $K_f = k_sI_f$ is the rotational stiffness of the footing, k_s the coefficient of subgrade reaction (N/m³ or kips/ft³), and I_f is the second moment of area of the contact interface between the spread footing and the soil. For a pile-supported footing, an equivalent rotational stiffness K_f can readily be derived from the axial stiffness and spatial distribution of the pile system as

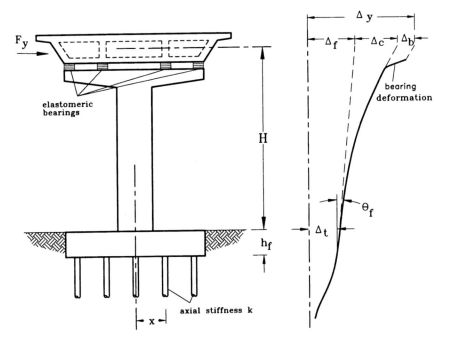

(a) Bent with flexible foundation and bearing pads   (b) Displacement profile at yield

FIG. 5.11 Influence of additional flexibility on yield displacement of a single-column bent.

$$K_f = \Sigma \, x_i^2 k_i \tag{5.18}$$

where k_i is the axial stiffness (N/m or kips/ft) of pile i at a distance x_i from the transverse axis of rotation through the centroid of the pile group.

With a pile-supported footing there may also be significant translational deformation Δ_t of the pile cap resulting from flexural deformation of the piles under the seismic shear force, as shown in Fig. 5.11(b). This deformation will contribute to Δ_f in terms of increased flexibility.

The additional elastic displacement at the level of the center of seismic force is thus

$$\Delta_f = \theta_f(H + h_f) + \Delta_t = \frac{M_f(H + h_f)^2}{K_f} + \Delta_t \tag{5.19}$$

If the superstructure is flexibly supported above the cap beam by elastomeric bearings of total stiffness K_b, with freedom of horizontal deformation in the direction considered, bearing shear deformations of

$$\Delta_b = \frac{F_y}{K_b}$$

will also be developed.

The displacement components due to footing and bearing flexibility have an identical effect on structural displacement ductility capacity to those resulting from cap-beam flexibility. The effects can thus be related to the rigid base ductility $\mu_{\Delta r}$ by

$$\mu_{\Delta f} = 1 + \frac{\mu_{\Delta r} - 1}{1 + (\Delta_f + \Delta_b)/\Delta_c} \tag{5.20}$$

where Δ_c is again the yield displacement at the center of seismic force resulting from column displacement alone.

(iii) Continuous Pile Shaft/column Designs. A construction form gaining considerable popularity in California and elsewhere involves the use of pile-shaft/column designs where the single pile, cast in a drilled hole, has the same section and reinforcement as the aboveground column. Advantages of this design approach were considered in Section 3.3.6. As shown in Fig. 5.12, the equivalent depth to fixity for displacements is typically three to five diameters below ground level, while maximum moment occurs closer to the surface,

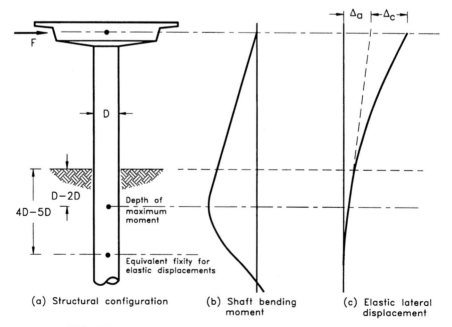

FIG. 5.12 Additional flexibility effects in a ductile pile shaft.

about one to two pile diameters down. Pile-shaft design is considered in more detail in Section 5.3.2(b)(iii). For the cantilever column shown, the yield displacement Δ_y at the center of mass can be much greater than the amount Δ_c contributed by structural deformation of the column above the level at which the plastic hinge forms. Since all plastic rotation is centered at the maximum moment location, the displacement ductility capacity is thus lower than it would be for a vertical cantilever rigidly fixed against elastic rotation and displacement at the position of maximum moment. Again a relationship between the ductility capacity of the actual column and an equivalent column rigidly supported at the position of maximum moment can be written as

$$\mu_{\Delta f} = 1 + \frac{\mu_{\Delta r} - 1}{1 + \Delta_a/\Delta_c} = 1 + \frac{\mu_{\Delta r} - 1}{1 + f_a} \qquad (5.21)$$

where Δ_a is the additional elastic displacement at yield caused by deflection components originating below the location of maximum moment, and $f_a = \Delta_a/\Delta_c$ may be considered an additional-flexibility coefficient. It will be appreciated that Eq. (5.21) is of general form, where Δ_a and f_a refer to additional elastic components of the yield displacement originating elsewhere than in the rigidly connected column and may be used for the cases described by Eqs. (5.19) and (5.20), with appropriate definition of Δ_a.

(b) Relationship Between Force-Reduction Factor and Displacement Ductility Factor. An early approach that recognized that different bridge structural systems have different inherent ductility capacity was the use by the California Department of Transportation (Caltrans) of adjustment factors for ductility and risk, Z (i.e., force-reduction factors) that were applied to member elastic force levels resulting from spectral modal analysis. The form of this adjustment factor is shown in Fig. 5.13 for the damage-control limit state. Different levels of period-dependent Z factors were specified for multicolumn bents, single-column bents, and wall-type piers, based primarily on experience with performance of different structural types in earthquakes. Since this experience was related to behavior of nonductile design details common to the pre-1971 (San Fernando earthquake) era, there is room for doubt about the applicability of this experience to modern, well-detailed bridges.

The decrease in Z with increasing period is contrary to trends predicted by theoretical considerations. As noted in Section 2.4.2, for long-period structures the equal-displacement approximation of the relationship between the displacements of elastic and ductile systems of equal elastic stiffness is appropriate and becomes increasingly conservative for very long period structures. Thus structural ductility and force-reduction factors may be considered equal. For short-period structures, dynamic analyses indicate that structure displacement ductility factors are larger than force-reduction factors, and approximate values are predicted by an equal-energy approach. Furthermore, short-period

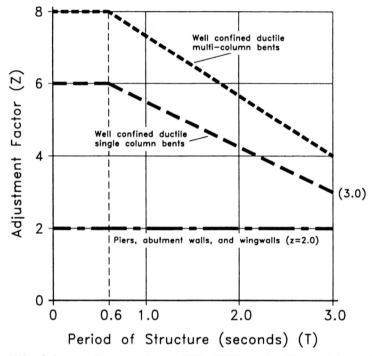

FIG. 5.13 Caltrans adjustment factor (1973–1994) for ductility and risk, Z [C1].

structures are also more likely to be subjected to a greater number of inelastic response cycles than long-period structures are for a given earthquake duration. As a consequence, the cumulative ductility factor, which is the summation of all displacement ductilities sustained during all yield excursions and which has some relevance as a measure of damage potential, increases as the period decreases. It is thus preferable to base design force-reduction factors on inelastic design acceleration spectra, where the relationship between ductility and force-reduction factors is nonlinear, as shown in the example of Fig. 2.25. Alternatively, the force reduction factor Z can conservatively be related to the structural displacement ductility capacity μ_Δ by the relationship [B2]

$$Z = 1 + 0.67(\mu_\Delta - 1)\frac{T}{T_0} \le \mu_\Delta \tag{5.22}$$

where T_0 is the period at peak elastic spectral response and T is the elastic first-mode period. Equation (5.22) provides a gradual variation from $Z = 1$ at $T = 0$, which is theoretically correct, through the equal-energy equation at about $T = 0.8T_0$, to the equal displacement equation ($Z = \mu_\Delta$) at $T = 1.5T_0$.

There are other factors, however, that produce trends which counteract, to some degree, the period dependency of the relationship between force-reduction factors and ductility. First, short-period structures may have reduced response as a consequence of increased damping resulting from soils effects. Also, as discussed in Section 5.2.3(b)(i), the effective plastic hinge length is reduced as a proportion of the column length, as the aspect ratio (height divided by diameter or height divided by depth) increases. As a consequence, the structure ductility capacity is reduced as the aspect ratio of the column increases, for a specified amount of confining reinforcement.

Figure 5.14 shows a typical relationship between theoretical structure displacement ductility capacity μ_Δ and column aspect ratio H/D for a fixed-base cantilever or multicolumn bent with rigid cap beam and foundation, including the foregoing trend, as line 1, based on 1994 Caltrans design equations for amount of confining reinforcement in plastic hinges [see Section 5.3.2(c)]. Also shown in Fig. 5.14 are curves for ductility capacity when additional flexibility due to cap beam and/or foundation flexibility increase the yield displacement, shown as a family of curves, marked 2, where f_a is the additional flexibility coefficient, defined in Eq. (5.21). These curves apply only for footing-supported columns. For the special case of pile shaft/column designs, the effective plastic hinge length is greater than for cases where the hinge forms against a constraining member [see Section 5.3.2(b)] and the flexibility coefficient is related uniquely to the column aspect ratio. Separate curves are thus given in Fig. 5.14 for two cases of pile shaft: cantilever pile shaft (as discussed in relation to Fig. 5.12) and fixed-top pile shaft, such as multicolumn pile shaft bents where the columns frame monolithically into the cap beam. For cantilever pile shafts, the increasing plastic hinge length/column height ratio compensates for increasing relative soil flexibility as aspect ratio is reduced,

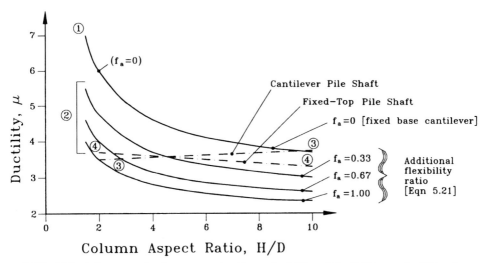

FIG. 5.14 Influence of column aspect ratio and additional flexibility on ductility of columns designed to 1994 Caltrans confinement requirements.

making displacement ductility rather insensitive to aspect ratio. In Fig. 5.14, column aspect ratio refers to distance from section of maximum moment (or ground level for pile shafts) to point of contraflexure.

It will be noted from Fig. 5.14 that the ductility capacity is expected to be reduced substantially as the aspect ratio of the column increases, and that multicolumn bents ($f_a > 0$) may have significantly lower ductility capacity than fixed-base cantilever columns ($f_a = 0$). It will also be noted that significantly lower ductility capacities than suggested by Fig. 5.13 can be expected in many cases, particularly when additional flexibility is important. In this context it should be noted that the curves of Fig. 5.14 relate to typical designs with low axial loads and a low longitudinal reinforcement ratio. For high axial loads or reinforcement ratios, theoretical ductility capacity is often less than implied by Fig. 5.14. As a consequence, it is suggested in Section 5.3.2(*c*) that levels of confining reinforcement be increased by one-third above the level currently specified by Caltrans and that a further increase be required for columns with high longitudinal steel ratio. Based on this increased level of confinement, a simplified relationship between ductility capacity and column aspect ratio suitable for force-based design of simple structures is presented in Fig. 5.15. In lieu of specific calculation of the additional flexibility coefficient f_a for multicolumn bents, a value of $f_a = 0.5$ may be adopted except for unusually wide or squat bents. Where foundation or bearing flexibility is expected to be significant, the value for f_a should always be calculated. Use of Fig. 5.15 in conjunction with Eq. (5.22) provides a suitable basis for force-

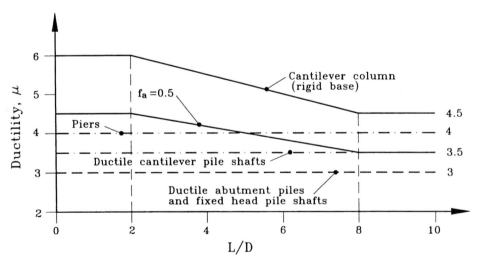

FIG. 5.15 Proposed design displacement ductility levels for bridges confined in accordance with Eq. (5.47). [*Note:* $f_a = 0.5$ is used for columns in double bending (multicolumn bents, longitudinal response) unless designed to Eq. (5.21).]

based seismic design, while recognizing some of the inherent displacement-based influences of structural form on seismic response that are automatically considered in the displacement-based design approach of the next section.

(c) *Displacement-Based Design.* For more complex bridges or bridges of special importance, the appropriate relationship embodied in the development of Fig. 5.15 may be unacceptably coarse. In such cases, reanalysis (see Section 1.5.5) after initial design using dynamic time-history analysis (see Section 4.5.4) or other means of analysis capable of determining member ductility demand directly should be used.

An alternative approach [K3,K8,C3] that shows considerable promise is to use a design approach based on displacement rather than force considerations. In this approach the design procedure attempts to provide the appropriate member detailing (member size and reinforcement content) to achieve a specified displacement at the center of seismic force (or a displacement profile, in more complex structures), under the design level seismic input. The reason for adopting this approach is that damage limit states can be rather closely related to strain limits, which in turn can be converted to equivalent displacements but cannot be directly related to force-level or displacement ductility factor. The procedure, then, attempts to provide a structure that as closely as possible achieves the required limit state under the design level of excitation.

To do this, some fundamental changes to the normal force-driven design procedure are made. First, initial stiffness and strength of members is considered to be of only secondary importance. Instead of characterizing the bridge stiffness by initial elastic properties, the secant properties pertaining at *maximum* displacement are used, as suggested in Fig. 5.16(b). Second, instead of using 5% elastic damping, which is appropriate for initial preyield response, a value appropriate for the hysteretic response expected is used. As shown in Fig. 5.16(c), the equivalent viscous damping is a function of the expected ductility level and the location of inelastic response (beam hinges dissipate more energy and hence have higher equivalent damping than column hinges). Third, the approach uses a design displacement spectra set for different levels of equivalent damping, as shown, for example, in Fig. 5.16(d). The approach, then, is based on the substitute structure method outlined in Section 4.4.3.

The method is most simply explained by reference to a 1-DOF example— say, a simple regular multispan bridge under transverse response [Fig. 5.16(a)]—and proceeds according to the following sequence of operations:

1. An initial estimate of the structural yield displacement Δ_y is made. Since final results are not particularly sensitive to the value assumed, an initial estimate based on a yield drift of 0.005 is recommended. Hence $\Delta_y = 0.005H$.

2. The limit to acceptable plastic rotation, θ_p, of the critical hinges (in this case, the column base hinge) is determined. This will be a function of

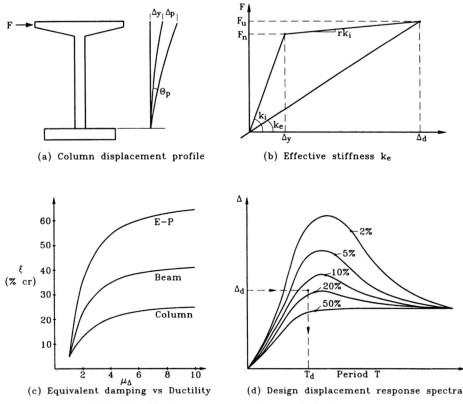

(a) Column displacement profile   (b) Effective stiffness k_e

(c) Equivalent damping vs Ductility   (d) Design displacement response spectra

FIG. 5.16 Displacement-based design of a bridge pier.

the design limit state considered and the geometry of the bridge [see Section 5.3.2(*b*) and (*d*)]. For the damage-control limit state, a value of about $\theta_p = 0.03$ will often be appropriate.

3. The design level of displacement Δ_d at the center of seismic force, corresponding to the plastic rotation limit of the most critical hinge, is found from considerations of mechanism deformation, based on the plastic collapse analysis method outlined in Section 4.5.2(*b*)(ii). Hence, for the example of Fig. 5.16(*a*),

$$\Delta_d = \Delta_y + \Delta_p = \Delta_y + \theta_p H$$

4. An estimate of effective structural damping is made based on the implied ductility level $\mu_\Delta = \Delta_u/\Delta_y$, using an appropriate relationship between ductility and damping [e.g., Fig. 5.16(*c*)]. More realistic relationships are given in Section 4.4.3.

5. With reference to the elastic displacement response spectra [Fig. 5.16(d)], the effective period T_d is found by entering at the required displacement Δ_d and intersecting the appropriate damping curve (follow the dashed line). The effective stiffness of the substitute structure at maximum response can thus be found from

$$T_d = 2\pi \sqrt{\frac{M}{K_e}}$$

as

$$K_e = \frac{4\pi^2}{T_d^2} M \qquad (5.23a)$$

and a first estimate of the required strength at maximum response by

$$F_u = K_e \Delta_d \qquad (5.23b)$$

The required nominal strength of the equivalent bilinear response, F_n, can thus be found from

$$F_n = \frac{F_u}{r\mu - r + 1} \qquad (5.24)$$

6. Preliminary estimates for member dimensions and reinforcement content can now be made, the elastic stiffness K_i can be calculated, and a refined estimate of the yield displacement can be obtained. In practice, a range of possible solutions may be generated.

7. The total displacement, structure ductility, and hence effective damping are thus revised, and steps 5 and 6 are repeated until a stable and satisfactory solution is obtained. Results for a typical set of analyses for simple cantilever bridge columns are shown in Fig. 5.17. The columns, of different heights, were designed to the Eurocode 8 response spectra, expressed in displacement form, for an earthquake characterized by a peak ground acceleration of 0.5g. A series of design options, each resulting in a plastic drift of 0.03, were developed for each column height, and the implications in terms of column diameter, design moment, longitudinal reinforcement content, ductility demand, and transverse reinforcement content evaluated.

As can be seen from Fig. 5.17, the results follow trends that might not be expected from force-based design. For a given height, the design moment decreases as the diameter increases [Fig. 5.17(a)]. This is because a greater displacement ductility results from the larger, hence stiffer, columns. Force-based design using initial stiffness as the prime indicator of response would imply *higher* design moments for the stiffer piers. As seen from Fig. 5.17(c),

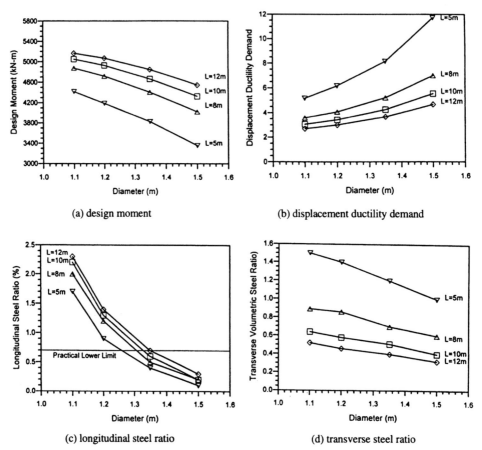

FIG. 5.17 Parameter study for displacement-based design of simple cantilevers to a target plastic drift of $\theta_p = 0.03$ (design spectra: Eurocode 8, 0.5g).

the reduced design moments results in greatly reduced longitudinal steel content for the larger-diameter piers. The consequence of the reduced design moment is increased displacement ductility demand [Fig. 5.17(b)] with increasing diameter, and the values might appear unacceptable in a force-based design philosophy. However, Fig. 5.17(d) shows that the volumetric ratio of confining reinforcement required to sustain these ductilities actually is reduced with increasing diameter, as a result of reduced neutral-axis depth, and hence increased ultimate curvature.

The use of displacement-based design for more complex structures, particularly those involving significant differences in design displacements at adjacent piers, is still in the experimental stage [C3]. Care is needed in assessing the global damping appropriate to different levels of ductility at adjacent piers. However, results appear promising. For multi-degree-of-freedom structures, elastic modal analyses based on the substitute structure are used to check that the displacement profile satisfies the drift constraints, and to define design parameters.

(d)   Design for Required Flexural Strength of Plastic Hinges. Although it is possible to conceive of bridge seismic designs where ductility will be provided by beam hinges, it is normally the columns that will be required to develop inelastic rotational capacity. Required flexural strength of the plastic hinges will be found from either the displacement- or force-based design approaches of preceding sections.

Current design philosophy in most codes or design recommendations would match a conservative estimate of flexural strength M_n, further reduced by a flexural strength reduction factor ϕ_f, to the required strength M_r, in accordance with

$$\phi_f M_n \geq M_r \tag{5.25}$$

Since the capacity design approach requires nonductile elements or modes of deformation to be protected by designing for force levels corresponding to maximum feasible (extreme) estimates of flexural strength developing at plastic hinge locations (i.e., $M^\circ = \phi^\circ M_n$), the required strength S_r of capacity-protected actions may be very much higher than that corresponding to the required strength of plastic hinges. Design strength for capacity-protected actions is thus dictated by

$$\phi_s S_n \geq S_r = S^\circ \tag{5.26}$$

where S° is the action corresponding to flexural overstrength moment M° of the plastic hinges, discussed further in Section 5.3.3, and ϕ_s is the strength reduction factor appropriate to action S. It should be noted that many codes do not specify sufficiently high capacity protection factors ($\phi^\circ = M^\circ/M_n$) to ensure that undesirable deformation modes, such as shear failure, are inhibited. If adequately high capacity protection factors are specified in conjunction with current estimates of design strength, seismic design forces for all parts of the structure, except for plastic hinge regions, will be much higher than those corresponding to the design level of seismic force. The economic consequences can be considerable, particularly for foundation design.

In the United States, nominal flexural strength is normally calculated by the ACI method, using specified minima material strength and material strains corresponding to an extreme fiber compression strain of 0.003. Some aspects relevant to this approach are discussed briefly next.

(i)   Reinforcement Yield Strength. Because of the wide variation in properties of grade 40 ($f_y = 40$ ksi $= 276$ MPa) reinforcement (see Section 5.2.3), grade 60 ($f_y = 60$ ksi $= 414$ MPa) is normally specified for seismic design. The permissible range for yield strength of A706 steel, which is preferred over A615 because of tighter control of carbon content and improved weldability properties, is

$$(414 \leq f_y \leq 534 \text{ MPa}) \qquad 60 \leq f_y \leq 78 \text{ ksi}$$

Thus the upper limit is 30% higher than the specified design value.

(ii) Reinforcement Strain Hardening. At curvature ductility levels corresponding to peak response, steel strains as high as 7% may occur. At this level the steel stress is typically 30 to 40% above the actual yield stress. High steel tensile strains occur particularly with columns subjected to low axial load and containing low longitudinal reinforcement ratios. In combination with high initial yield stress it is thus possible for stress in some reinforcing bars to exceed the nominal 60 ksi (414 MPa) by 70% or more.

(iii) Concrete Compression Strength. The specified 28-day compression strength is a low estimate of the strength expected in the field. Conservative mix design and the requirements for concrete suppliers to guarantee specified strength result in mean 28-day strengths about 20 to 25% higher than the specified strength. Concrete continues to gain strength after 28 days [see Section 5.2.1(a)] with strength gain in the first year typically being about 20% above the 28-day strength. After 30 years, compression strength often exceeds the specified strength by 100% or more. As noted earlier, cores taken in the 1990s from California bridges built in the 1950s and 1960s have consistently tested out with compression strengths in the range 38 to 62 MPa (5500 to 9000 psi) for nominal 22.5-MPa (3250-psi) concrete. Since the design-level earthquake has an extremely low probability of occurring by an age of 28 days, at which stage the bridge will probably still be under construction, a higher compression strength is needed in assessing overstrength potential.

Concrete compression strength is further enhanced by passive confinement provided to plastic hinge regions by transverse reinforcement, as noted in Section 5.2.2. With levels of confinement common in modern bridge designs, compression strength enhancement about 50% above actual unconfined strength can be expected.

(iv) Maximum Compression Strain. The use of an extreme fiber compression strain of 0.003 for flexural strength calculations does not reflect ultimate conditions, when extreme fiber strains as high as 0.02 may develop. It is also a very conservative estimate of strains at which crushing and spalling first develop. When the critical section is confined by an adjacent member (footing or cap beam), first signs of crushing generally are visible at extreme fiber strains in the range 0.006 to 0.010 [M6]. Figure 5.18 shows a typical relationship between extreme fiber compression strain and moment for a typical circular bridge column. The use of $\varepsilon_c = 0.003$ and nominal material strengths results in a

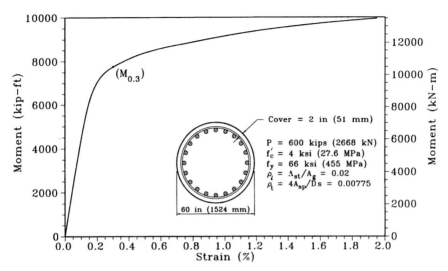

FIG. 5.18 Moment–extreme fiber compression strain relationship for a typical circular column.

rather conservative estimate of peak strength, even when nominal material strengths are used for predicting the moment–strain curve. When it is considered that yield strengths may greatly exceed the specified minimum, as noted above, the conservatism is severe. This is examined further in Fig. 5.19, which plots the relationship between overstrength ratio $\phi°$ for circular and rectangular columns at maximum ductility capacity with different axial load ratios and longitudinal reinforcement ratios ρ_l related to ACI design assumptions for flexural strength. Nominal and overstrength material properties assumed were $f_y = 60$ ksi (414 MPa), $f_{yo} = 78$ ksi (538 MPa), $f_c = 4$ ksi (27.6 MPa), $f_{co} = 1.7f_c = 6.8$ ksi (46.9 MPa). It is seen from Fig. 5.19 that $\phi°$ is strongly dependent on ρ_l and $P/f_c'A_g$ and that values of $\phi° = 1.6$ and 1.7 would be needed to provide adequate protection over the common range of parameter values for circular and rectangular sections, respectively.

(v) Flexural Strength Reduction Factors. Most seismic design codes specify the use of flexural strength reduction factors to obtain a dependable strength value which is then equated to the required strength in accordance with Eq. (5.25). For example, 1992 AASHTO requirements [A4] specify a flexural strength reduction factor ϕ_f of

$$0.9 \ge \left(\phi_f = 0.9 - \frac{2P}{f_c'A_g} \right) \ge 0.5$$

In conjunction with the overstrength factors of Fig. 5.19, it is thus possible for extreme flexural strength to exceed the required strength by 200 to 250%

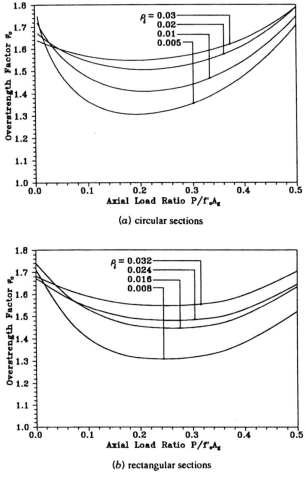

FIG. 5.19 Relationship between maximum feasible flexural strength and ACI nominal flexural strength [$f_c' = 4$ ksi (27.6 MPa), $f_y = 60$ ksi (414 MPa)].

even if no excess dependable strength is provided. It should, however, be noted that AASHTO seismic design recommendations were under review at the time of preparing this book, and less conservative approaches are expected to be adopted in the revised version.

Flexural strength reduction factors specified by ACI are similar to those of AASHTO but with a lower limit of 0.7. Caltrans has for many years used an approach where $\phi_f = 1$ is a lower limit. In New Zealand a uniform value of $\phi_f = 0.9$ is adopted, regardless of axial load level. Japan adopts $\phi_f = 1.0$, and in Europe the dependable strength is determined by use of low estimates for f_c' and f_y rather than by specifying a flexural strength reduction factor.

(vi)   Consequences of Unconservatism in Design Strength. It is clear that current flexural design approaches are deliberately conservative, and rather inconsistent, and that the consequence is a requirement for high overstrength factors if capacity design objectives are to be met. It would seem that the consequences of a less conservative flexural strength design approach might be beneficial since overstrength factors could be reduced and since the possibility of actual flexural strength begin less than the required strength does not have the same significance to ductile flexural design as it does to gravity load design. With reference to Fig. 5.20 it is seen that if actual flexural strength is (say) 10% below required strength, the prime structural consequence is that flexural yielding may occur slightly earlier than expected, resulting in a corresponding (11%) increase in the curvature ductility demand on the section. Considering the conservatism and lack of precision inherent in design for ductility, this is hardly significant. It will be noted that the situation is very different from gravity-load design, where it is essential that an adequate margin be maintained between strength and applied loads, to avoid failure. For seismic design the strength is expected to be developed at a fraction of the design response, so the concept of maintaining a strength margin is meaningless. It follows that for seismic design, the consequences of *over*prediction of the design flexural strength of a plastic hinge are less significant that those of *under*predicting the overstrength capacity.

(vii)   Design Recommendations. Based on the arguments above, it will be seen that there is a strong case for less conservative flexural design of plastic hinge regions. The following design recommendations are thus proposed.

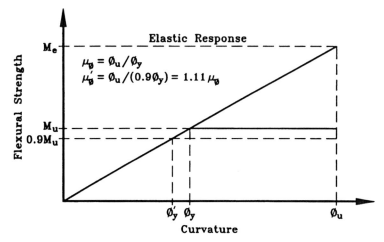

FIG. 5.20   Influence of low flexural strength on ductility demand, based on an equal-displacement approximation.

1. Design flexural strength of plastic hinges should be computed based on characteristic material strengths, corresponding approximately to the lower 5% limit at a characteristic age for occurrence of the seismic design event. For U.S. design, the following values are recommended:

    Characterisic yield strength: $\qquad\qquad f_{ye} = 1.1 f_y$

    Characteristic compression strength: $\quad f'_{ce} = 1.3 f'_c$

2. Design extreme fiber compression strain $\varepsilon_c = 0.004$.
3. An equivalent rectangular stress block (or other appropriate shape, justified by experimental results) may be assumed to represent the concrete stress–strain relationship in compression.
4. A flexural strength reduction factor of $\phi_f = 1.0$ should be adopted for dependable flexural strength.

(viii) Moment–Curvature Analysis. As an alternative to design using an equivalent stress block, the design flexural strength may be based on a moment–curvature analysis of the section. It is envisaged that this will be used where moment–curvature analyses are used to predict the expected overstrength values at design plastic rotation levels. Where moment–curvature analysis is used, the design flexural strength should correspond to conditions when the extreme fiber compression strain reaches 0.004 or when the tensile strain in the extreme tension reinforcement reaches 0.015, whichever occurs first. The steel stress–strain curve should include allowance for strain hardening. For grade 60 reinforcement ($f_y = 414$ MPa $= 60$ ksi), used in the United States, the curve given in Fig. 5.21 may be adopted if local test results are not available. In Fig. 5.21, the strain-hardening part of the stress strain curve may be represented by the following equation:

$$f_s = f_{ye} \left[1.5 - 0.5 \left(\frac{0.12 - \varepsilon_s}{0.112} \right)^2 \right] \tag{5.27}$$

Equation (5.27) assumes $\varepsilon_{sh} = 0.008$ and $\varepsilon_{su} = 0.12$.

Note that it is important that the concrete stress–strain relationship distinguish between cover concrete and confined concrete, with allowance for the different compression strength and strain capacities of these two components as discussed in Section 5.2.2, when moment–curvature analysis is carried out. Thus, with reference to Fig. 5.22(a), the moment–curvature curve for a circular column may be generated for specified values of extreme fiber compression strain ε_c by considerations of axial and moment equilibrium. From considerations of axial equilibrium,

$$P = \int_{x=(D/2)-c}^{D/2} [b_{c(x)} f_c(\varepsilon_x) + (b_{(x)} - b_{c(x)}) f_{cu}(\varepsilon_x)] \, dx + \sum_{i=1}^{n} A_{si} f_{s(\varepsilon_{xi})} \tag{5.28}$$

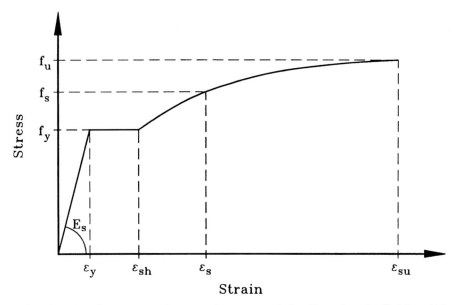

FIG. 5.21 Tensile stress–strain curve for monotonic loading of grade 60 (f_y = 414 MPa nominal) reinforcement.

where

$$\varepsilon_x = \frac{\varepsilon_c}{c} (x - 0.5D + c)$$

From consideration of moment equilibrium,

$$M = \int_{x=(D/2)-c}^{D/2} [b_{c(x)} f_{c(\varepsilon_x)} + (b_{(x)} - b_{c(x)}) f_{cu(\varepsilon_x)}] x \, dx + \sum_{i=1}^{n} A_{si} f_{s(\varepsilon_{xi})} x_i \quad (5.29)$$

and the curvature is

$$\phi = \frac{\varepsilon_c}{c} \quad (5.30)$$

In Eqs. (5.28) and (5.29), $f_{c(\varepsilon)}$, $f_{cu(\varepsilon)}$, and $f_{s(\varepsilon)}$ are the stress–strain relationships for confined concrete, unconfined concrete, and reinforcing steel, respectively, and A_{si} is the area of a reinforcing bar with distance x_i from the centroidal axis. Other nomenclature is defined in Fig. 5.22(a).

Equation (5.28) is solved for c by trial and error using the known axial load level P and the specified extreme fiber compression strain. This enables the moment M and curvature ϕ to be calculated directly from Eqs. (5.29)

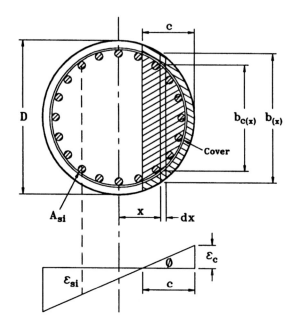

(a) Circular Column

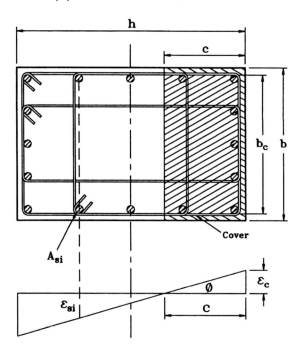

(b) Rectangular Column

FIG. 5.22 Moment-curvature analysis of columns sections.

and (5.30). The entire moment–curvature curve is generated by specifying a sequence of ε_c values up to the ultimate compression strain, as given by Eq. (5.14). Substituting $b_{(x)} = b$ and $b_{c(x)} = b_c$, Eqs. (5.28) to (5.30) also apply to rectangular sections, using the nomenclature of Fig. 5.22(b).

The moment–curvature calculations described above are suitable for inclusion in simple computer codes [S12] and are used as the basis of most analysis and design examples of this book, using the models for stress–strain curves of confined and unconfined concrete and reinforcing steel described in Section 5.2.

(ix) Limits to Longitudinal Reinforcement. It is common for design codes to specify lower and upper limits to the area ratios of longitudinal reinforcement permitted in column sections. There is, however, remarkable variation in both codified limits and common design practice between different countries. In the United States, longitudinal reinforcement ratios for columns between 1 and 8% are permitted. In New Zealand the permitted range is 0.8 to 8.0%. In Japan much lower reinforcement ratios are permitted, and values as low as 0.5% are common. It is thus useful to examine briefly the purposes of specifying reinforcement limits.

LOWER LIMIT. It is important that the column flexural strength should exceed the cracking strength by an adequate margin. A low margin of flexural strength over cracking moment could result in only one or two cracks forming in the plastic hinge region of ductile columns, as indicated in Fig. 5.23. If the tensile

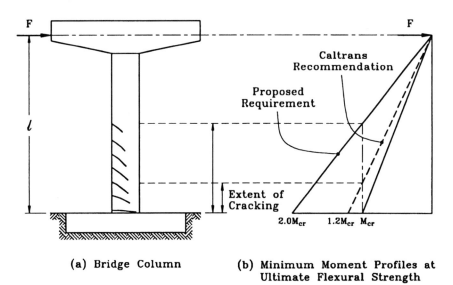

(a) Bridge Column   (b) Minimum Moment Profiles at
                          Ultimate Flexural Strength

FIG. 5.23 Spread of flexural cracking for columns with low longitudinal reinforcement ratios.

strength of the concrete is higher than anticipated, which is probable for seismic rates of loading, only one crack may form at the column base. The result will be a reduced effective plastic hinge length and increased curvature ductility demand. This, coupled with the low compression zone depth probable for low reinforcement ratios, may result in excessive reinforcement tensile strains, possibly leading to fracture of reinforcement. This can be anticipated particularly for small-diameter bars, where strain penetration from the critical section will be small or where the reinforcing steel has a low ultimate tensile strain.

A common code provision requires a margin of 20% of ultimate strength over cracking strength. This is inadequate for seismic response. To ensure satisfactory performance, it is recommended that the flexural strength of the column section should exceed the probable cracking moment by at least 100%. Figure 5.24 plots the flexural strength/cracking moment ratios for circular and rectangular columns with different axial load and longitudinal reinforcement ratios. Flexural strength is based on moment–curvature analyses with $f'_{ce} = 4.2$ ksi (29 MPa) and $f_{ye} = 66$ ksi (455 MPa), in accordance with Section 5.3.1(d) (vii), based on $f'_c = 3.25$ ksi (22.4 MPa) and $f_y = 60$ ksi (414 MPa). Figure 5.24 indicates that lower limits of $\rho_l = 0.005$ for circular columns and 0.008 for rectangular columns would be appropriate. The lower limit of $\rho_l = 0.005$ for circular columns is confirmed by the excellent ductile behavior exhibited by a circular column reinforced to this limit, whose hysteretic performance

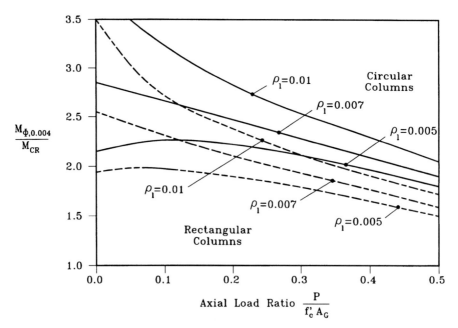

FIG. 5.24 Relationship between flexural strength and cracking moment.

is plotted in Fig. 5.25 [P9]. This column developed displacement ductility factors in excess of $\mu_\Delta = 10$ before failing in shear.

UPPER LIMIT. It is uncommon to use steel ratios anywhere approaching the permissible limit of 8% in current designs, as a consequence of practical difficulties in placing and confining large steel quantities. Other reasons for placing an upper limit on steel ratios in columns include increased sensitivity to $P-\Delta$ effects, increased difficulty in restraining large areas of compression reinforcement against buckling, particularly for circular columns, and excessive joint shear stresses developed in column–cap beam and column–footing joints when high reinforcement ratios are adopted. The later point will almost always require longitudinal reinforcement ratios to be less than 4% if joint shear stresses are to be limited to $0.25 f'_c$, as recommended elsewhere [P4]. Consequently, the following limits are recommended:

$$0.005 \le \rho_l \le 0.04 \quad \text{(circular columns)} \quad (5.31a)$$

$$0.008 \le \rho_l \le 0.04 \quad \text{(rectangular columns)} \quad (5.31b)$$

Practical designs will normally be found in the range $0.01 \le \rho_l \le 0.03$.

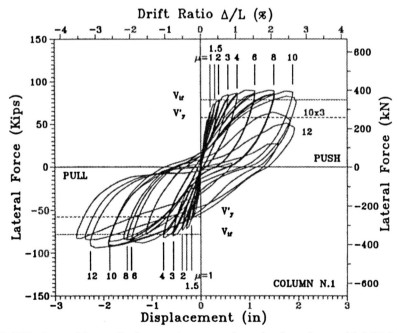

FIG. 5.25 Lateral force–displacement response for a circular column with 0.5% longitudinal steel ratio [P9].

(x) Design Axial Force on Columns. When designing column sections, it is essential that the axial force on the column corresponds to the level of lateral response considered. This is satisfied by ensuring that all internal and external forces are in equilibrium. In the case of the simple cantilever column of Fig. 5.26 under transverse response this will be trivial, provided that the superstructure is straight, so that effects of horizontal curvature do not result in seismically induced axial forces. The design axial force will then be

$$P = kP_D \tag{5.32}$$

where k represents the expected variation in axial dead load resulting from vertical acceleration effects. Because this effect is generally not particularly dominant, it is common to take $k = 1$. If variations due to vertical acceleration are to be considered, this should strictly result from a vertical response analysis. However, since peak vertical and horizontal response are unlikely to occur at the same time, some reduction in the vertical response is appropriate. It is suggested that unless the columns are felt to be unusually sensitive to vertical response effects, a value of

$$k = 1 \pm 0.5\text{PGA} \tag{5.33}$$

could be considered, where PGA is the design peak ground horizontal acceleration, expressed as a ratio to g. Equation (5.33) is based on an assumed ratio of horizontal to vertical PGA of 0.67. In structures sensitive to vertical acceleration the combinations of vertical and horizontal spectral responses in accordance with recommendations of Section 4.5.3 should be adopted, or inelastic time-history analyses in accordance with Section 4.5.4 should be used to determine appropriate combinations.

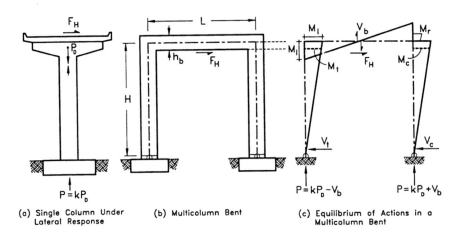

(a) Single Column Under Lateral Response (b) Multicolumn Bent (c) Equilibrium of Actions in a Multicolumn Bent

FIG. 5.26 Design axial forces for ductile columns.

For multicolumn bents such as depicted in Fig. 5.26(*b*), or when considering longitudinal response, the axial force must be determined from equilibrium considerations. Thus for the direction of seismic force shown in Fig. 5.26(*b*), the column axial forces are

$$P = kP_D \pm V_b \tag{5.34}$$

where the seismically induced cap beam shear force V_b is given by

$$V_b = \frac{M_l + M_r}{L} \tag{5.35}$$

and where

$$M_l = \frac{M_t}{\left(1 - \dfrac{0.5h_b}{H}\right)} \tag{5.36a}$$

$$M_r = \frac{M_c}{\left(1 - \dfrac{0.5h_b}{H}\right)} \tag{5.36b}$$

Since the moments M_t and M_c developed in the plastic hinge regions at the top of the tension and compression columns, respectively, are influenced by the axial forces on the columns, an iterative procedure is required in design to satisfy the design requirement that

$$V_t + V_c = \frac{M_t + M_c}{H} \geq F_H \tag{5.37}$$

(xi) *Moment Redistribution of Design Actions.* Equation (5.37) implies that the equilibrium of internal actions and seismic force should be carried out in global rather than local terms. This may imply some redistribution of elastic actions when force-based design is adopted. As shown in the example of Fig. 5.27, the combination of gravity and seismic actions may result in considerable inequality in the design moments for different columns. In the example shown, this will typically be partly offset by reduced moment capacity for the column with tensile seismic axial force, but the required moment capacity of the compression column is likely to dominate design. Since both columns (in a symmetrical bent) would be provided with the same reinforcement, to cope with the reversed direction of seismic response, the moment capacity of the tension column would end up being larger than required. Hence the overall lateral strength provided would exceed the required level. In the capacity

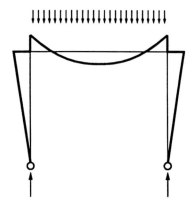

(a) Gravity Loads, D

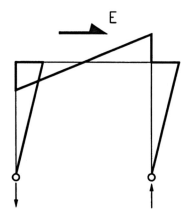

(b) Seismic Force, $\vec{E}$

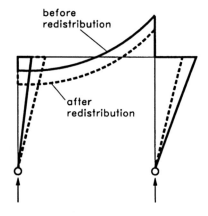

(c) Combined Action, $D+\vec{E}$

FIG. 5.27 Moment redistribution of design actions.

design process, this would result in higher design forces for capacity-protected members and actions, with consequent undesirable economic consequences.

To avoid this unnecessary extra strength, redistribution of the bending moment diagram may be permitted, as shown in Fig. 5.27(c) in such a way that the design capacity is optimized, and Eq. (5.37) becomes an equality. In this example, this requires reducing the design moment at the top of the compression column and increasing the design moment at the top of the tension column by an equal amount. The consequence will be slightly earlier onset of inelastic action at the compression column.

In determining the limits to redistribution, it is necessary to ensure that service load response under gravity-load effects is not impaired. Also, to minimize extra ductility demand on the compression column, it is recommended that no critical moment be reduced by more than 30%, in accordance with recommendations made for redistribution of actions in building frames [P4].

5.3.2 Flexural Ductility and Inelastic Rotation

(a) Required Ductility. When designing to the force-based design approach to Section 5.3.1(b), the structural ductility capacity must be known to determine the appropriate force-reduction factor. As explained in that section, geometrical considerations, including the effects of foundation flexibility, influence the relationship between structural displacement ductility factor and member ductility factor, which may be expressed in curvature, rotation, or displacement units. Equation (5.21) is formulated as a basic equation relating member to structure displacement ductility capacity. Using the displacement-based design procedure of Section 5.3.1(c), it is the plastic rotation of potential plastic hinges that is of greatest interest. From this, the permitted displacements are calculated, and hence the basic force requirements established. In both design approaches it is thus necessary to be able to determine the inelastic rotation and ductility capacity, since these are interrelated, of individual members.

(b) Assessment of Member Inelastic Rotation and Ductility Capacity. The available plastic rotation capacity, and hence the member ductility capacity, depend on section geometry and the amount and distribution of transverse reinforcement within the plastic hinge region. Transverse reinforcement provides the dual functions of confining the core concrete, thus enhancing its compression strength and enabling it to sustain higher compression strains, and restraining the longitudinal compression reinforcement against buckling. These two actions interact in a complex fashion not yet fully understood.

(i) Plastic Rotation Capacity. A bilinear approximation to the moment–curvature relationship for the critical section is required, as shown in Fig. 5.28. The equivalent yield curvature ϕ_y is found by extrapolating the line joining

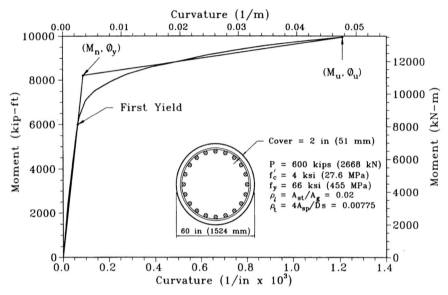

FIG. 5.28 Bilinear approximation of a column moment–curvature relationship.

the origin and conditions at first yield, to the nominal moment capacity M_n, determined in accordance with the recommendations of Section 5.3.1(d) (vii) or (viii). The plastic curvature capacity ϕ_p is the difference between ultimate curvature ϕ_u corresponding to the limit compression strain ε_{cu}, and the yield curvature. Thus

$$\phi_p = \phi_u - \phi_y \tag{5.38}$$

This plastic curvature is assumed to be constant over the equivalent plastic hinge length L_p, which is calibrated to give the same plastic rotation ϕ_p as occurs in the real structure.

From analyses and test results, a reasonable estimate for the plastic hinge length when the plastic hinge forms against a supporting member, such as the footing in Fig. 5.29, is given by

$$L_p = \begin{cases} 0.08L + 0.022f_{ye}d_{bl} \geq 0.044f_{ye}d_{bl} & (f_{ye} \text{ in MPa}) \\ 0.08L + 0.15f_{ye}d_{bl} \geq 0.3f_{ye}d_{bl} & (f_{ye} \text{ in ksi}) \end{cases} \tag{5.39}$$

In Eq. (5.39), L is the distance from the critical section of the plastic hinge to the point of contraflexure and d_{bl} is the diameter of the longitudinal reinforcement. The second term in Eq. (5.39) makes allowance for additional rotation at the critical section resulting from strain penetration of the longitudi-

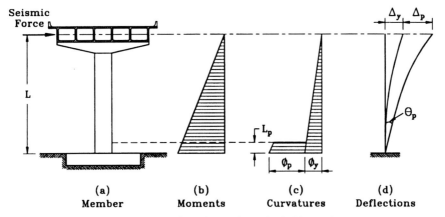

FIG. 5.29 Inelastic deformation of a bridge column.

nal reinforcement into the supporting element, in this case the footing. The plastic rotation is therefore

$$\theta_p = L_p \phi_p = L_p(\phi_u - \phi_y) \tag{5.40}$$

The limit curvatures ϕ_y and ϕ_u can be found without recourse to a full moment–curvature analysis. Using the design charts of Fig. 4.9 for effective stiffness I_e of cracked sections, the yield curvature may be approximated as

$$\phi_y = \frac{M_n}{E_c I_e} \tag{5.41}$$

From section analyses of the critical section at the ultimate extreme fiber compression strain ε_{cu},

$$\phi_u = \frac{\varepsilon_{cu}}{c_u} \tag{5.42}$$

where c_u is the neutral-axis depth and ε_{cu} is given by Eq. (5.14).

(ii) Member Ductility Capacity. The section curvature ductility capacity is defined by

$$\mu_\phi = \frac{\phi_u}{\phi_y} \tag{5.43}$$

Excluding all additional flexibility effects [see Section 5.3.1(a)], the member yield displacement, referring again to Fig. 5.29, may be approximated by

$$\Delta_y = \frac{\phi_y L^2}{3} \tag{5.44}$$

The plastic displacement Δ_p includes the component due to the plastic rotation θ_p and additional elastic displacement resulting from the increase in moment from M_n to M_u (see Fig. 5.28). Thus

$$\Delta_p = \left(\frac{M_u}{M_n} - 1\right)\Delta_y + L_p(\phi_u - \phi_y)(L - 0.5L_p) \tag{5.45}$$

Hence the member displacement ductility factor μ_Δ is given by

$$\mu_\Delta = \frac{\Delta_u}{\Delta_y} = 1 + \frac{\Delta_p}{\Delta_y} \tag{5.46}$$

$$= \frac{M_u}{M_n} + 3(\mu_\phi - 1)\frac{L_p}{L}\left(1 - 0.5\frac{L_p}{L}\right)$$

EXAMPLE. Figure 5.28 shows the moment–curvature curve computed for a 60-in. (1524-mm)-diameter column confined with No. 4 (12.7 mm diameter) spirals of $f_y = 66$ ksi (455 MPa) at 4-in. (102-mm) centers. Longitudinal reinforcement is No. 14 bars ($d_{bl} = 1.69$ in. $= 43$ mm) with $f_{ye} = 66$ ksi (455 MPa). The key results for the bilinear approximation, shown in Fig. 5.28, are $M_n = 8118$ kip-ft (11.0 MNm), $\phi_y = 0.0000635$/in. (0.0025/m), $M_u = 9950$ kip-ft (13.5 MNm), and $\phi_u = 0.00121$/in. (0.0476/m).

If the section is representative of the bridge column of Fig. 5.29, with $L = 30$ ft (9.15 m), calculate plastic rotation capacity, yield and ultimate displacements, and member ductility factors.

From Eq. (5.39), $L_p = 0.08 \times 360 + 0.15 \times 66 \times 1.69 = 45.5$ in. (1156 mm)
From Eq. (5.40), $\theta_p = 45.4(0.00121 - 0.0000635) = 0.052$
From Eq. (5.43), $\Delta_y = 0.0000635 \times \dfrac{360^2}{3} = 2.74$ in. (69.7 mm)

From Eq. (5.44), $\Delta_p = \left(\dfrac{9950}{8118} - 1\right)2.74 + 0.052(360 - 22.75) = 18.16$ in. (461 mm)

Therefore,

$$\Delta_u = 2.74 + 18.16 = 20.9 \text{ in. (531 mm)}$$

Hence, from Eq. (5.46),

$$\mu_\Delta = \frac{20.9}{2.74} = 7.6$$

Note, from Eq. (5.43),

$$\mu_\phi = \frac{0.00121}{0.0000635} = 19.1$$

The structural displacement ductility capacity will be less than the member displacement ductility capacity of $\mu_\Delta = 7.6$ by any additional flexibility influences, in accordance with Eq. (5.21).

(iii) *Pile-Shaft Designs with In-Ground Hinge.* As noted in Section 3.3.6(*c*), in-ground hinges of integral pile shaft/columns are comparatively long as a consequence of the gradual variation in moment close to the critical section. Figure 5.30 plots the theoretical plastic hinge length as a fraction of pile diameter for cantilever pile shafts with different heights *H* from ground level to the aboveground point of contraflexure, and different subgrade reaction moduli, assumed to increase linearly with depth and to represent the full soil resistance over the pile diameter [B3]. Results are expressed in terms of the

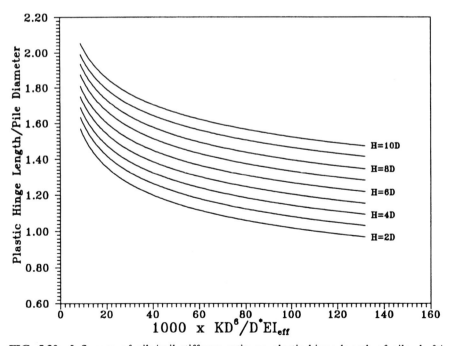

FIG. 5.30 Influence of pile/soil stiffness ratio on plastic hinge length of pile shaft/column designs [B3].

stiffness parameter KD^6/D^*EI_e, where EI_e is the effecive stiffness of the pile cracked section at first yield, given by Fig. 4.9, or directly from moment–curvature analysis, and subgrade modulus K. The ratio D/D^*, where D^* is a reference diameter of 6 ft (1.83 m) reflects the influence of variable column diameter on subgrade modulus. Thus, for example, a subgrade modulus of $K = 50$ tons/ft^3 (15.7 MN/m^3) represents a total soil stiffness for a 1-ft (305-mm) height of pile at depth 10 ft (3.05 m) of 500 tons/ft (14.4 MN/m). Values of plastic hinge length vary between D and $2D$, with highest values occurring for greater heights and more flexible soils.

The depth at which the plastic hinge forms below ground level also depends on the cantilever height of the column and the subgrade reaction modulus, as shown in Fig. 5.31. The depths indicated in Fig. 5.31 are rather less than commonly assumed and result from extensive inelastic analyses. These analyses showed that as plasticity developed in the piles, the location of maximum moment migrated upward toward the ground surface, with a final depth of approximately 70% of that predicted by elastic analysis [B3].

(c) Confinement for Plastic Hinges

(i) Concrete Confinement Requirements. With displacement-based design the volumetric ratio of transverse reinforcement required to provide adequate ultimate compression strain may be found from the required plastic rotation θ_p,

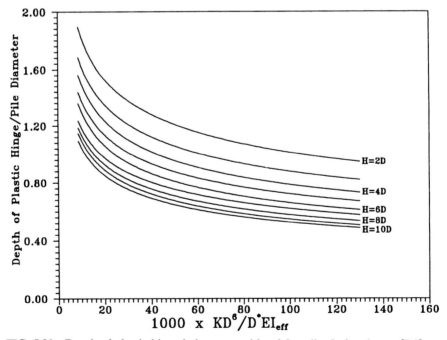

FIG. 5.31 Depth of plastic hinge below ground level for pile-shaft columns [B3].

the relationship between plastic rotation and ultimate curvature [Eq. (5.40)], between ultimate curvature and ultimate compression strain [Eq. (5.41)], and between ultimate compression strain and volumetric ratio of transverse reinforcement [Eq. (5.14)].

Where force-based seismic design is adopted, using the force-reduction factors recommended in Fig. 5.15, standard prescriptive requirements for confinement may be used. The following equations have been modified from Caltrans design practice to ensure adequate ductility for the force-reduction factors in Fig. 5.15.

Circular columns or rectangular columns with interlocking spirals:

$$\rho_s = \frac{4A_{sp}}{D'_s} \geq 0.16 \frac{f'_{ce}}{f_{ye}} \left(0.5 + \frac{1.25P}{f'_{ce}A_g}\right) + 0.13 \, (\rho_l - 0.01) \qquad (5.47)$$

where ρ_l is the longitudinal steel ratio.

Rectangular columns with rectangular distribution of longitudinal reinforcement:

$$A_{sh} = 0.12sh_c \frac{f'_{ce}}{f_{ye}} \left(0.5 + \frac{1.25P}{f'_{ce}A_g}\right) + 0.13(\rho_l - 0.01) \qquad (5.48)$$

where h_c is the core width perpendicular to the direction of placement of A_{sh} and s is the vertical spacing. Equation (5.48) provides the same confining effect as Eq. (5.47), assuming that the effectiveness of the confinement is 12% less, on average for rectangular compared to circular columns.

The amount of confinement provided by Eqs. (5.47) to (5.48) is somewhat higher for low axial load ratios than has recently been suggested based on theoretical research in New Zealand [P10]. However, conflicting evidence is available from tests of circular columns in the United States [I5, M5], particularly when columns have high aspect ratios (L/D), leading to long plastic hinge regions and hence reduced confinement influence from the supporting member (footing or cap beam) or when the column is subjected to bidirection loading. Pending clarification of these points, the equations above, which may in some cases be excessively conservative, are recommended.

For compression members of smaller size [say, $D \leq 36$ in. (914 mm)] subjected to high compressive loads, such as may be the case with piles, it is recommended that the ACI-318 requirement [A5] that

$$\rho_s \geq 0.45 \left(\frac{A_g}{A_c} - 1\right) \frac{f'_c}{f_{yh}} \qquad (5.49)$$

also be satisfied. This requirement is based on maintaining the axial load capacity of the entire section of the column, assuming it to be unconfined, by the confined strength of the core after spalling of all cover concrete.

(ii)   Antibuckling Requirements. The transverse reinforcement in the plastic hinge region must also be capable of restraining longitudinal compression reinforcement against buckling. Two possible buckling modes should be considered: buckling between layers of transverse reinforcement, and buckling over a longer length, involving yield, and ultimately fracture of one or more layers of transverse reinforcement. These modes are illustrated in Fig. 5.32(*a*) and (*b*), respectively. The spacing between layers of transverse reinforcement to avoid the first mode [Fig. 5.32(*a*)] depends on the effective modulus of elasticity of the longitudinal reinforcement in the strain-hardening range, and the peak compression strain expected of the longitudinal reinforcement. A common requirement has been that the spacing of transverse reinforcement layers should not exceed $6d_{bl}$ [C1]. Although this has been shown to be adequate for columns reinforced with steels whose ultimate strength is approximately 50% higher than yield strength, it has been shown to be inadequate for support of steels whose f_u/f_y ratio is rather less, as is common with the new European tempcore steels. Although extensive data are not yet available, it is suggested that the maximum spacing of lateral support to the longitudinal reinforcement provided by transverse reinforcement should be

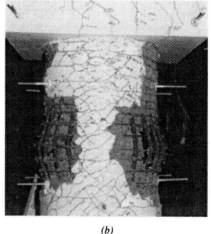

(*a*)   (*b*)

FIG. 5.32   Buckling of longitudinal reinforcement in plastic hinges. (*a*) Buckling between layers of transverse reinforcement; (*b*) Buckling involving several layers of transverse reinforcement.

$$s \leq \left[3 + 6 \left(\frac{f_u}{f_y} - 1 \right) \right] d_{bl} \tag{5.50}$$

Equation (5.50) results in the familiar $s \leq 6d_{bl}$ [S2,C1] for $f_u/f_y = 1.5$ but requires closer spacing for steels with low f_u/f_y ratios.

The second requirement of avoiding buckling over several multiples of hoop spacing [Fig. 5.32(b)] has traditionally been met for rectangular sections by requiring a restraint force of $\frac{1}{16}$ of the yield capacity of the restrained bar. For longitudinal and transverse reinforcement of nominally equal yield strength, this requires that the tie diameter should be at least one-fourth that of the longitudinal bar, provided that each longitudinal bar is restrained by a tie or hoop leg parallel to the potential direction of buckling. Although this requirement has in the past been related to a spacing of 4 in. (102 mm) between lateral supports [S2], it is clearly more reasonable to relate the force to a dimensionless distance. It is thus suggested that the requirement be related to a spacing of $6d_{bl}$. This implies that the required area A_{tr} at spacing s to restrain longitudinal bars of total area, ΣA_l and diameter d_{bl} is

$$A_{tr} = \frac{\Sigma A_l}{16} \frac{s}{6d_{bl}} \frac{f_y}{f_{yh}} \tag{5.51}$$

$$\approx \frac{\Sigma A_l s}{100 d_{bl}} \frac{f_y}{f_{yh}}$$

The significance of ΣA_l is explained with reference to Fig. 5.33. The corner bar A_{l1} is restrained in all directions by the peripheral hoop. Thus in computing A_{tr1}, Eq. (5.51) applies with $\Sigma A_l = A_{l1}$. The inner hoops of Fig. 5.33(a) restrain three longitudinal bars against outward buckling; hence A_{tr2} is calculated based on $\Sigma A_l = A_{l2} + \frac{1}{2} A_{l3}$. In Fig. 5.33(b), an octagonal hoop is used to reduce congestion, but this results in reduced restraint of outward buckling, since the maximum restraint force perpendicular to the column face is $A_{tr2} f_{yh}/\sqrt{2}$. It thus follows that since this must restrain $A_{l2} + 0.5A_{l3}$,

$$A_{tr2} = \frac{\sqrt{2} (A_{l2} + 0.5A_{l3})s}{100 d_{bl}} \frac{f_y}{f_{yh}} \tag{5.52}$$

Although it is clear that adequate restraint should be provided against buckling of longitudinal reinforcement in circular columns, this requirement has not in the past been quantified, with the consequence that bar buckling has been common in tests of circular columns, as shown in the example of Fig. 5.32(b). Adopting the same requirements as for rectangular columns, that an equivalent restraint force of $\frac{1}{16} A_l f_y$ at a spacing of $6d_{bl}$ be provided, leads

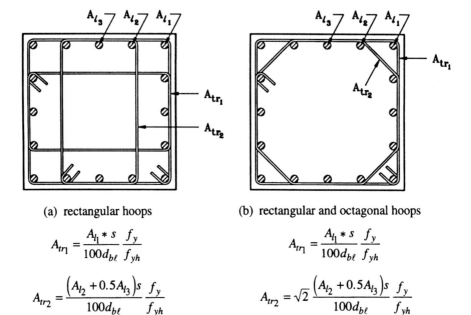

FIG. 5.33 Restraint of longitudinal bar buckling in rectangular columns.

to the requirement that the volumetric ratio of transverse confining steel should be at least

$$\rho_s = \frac{0.0052\,\rho_l D}{d_{bl}}\,\frac{f_y}{f_{yh}} \tag{5.53}$$

It will, however, be recognized that the provision of a restraint force of $\frac{1}{16}$ $A_l f_y$ at $6d_{bl}$ is somewhat arbitrary and cannot be directly related to the mechanics of buckling. A more rigorous approach can be developed by considering the restraint needed to avoid buckling over a critical length involving several hoops of a longitudinal bar in the strain hardening range of axial compression. If the longitudinal bar stiffness is based on a secant modification [M6] to the double-modulus approach, the required level of transverse reinforcement is approximated by

$$\rho_s = \frac{0.45 n f_s^2}{E_{ds} E_t} \tag{5.54}$$

where $E_{ds} = 4E_s E_i/(\sqrt{E_s} + \sqrt{E_i})^2$ is the double modulus of the longitudinal reinforcement at f_s, the axial stress in the bar at buckling, E_t the modulus of elasticity of the transverse reinforcement, E_i the elastic modulus of the longitudinal reinforcement, and E_s the secant modulus from f_s to f_u, the ultimate stress. For grade 60 reinforcement, taking $f_s = 74$ ksi (510 MPa) corresponding to an axial compression strain of 4% based on f_{ye} = 66 ksi (455 MPa) and assuming that E_t is the elastic value of 29,000 ksi (200 GPa), E_{ds} is found to be 657 ksi (4530 MPa) and Eq. (5.54) reduces to

$$\rho_s = 0.00013n \tag{5.55}$$

That is, the only significant parameter is the number of longitudinal bars, n.

Equations (5.54) and (5.55) are based on equilibrium of the longitudinal bar under the P–Δ effects of an assumed displaced shape and the elastic restraining forces of the hoops. Thus the effect of the outward pressure of the confined core to increase the tendency of the longitudinal bar to buckle has not been considered. It is thus apparent that the full yield strength of the hoops cannot simultaneously be used to provide confinement to the concrete core and to restrain the longitudinal steel against buckling. As a consequence, in the absence of definitive analyses describing the interaction between confinement and buckling restraint, it is recommended that the amount of transverse reinforcement required to restrain buckling be increased by 50% above that given by Eq. (5.54) or (5.55). For grade 60 reinforcement or the equivalent, it is thus recommended for design purposes that

$$\rho_s \geq 0.0002n \tag{5.56}$$

However, in members with low aspect ratio, it appears that the decrease in compression stress in the longitudinal bars with distance from the critical section is too rapid for the buckling mode represented by Fig. 5.32(b) to develop. Consequently, the requirements of Eq. (5.56), which can be onorous when large numbers of reinforcing bars are present, need not be applied to columns with aspect ratio of $M/VD < 4$.

The amount of transverse reinforcement required for confinement of the concrete core, defined by Eq. (5.47), has already been increased to account for the foregoing interaction between confinement and antibuckling. Thus for circular columns it is recommended that transverse reinforcement should satisfy Eq. (5.47) [or the amount required by Eqs. (5.40), (5.42), and (5.14)] and also the amount required by Eq. (5.56).

(iii)   Extent of Confinement: Plastic End Regions. There has traditionally been confusion between the equivalent plastic hinge length and the plastic end region. The former is the mathematical approximation over which plastic curvature is assumed constant when calculating plastic rotation [e.g., Eq. (5.39)]. The latter is the length over which special detailing requirements such as enhanced confinement should extend. The plastic end region depends on the axial load ratio and the length of column subjected to inelastic action. It is essential that the full length of column over which cover spalling might develop should be subject to full confinement. The following conservative rules included in the New Zealand Concrete Code [S2] are thus recommended:

- For axial load ratios of $P/f'_{ce}A_g \leq 0.3$, the plastic end region shall be the greater of (1) the section dimension in the direction considered, or (2) the region over which the moment exceeds 80% of the maximum moment. The meaning of the requirements above is illustrated in Fig. 5.34.
- For axial load ratios $P/f'_{ce}A_g \geq 0.3$ the plastic end region defined above should be increased by 50%.

Where shear requirements govern transverse reinforcement design for plastic end regions, a longer end region is required, in accordance with Section 5.3.4(b)(iii).

(iv) Detailing Requirements for Confinement. Special care must be taken with detailing transverse confinement reinforcement. Since cover concrete will be expected to spall off during design-level seismic response, it is essential that the cover not be relied upon for anchorage. Consequently, hoops (circular or retangular) should be closed by full-strength lap welds or by overlapping and bending the ends around longitudinal bars and back into the core with at least a 135° hook and a straight bar extension at the end of the hook of at least $8d_{bh}$, where d_{bh} is the diameter of the hoop reinforcement. Laps of spiral confinement should be treated similarly.

Spacing of hoop sets or spirals up the column axis should satisfy Eq. (5.50) but should not exceed $D/5$ (or $h/5$ for rectangular columns). With larger spacing, the effectiveness of confinement of the column core will be reduced. However, it will generally be found that with typical transverse reinforcement bar diameters, much smaller spacings will be required to satisfy Eqs. (5.47), (5.48), and (5.56).

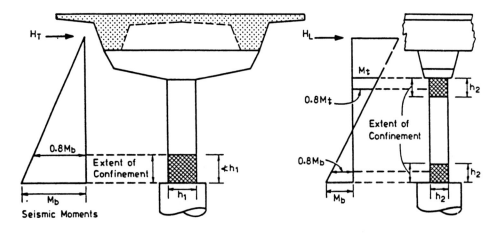

(a) Transverse Response   (b) Longitudinal Response

FIG. 5.34 Plastic end regions for confinement of columns [S2].

Column longitudinal reinforcement should be distributed essentially uniformly around the section, to aid confinement. Where a rectangular distribution of reinforcement is used, a bend of at least 45° in a transverse hoop should restrain at least every other longitudinal bar (see Fig. 5.33), and bars not directly restrained by a hook should not be farther than 150 mm (6 in.) from a hook.

An exception to the desirable recommendations above needs to be made for rectangular piers, where the section shape approximates that for structural walls (see Fig. 3.17, section G–G, for example). In such piers, ductility capacity in the longitudinal (i.e., weak) direction will typically be adequate without special transverse ties. Tests [P15] have indicated displacement ductility capacities of about $\mu_\Delta = 6$ before spalling and strength degradation occurs. Consequently, in the central portion of the pier, only nominal transverse reinforcement, linking the reinforcement layers through the pier thickness, needs to be provided.

In the transverse direction, special confinement reinforcement will be required at each end of the pier to provide the required ductility. The length over which this extends should not be less than the calculated depth of the compression zone. Although axial load levels on piers are typically low, it is clearly inappropriate to use Eq. (5.48) directly for the amount of transverse reinforcement needed perpendicular to the major principal axis when the extent of the confinement being considered is the compression zone rather than the full wall length. Consequently, within the compression zone it is recommended that Eq. (5.48) be used, with an effective axial load ratio of $P/f'_c A_g = 0.4$, regardless of the actual value. This would imply similarity in the end regions of the pier to a rectangular section about 2.5 times the compression zone depth. Alternatively, the required volumetric ratio of confinement can be found from the direct approach, related to required plastic rotation θ_p, described in Section 5.3.2(c)(i), and placing at least 50% of this requirement both transversely and longitudinally in the pier section. Where interlocking spirals are used, the center-to-center spacing of the spirals should not exceed $0.75D'$, to avoid large areas of unconfined cover concrete.

EXAMPLE. A circular column with $D = 72$ in. (1860 mm) sustains a design axial compression force of $P = 4232$ kips (18.8 MN) and is reinforced with 36 No. 14 bars ($d_{bl} = 43.0$ mm) of grade 60 reinforcement ($\rho_l = 0.0199$). Specified concrete strength is $f'_c = 4.0$ ksi (27.6 MPa). Calculate the amount of transverse reinforcement required by a force-based design, and compare with the requirements for a displacement-based design using a maximum plastic rotation of 0.035 rad and a plastic hinge length of 50 in. (1270 mm).

Force-based (prescriptive) design: design material strengths are

$$f'_{ce} = 1.3 f'_c = 5.2 \text{ ksi (35.9 MPa)}$$

$$f'_{ye} = 1.1 f_y = 66 \text{ ksi (455 MPa)}$$

The axial load ratio is

$$\frac{P}{f'_{ce}A_g} = \frac{4232}{5.2 \times 4069} = 0.20$$

From Eq. (5.47),

$$\rho_s \geq 0.16 \times \frac{5.2}{66}(0.5 + 1.25 \times 0.2) + 0.13(0.0199 - 0.01) = 0.0107$$

is required for concrete confinement. From Eq. (5.56), $\rho_s \geq 0.0002n = 0.0002 \times 36 = 0.0072$ is required for antibuckling, with a maximum spacing of $s = 6 \times 1.69 = 10.1$ in. (258 mm). Hence concrete confinement governs. Now $\rho_s = 4A_{sp}/D's$. Using a No. 7 (22.2 mm diameter) spiral, $A_{sp} = 0.601$ in^2 (388 mm^2), the required spacing is $s = 4 \times 0.601/(70.88 \times 0.0107) = 3.2$ in. < 10.1 in. Therefore, use a No. 7 (22.2 mm) spiral at $s = 3.2$ in. (89 mm).

Displacement-based design for $\theta_p = 0.035$: Plastic curvature

$$\phi_p = \frac{\theta_p}{L_p} = \frac{0.035}{50} = 0.0007 \text{ in}^{-1} (0.0276 \text{ m}^{-1})$$

Assume that $\phi_y \approx 0.05\phi_p$; therefore, $\phi_u = 0.000735$ in^{-1} (0.0289 m^{-1}). From a trial computer section analysis with $\rho_s \approx 0.008$, the neutral-axis depth from the outside of the confined core is $c_u = 21.9$ in. (556 mm). Hence the required extreme fiber compression strain, from Eq. (5.42), is

$$\varepsilon_{cu} = 0.000735 \times 21.9 = 0.0161$$

From Eq. (5.14), the required transverse reinforcement for concrete confinement is

$$\rho_s = (\varepsilon_{cu} - 0.004)\frac{f'_{cc}}{1.4 f_{yh}\varepsilon_{su}}$$

With $f_{yh} = 66$ ksi (455 MPa), $\varepsilon_{su} = 0.12$ for grade 60 reinforcement, and $f'_{cc} = 1.3 f'_{ce}$ [from Eqs. (5.6) and (5.4), with a trial value of $\rho_s = 0.008$],

$$\rho_s = (0.0161 - 0.004) \times \frac{1.3 \times 5.2}{1.4 \times 66 \times 0.12}$$

$$= 0.0074$$

This is marginally more than the $\rho_s = 0.0072$ required by Eq. (5.56) for antibuckling but 33% less than required by the prescriptive approach. This can be satisfied using No. 7 (22.2 mm) spirals at s = 4.58 in. (say, 4.5 in. = 114 mm).

(d) Serviceability Considerations. There is a trend with seismic codes toward specification of a two-level design process, with consideration of a serviceability limit state as well as the damage control limit state. The serviceability limit state may be of particular importance for lifeline structures, where a high degree of assurance of continued functionality is required in the immediate post-earthquake rescue phase. There is also, however, growing concern that "normal" structures should not need repair after moderate earthquakes that might be expected to occur once or twice in the design life of the structure.

Various definitions of what physically constitutes the serviceability limit state have been used. Perhaps the most common is to restrict reinforcement strains to be less than or equal to the yield strain, and also place an upper limit, typically about ε_c = 0.002, on concrete strains. Although this may be computationally convenient in that a simple elastic strength check is carried out without considerations of ductility being applied, there are serious defects to this definition. First, it should be appreciated that the limits, $\varepsilon_s \le \varepsilon_y$ and $\varepsilon_c \le 0.002$, do not constitute the onset of damage. A bridge structure exercised by seismic response to these limits would develop seismically induced crack widths that would probably be undetectably to the naked eye after the earthquake. Significantly greater peak response could be sustained before the initiation of concrete spalling or the development of unacceptable residual crack widths.

Second, elastic design to limited strains is meaningless as a result of strain-induced stresses due to such effects as settlement, creep, and shrinkage, particularly for compression members, and redistribution of gravity-load effects due to shakedown. Since these effects will not normally be included in the serviceability assessment, it follows that the accuracy with which a limit state based on first yield can be identified will be extremely low.

Third, specification of limited elastic strains provides uneven protection against damage in terms of probability of occurrence. This is due to the influence of additional flexibility aspects, discussed in Section 5.3.1(b). In a structural system with no additional flexibility, the onset of damage would typically be at seismic excitation levels about 2.5 to 4 times that corresponding to the elastic limit, while a structure with high additional flexibility may suffer significant spalling of cover concrete at an excitation level 1.5 times that corresponding to the elastic limit.

Fourth, setting the criteria for the serviceability limits state artificially low requires that the level of excitation corresponding to the serviceability earthquake must also be set artificially low to avoid serviceability aspects dominating design. Thus the probability of occurrence of the serviceability-level earthquake will not correspond to the probability of reaching the true serviceability limit state. This has significance to seismic risk analyses.

A more reasonable approach to serviceability design would be to use more realistic limit states. There should be no crushing of the concrete, and crack widths should remain acceptably small, not requring remedial action in the form of, for example, epoxy injection grouting. Calculations based on an acceptable residual crack width of 0.04 in. (1 mm) after the earthquake suggest that at maximum response, the tensile steel strain should not exceed 0.015. For allowable concrete strain, a value of $\varepsilon_c = 0.004$ at peak response may conservatively be accepted. It will be noted that these correspond to suggested design strength criteria in Section 5.3.1(d)(viii). The plastic rotations associated with these limit states can readily be determined by moment–curvature analysis and hence be incorporated in a plastic mechanism analysis. Consequently, the displacement and hence structural ductility corresponding to the serviceability limit state can thus be identified. This can then be related to the serviceability excitation level either directly through the displacement spectra, if displacement-based design is used, or by use of appropriate relationships between ductility and force-reduction factors [Eq. (5.22)] and relating response to the serviceability acceleration spectrum, if force-based design is used.

It will be clear that checking serviceability will require little extra work if a plastic mechanism analysis is carried out for the damage control limit state. It is suggested that serviceability aspects need to be considered only for important structures, and for these it will be advisable to carry out a plastic mechanism analysis regardless of whether force- or displacement-based design is adopted.

5.3.3 Required Strength of Capacity-Protected Actions

To ensure that inelastic deformations occur only in designated and properly detailed plastic end regions, it is necessary to determine the maximum feasible moment capacity of the plastic hinges and to design the rest of the structure for the actions corresponding to gravity loads plus this flexural overstrength. Generally, the basis for this is that all plastic hinges develop their full plastic capacity, although more realistic values will naturally result from a plastic collapse mechanism analysis or inelastic time-history analyses. In these analyses, strain-hardening effects and material overstrength can be modeled directly if desired. Once the demand on the capacity-protected members has been determined, the sections are designed conservatively to ensure essentially elastic response results. By this means, unexpected plastic hinge formation and undesirable inelastic deformation modes, such as shear failure of members or joints, are effectively proscribed.

(a) Standard Prescriptive Force-Based Design. For routine structures designed in accordance with the force-based approach of Section 5.3.1(*a*), it is desirable that overstrength computations be as simple as possible. Recent Caltrans practice has been to design for actions corresponding to a plastic

mechanism with all potential plastic hinges at an overstrength moment capacity of

$$M_p = \phi° M_{ACI} \tag{5.57}$$

where the flexural overstrength factor is uniformly taken as $\phi° = 1.3$. As discussed in relation to Fig. 5.19, this overstrength factor will often provide insufficient protection, and a value of about $\phi° = 1.7$ would be needed to provide protection for the full range of axial load and reinforcement ratios.

Using the design recommendations for flexural strength given in Section 5.3.1(d)(vii), lower overstrength factors are appropriate. Figure 5.35 plots

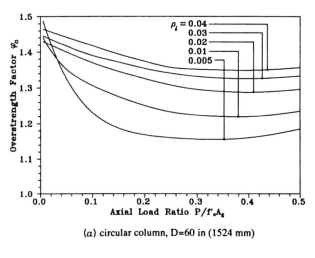

(a) circular column, D=60 in (1524 mm)

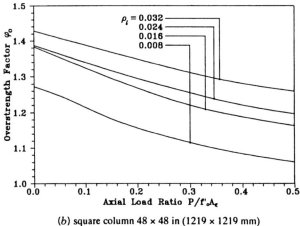

(b) square column 48 × 48 in (1219 × 1219 mm)

FIG. 5.35 Maximum overstrength factor $\phi°$ for columns designed to Section 5.3.1(d)(vii) using Grade 60 reinforcement.

computed overstrength factors for circular and rectangular columns reinforced with grade 60 reinforcement (f_y nominal = 414 MPa), based on a specified compression strength of f'_c = 4 ksi (27.6 MPa). In accordance with Section 5.3.1(d)(vii), design material strengths are f'_{ce} = 1.3 × 4 = 5.2 ksi (35.9 MPa), f_{ye} = 66 ksi (455 MPa), and ε_c = 0.004, Flexural overstrength has been computed at maximum ductility capacity, using f'_{co} = 1.7 × 4 = 6.8 ksi (46.9 MPa) and f_{yo} = 78 ksi (538 MPa). Where appropriate, strain hardening was included in both design strength and overstrength.

It is apparent from Fig. 5.35 that the maximum overstrength factor is strongly influenced by both axial load and reinforcement ratio. For standard design a value of $\phi°$ = 1.45 could be adopted uniformly, but it is clear that this would be overly conservative for many practical conditions. Thus Fig. 5.35 could be used directly or results from column-specific moment–curvature analyses adopted.

Although Fig. 5.35 has been computed for specific column dimensions, and f'_c = 4.0 ksi (27.6 MPa), the results for $\phi°$ are insensitive to column size or f'_c for normal variations and may thus be used for other columns. The results are, however, strongly influenced by the stress–strain characteristics of the reinforcing steel and would not be appropriate for (say) tempcore steel.

It should be emphasized that the values of Fig. 5.35 are maximum values that will not necessarily be achieved in every case. Lower values may be justified based on the approaches outlined in Sections 5.3.3(b) and (c).

Member forces must be determined based on appropriate combination of seismic response and gravity loads. It may be necessary to consider a reasonable range of gravity loads to ensure that maximum feasible conditions are met. For example, Fig. 5.36 shows moment profiles in the cap beam of a two-column bent based on dead load and on dead load plus live load. Although the peak moments in the cap beam will probably be influenced only slightly

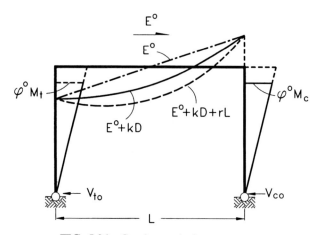

FIG. 5.36 Cap beam design moments.

by the live load, the moment pattern, and particularly the points of contraflex-ure, may be affected significantly. This must be considered when locations of termination of reinforcement are decided.

It is also appropriate to include some variation in dead load as a conse-quence of uncertainties in absolute dead-load values and because of the possi-bilities of vertical acceleration response. In line with the recommendations of Section 5.3.1(d)(x), it is suggested that the gravity loads to be considered in combination with seismic forces corresponding to plastic moment capacity be in accordance with the following ultimate load combinations:

$$U_1 = E° + (1 - 0.5\text{PGA})D \qquad (5.58a)$$

$$U_2 = E° + (1 + 0.5\text{PGA})D + rL \qquad (5.58b)$$

where $E°$ are the actions corresponding to overstrength (plastic) capacity at the hinges and r is a reduction factor applied to the design live load to provide a reasonably conservative estimate of expected live load occurring in conjunction with seismic actions. The appropriate value of r will depend on span length, traffic density, number of lanes, and importance of the structure. In the absence of specific guidelines in local codes, it is recommended that $r = 0.2$ be adopted.

It must be appreciated that the actions induced by dead load and live load do not correspond to those predicted by elastic analysis under gravity loads alone. The forces are those corresponding to gravity load applied to the base structure modified by removing rotational constraints at positions of seismic plastic hinging. This is illustrated in Fig. 5.36, where formation of plastic hinges at the column tops results in gravity loads being applied to the cap beam as though it were simply supported. That is, gravity loads do not influence the column moments.

A common error in cap beam design may also be illustrated in relation to Fig. 5.36. Many designers ignore the seismically induced axial forces in cap beams. Equilibrium requirements must, however, be met, and hence the critical cap beam section at the face of the tension (left) column will be subjected to a tension force of $T = V_{to}$, where V_{to} is the column shear corresponding to overstrength moment capacity at the left column plastic hinge. Similarly, an axial compression force of $C = V_{co}$ will be induced at the right end of the cap beam. These axial forces, which are commonly ignored, may increase or decrease the moment capacity by as much as 20% in typical designs, and must be considered.

If the columns are not located, as shown in Fig. 5.36, at the ends of the cap beam, the axial force in the cap beam may require more analysis to determine the appropriate value. This may be done simply by assuming the total shear force $(V_{to} + V_{co})$ to be uniformly distributed over the length of the cap beam as a shear flow $v = (V_{to} + V_{co})/L$ and taking normal considerations of equilibrium.

(b) *Actions Determined by Displacement-Based Design.* Where displacement-based design is adopted, the plastic rotations of critical plastic hinges will be determined by a plastic mechanisms analysis, as outlined in Section 4.5.2(*b*)(ii). Such an analysis would normally be carried out using design-level estimates of material strength and would thus not be directly applicable to determination of required strengths of actions to be protected by application of capacity design principles. The analyses would, however, incorporate actual expected overstrength resulting from strain hardening at the expected plastic hinge rotations. As a consequence, a lower overstrength factor is appropriate than that suggested for force-based design in the preceding section. Thus, when design is based on expected material strengths in accordance with Section 5.3.1(*d*)(vii), and strain hardening is included in the moment–curvature analyses, it is recommended that $\phi_0 = 1.15$ be used to determine maximum feasible seismic actions. These would then be combined with gravity loads in accordance with Eq. (5.58).

More consistent results will be obtained if additional plastic collapse analyses are run based on maximum expected material properties, using $f'_{co} = 1.7 f'_c$ and $f_{yo} = 1.3 f_y$, for grade 60 reinforcement. Two analyses should be run, corresponding to the two load combinations of Eq. (5.58). With this approach, maximum overstrength actions are obtained directly and no further overstrength factors are needed.

(c) *Actions Determined by Inelastic Time-History Analyses.* A time-history analysis, if used, is likely to be carried out with the purpose of determining maximum plastic hinge rotations and structure displacements and hence will be based on lower estimates of material strength. Simple multiplication by an overstrength factor and combination with gravity-load forces would seem inappropriate, since peak plastic moments at different hinges may not occur at the same time instant. Consequently, if capacity actions are to be determined by time-history analysis, a second analysis should be run, using section strengths at plastic hinges based on $f'_{co} = 1.7 f'_c$ and $f_{yo} = 1.3 f_y$, for the two ultimate load combinations represented by Eq. (5.58).

5.3.4 Design Strength of Capacity-Protected Actions

In Section 5.3.1(*d*), the basic capacity design equation for capacity-protected actions was given by Eq. (5.26) as $\phi_s S_n \geq S_r = S^\circ$. That is, the dependable strength of the protected action should exceed the maximum demand corresponding to flexural overstrength in the plastic hinges. Traditionally, capacity design practice, as developed and implemented in New Zealand [P3], has adopted $\phi_s = 1.0$ in Eq. (5.26) on the basis that adequate conversatism exists in the process used to determine the overstrength action, S°. It is, however, clearly more consistent to adopt appropriate values for ϕ_s when there is any uncertainty associated with the ideal strength. This approach is adopted in the following.

(a) Flexural Strength. It could reasonably be suggested that the approach advanced in Section 5.3.1(d)(vii) for design flexural strength of plastic hinges could also be adopted for design flexural strength of sections to be protected against hinging, provided that a flexural strength reduction factor of $\phi_f = 0.9$ is incorporated. This implies the use of material strengths exceeding specified minima but would produce design strengths essentially equal to the minimum ideal capacity corresponding to specified minima material strengths, as a result of factoring down by the strength reduction factor.

For members, such as columns containing plastic hinges, this would clearly by appropriate, since the same reinforcing bars are generally present in the plastic hinge zone as in the remainder of the member, and it would be inappropriate to assign two different yield strengths to the same bar. Also, the consequences of unexpected flexural yielding at sections not designated as plastic hinges will not be serious. Since design actions are based on overstrength in plastic hinges, these will occur only at the limits of plastic deformation, if at all, and ductility demand on the unanticipated hinges will be minimal.

However, consequences of adopting a higher-than-minima material strengths for capacity-protected actions would be more serious when brittle actions are considered, particularly shear strength, which is rather strongly dependent on concrete compression strength. It might thus be unwise to rely on $f'_{ce} = 1.3 f'_c$. It is desirable that a consistent approach be adopted for all capacity-protected actions. It is also desirable that the same procedures be used for determining design strengths of capacity-protected members for actions resulting from seismic and nonseismic (e.g., gravity) load combinations.

Consequently, it is recommended that specified material strengths be adopted for determining design flexural strength of capacity-protected members. It will be noted that in many countries, the specified material strengths are characteristic lower 5 percentile values, and the flexural strength reduction factors adopted should reflect the possible material strength variations. It is suggested that an ultimate compression strain of $\varepsilon_{cu} = 0.004$ be used for design strength of all flexural actions.

(i) Columns with Plastic Hinges. For columns with designated plastic hinges, special consideration is required. Clearly, it is not feasible to require the design strength immediately adjacent to plastic hinges to satisfy Eq. (5.26), particularly when tension-shift effects are considered, since this would imply that a requirement for additional longitudinal reinforcement exists outside the hinge region, whereas the designer might naturally consider terminating reinforcement because of reduced moment.

This conflict is resolved in Fig. 5.37, where design moment profiles for simple vertical cantilevers and columns subjected to double bending are considered. Since the same reinforcing steel is used within and adjacent to the plastic hinge, it is only necessary to consider overstrength resulting from strain hardening. Conservatively, this may be taken as 20% for steels with $f_u \approx 1.5f_y$. Moment profiles 1 and 2 in Fig. 5.37 thus correspond to design strength

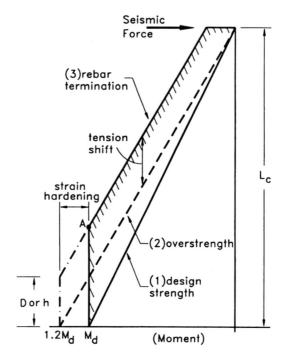

(a)   Simple   cantilever

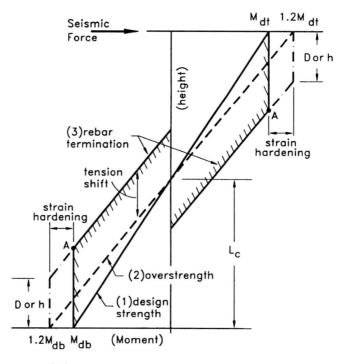

(b)   Reversed   bending

FIG. 5.37 Moment profiles for termination of column longitudinal reinforcement.

and overstrength, respectively, based on the *actual* yield strength of column reinforcement.

Inclined flexure–shear cracking in columns results in tension shift, where tensile reinforcement stress exceeds that corresponding to the section moment based on the plane-sections hypothesis. The extent of tension shift depends on the angle of the inclined flexure–shear cracks to the column axis and the amount of transverse reinforcement within the hinge region. A full discussion of tension-shift effects is given in [P4].

For columns, the angle of flexure–shear cracking to the column axis may be taken as 30°, as discussed in Section 5.3.4(*b*), and the tension shift may be taken as $0.5D\cot 30°$ or $0.5h\cot 30°$, where D is the column diameter and h the section depth for a rectangular column. This results in a tension shift of $0.87D$ or $0.87h$. In Fig. 5.37 this is conservatively rounded up to D or h, and the moment profile corresponding to overstrength is thus displaced vertically by D or h to provide profile 3, which may be used for termination of column longitudinal reinforcement. Thus all reinforcement is required at full strength up to heights designated as A, by the intersection of the design strength M_d and the tension-shift profile 3. Any reinforcement terminated in the column must extend past profile 3 by at least the development length l_d. This implies that the minimum length for terminating column reinforcement is a distance $l_{t,min}$ from the critical section, where

$$l_{t,\text{min}} = 0.167L_c + (D \text{ or } h) + l_d \qquad (5.59)$$

where L_c is the distance from the critical section to the point of contraflexure.

The moment profiles of Fig. 5.37(*b*) are based on the assumption that the cap beam or superstructure, for longitudinal response, is relatively rigid. A similar profile may develop for transverse response of a cantilever column with rigid connection to a horizontally curved superstructure. In this case the point of contraflexure may move up and down the column depending on the amount of moment contributed by different modes of vibration. In such cases a variation in the position of contraflexure should be considered when terminating reinforcement. In all cases of reversed bending it is recommended that the longitudinal reinforcement ratio not be reduced below 50% of the value at the critical sections. In many cases it will not be found to be economical to reduce reinforcement at all, because of the large development lengths required.

Very tall columns may have significant self-weight moments induced as a consequence of inertial response of the column vibrating as a free beam of constant distributed mass. These effects should be considered in determining the design strength envelope and hence the points of rebar termination based on overstrength and tension shift.

(ii)   Other Members. Members other than columns with designated plastic hinges should also be designed for tension-shift effects by carrying longitudinal reinforcement a distance $d + l_d$ past the location where it is required at full

strength, but not less than a distance of $12d_b$ past the location where it is required to provide any contribution to flexural strength, in accordance with standard provisions incorporated in most codes, such as [A5,S2].

As discussed in Section 5.3.3(a), design of cap beams should incorporate the effects of seismic axial forces, since these have a significant influence on longitudinal reinforcement requirements. Consideration should also be given to the way in which longitudinal cap beam reinforcement is distributed through the section. Traditionally, this has been achieved by concentrating the reinforcement in two groups, each at maximum distance from the centroidal axis, as shown in Fig. 5.38(a). This frequently has the effect of requiring several layers of rebar at the top and bottom of the section, and of creating considerable congestion. There is absolutely no reason for choosing the distribution of Fig. 38(a), which is based solely on our experience with one-way bending of beam sections, which implies greatest design efficiency from reinforcement

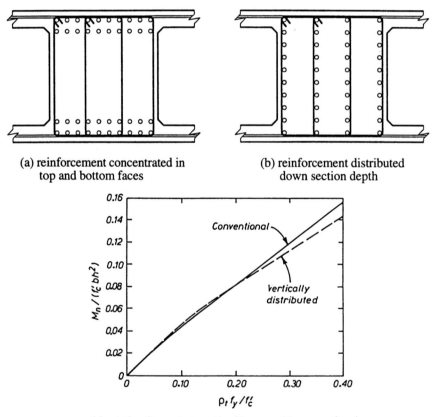

(a) reinforcement concentrated in top and bottom faces

(b) reinforcement distributed down section depth

(c) relative flexural strength of beams with conventional and vertical distribution longitudinal reinforcement [W4]

FIG. 5.38 Influence of distribution of longitudinal reinforcement on flexural strength of cap beam.

placed at maximum distance from the compression face. When the section is subjected to reversal of bending moment, requiring reinforcement in opposite faces, the compression reinforcement has very little effect on the bending moment capacity. If the total longitudinal reinforcement of Fig. 5.38(a) is distributed down the sides of the section, as shown in Fig. 5.38(b), the bending moment capacity is almost identical to that for the concentrated case of Fig. 5.38(a). Figure 5.38(c) compares theoretical flexural strength for the two cases as a function of the mechanical longitudinal reinforcement ratio $\rho f_y / f'_c$ [W4]. As will be seen, the difference is negligible.

Choosing the distributed reinforcement option of Fig. 5.38(b) has the advantage of reducing congestion, controlling side face shrinkage cracking, improving member shear strength and shear resistance through column–beam intersections, and is preferred over the more conventional concentrated reinforcement. In practice, a solution intermediate between the two cases of Fig. 5.38 is recommended, with more reinforcement in top and bottom faces, as this has the further advantage of enhanced torsional resistance.

(iii) *Summary: Flexural Strength.* Except for columns containing designated plastic hinges, design flexural strength of capacity-protected sections should be based on the following requirements:

1. Required strength should be based on Eq. (5.58).
2. Nominal strength should be based on nominal (specified) material strengths f'_c and f_y and an ultimate compression strain of $\varepsilon_u = 0.004$.
3. A flexural strength reduction factor of $\phi_f = 0.9$ should be used to determine dependable (design) flexural strength of capacity-protected members.
4. Axial forces must be included in the design.
5. Longitudinal reinforcement should be distributed through the section.

For columns containing designated plastic hinges, reinforcement termination should be based on considerations of strain hardening and tension shift, in accordance with Fig. 5.37.

(b) *Shear Strength.* There is still a wide divergence of opinions, design approaches, and code equations related to the shear strength of reinforced concrete members. This is particularly the case when the influences of axial load and flexural ductility are considered. Axial load is often considered quite differently when it results from gravity effects or from prestress, even within the same code, such as ACI 318-89 [A5], although it is difficult to rationalize any significant difference in the effects on the concrete member. It was mentioned in Chapter 1 that evidence from bridge columns failing in shear under seismic attack indicates that shear strength within plastic end regions is reduced with increasing flexural ductility. This effect is frequently ignored in codified

design equations or is at best treated in simplistic fashion, by discounting the concrete contribution to shear strength in plastic end regions when the axial load is less than $P = 0.05f'_cA_g$ [A5] or $0.1f'_cA_g$ [S2].

The divergence between different shear strength design equations is a result of the considerable scatter between predicted and observed shear strengths for all the methods. Each can claim as good (or bad) a fit with the experimental data as the others, even though each may result in markedly different estimates in strength for a specific design case. A consequence of this lack of fit with experimental data is that conservative lower limits must be adopted, which may, on occasions, be excessively conservative.

Recently, however, there have been significant improvements in the understanding of shear in reinforced concrete members. Integrated design approaches such as compression field theory [C4] seek to avoid the artificial separation of flexure and shear adopted for convenience of design in most current design approaches. The influence of flexural ductility in reducing aggregate interlock shear transfer across wide cracks in the plastic hinge regions of members is also much better quantified now, after extensive experimentation [A6,W3,P11].

The approach taken in this section is to present the most recently developed design approaches for shear strength of reinforced concrete members and to compare with more commonly accepted approaches. Clearly, a complete analysis of existing and new approaches is beyond the scope of this book and the interested reader is referred elsewhere to more complete presentations [P11,P12]. Although compression field theory shows much promise, it has not yet been extended successfully to cope with degradation of shear strength in plastic end regions and is not considered in this book. It is emphasized, however, that compression field theory can be used with confidence for members that do not contain plastic hinges.

(i)   ASCE-ACI 426 Shear Strength Approach.   Committee 426, a joint ASCE–ACI committee on shear strength of concrete members, has produced design equations based on the additive model

$$V_d = \phi_s(V_c + V_s) \qquad (5.60)$$

where V_d is the dependable (design) shear strength, V_c and V_s the nominal strength of concrete and transverse reinforcement shear resisting mechanisms, respectively, and $\phi_s = 0.85$ is the shear strength reduction factor. The ASCE–ACI design equations [A7] for V_c and V_s have not been fully adopted by ACI 318 [A5] but have been incorporated with some modification to cope with strength reduction in plastic hinge zones into the New Zealand concrete code [S2] and are widely respected as a conservative basis for design.

CONCRETE CONTRIBUTION.   The strength of concrete shear resisting mechanisms is related to a nominal shear stress v_c by the equation

$$V_c = v_c A_e \qquad (5.61)$$

where A_e is the effective shear area, taken as

$$A_e = b_w d \leq 0.8 b_w h \qquad (5.62)$$

for rectangular sections of web width b_w, effective depth d, and overall depth h. The ASCE–ACI 426 recommendations imply that $A_e = 0.8D^2$ for circular columns of diameter D. This exceeds the gross area, and it is recommended that $A_e = 0.8A_{\text{gross}} = 0.628D^2$ be adopted in conjunction with these recommendations.

The nominal shear stress is

$$v_b = \begin{cases} (0.066 + 10\rho_t)\sqrt{f_c'} \leq 0.2\sqrt{f_c'} & \text{(MPa)} & (5.63) \\[2mm] (0.8 + 120\rho_t)\sqrt{f_c'} \leq 2.4\sqrt{f_c'} & \text{(psi)} & (5.64) \end{cases}$$

where $\rho_t = A_s/b_w d$ is the tension steel ratio, taken as $0.5\rho_t$ for columns.

The presence of axial load modifies v_c such that

$$v_c = \begin{cases} v_b\left(1 + \dfrac{3P}{f_c' A_g}\right) & \text{(axial compression)} & (5.65) \\[4mm] v_b\left(1 + \dfrac{P}{500 A_g}\right) & \text{(axial tension) (psi)} & (5.66) \end{cases}$$

where P is taken negative for tension.

Within plastic end regions, the New Zealand Concrete Code, which adopts the foregoing approach, reduces v_c in accordance with the following provisions:

$$v_c = \begin{cases} 0 & \text{(for } P \leq 0.1 f_c' A_g) & (5.67a) \\[4mm] 4v_b\sqrt{\dfrac{P}{f_c' A_g} - 0.1} & \text{(for } P \geq 0.1 f_c' A_g) & (5.67b) \end{cases}$$

TRANSVERSE REINFORCEMENT CONTRIBUTION (TRUSS MECHANISM). It is assumed that transverse reinforcement contributes to stabilizing diagonal compression struts at $\theta = 45°$ to the member axis to produce a strength

$$V_s = \frac{A_v f_{yh} d}{s} \qquad (5.68)$$

where A_v is the total area of transverse reinforcement in a layer in the direction of the shear force, s the spacing along the member axis of the layers of stirrups or hoops, and d the effective depth, again taken as $0.8h$ for columns with longitudinal reinforcement distributed around the section.

The appropriate definition of A_v for rectangular columns may be determined in relation to Fig. 5.39(a), which shows a column reinforced with 12 longitudinal bars and two possible arrangements of transverse reinforcement. Both include a peripheral hoop, but in one alternative the internal eight longitudinal bars are confined by an octagonal hoop, whereas in the other they are confined by two independent rectangular hoops, shown by dashed lines.

Considering a flexure–shear crack inclined at $\theta = 45°$ to the column axis, as shown, the resisting force crossing the crack depends on whether the crack crosses the hoop layer in the outer regions, indicated by lines 1–1, or in the center region, indicated by line 2–2. The peripheral hoop contributes fully at all sections, but in the outer sections the yield force in the octagonal hoop is

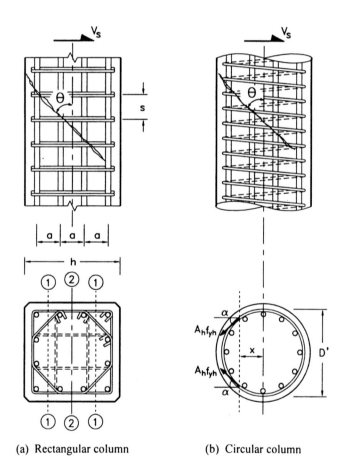

(a) Rectangular column   (b) Circular column

FIG. 5.39 Effectiveness of transverse reinforcement for shear resistance of columns.

at 45° to the direction of applied shear force and must be resolved back to the line of action of V_s. Thus, considering first the case where the section is reinforced by the peripheral and octagonal hoops, the resisting force F_v parallel to the applied shear force will be
Octagonal case, section 1–1:

$$F_v = \left(2 + \frac{2}{\sqrt{2}}\right) A_h f_y = 3.41 A_h f_{yh} \tag{5.69a}$$

Octagonal case, section 2–2:

$$F_v = (2 + 2) A_h f_y = 4.0 A_h f_{yh} \tag{5.69b}$$

where A_h is the area of one leg of the hoops.

Similarly, for the case with peripheral hoop and overlapping internal hoops, it is seen that four legs of hoop are crossed by the crack at section 1–1 and six legs at section 2–2. Hence:
Rectangular case, section 1–1:

$$F_v = 4 A_h f_{yh} \tag{5.70a}$$

Rectangular case, section 2–2:

$$F_v = 6 A_h f_{yh} \tag{5.70b}$$

Since an inclined crack will cross approximately twice as many hoop sets in the outer (1–1) than the inner (2–2) section, the average effective area of transverse reinforcement for the two cases is thus:
Octagonal case:

$$A_v = \left(\frac{2 \times 3.41 + 1 \times 4}{3}\right) A_h = 3.61 A_h \tag{5.71a}$$

Rectangular case:

$$A_v = \left(\frac{2 \times 4 + 1 \times 6}{3}\right) A_h = 4.67 A_h \tag{5.71b}$$

It can readily be shown that the efficiency of the two alternatives (shear strength divided by volume of steel) is the same. The octagonal case will generally be preferable because of reduced congestion in the core.

The same approach may be adopted to determine the shear strength provided by the two hoop alternatives under a diagonal response. In this case, most of the hoop forces for the octagonal case and all of the hoop forces for rectangular case must be resolved 45° to be parallel with the applied shear force. It is thus found that the effective areas of transverse reinforcement to be used in Eq. (5.68) are $A_v = 2.55A_h$ and $3.31A_h$ for octagonal and rectangular cases, respectively. In each case, these are $1/\sqrt{2}$ times the areas for attack parallel to one of the principal axes. However, since the effective depth d is increased by $\sqrt{2}$ for the diagonal case, the shear strength given by Eq. (5.68) will be the same for diagonal and principal axes. Since flexural strength is typically slightly lower for diagonal response, it follows that in a capacity design approach it is sufficient to check the shear strength for the principal axes. This conclusion applies for rectangular as well as square sections.

In applying Eq. (5.68) to circular sections reinforced with spirals or circular hoops, codes have generally recommended taking $A_v = 2A_h$ and $d = 0.8D$ [A5,S2]. However, it is clear from the discussion above related to resolving hoop forces parallel to the applied shear force that this is inappropriate. In Fig. 5.39(b), the component of hoop force parallel to the applied shear force, exposed by a diagonal flexure–shear crack is

$$F_v = A_h f_{yh} \cos \alpha \tag{5.72}$$

where α increases from 0 to 90° as the distance x from the column axis perpendicular to the applied shear force increases from 0 to $0.5D'$, and D' is the diameter of the spiral or hoop. It can readily be shown [A6] that the total shear resistance, assuming a 45° crack inclination, is thus

$$V_s = \frac{\pi}{2} \frac{A_h f_{yh} D'}{s} \tag{5.73}$$

Although this assumes that the crack traverses the full diameter D' of the core, the error resulting from a compression zone extending into the core is negligible, since the contribution of hoop force to V_s from the regions where x is a maximum is very small.

(ii) Approach of Priestley et al. for Columns. It has been found that the ASCE–ACI 426 approach for shear strength does not provide a particularly good estimate of the shear strength of columns. For low ductility levels, the approach tends to be excessively conservative, while at high ductility levels it is in some cases nonconservative. Figure 5.40 compares the measured [A6,W-3,P11,J1,M7] and predicted shear strength of circular and rectangular columns, where predicted strength is based on the equations above, using measured material strengths and $\phi_s = 1.0$, and the data are organized according to displacement ductility level, axial load level, and column shear span/section depth ratio (M/VD, or aspect ratio). The decreasing conservatism with increasing ductility is apparent, as is the general conservatism and high scatter.

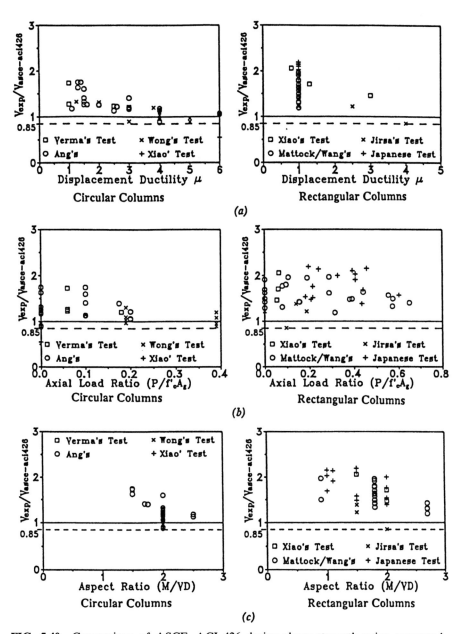

FIG. 5.40 Comparison of ASCE–ACI 426 design shear strength using measured material properties with column test results [P12].

A recently developed approach [P12] provides greatly improved predictions. The ideal shear strength is given by

$$V_d = V_c + V_s + V_p \tag{5.74}$$

where

$$V_c = k \sqrt{f'_c} A_e \tag{5.75}$$

$A_e = 0.8 A_{\text{gross}}$, and k, within plastic end regions, depends on the member displacement ductility μ_Δ, reducing from 3.5 in psi units (0.29 in MPa units) for $\mu \leq 2$ to 0.6 in psi units (0.05 in MPa units) for $\mu \geq 8$, as shown in Fig. 5.41. This implies a considerable increase in V_c for low ductility levels over that permitted by ASCE–ACI 426 or ACI 318-89. For columns subjected to ductile response in two orthogonal axes, the reduction in k begins earlier than under uniaxial response [W3]. For regions of columns outside plastic end regions, the initial value for k pertains.

The truss mechanism strength for circular columns is given by

$$V_s = \frac{\pi}{2} \frac{A_h f_{yh} D'}{s} \cot \theta \tag{5.76a}$$

and for rectangular columns by

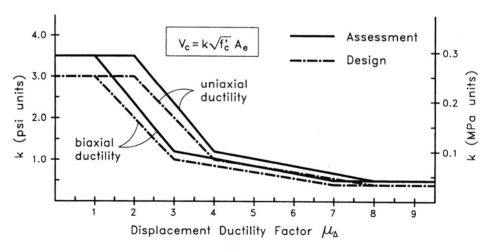

FIG. 5.41 Relationship between ductility and strength of concrete shear-resisting mechanisms [P9].

$$V_s = \frac{A_v f_y D'}{s} \cot \theta \qquad (5.76b)$$

where D' is the core dimension, from center to center of peripheral hoop, for both circular and rectangular columns, as defined in Fig. 5.42. In Eq. (5.76), the angle of the critical inclined flexure shear cracking to the column axis is taken as $\theta = 30°$, unless limited to larger angles by the potential corner-to-corner crack. Equation (5.76a) differs from Eq. (5.73) only by the inclusion of $\cot \theta$, while the effective depth for shear in Eq. (5.76b) is increased from Eq. (5.68), to be consistent with the approach for circular columns.

The development of steeper angles of cracking than the $\theta = 45°$ assumed by the ASCE–ACI 426 approach is well supported by experimental results (see Fig. 5.43, for example), and is accepted in the variable-angle truss mechanism design approach used in some European countries.

The shear strength enhancement resulting from axial compression is considered as an independent component of shear strength, resulting from a diagonal compression strut, as shown in Fig. 5.44, given by

$$V_p = P \tan \alpha \qquad (5.77a)$$

For a cantilever column, α is the angle formed between the column axis and the strut from the point of load application to the center of the flexural compression zone at the column plastic hinge critical section. For a column in reversed or double bending, α is the angle between the column axis and the line joining the centers of flexural compression at the top and bottom of the column. The justification for the foregoing approach is the simple observa-

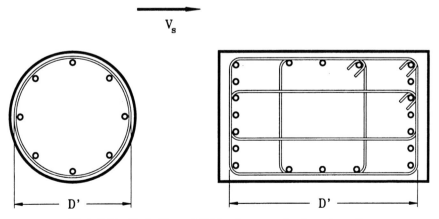

FIG. 5.42 Definition of D' for truss mechanism strength.

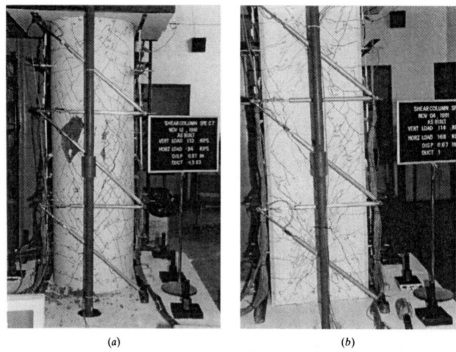

(*a*)	(*b*)

FIG. 5.43 Inclined flexure–shear cracking in columns under cyclic reversals of deformation. (*a*) Circular column; (*b*) Rectangular column.

tion that the axial load must effectively form a compression strut at an angle to the column axis since it must be transmitted through the flexural compression zone, and that the direction of the horizontal component of the force resists the applied shear force.

Equation (5.77) implies that the shear strength of squat axially loaded columns should be greater than that of more slender columns. This is well known from test results. It also implies that as the axial load increases, and hence the depth c of the flexural compression zone increases, the increase in shear strength will become less significant, since α will be reduced. This also agrees with test results. Equation (5.77) may also be used for resultant tensile forces on columns where P is taken negative for tension.

The experimental data used to assess the ASCE–ACI 426 approach in Fig. 5.40 is compared with predictions from Eqs. (5.74) to (5.77) in Fig. 5.45. Considerably improved representation of actual performance is apparent compared with Fig. 5.40. The use of a shear strength reduction factor of $\phi_s = 0.85$ effectively provides a lower limit to the data. It is suggested that the approach above may also be used to assess the expected increase in shear strength imparted to members by prestressing, where again, the angle α is based on distance between critical sections and eccentricity of the centroid of the flexural compression zone.

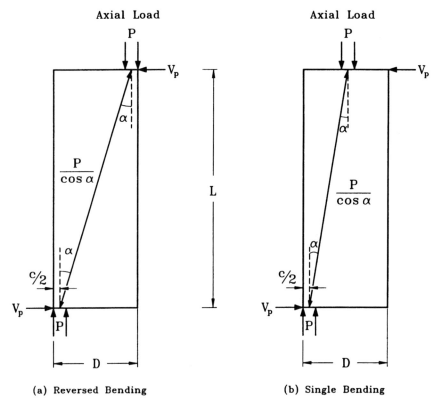

(a) Reversed Bending   (b) Single Bending

FIG. 5.44   Contribution of axial force to column shear strength [P12].

(iii)   Design Recommendations

COLUMNS. Equations (5.74) to (5.77) have been developed as predictive equations, which although showing very small scatter when compared with experimental results nevertheless overpredict shear strength in a small number of cases. Thus although they are appropriate for estimated shear strength of existing bridges, extra conservatism is appropriate for design. As a consequence, it is recommended that an additional 15% conservation be provided by using the following equations for the components of Eq. (5.74).

CONCRETE CONTRIBUTION FOR COLUMNS. For normal-weight concrete, use Eq. (5.75), with a maximum value of $k = 3.0$ (psi units) (0.25 MPa units), reducing with ductility as indicated in Fig. 5.41 by the lines marked "design." Limited test results [K4] indicate that these values should be reduced a further 25% for lightweight concrete columns.

It is emphasized that the displacement ductility of Fig. 5.41 is the member displacement ductility and will generally be somewhat more than the structural

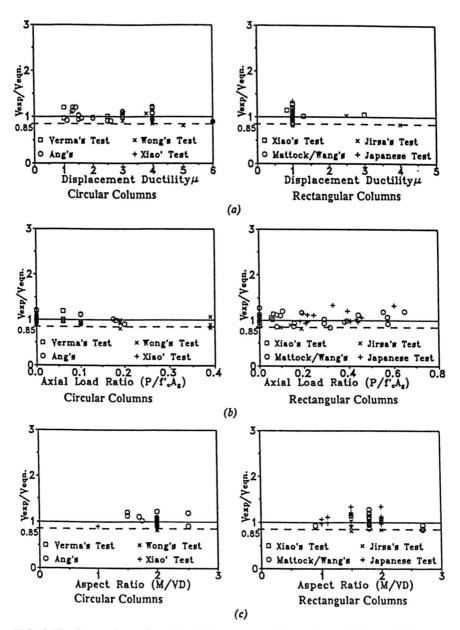

FIG. 5.45 Comparison of predicted shear strength from Eqs. (5.74) to (5.77) with column test results [P12].

displacement ductility factor. The relationships developed in Section 5.3.1(b) may be used to convert from structure to member ductility demand when using force-based design. When using displacement-based design it will generally be more convenient to assess member ductility demand in terms of curvature ductility factor. It would also appear that curvature ductility factor, which is a more meaningful indicator of crack widths, and hence loss of aggregate interlock capacity, would be a more appropriate base than the displacement ductility factor in determining the reduction in shear strength of the concrete shear-resisting mechanism. Consequently, Fig. 5.46 presents the design model of Fig. 5.41 in terms of curvature ductility rather than displacement ductilities. This has been developed from Fig. 5.41, based on the typical M/VD ratio of 2 used in most of the experimetnal data and typical bar diameters. Again, the design strengths of Fig. 5.46 should be reduced by 25% for lightweight concrete columns.

Truss mechanism contribution: Use Eq. (5.76) with $\theta = 35°$.
Axial load contribution: Use

$$V_p = 0.85 P \tan \alpha. \qquad (5.77b)$$

Shear strength reduction factor: Use $\phi_s = 0.85$.
Test results indicate that this will generally result in more economical designs than the ASCE–ACI approach but will also provide more consistent protection against shear failure.

For columns reinforced with interlocking spirals (see Fig. 3.17, section F–F), theoretical considerations indicate that the shear strength of the truss mechanism provided by the spirals should be the sum of the shear strength

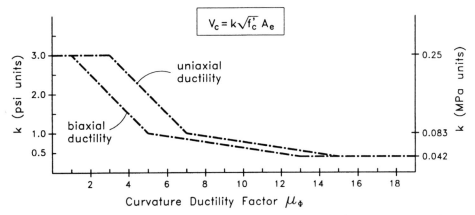

FIG. 5.46 Design strength of concrete shear-resisting mechanisms based on section curvature ductility.

of each of the spirals considered independently. The few data from relevant tests support this. As noted in Section 5.3.2(c)(iv), the center-to-center spacing of interlocking spirals should not exceed $0.75D'$.

BEAMS. The shear strength of concrete shear-resisting mechanisms appears to be lower in beams than in columns, primarily because columns benefit from the uniform distribution of longitudinal reinforcement around the section, helping to enhance dowel action and confinement. In the absence of further examination of the vast body of test data, it is recommended that the ASCE–ACI equation for basic shear strength be used, in combination with the independent axial load contribution described by Fig. 5.44. This should be used for axial forces due to prestress and also for axial forces induced in cap beams by column shear forces.

CONCRETE CONTRIBUTION FOR BEAMS. Use Eqs. (5.63) and (5.65) for the concrete contribution for normal-weight concrete, with an effective area of $A_e = 0.8A_{gross} = 0.8b_w h$. Reduce by 25% for lightweight concrete.

Truss mechanism contribution for beams: Use Eq. (5.76b) with $\theta = 35°$.

Axial load contribution: Use $V_p = 0.85P \tan \alpha$, where P is the net axial force on the section, resulting from prestress and other actions.

The use of this approach for cap beams is illustrated by the twin-column bent of Fig. 5.47. The angle α is found from the inclination of the strut between the centers of flexural compression zones at sections 1–1 and 2–2, and the axial forces to be used at these sections are

At section 1–1:

$$P = F_p - V_{col\,1}; \text{ hence } V_p = 0.85(F_p - V_{col\,1}) \tan \alpha \qquad (5.78a)$$

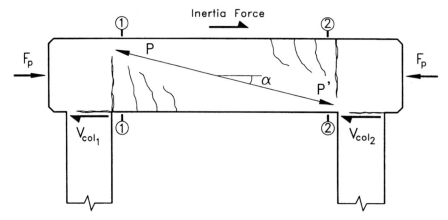

FIG. 5.47 Axial force contribution to shear strength of cap beam.

At section 2–2:

$$P' = F_p + V_{col2}; \text{ hence } V_p = 0.85(F_p + V_{col2}) \tan \alpha \qquad (5.78b)$$

(iv) Limit to Shear Stress. To avoid high diagonal compression stress levels in members, required shear stress levels V_r/A_e should not exceed $0.2f'_c$.

(v) Extent of End Region for Shear. The reduction in V_c implied by Figs. 5.41 and 5.46 applies only within plastic end regions, where the increase in width of flexure–shear cracks with ductility reduces the effectiveness of aggregate interlock. Since inclined shear cracking is expected at angles close to 30° to the column axis, the critical cracks can be expected to extend almost twice the member depth from the critical section. Consequently, the region over which the reduced V_c component applies should be taken as $2D$ or $2h$ from the critical section for circular and rectangular columns, respectively.

(c) Torsional Strength. In designing for torsional resistance of bridge members it is important to distinguish between torsion induced by equilibrium requirements and torsion induced by compatibility requirements. In the former, torsional resistance is essential to complete the load path from action to support (i.e., ground). In the second, torsional resistance is not essential to satisfy equilibrium requirements but is induced by requirements of rotational compatibility. In equilibrium-induced torsion, the magnitude of the torque is normally independent of the torsional rigidity, but in compatibility-induced torsion, the torque is directly dependent on the stiffness.

The difference is best illustrated by example. Figure 5.48 shows two situations where seismic actions induce torque in members. In Fig. 5.48(a), the bent shown has an outrigger column on one side as a result of geometric constraints below. Under longitudinal response, the column, which is connected monolithically to the cap beam outrigger, is expected to develop plastic

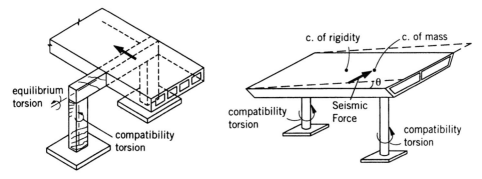

(a) Longitudinal Response with Outrigger Bent   (b) Torsion under Tranverse Response

FIG. 5.48 Equilibrium and compatibility torsion under seismic response.

hinges at top and bottom. The column top moment, extrapolated to the cap beam centerline, must be equilibrated by a cap beam torque of equal magnitude. Capacity design requires that the torsional strength of the cap beam must equal or exceed the torque corresponding to development of the column hinge plastic moment capacity. Unless the bent is stabilized by transverse strength of adjacent bents, the torsional strength of the cap beam is essential to equilibrium of the structure. This is an example of equilibrium torque. In fact, with consideration of adjacent bent transverse strength, it will be found that the cap beam torsional strength is not essential, but lack of adequate torsional strength will translate into reduced column moment, resulting in the intended plastic hinge not forming.

The same example induces compatibility torque in the column. Due to flexural deformation of the cap beam under the shear force necessary to induce the column moments, the column is twisted, inducing torque. The magnitude of this torque will depend on the torsional stiffness J of the column. If this is zero, the torque will be zero. In fact, with plastic hinges forming at the top and bottom of the column, the column torsional stiffness will be very low and the requirement for satisfactory response will be the ability of the column plastic hinges to develop ductile response while sustaining minor torsional rotations.

A second example illustrating compatibility torque is shown in Fig. 5.48(b), where a segment of bridge superstructure between movement joints contains one short column and one long column. As a consequence, the center of rigidity is displaced from the center of mass toward the shorter column. Under transverse response, a "swinging" mode of deformation develops, as suggested by the dashed lines of Fig. 5.48(b), with a rotation θ of the superstructure about the vertical axis. This rotation implies torque in the columns, but the magnitude is proportional to the column torsional stiffness, and with $J = 0$ the system is still stable.

Design for equilibrium and compatibility torques requires different approaches. In the former, torsional strength must be assured, while in the second, it is sufficient to ensure that rotational capacity is adequate. Equilibrium torque should always be resisted by elastic actions. Torsion is not a suitable mode of inelastic response. Standard approaches are available in specialist texts [C5] and in codes [A5,S2,E5] and will not be repeated here. The requirements include the use of longitudinal reinforcement distributed around the section, and special full peripheral hoops whose ends are anchored by 135° hooks bent back into the core. The area of these hoops is additional to any requirements for direct shear force.

For compatibility-induced torsion, numerical calculations are generally not necessary. It is sufficient to ensure that well-distributed longitudinal and transverse hoop reinforcement is placed. Tests of plastic hinges in columns under combined flexural displacement and torsional rotation [S6] have shown that plastic hinge capacity of typical circular columns is not affected adversely by torsional rotations up to 5%.

As noted in Section 4.4.2, torsional stiffness is greatly reduced after the onset of torsional cracking. However, the anticipated mode of cracking envisaged by typical design equations involves a series of spiral cracks forming around the section, which typically require a length of member several times the section width or depth to be fully developed. For shorter members, as will often be the case for outrigger cap beams, full torsional spiral cracks cannot develop, and the design equations are conservative and of doubtful validity. For such cases, such as when the freè length of the outrigger to the inside face of the column is less than 1.5 times the lesser transverse section dimension of the cap beam, it is more appropriate to consider a shear friction approach to resistance under the combined effects of shear and torsion.

The approach, based on a constant plastic shear friction stress, is illustrated in Fig. 5.49. The section is subjected to vertical and horizontal shear forces V_v and V_L, a torque T, and axial clamping force P. It is conceptually divided into four unequal quadrants of areas, A1 to A4, as shown in Fig. 5.49, within each of which the direction of shear friction resistance is taken as parallel to the outer edge, and the shear friction stress is taken as $\tau = \mu P/A$, where A is the total section area and μ is the coefficient of friction over the interface. Equilibrium under the external torsion T, longitudinal shear force V_L, and vertical shear force V_v requires that

$$V_v = F_1 - F_3 = \frac{\mu P}{A}(A_1 - A_3) \tag{5.79a}$$

$$V_L = F_2 - F_4 = \frac{\mu P}{A}(A_2 - A_4) \tag{5.79b}$$

$$T = F_1 x_1 + F_2 y_2 + F_3 x_3 + F_4 y_4$$

$$= \frac{\mu P}{A}(A_1 x_1 + A_2 y_2 + A_3 x_3 + A_4 y_4) \tag{5.79c}$$

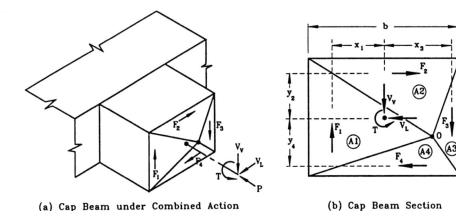

(a) Cap Beam under Combined Action   (b) Cap Beam Section

FIG. 5.49 Plastic shear friction design of short outrigger cap beam.

where x_i and y_i are the distances from the quadrant centroids to the section centroid. The components of Eq. (5.79) may be solved by trial and error, dividing the section into quadrants until all three equations are satisfied and then checking the implied value for μ, or by utilizing a limiting design value for μ (say, $\mu = 1.4$), then chosing F_1 to F_4 to satisfy Eqs. (5.79a) and (5.79b), and finally, checking that the torque predicted by Eq. (5.79c) exceeds the applied torque T.

The clamping force P in Eq. (5.79) should include the cap beam prestress force F (if present), axial force due to equilibrium with the column transverse shear force V_T (see Fig. 5.53), and clamping provided by the longitudinal reinforcement in the section. Tests on this mechanism on a large-scale model of a retrofit solution for the San Francisco double-deck viaducts following the 1989 Loma Prieta earthquake [P13,P14] (discussed further in Section 8.5.2) indicated that the effective clamping force at shear friction failure of the cap beam was less than that corresponding to yield of the cap beam reinforcement and implied an effective strain of 0.0006. Thus it is recommended that P in Eq. (5.79) be taken as

$$P = F + V_T + 0.0006E_sA_{st} \tag{5.80}$$

where A_{st} is the total area of cap beam longitudinal reinforcement at the section considered. These tests [P13,P14] indicated that torsional failure involing shear friction degraded only slowly with increasing torsional rotation of the failure surface, despite the very considerable physical degradation at the outrigger–superstructure connection, visible in Fig. 5.50, and that more than 80% of initial torsional capacity was maintained for torsional rotations as high as 2%.

5.4 DESIGN OF BEAM–COLUMN JOINTS

The 1989 Loma Prieta earthquake alerted the bridge design community to the need for special attention to provision of adequate mechanisms for force

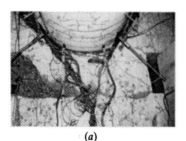 

(a)   (b)

FIG. 5.50 Shear friction failure under torsion of a short cap beam outrigger. (a) Looking into column; (b) Looking away from column.

transfer through connections between columns and cap beams, and columns and footings. Examples of damage of joints were discussed in Section 1.2.5.

Beam–column joints have received special attention in reinforced concrete frame design for some time. An extensive and detailed presentation of the performance and design of building frame joints is presented in [P4], and many codes have specific requirements for joint reinforcement to ensure satisfactory performance under seismic attack. However, until very recently, little attention has been paid to the performance or design of joints in bridge structures. Since the situation with bridges is typically different from that of building frames, with more reliance being placed on the ductile performance of columns than of beams, and with knee joints and vertical tee joints being more common than full intersections of columns and beams as occur at interior joints of building frames, a rather extensive treatment is useful here. Nevertheless, constraints on the size of this book preclude full coverage of the complexities of the subject herein and the reader is referred elsewhere [P30] for further details.

5.4.1 Shear Force in Beam–Column Joints

The traditional approach for investigating force transfer in beam–column joints has been based on an assessment of the joint shear force developed from equilibrium considerations of the member forces acting at the joint boundary. Consider the knee joints and tee joints shown in Fig. 5.51, where column moment is at overstrength, corresponding to plastic moment capacity in accordance with Section 5.3.3. In Fig. 5.51 the column is considered as an independent member extending to the top of the joint, with the influence of the beam or beams represented by the forces T, C, and V_B applied to this independent member. The moment profile up this member is shown in Fig. 5.51(c). The overstrength moment $M°$ continues to increase above the level of the beam soffit until the line of action of the beam force C (for the knee joint) or $C_r + T_l$ (for the tee joint). The

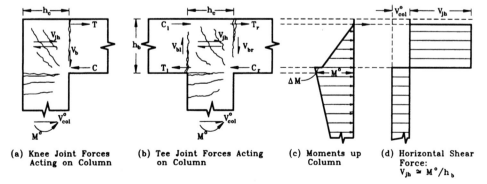

(a) Knee Joint Forces Acting on Column   (b) Tee Joint Forces Acting on Column   (c) Moments up Column   (d) Horizontal Shear Force: $V_{jh} \simeq M°/h_b$

FIG. 5.51 Horizontal shear force in knee and tee joints.

moment slope reverses under this force, decreasing to zero at the height of the upper stress resultant T (or $T_r + C_l$), as shown. Note that an incremental moment decrease ΔM is shown at the level of the beam lower stress resultant due to moment provided by the beam shear. That is,

$$\Delta M = \begin{cases} \dfrac{V_b h_c}{2} & \text{(knee joint)} & (5.81a) \\[3mm] \dfrac{(V_{bl} + V_{br})h_c}{2} & \text{(T joint)} & (5.81b) \end{cases}$$

This moment, which exists as a consequence of taking moments on the line of the column axis, should theoretically be distributed up the height of the joint, but has been concentrated, for convenience, at the level of the beam lower stress resultants, as shown.

The distribution of column horizontal shear force corresponding to the moments of Fig. 5.51(c) is shown in Fig. 5.51(d). Below the joint the shear is, of course V_{col}°, the shear corresponding to flexural overstrength. However, because of the very high moment gradient within the joint, the shear force V_{jh} within the joint is much larger than V_{col}°, and may be found, approximately, from the relationship

$$V_{jh} = \frac{V_{col}^\circ l}{h_b} \tag{5.82}$$

where l is the distance from the beam soffit to the column point of contraflexure. Alternatively, V_{jh} may be calculated with adequate accuracy as

$$V_{jh} = \frac{M^\circ}{h_b} \tag{5.83}$$

which follows directly from Eq. (5.82). Two compensating approximations are made in Eqs. (5.82) and (5.83): (1) the incremental moment decrease ΔM is ignored, and (2) the lever arm between beam stress resultants T and C is approximated by h_b. As the horizontal joint shear force will be used primarily as an indicator of joint stress level, this approximation is adequate.

In similar fashion, the forces acting on the beam, considered to extend through the joint, are shown in Fig. 5.52. The column tension stress resultant is distributed across the section and cannot properly be considered a single force. As a consequence, the moment profile through the joint is nonlinear, as shown, implying a variation in vertical shear force through the joint. Since the absolute maximum value is of little interest, an average value

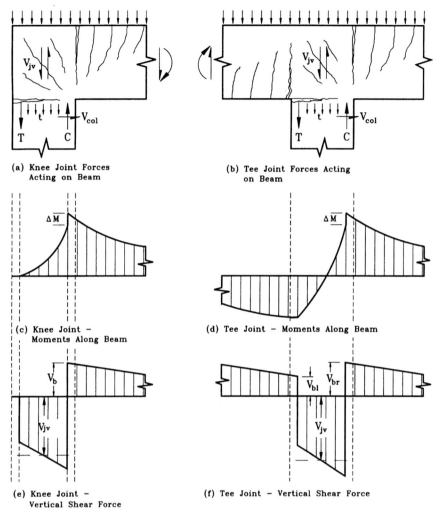

(a) Knee Joint Forces
    Acting on Beam

(b) Tee Joint Forces Acting
    on Beam

(c) Knee Joint –
    Moments Along Beam

(d) Tee Joint – Moments Along Beam

(e) Knee Joint –
    Vertical Shear Force

(f) Tee Joint – Vertical Shear Force

FIG. 5.52 Vertical shear force in knee and tee joints.

V_{jv} is shown in Fig. 5.52(e) and (f). Since average shear stress in the joint region should be equal in orthogonal directions, we can express V_{jv} directly as

$$V_{jv} = \frac{V_{jh}h_b}{h_c} \tag{5.84}$$

Joint shear forces can also be directly calculated, with greater accuracy than in the foregoing development, by equilibrium considerations. Although it is felt that the approach decribed above has adequate accuracy, the more rigorous approach is outlined in the following, since it requires careful consideration of joint equilibrium, which is often ignored by designers.

(a) **Knee Joints.** Knee joints are the most common type of joint occurring in multicolumn bridge bents when transverse response is considered. Equilibrium conditions under closing and opening moment are represented in Figs. 5.53(a) and (b), respectively. In these figures, the beam tensile, compressive, and shear stress resultants are indicated by T_b, C_b, and V_b, with T_c, C_c, and V_{col} being the corresponding forces for the column. Axial forces P_c and P_b are present in the column and beam, respectively, and a prestress force F is shown, which will, of course, be zero if the cap beam is reinforced conventionally. Moments M_b and M_c on the joint boundaries induce the flexural stress resultants noted above. Equilibrium equations governing the relationships between the various forces are summarized below.

Action	Closing Joint	Opening Joint	
Beam moment	$M_b = T_b\left(d - \dfrac{a_b}{2}\right)$ $+ P_b\left(\dfrac{h_b}{2} - \dfrac{a_b}{2}\right)$	(same)	(5.85a)
Beam axial force	$P_b = (F) + V_{col}$	$P_b = (F) - V_{col}$	(5.85b)
Beam compressive force	$C_b = T_b + P_b$	(same)	(5.85c)
Column moment	$M_c = M_b + V_b\dfrac{h_c}{2} - V_{col}\dfrac{h_b}{2}$ $\approx T_c\left(0.7h_c - \dfrac{a_c}{2}\right)$ $+ P_c\left(\dfrac{h_c}{2} - \dfrac{a_c}{2}\right)$	(same) (same)	(5.85d) (5.85e)
Column compressive force	$C_c = T_c + P_c$	(same)	(5.85f)
Horizontal joint shear force	$V_{jh} = T_b(+0.5F)$	$V_{jh} = C_b(-0.5F)$	(5.85g)
Vertical joint shear force	$V_{jv} \approx \dfrac{V_{jh}h_b}{h_c}$	(same)	(5.85h)

In the equations above, a_b and a_c are the depths of the beam and column

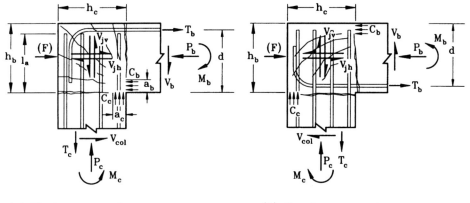

(a) Closing moment   (b) Opening moment

FIG. 5.53   Forces acting on knee joints.

compression zones, and it is assumed that the beam prestress F, if present, is distributed uniformly over the depth of the beam, and joint shear is calculated at the joint centroid.

Column moment is found in Eq. (5.85d) from the requirement that beam and column moments must be in equilibrium at the joint centroid. In normal capacity design calculations, it will be the column moment that is known and which will correspond to flexural overstrength of the column plastic hinges (see Section 5.3.3). Equations (5.85) can thus be inverted to find M_b. The beam tension force T_b can then be found from the known beam moment and beam axial force [from Eq. (5.85b)] using Eq. (5.85a). This enables the beam horizontal and vertical joint shear forces to be calculated from Eqs. (5.85g) and (5.85h).

It should be noted that it is more convenient to calculate the joint shear forces in this fashion than directly from the column stress resultants T_c and C_c. These are not normally known by the designer, particularly for circular columns and where the column longitudinal reinforcement is distributed around the section, as is the usual case. Although approximate equations for the relationship between column moment and internal stress resultants given in Eq. (5.85e) could be used for this purpose, this will result in the maximum rather than the average joint vertical stress [see Fig. 5.52(e)] which is inappropriate for design.

(b)   Tee Joints.   Tee joint conditions occur at internal columns of multicolumn (i.e., three or more columns) bents under transverse seismic response, and also for all columns of bents with monolithic column/superstructure connection details under longitudinal response. Figure 5.54 shows forces acting on a typical tee joint. Equilibrium relationships are based on horizontal and vertical equilibrium of forces and moment equilibrium at the joint centroid, and are summarized below:

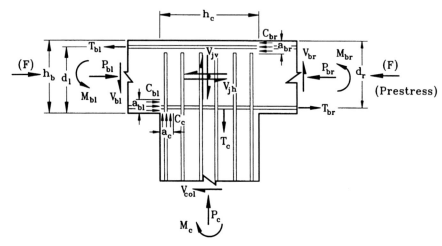

FIG. 5.54 Forces acting on tee joints.

Joint moment equilibrium	$M_c + V_{col} \dfrac{h_b}{2} = M_{bl} + M_{br} + (V_{bl} + V_{br}) \dfrac{h_c}{2}$	(5.86a)
Beam moment (left)	$M_{bl} = T_{bl}\left(d_l - \dfrac{a_{bl}}{2}\right) + P_{bl}\left(\dfrac{h_b}{2} - \dfrac{a_{bl}}{2}\right)$	(5.86b)
Beam moment (right)	$M_{br} = T_{br}\left(d_r - \dfrac{a_{br}}{2}\right) + P_{br}\left(\dfrac{h_b}{2} - \dfrac{a_{br}}{2}\right)$	(5.86c)
Beam axial force (left)	$P_{bl} \approx F + 0.5 V_{col}$	(5.86d)
Beam axial force (right)	$P_{br} \approx F - 0.5 V_{col}$	(5.86e)
Horizontal joint shear force	$V_{jh} = T_{bl} + C_{br}$	(5.86f)
Vertical joint shear force	$V_{jv} = \dfrac{V_{jh} h_b}{h_c}$	(5.86g)

Again, the prestress force F, if present, is assumed to be uniformly distributed over the beam depth. Since average rather than maxima values for shear force are required for design purposes, it will be more convenient to calculate V_{jh} directly from Eq. (5.83) than from Eq. (5.86g).

5.4.2 Nominal Shear Stress

The nominal shear stress level in beam–column joints can be found directly from the joint shear forces as

$$v_{jh} = \frac{V_{jh}}{b_{je}h_c} = v_{jv} = \frac{V_{jv}}{b_{je}h_b} \qquad (5.87)$$

where b_{je} is the effective width of the joint, defined in Fig. 5.55. In Fig. 5.55, the effective width is taken at the center of the column section, allowing a 45° spread from boundaries of the column section into the cap beam. Thus under longitudinal response, the effective width for the cases shown in Fig. 5.55 measured along the cap beam axis is

$$b_{je} = \begin{cases} \sqrt{2}D & \text{(circular column)} & (5.88a) \\ h_c + b_c & \text{(rectangular column)} & (5.88b) \end{cases}$$

Under transverse response, the effective width will be the smaller of the value given by Eq. (5.88) and the cap beam width b_b. In the example shown in Fig. 5.55(b), $b_{je} = b_b$ governs. Note that an end column of a cap beam under longitudinal response will have a smaller effective width than given by Eq. (5.88) if the cap beam does not extend beyond the column exterior face.

The effective width defined above is largely based on engineering judgment and is somewhat larger than recommended for building frames [P4]. However, in conjunction with stress limitations recommended in the following section, it appears, based on test results, to be adequately conservative.

5.4.3 Principal Stress Levels in Joints

In design of beam/column joints of building frames it has been the practice to set shear stress limitations based on the nominal shear stress calculated in accordance with Eq. (5.87). Thus for typical knee and tee joint conditions, ACI 318-92 [A5] would limit shear stress to $12\sqrt{f_c'}$ psi ($1.0\sqrt{f_c'}$ MPa) and $15\sqrt{f_c'}$ psi ($1.25\sqrt{f_c'}$ MPa), respectively.

It can be argued that these limitations, which are based on the performance of joints designed without specific consideration of force transfer mechanisms, are somewhat arbitrary in that well-designed joints have been shown to sustain significantly higher shear stress levels without distress (see, e.g., [B4]). It would appear that more basic and relevant information could be obtained from examination of nominal principal tension and compression stresses in the joint. If the average principal tension stress is less than the cracking strength of the concrete, it is unlikely that the joint will exhibit distress, because any joint cracking that develops should be minimal and related to local stress concentrations. In such cases, special joint reinforcement is unlikely to be

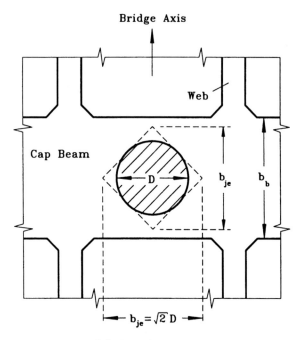

(a) Circular Column

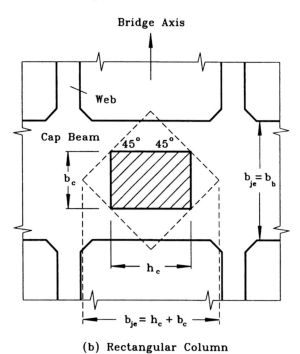

(b) Rectangular Column

FIG. 5.55 Effective joint width for joint shear stress calculation.

needed, although nominal minima values should be supplied. On the other hand, if nominal principal compression stresses exceed the crushing strength of the concrete, special joint reinforcement is unlikely to be particularly effective. In this context it should be noted that crushing must be expected at a nominal compression stress significantly less than f_c', since the average stress will be less than the maximum and compression strength is reduced by transverse tensile strain [C4] in the joint region.

A simple Mohr's circle analysis for stress shows that the nominal principal stresses in the joint region are given by

$$p_c, p_t = \frac{f_v + f_h}{2} \pm \sqrt{\left(\frac{f_v - f_h}{2}\right)^2 + v_j^2} \qquad (5.89)$$

where p_c and p_t are nominal principal compression and tension stresses, respectively; v_j is the joint shear stress, from Eq. (5.87); and f_v and f_h are average axial stresses in the vertical and horizontal directions.

In a typical joint, f_v is provided by the column axial force P_c, including the seismic component, which should be the appropriate value for M° if overstrength conditions are considered. An average stress at midheight of the cap beam should be used, assuming a 45° spread up from the boundaries of the column in all directions. Given the nonuniform nature of column axial force introduction into the cap beam, this is clearly a rather coarse approximation but is likely to provide a conservatively low value for f_v. The horizontal axial stress f_h is based on the mean axial force at the center of the joint. Under transverse response this will be the force P_b from Eq. (5.85b) for knee joints, and the average of P_{bl} and P_{br} (i.e., $= F$) from Eqs. (5.86d) and (5.86e) for tee joints. Under longitudinal response, $f_h = f_p$ for a prestressed solid section superstructure and $f_h = 0$ for a box section, even if prestressed, since there is unlikely to be any significant longitudinal axial stress transverse to the cap beam at midheight unless specially provided for that purpose.

In the transverse direction the average stress can be approximated by

$$f_h = \frac{P_b}{b_b h_b} \qquad (5.90)$$

where P_b is given by Eq. (5.85c).

(a) *Principal Tension Stress.* Experimental evidence indicates that diagonal cracking is initiated in the joint region when the principal diagonal tension stress is approximately $3.5\sqrt{f_c'}$ psi ($0.29\sqrt{f_c'}$ MPa). Although alternative concrete mechanisms and residual tension strength enable joints without special reinforcement to resist higher tension stresses before joint failure occurs, as discussed in Section 7.4.9, these should not be relied on in new design.

Examination of Eq. (5.89) indicates that for a given shear stress v_j, the magnitude of the principal tension stress, which has a negative sign in Eq. (5.89) if a compression-positive sign convention is adopted, is reduced by both column axial compression and cap beam prestress. The latter is a very powerful means for reducing potential problems in joint regions.

(b) Principal Compression Stress. Current code limitations on joint shear stress are intended to keep principal compression stress levels to acceptable values. As such, they will be seen to be rather ineffective, since principal compression stress is frequently more influenced by the axial stress levels than by the shear stress levels. It is consequently recommended that the joint principal compression stresses given by Eq. (5.89) should be limited to

$$p_c \leq 0.3f_c' \tag{5.91}$$

For average column compression stress of $f_v = 0.1f_c'$ and $f_h = 0$, this requirement corresponds closely to limits suggested for building frames [P4], $v_j \leq 0.25f_c'$, and to the current ACI-318 limits noted above. Although higher compression stresses can be sustained by well-confined joints, the conservative recommendation of Eq. (5.91) is favored in light of the comparative sparcity of data.

(c) Example of Principal Stress Computation. The knee joint shown in Fig. 5.56 develops a plastic moment capacity at an overstrength of $M_c^\circ = 20,600$ kip-ft (27.9 MNm) under closing moments and a corresponding shear force of $V_{col}^\circ = 688$ kips (3060 kN). The column axial force is 2040 kips (9074 kN), including seismic effects. Calculate the joint principal stresses (a) if the cap beam is conventionally designed, and (b) if 50% of the cap beam flexural strength is provided by a centrally located prestress force of 3750 kips (16,670 kN). The cap beam and joint specified compression strength is $f_c' = 4$ ksi (27.6 MPa) for the conventional cap beam design and $f_c' = 5$ ksi (34.5 MPa) for the prestressed cap beam.

From Eq. (5.83), the horizontal joint shear force is given by

$$V_{jh} = \frac{M^\circ}{h_b} = \frac{20,600}{6} = 3433 \text{ kips (15,270 kN)}$$

Note that this is about five times the column shear force of 688 kips.

Joint Shear Stress. From Fig. 5.55, the effective joint width is clearly the cap beam width [i.e., $b_{je} = 84$ in. (2134 mm)]. Hence, from Eq. (5.87),

$$v_j = \frac{V_{jh}}{b_{je}h_c} = \frac{3433}{84 \times 72} = 568 \text{ psi } (9.0\sqrt{f_c'} \text{ psi} = 0.75\sqrt{f_c'} \text{ MPa})$$

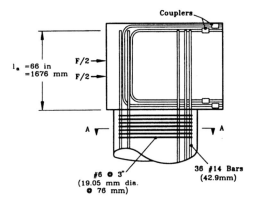

Elevation

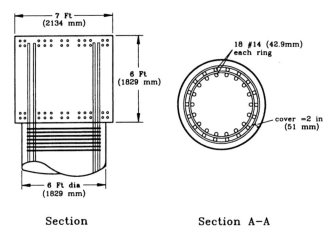

Section   Section A–A

FIG. 5.56   Dimensions for knee joint example of Section 5.4.3(*c*).

CASE (a):   CONVENTIONAL CAP BEAM DESIGN. The vertical axial stress in the joint, allowing 45° spread into the cap beam, is

$$f_v = \frac{P_c}{b_{je}(h_c + 0.5h_b)} = \frac{2040}{84(72 + 36)} = 225 \text{ psi } (1.55 \text{ MPa})$$

The horizontal axial stress (compressive) is found from $P_b = V_{col}$ [Eq. (5.85b)] as

$$f_h = \frac{688}{84 \times 72} = 114 \text{ psi } (0.78 \text{ MPa})$$

Therefore, principal stresses, from Eq. (5.89), are

$$p_c, p_t = \frac{225 + 114}{2} \pm \sqrt{\left(\frac{225 - 114}{2}\right)^2 + 568^2}$$

That is,

$$p_t = 401 \text{ psi } (6.3\sqrt{f_c'} \text{ psi} = 0.53\sqrt{f_c'} \text{ MPa})$$

$$p_c = 740 \text{ psi } [0.185f_c', \text{ based on } f_c' = 4 \text{ ksi (27.6 MPa)}]$$

CASE (b): PRESTRESSED CAP BEAM DESIGN. v_j and f_v remain unchanged, but the horizontal axial stress is now

$$f_h = \frac{3750 + 688}{84 \times 72} = 734 \text{ psi (5.06 MPa)}$$

Therefore, principal stresses, from Eq. (5.89) are

$$p_c, p_t = \frac{(225 + 734)^2}{2} \pm \sqrt{\left(\frac{225 - 734}{2}\right)^2 + 568^2}$$

that is,

$$p_t = 143 \text{ psi } (2.0\sqrt{f_c'} \text{ psi} = 0.17\sqrt{f_c'} \text{ MPa})$$

$$p_c = 1102 \text{ psi } [0.22f_c', \text{ based on } f_c' = 5 \text{ ksi (34.5 MPa)}]$$

Note that for the conventional cap beam design, the principal tension stress of $6.3\sqrt{f_c'}$ psi ($0.53\sqrt{f_c'}$ MPa) implies extensive cracking, and in accordance with the following sections, special joint reinforcement would be needed. With the prestressed cap beam, principal tension stress is reduced to $2.0\sqrt{f_c'}$ psi ($0.17\sqrt{f_c'}$ MPa) and no joint shear cracking is expected. In both cases, principal compression stresses are well within the limit of $0.3f_c'$ required by Eq. (5.91). The joint should also be checked under opening moments.

5.4.4 Mechanisms of Force Transfer in Cracked Joints

(a) *Knee Joints.* When principal tension stresses exceed the joint tension strength, cracking occurs and the force transfer from beam to column implied by equilibrium considerations can no longer be based on assumptions of isotropic material performance. Figure 5.57 shows typical patterns of cracks developed in knee joints under closing and opening moments.

(i) *Closing Moments.* Under closing moment, a fan-shaped pattern of cracks develops, radiating from the outer surfaces of beam and column toward the

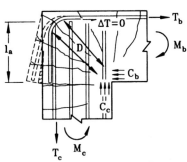
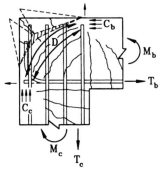

(a) Closing Moments. Beam Bottom
    Steel Omitted for Clarity

(b) Opening Moments. Beam Top
    Steel Omitted for Clarity

FIG. 5.57   Cracking and joint failure of knee joints.

inside corner, as shown in Fig. 5.57(*a*). If there is no vertical joint reinforcement clamping the beam top reinforcement into the joint, the entire beam tension force T_b is transferred to the back of the joint since there is no mechanism available to resist the moment at the base of the wedge-shaped concrete elements caused by tension transfer to the concrete by bond. If the beam top reinforcement is anchored by a straight extension to the back of the joint, it is apparent that failure will occur by opening up of the corner-to-corner crack or beam bar bond failure.

If the beam reinforcement is anchored by bending down to the bottom of the beam, better conditions for force transfer exist. If the lap length l_a provided with the column reinforcement is adequate, continuity of tension force capacity around the corner is provided, and a diagonal strut D can form from the bend of the beam bars to the inner corner, where it is resisted by the resultant of the beam and column compression forces C_b and C_c. However, resistance to the diagonal strut D at the hook creates lateral concrete tension stresses below the hook, which are exacerbated by the tension forces associated with lap splice action between the beam bar tails and the column longitudinal reinforcement. If splitting occurs, as is likely, the hook will tend to straighten and failure will occur, as suggested in Fig. 5.57(*a*).

(*ii*) *Opening Moments.* The situation under opening moments is illustrated in Fig. 5.57(*b*). A series of arch-shaped cracks tends to form between the compression zones at the outside of the column and the top of the beam. The curved nature of these cracks is a function of the distributed column reinforcement, as discussed subsequently, in relation to Fig. 5.61. The intersection of the arch strut D and the flexural compression zones at the top of the beam and the back of the column create outward-acting resultant forces. If the beam bottom reinforcement is anchored by straight bar extension, as shown by the dashed profile in Fig. 5.57(*b*), there is nothing to resist the horizontal tensile resultant, and vertical splitting will result, reducing competence of the anchorage of outer column bars and of beam top bars bent down, as in Fig. 5.57(a). At the intersection of D with C_b, the vertical resultant will tend to form a horizontal crack at the top of the column reinforcement. Flexural compression force(s) then tend to wedge-off the outside corner of the joint, aided by the diagonal fan cracks developed under closing moments. Anchorage of the column steel in the joint is lost, and joint failure occurs.

The result of either of these mechanisms of failure can be sudden joint failure at moments less than those corresponding to the flexural strength of the column. Figure 5.58 shows the results of a model knee joint representing the I-980 bent 35 joint, which failed in the 1989 Loma Prieta earthquake [Fig. 1.27(a)]. Similar cracking is observed in the model [Fig. 5.58(a)], and as seen from the lateral force–displacement response of Fig. 5.58(b), failure occurred under the closing moment at a level considerably below that corresponding to the ideal flexural strength F_i, or even strength F_y, corresponding to first yield. Joint shear cracking in the test unit first developed at a nominal principal tension stress of about $3.5\sqrt{f_c'}$ psi ($0.29\sqrt{f_c'}$ MPa), in close agreement with suggestions of Section 5.4.3(a), but failure did not occur until a nominal principal tension stress of about $5.6\sqrt{f_c'}$ psi ($0.47\sqrt{f_c'}$ MPa).

It is apparent that satisfactory performance of knee joints with ductile columns will require specially detailed reinforcement to enable the diagonal struts of Fig. 5.57 to develop. For closing moments, the tension force must be transferred around the outside of the joint, which involves restraint of the tails of the bent beam top reinforcement and adequate anchorage of the column longitudinal reinforcement. If this can be assured, the strut D of Fig. 5.57(a) can effect the required force transfer, provided that the compression stresses within the strut are acceptably low. This will depend, in part, on the magnitude of D and the radius of the bend of the beam bars.

Three possible details are shown in Fig. 5.59. In Fig. 5.59(a), the common detail involving bending the beam steel down and up inside the column cage is shown. Although this provides good restraint against straightening of the beam bar tails, provided that horizontal hoop reinforcement is provided around the column bars within the joint region, as shown, anchorage details for the outer column bars are very poor. Unless a large clamping force is provided by the hoop reinforcement, anchorage of the outer column bars is likely to degrade rapidly under cyclic response involving alternate tensile and compressive yield of the column bars, as suggested by the crack pattern of Fig. 5.59(a).

The amount of transverse hoop reinforcement required to provide anchorage of the column reinforcement after the splitting cracks develop can be calculated by a shear friction approach. Tests indicate [P19] that the stress in the hoop reinforcement should not exceed that corresponding to a strain of $\varepsilon_s = 0.0015$, since higher strains appear to result in excessive dilation of the circumferential crack with reduced efficiency of the shear friction mechanism. Based on a coefficient of friction of $\mu = 1.4$ across the circumferential crack

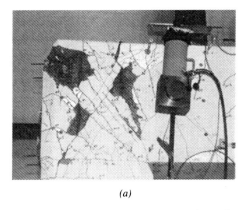

(a)

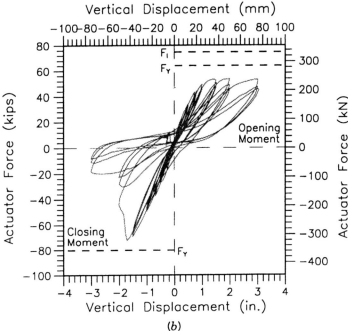

(b)

FIG. 5.58 Failure of a knee joint without joint reinforcement.

surface, it can be shown (see Section 5.5.3) that the required volumetric ratio of transverse reinforcement to avoid anchorage failure is

$$\rho_s = \frac{4A_h}{D's} = \frac{0.46A_{sc}}{D'l_a}\frac{f^o_{yc}}{f_{sh}} \tag{5.92}$$

where A_h and s are the area and spacing of the hoop reinforcement, A_{sc} is

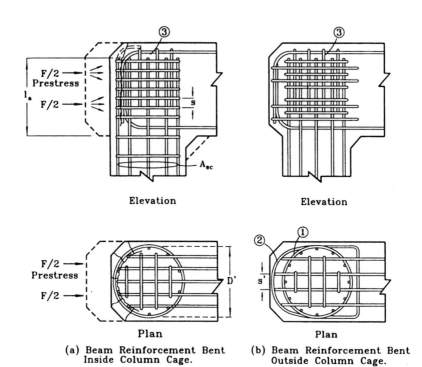

Elevation   Elevation

Plan   Plan

(a) Beam Reinforcement Bent   (b) Beam Reinforcement Bent
    Inside Column Cage.   Outside Column Cage.

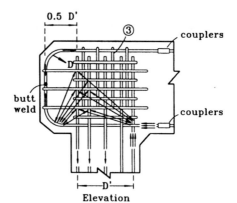

Elevation

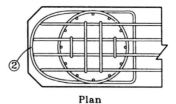

Plan

(c) Beam Stub with Continuous Top
and Bottom Reinforcement.

FIG. 5.59 Knee joint reinforcement for closing moments.

the total area of column longitudinal reinforcement, f_{yc}^o is the overstrength stress in the column reinforcement, including strain hardening and yield over-strength (i.e., $f_{yc}^o \approx 1.4f_y$ for grade 60 rebar), l_a is the anchorage length in the joint [Fig. 5.59(a)], and $f_{sh} = 0.0015E_s$ is the stress permitted in the hoop reinforcement.

As seen in the next example, the ρ_s value required by Eq. (5.92) can be quite high. An improvement to the force transfer mechanism can be achieved by prestressing the cap beam, with the anchorages contained within an extension to the cap beam, indicated by the dashed lines of Fig. 5.59(a). Equation (5.92) can be reinterpreted as requiring a clamping pressure of

$$ f_l = 0.5\rho_s f_{sh} = \frac{0.23A_{sc}f_{yc}^o}{D'l_a} \tag{5.93} $$

If the cap beam prestress f_p is used to provide part of the beam moment capacity required, no special confining hoop reinforcement will be required, provided that $f_p > f_l$. In this case, nominal joint horizontal reinforcement may be provided, in accordance with Eq. (5.96). If $f_l > f_p$, reduced hoop reinforcement may be provided, to satisfy

$$ \rho_s = \frac{2}{f_{sh}} \left(\frac{0.23A_{sc}f_{yc}^o}{D'l_a} - f_p \right) \tag{5.94} $$

Again, the minimum requirement of Eq. (5.96) should be satisfied. A further possibility involves bending the outer column bars over the joint, shown dashed in Fig. 5.59(a). This is likely to result in unacceptable congestion.

In Fig. 5.59(b), the beam reinforcement passes through and is anchored by hooks beyond the column cage. As shown in the plan view of Fig. 5.59(b), the tails should be confined within circular hoops, marked 2, to restrain straightening. Alternatively, the tails of each bar may be restrained by individual hair pins, or U bars, providing a restraining force of not less than $0.033A_b f_y$ centered $6d_b$ down from the end of the bend, where A_b and d_b are the area and diameter of the bent-down (or bent-up) beam reinforcement. This force is sufficient to develop the plastic moment capacity of the bar in bending. It would seem superfluous to provide more.

When the tails are confined by circular hoops, as suggested in Fig. 5.59(b), the required volumetric ratio of confining hoop reinforcement is given by [P30]

$$ \rho_s' = \frac{0.0088d_b}{s'} \frac{f_y}{f_{yh}} \tag{5.95} $$

where s' is the spacing of the beam bar tails around the hoop circumference. The requirements of Eq. (5.95) are generally much less onerous than those required by Eq. (5.92). The clamping capacity of the tail restraint also provides some clamping action across the circumferential anchorage splitting crack, which may occur inside the outer column bars in a fashion similar to that illustrated for Fig. 5.59(a). Transverse hoop requirements for hoop 1 around the column steel can thus be relaxed somewhat.

A third and preferred alternative is shown in Fig. 5.59(c). In this case, most or all of the beam reinforcement is provided in the form of continuous U bars. This can be effected either by butt welding top and bottom beam bars at the back of the joint or by placing short U bars through the joint and connecting to the beam reinforcement by approved mechanical couplers, capable of transferring the full yield strength of the bars. The tails of the beam steel are located approximately $0.5D'$ beyond the column cage and are restrained in accordance with Eq. (5.95). Anchorage of the column bars is now provided by struts that form outward to the bottom beam bar bends and inward to the compression zone at the inside corner, as shown in Fig. 5.59(c). The horizontal components of these struts provide the clamping action necessary to anchor the column steel. The vertical component of the outwards struts is transferred up to the top bend, where it is resolved into the major diagonal strut D. With this mechanism, nominal horizontal joint hoop reinforcement around the column cage, in accordance with Eq. (5.96), is sufficient.

In all cases, sufficient hoop reinforcement should be provided to transfer at least 50% of the principal tension strength of $3.5\sqrt{f_c'}$ psi ($0.29\sqrt{f_c'}$ MPa) by a lateral clamping pressure $f_l = 0.5\rho_s f_{yh}$. This thus requires that

$$\rho_{s,min} = \begin{cases} \dfrac{3.5\sqrt{f_c'}}{f_{yh}} & \text{(psi)} \\[2em] \dfrac{0.29\sqrt{f_c'}}{f_{yh}} & \text{(MPa)} \end{cases} \tag{5.96}$$

A minimum amount of vertical joint reinforcement (marked 3 in Fig. 5.59) should also be provided to assist with bond transfer of the top beam reinforcement and to assist in the development of the diagonal strut under opening moments, as discussed in Section 5.4.4(b).

(ii) Opening Moments. Three mechanisms are illustrated in Fig. 5.60 to avoid the potential failures under opening moment, described in relation to Fig. 5.57(b), where the plug of concrete encasing the column reinforcement pulls out of the joint, separating along a horizontal crack at the top of the column rebar. This behavior has been observed in testing of knee joints without vertical joint shear reinforcement but with considerable hoop reinforcement [I5]. To avoid this type of failure, a means must be provided to redirect the beam compression force C_b into the diagonal strut D toward the column compression force C_c. The inside column reinforcement could be bent over the joint, as shown in Fig. 5.60(a), to provide the necessary vertical component

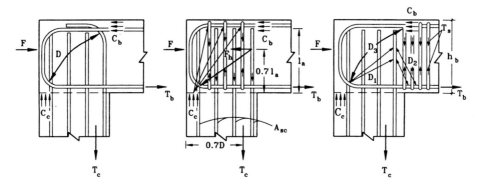

(a) Bent inside column bar   (b) Vertical stirrups   (c) External stirrups

FIG. 5.60   Knee joint reinforcement for opening moments.

to C_b. Vertical joint reinforcement could be placed to transfer part of the column tension force T_c by bond up to the top of the joint, where diagonal struts can form as in Fig. 5.60(b). A third alternative involves providing additional vertical reinforcement *outside* the joint in the beam, to encourage the formation of a diagonal compression strut initiating outside the joint, as indicated in Fig. 5.60(c).

The solution proposed in Fig. 5.60(a) is likely to cause unacceptable congestion and would require each of the tails of the column bars to be anchored with a restraining force of not less than $0.033A_b f_y$, as identified in discussion of Fig. 5.59(b). In Fig. 5.60(b), a proportion of the column tension force T_c is transferred directly by bond to the vertical stirrups, with the remainder transferred by horizontal hoop reinforcement, represented by the force F_h. It is recommended that with this design, 50% of T_c be transferred by vertical stirrups and 50% by the hoops. It is reasonable to put $T_c = 0.5A_{sc}f_{yc}^\circ$, even though the actual tension force could be higher. This is because conditions for bond transfer for the vertical reinforcement on the compression side of the centroidal axis are favorable, since the bars are well anchored in the diagonal compression strut. Special joint reinforcement is thus not needed for these bars. Thus

$$F_v = 0.25A_{sc}f_{yc}^\circ$$

and the total area of vertical stirrup reinforcement required is

$$A_{jv} = 0.25A_{sc}\frac{f_{yc}^\circ}{f_{yv}} \tag{5.97}$$

where f_{yv} is the yield strength of the joint vertical reinforcement. If $f_{yc}^\circ = 1.4f_{yc}$

and $f_{yc} = f_{yv}$, as for grade 60 rebar design, this requires that $A_{jv} = 0.35A_{sc}$. Placing this amount of joint vertical reinforcement can be difficult.

Horizontal hoops are needed to carry the remainder of T_c, via the clamping force F_h in Fig. 5.60(b). Assuming this force of $0.5T_c$ to be centered $0.7l_a$ into the joint to allow for strain penetration, and further assuming T_c to be centered $0.7D$ from the compression face, it can be shown [P30] that the amount of hoop reinforcement required is given by

$$\rho_s = \frac{3.3}{Df_{yh}l_a} \left(\frac{0.18A_{sc}f^\circ_{yc}D}{l_a} - F \right) \tag{5.98}$$

where F is the cap beam prestress, if present. If no prestress is provided, Eq. (5.98) can be simplified, with minor rounding up, to

$$\rho_s = \frac{0.6A_{sc}f^\circ_{yc}}{l_a^2 f_{yh}} \tag{5.99}$$

The mechanism of Fig. 5.60(b) is probably the most dependable but requires considerable amounts of both horizontal and vertical joint reinforcement.

The mechanism of Fig. 5.60(c) has similarities to that of Fig. 5.59(c). Anchorage to the column bars closest to the cap beam is provided by struts D_1, directed toward the column compression resultant C_c, and D_2 directed outward, into the beam. The vertical component of D_2, namely T_s, is provided by beam stirrups close to the joint. Transfer of this tension force to the top of the beam provides the necessary force to incline the beam compression force C_b into the major compression arch D_3. Horizontal components of D_1 and D_2 approximately balance each other, reducing the need for hoop reinforcement.

It is recommended that $0.5T_c$ (that portion closest to the inner face of the column) be transferred by this mechanism. Allowing 50% of this force to be transferred into the joint via D_1 and 50% outward via D_2, the tension force T_s in the external beam stirrups is $T_s = 0.25T_c$. Again, approximating $T_c = 0.5A_{sc}f^\circ_{yc}$, the required amount of vertical beam stirrup reinforcement is

$$A_{jv} = 0.125A_{sc} \frac{f^\circ_{yc}}{f_{yv}} \tag{5.100}$$

This reinforcement should be placed over a length not greater than $0.5h_b$ from the column face and is additional to shear reinforcement required for conventional shear transfer in the beam. Note, however, that design of beam shear reinforcement close to the column face will normally be governed by the closing moment condition, where gravity and seismic shear forces are additive. Under opening moments, gravity and seismic shear forces normally oppose each other, and beam stirrups are typically underutilized for this case. As a consequence, much of the reinforcement required by Eq. (5.98) may be provided by existing beam shear reinforcement.

The strut D_2 imposes additional tension force in the beam bottom flexural reinforcement, as is apparent from equilibrium of forces under D_2 and T_s. Assuming the special vertical reinforcement to be placed over a length $h_b/2$ from the column face, the additional horizontal force to be resisted by the bottom beam reinforcement will be approximately $0.5T_s$. The additional area of beam bottom reinforcement required is thus

$$\Delta A_{sb} = 0.0625 A_{sc} \frac{f^o_{yc}}{f_{yb}} \tag{5.101}$$

This additional reinforcement must be added to that required for capacity resistance at column overstrength, since this mechanism can develop only if the beam remains elastic. Test results [I5] clearly show significantly increased stress in cap beam bottom reinforcement as a result of this mechanism.

With the diagonal strut D_2 thus properly anchored, the remaining 50% of the column tension force can be transferred from the inner bars by successive increase in the inclination of the diagonal strut, forming the arch-type strut suggested in Fig. 5.61. This implies transfer of the column bar tension force high up the bar in some cases, implying very high bond stresses. Measurements of column bar strains in knee joint experiments [I5] have indicated that bond stresses as high as $30\sqrt{f'_c}$ psi $(2.5\sqrt{f'_c}$ MPa) can be sustained. These bond

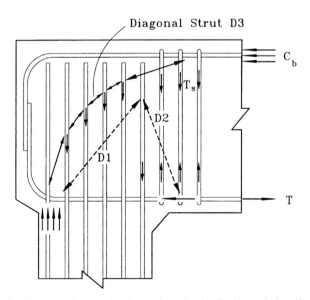

FIG. 5.61 Anchorage of central column bars by inclination of the diagonal strut.

stress levels, which have also been consistently recorded in other confined conditions, such as building frame joints [P4,E7] imply that reinforcement with yield stress of 60 ksi (414 MPa) can develop yield in as little as $9d_b$ from the free end of the bar. However, it will be apparent that for the internal bars to be effectively clamped by the diagonal strut, they must be carried as high as possible toward the top of the joint. This is discussed further in Section 5.5.3.

To provide assistance in bond transfer of the top reinforcement and to avoid the total beam tension force being transferred across to the hook, it is recommended that vertical stirrups, inside the column core, be provided for a vertical resistance equal to $0.5T_s$. This requires an internal vertical joint stirrup area of

$$A_{vi} = 0.0625A_{sc} \frac{f_{yc}^{\circ}}{f_{yv}} \tag{5.102}$$

Since this clamping action occurs at the top of the joint, these stirrups need not extend to the base of the joint. It is, however, recommended that they extend at least two-thirds of the cap beam depth.

Although the mechanism of Fig. 5.60(c) would not appear to require horizontal stirrups for stability, the inclination of struts D_1 and D_2 are not equal, implying an outward thrust. It is thus recommended that hoop reinforcement capable of providing at least 50% of the clamping force provided by Eqs. (5.98) and (5.99) be provided. This requires that

$$\rho_s = \frac{3.3}{Df_{yh}l_a} \left(\frac{0.09A_{sc}f_{yc}^{\circ}D}{l_a} - F \right) \tag{5.103}$$

which for $F = 0$ simplifies to

$$\rho_s = \frac{0.3A_{sc}f_{yc}^{\circ}}{l_a^2 f_{yh}} \tag{5.104}$$

Again, Eq. (5.96) should be considered a lower limit for joint hoop reinforcement.

(iii) Joint Reinforcement Requirements for Low Principal Tension Stress. When principal tension stress is less than $p_t = 3.5\sqrt{f_c'}$ psi $(0.29\sqrt{f_c'}$ MPa), no vertical joint reinforcement is needed, and only nominal transverse hoop reinforcement in accordance with Eq. (5.96) is required. The requirements of the preceding two sections of this chapter, including Eqs. (5.92) to (5.95) and Eqs. (5.97) to (5.104), should be fully implemented when principal tension stresses exceed $p_t = 5\sqrt{f_c'}$ psi $(0.42\sqrt{f_c'}$ MPa). For principal tension stresses in the range $3.5\sqrt{f_c'}$ psi $\leq p_t \leq 5\sqrt{f_c'}$ psi $(0.29\sqrt{f_c'}$ MPa $\leq p_t \leq 0.42\sqrt{f_c'}$ MPa)

linear interpolation between the full requirements of these sections and the nominal requirements of Eq. (5.96) may be adopted.

(iv) Summary of Knee Joint Design Requirements

1. Principal tension stress should be calculated in accordance with Eq. (5.89). If $p_t \leq 3.5\sqrt{f_c'}$ psi $(0.29\sqrt{f_c'}$ MPa), joint shear cracking is not expected, and nominal joint reinforcement, in the form of hoops satisfying Eq. (5.96), should be provided. For principal tension stress $p_t >$ $5.0\sqrt{f_c'}$ psi $(0.42\sqrt{f_c'}$ MPa), the full requirements of this section should be followed. For intermediate stess levels, linear interpolation between nominal and full design values can be used.

2. Principal compression stress, calculated in accordance with Eq. (5.89), should not exceed $0.3f_c'$.

3. Mechanisms for resistance of closing moments are shown in Fig. 5.59. When beam steel is bent into the joint within the column cage [Fig. 5.59(a)], transverse hoop reinforcement should satisfy Eq. (5.92), or if the cap beam is prestressed, Eq. (5.94). When the beam reinforcement is anchored beyond the column cage, as in Fig. 5.59(b) or Fig. 5.59(c), the tails should be restrained against opening in accordance with Eq. (5.95). The preferred design, shown in Fig. 5.59(c), involves continuity between top and bottom beam bars and enables mechanisms to develop requiring minimal transverse reinforcement, in accordance with Eq. (5.96), which is considered a minimum for all joints. Beam prestress results in a considerable reduction in required transverse joint reinforcement.

4. Mechanisms for resistance of opening moments are shown in Fig. 5.60. The arrangement of Fig. 5.60(b) requires internal vertical stirrups in accordance with Eq. (5.97) and transverse hoops in accordance with Eq. (5.98) or (5.99). Transverse hoops provided should satisfy the more onorous of opening and closing requirements. Although the mechanism of Fig. 5.60(b) is probably the most dependable, the mechanism of Fig. 5.60(c), involving beam vertical stirrups external to the joint, is generally most practical. This involves placing stirrups satisfying Eq. (5.100) within a distance $h_b/2$ from the joint, additional beam bottom reinforcement satisfying Eq. (5.101), and nominal vertical stirrups within the joint satisfying Eq. (5.102). Transverse hoop reinforcement within the joint should satisfy the more severe of Eqs. (5.96) and (5.103) [or Eq. (5.104), if $F = 0$].

5. Because the mechanisms described in previous sections have been shown by experiments to be slightly conservative and since consequences of joint failure are rarely catestrophic, it is recommended that a strength reduction factor of $\phi = 1.0$ be used for design of joints.

Knee joints designed in accordance with these provisions have performed very well. Figures 5.62 and 5.63 show crack patterns and force–displacement hysteresis loops for conventionally reinforced and prestressed knee joints,

(a)

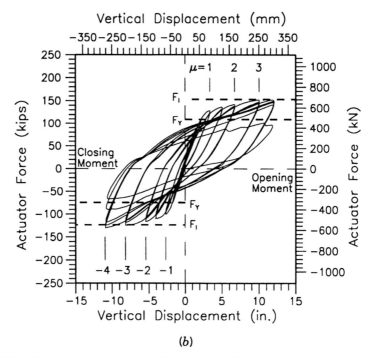

(b)

FIG. 5.62 Behavior of knee joint with conventionally reinforced cap beam designed to provisions of Section 5.4.1, subjected to simulated seismic loading [15]. (a) Crack pattern at maximum response; (b) Lateral force–displacement response.

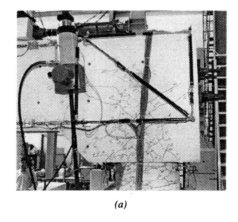

(a)

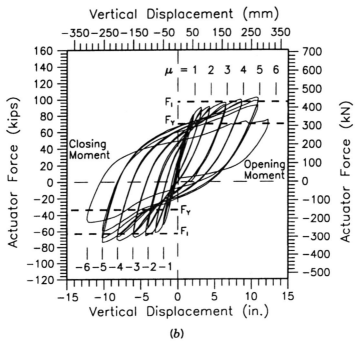

(b)

FIG. 5.63 Behavior of knee joint with prestressed cap beam designed to provisions of Section 5.4.1, subjected to simulated seismic loading [I5]. (a) Crack patterns at maximum response; (b) Lateral force–displacement response.

respectively. In both cases the joint regions remained essentially elastic although stirrup strains approached yield, and ductility was confined to the column plastic hinge, as intended by the design philosophy.

(v) Rectangular Columns. The design approaches developed above can be applied directly to rectangular columns reinforced with longitudinal bars con-

tained within interlocking spirals, as shown in Fig. 3.17, section F–F. Where the column bars are disposed in rectangular fashion around the column perimeter, adequate anchorage of the outer layer of bars is difficult to provide without excessive congestion by large numbers of transverse hoops through the joint, poviding clamping force to each column bar in accordance with the reinforcement configuration of Fig. 3.17, section E–E. This is satisfactory only when the total number of column bars is small and the longitudinal reinforcement ratio is low. If this is not the case, it is recommended that the cap beam be prestressed to provide adequate anchorage or that the outer column bars be bent in, as in the dashed profiles of Fig. 5.59(b), and restrained vertically. The mechanism of Fig. 5.60(c) can be employed to advantage for opening-moment resistance of rectangular columns. In this case the tension forrce to be anchored by the opposing struts D_1 and D_2 should include all column reinforcement with a distance of $0.2h_c$ from the inner column face. This will generally exceed the $0.5T_c$ required for circular columns.

(vi)   Example of Knee Joint Design. The knee joint shown in Fig. 5.56 and used for the example in Section 5.4.3(c) is now designed for joint reinforcement requirements using a number of different design alternatives, as outlined in previous sections. Column longitudinal reinforcement is provided by 36 No. 14 bars (d_b = 42.9 mm) arranged in two concentric rings of 18 bars each. As discussed later, this is to enhance force transfer to the bent beam bars by bond. A steel design yield strength of f_{ye} = 66 ksi (455 MPa) is used for column design, and an overstrength stress of f^o_{yc} = $1.3f_{ye}$ = 85.8 ksi (592 MPa) is used for determining joint requirements. Joint reinforcement nominal yield stress is f_y = 60 ksi (414 MPa), and a strength reduction factor of ϕ = 1 is used.

JOINT REINFORCEMENT FOR CLOSING MOMENTS. Consider the mechanism of Fig. 5.59(a). If the column reinforcement is in a single layer and the cap beam is conventionally reinforced, Eq. (5.92) requires, with D' = 68 in. (1727 mm), l_a = 66 in. (1676 mm), and f_{sh} = 0.0015 × 29,000 = 43.5 ksi (300 MPa):

$$\rho_s = \frac{0.46 \times 81.0}{68 \times 66} \left(\frac{85.8}{43.5}\right) = 0.0164$$

with $\rho_s = 4A_h/D's$, and using No. 9 hoops (d_{bh} = 28.6 mm), this requires a spacing of s = 3.5 in. (88.9 mm). Although feasible, the requirement is severe.

However, if the column steel is in a double ring, as shown in Fig. 5.56, two potential concentric fracture surfaces form and the clamping force is effective across both cracks. Hence, since the bar spacing along the circumference of each ring is double, ρ_s is halved. Thus ρ_s = 0.0082, which can be provided by No. 7 bars at s = $4\frac{1}{4}$ in. (22.2 mm diameter at 108 mm).

For the prestressed cap beam, the previous example established that the principal tension stress was ρ_t = $2.0\sqrt{f'_c}$ < $3.5\sqrt{f'_c}$; therefore, only nominal hoops are required in accordance with Eq. (5.96). Thus

$$\rho_s = \frac{3.5\sqrt{f'_c}}{f_{yh}} = \frac{3.5\sqrt{4000}}{60,000} = 0.00369$$

This can be provided by No. 5 hoops at 5-in. centers up the joint (15.9 mm diameter at 127 mm). If the cap beam is extended approximately 36 in. beyond the column, the mechanism of Fig. 5.59(c) can be employed, and again, nominal hoops (No. 5 hoops at 5-in. centers) are sufficient for the closing moment. In accordance with Eq. (5.95), the beam bar tails should be restrained by hoops providing

$$\rho'_s = \frac{0.0088 \times 1.69}{8.2} \left(\frac{60}{60}\right) = 0.0018$$

This can be provided by No. 5 hoops at 10-in. centers (15.9 mm diameter at 254 mm).

JOINT REINFORCEMENT FOR OPENING MOMENT. The solution adopted is to provide external reinforcement in accordance with Fig. 5.60(c). From Eq. (5.100), the amount of vertical reinforcement to be placed over a length not greater than $0.5h_b$ from the column face is

$$A_v = 0.125 \times 81.0 \times \frac{85.8}{60} = 14.5 \text{ in}^2 \ (9340 \text{ mm}^2)$$

This can be provided by 36 No. 6 (19.1 mm) stirrup legs, arranged in six layers of six legs each, with a spacing along the beam axis of 6 in. (152 mm). This reinforcement could be provided at least partially by beam shear reinforcement underutilized for the opening moment. An additional 18 No. 6 legs must be provided within the joint region in accordance with Eq. (5.102). Transverse reinforcement is required in accordance with Eq. (5.104) for the conventionally designed cap beam. Hence

$$\rho_s = \frac{0.3 \times 81.0 \times 85.8}{66^2 \times 60} = 0.0080$$

s essentially the same as that required for bond transfer under a closing moment, using Fig. 5.59(a). Note that this governs transverse hoop requirements for the design alternative utilizing Fig. 5.59(c). The $\rho'_s = 0.0018$ for beam bar tail restraint can be substracted, requiring $\rho'_s = 0.0080 - 0.0018 = 0.0062$, which can be provided by No. 7 hoops at $5\frac{1}{2}$ in. (19.9 mm diameter at 140 mm).

Equation (5.101) requires additional beam bottom longitudinal reinforcement of area $\Delta A_{sb} = 0.0625 \times 81 \times 85.8/60 = 7.25 \text{ in}^2 \ (4676 \text{ mm}^2)$. This could be provided by an extra 4 No. 14 bars ($A_b = 1451 \text{ mm}^2$).

For the prestress design, the principal tension stress under the opening moment exceeded that under the closing moment as a result of reduced column axial load, and was found to be $4.1\sqrt{f'_c}$ psi (not calculated in the previous example). Hence, in accordance with the recommendations of Section 5.4.4(a)(iii), external vertical reinforcement can be reduced by interpolation between the amount required for $p_t = 3.5\sqrt{f'_c}$ psi ($0.29\sqrt{f'_c}$ MPa), that is, $A_{jv} = 0$, and the full amount given by Eq. (5.100). Thus an amount $(4.1 - 3.5)/(5 - 3.5) = 0.4$ times the full amount is required. Thus $A_{jv} = 0.4 \times 14.5 = 5.8$ in² (3740 mm²); this requires 14 No. 6 legs but could almost certainly be provided by underutilized beam stirrups. Within the joint, 10 No. 5 legs are needed. The small amount of additional beam steel [also 2.9 in² (1870 mm²)] could be satisfied by a small increase of prestress force to provide the same tension capacity of $2.9 \times 60 = 435$ kips, requiring $F = 3750 + 435 = 4185$ kips (18.6 MN).

Figure 5.64 summarizes the various options for the joint design. Bending of the beam reinforcement into the column cage in Fig. 5.64(a) is likely to cause congestion at the back of the joint, and the detail of Fig. 5.64(b) is to be preferred. The beam bars are bent down to form a circle in plan, and the No. 5 hoops extended back across the joint, where they can be anchored by

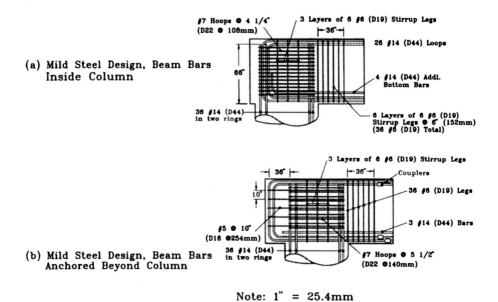

FIG. 5.64 Alternative joint details for knee joint design example.

lap splices at the inside of the column, for ease of construction. Figure 5.64(c) shows the very light reinforcement required for the prestressed design.

(b) Tee Joints. Like knee joints tee joints suffer joint shear failure if principal tension or compression stresses exceed the joint capacity. Figure 5.65 shows failure of a typical tee joint under transverse response. This joint, which was tested inverted for convenience and which represented a large-scale (three-

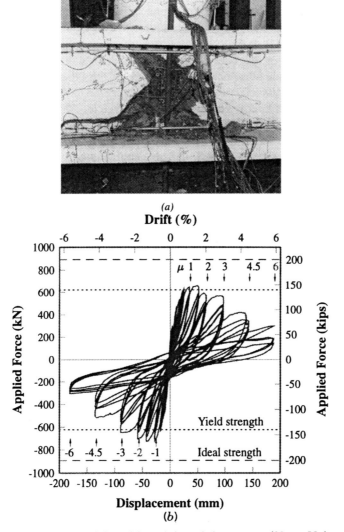

(a)

(b)

FIG. 5.65 Failure of a tee joint without joint reinforcement. (*Note:* Unit was tested inverted.) (*a*) Joint failure; (*b*) Lateral force–displacement response.

fourths size) model of an actual bridge detail constructed in the 1960s, contained no joint reinforcement and had inadequate development length of the column longitudinal reinforcement to satisfy the requirements of Section 5.5. Joint failure occurred at a principal tension stress of about $4.6\sqrt{f_c'}$ psi $(0.38 \sqrt{f_c'}$ MPa), at less than the lateral force corresponding to the column or cap beam moment capacity, with unsatisfactory lateral force displacement characteristics, as shown in Fig. 5.65(b).

As with knee joints, special transverse reinforcement is required to ensure that joint failures such as that shown in Fig. 5.65 do not occur. Figure 5.66 illustrates three different design approaches that can be used to design the joint reinforcement. The mechanism shown in Fig. 5.66(a) is essentially equivalent to the procedure that would be used for building frames. It is assumed that the beam and column compression forces C_{bl}, C_{br}, and C_c contribute directly to the major diagonal compression strut D. This requires a force to "bend" the right-side beam compression force C_{br} into the strut. For a building frame, which Fig. 5.66(a) could be taken to represent if the joint was rotated 90° and the roles of beams and columns thus reversed, this would be effected by bending the beam steel (i.e., column steel in our example) across the back of the joint. In the example shown, this is not provided, due to difficulty in bending the column bars inward, and as usual, the column steel terminates by straight bar extension as close as possible to the top of the joint. The beam tension force T_{bl} and the column tension force T_c are transferred by bond to horizontal and vertical stirrups, respectively, and are assumed to produce a set of diagonal cracks parallel to the potential corner-to-corner strut. The lower beam tension force T_{br} is considered noncritical since it can be transferred in the excellently confined region of the lower left compression corner, if necessary. Thus the joint reinforcement required for a nonprestressed joint would be

$$A_{jh} = \frac{T_{bl}}{f_{yh}} \tag{5.105}$$

$$A_{jv} = \frac{T_c}{f_{yv}} \tag{5.106}$$

The force T_{bl} can be found from the known column plastic moment using the equilibrium equation (5.86). As with knee joints, it is sufficiently conservative to put $T_c = 0.5 A_{sc} f_{yc}^o$, so that Eq. (5.106) can be rewritten as

$$A_{jv} = \frac{0.5 A_{sc} f_{yc}^o}{f_{yv}} \tag{5.107}$$

The vertical joint reinforcement provides the force necessary to transfer the beam compression force C_{br} from the horizontal to diagonal inclination and

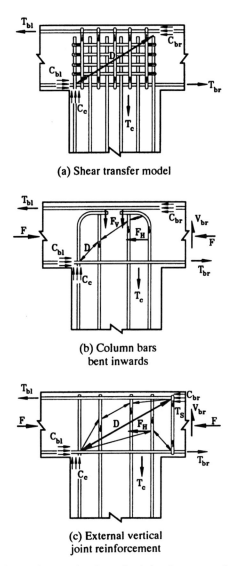

(a) Shear transfer model

(b) Column bars
bent inwards

(c) External vertical
joint reinforcement

FIG. 5.66 Alternative mechanisms for joint force transfer in tee joints.

also helps to transfer column tension force in reinforcement by bond force transfer.

It will be recognized immediately that Eqs. (5.105) and (5.107) will require unacceptably large amounts of joint reinforcement. For example, with $f_{yc}^\circ = 1.4 f_{yv}$, Eq. (5.107) would require vertical joint reinforcement equal to 70% of the total column longitudinal steel area to be placed within the joint region. This would be practically impossible to place.

The alternative mechanism of Fig. 5.66(b) involves all of the outer column reinforcement being anchored by bending in toward the column center. If this is anchored properly, a diagonal arch D can develop, in conjunction with the beam compression force C_{br}. The situation is now very similar to that of a ductile beam framing into an exterior column but rotated 90°. Note that the alternative option of bending the hooks outward, away from the column core, which is sometimes adopted by bridge engineers to improve column bar anchorage *should never be used,* as it directs the anchorage force away from the joint, creating additional tension stress within the joint region.

As a consequence of the inward bend of the column reinforcement, the amount of transverse joint reinforcement can be reduced considerably. A force F_v, shown in Fig. 5.66(b), is required to stabilize the bent ends of the vertical bars. As before, this is provided by anchoring each bent bar by a vertical bar of at least one-fourth the diameter of the longitudinal bar, at a distance of approximately $6d_b$ from the bend. It is recommended that horizontal reinforcement should be provided to transfer at least 25% of T_c. Thus hoop reinforcement satisfying Eq. (5.103) or (5.104) is required. As with knee joints, cap beam prestressing will greatly reduce the amount of transverse hoop reinforcement required. In all cases, however, some hoop reinforcement must be provided, in accordance with Eq. (5.96).

A third alternative, utilizing stirrup reinforcement outside the joint, shown in Fig. 5.66(c) is similar to that of Fig. 5.60(c) for knee joints and involves similar calculation. Thus vertical reinforcement should be placed over a distance of $h_b/2$ from the column face, in accordance with Eq. (5.100) on *each* side of the column. A further amount of vertical reinforcement equal to half of this [i.e., Eq. (5.102)] should be placed within the joint confines to help stabilize the top beam reinforcement and assist in transfer of the column tension force by bond. The area of beam bottom reinforcement must be increased in accordance with Eq. (5.101) and carried a sufficient distance to develop its yield strength a distance $h_b/2$ from the column face. Horizontal hoop reinforcement to carry a force of $0.25T_c$, in accordance with Eqs. (5.103) or (5.104), must also be provided, with a minimum as given by Eq. (5.96) in all cases.

It will be recognized that many other mechanisms are possible to assist in joint transfer. These include bending beam top longitudinal reinforcement down over the joint to lap splice the column rebar, passing beam reinforcement diagonally through the joint, and the use of headed reinforcement for the column rebar, to provide positive anchorage. These approaches, which are not developed further here because of a lack of relevant test data are illustrated in Fig. 5.67 and are all expected to be structurally competent, although the mechanism involving diagonal reinforcement could create unacceptable congestion.

Of the mechanisms proposed, that represented by Fig. 5.66(c) is the most constructable. Results of a test of a tee joint redesigned from that shown in Fig. 5.65 are given in Fig. 5.68. The test unit was essentially a 66% scale

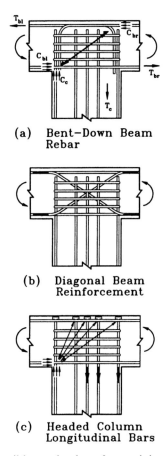

(a)   Bent–Down Beam
       Rebar

(b)   Diagonal Beam
       Reinforcement

(c)   Headed Column
       Longitudinal Bars

FIG. 5.67   Possible mechanisms for tee-joint force transfer.

model of the earlier test and contained identical (though scaled) column reinforcement. Beam reinforcement was redesigned to ensure that the plastic hinge would form in the column, and joint reinforcement complying with Eqs. (5.100) to (5.104) was provided. Unlike the unit without joint reinforcement, this test was very satisfactory, with predicted strength based on column plastic hinging being achieved up to displacement ductility factors of $\mu_\Delta = 6$ and drift angles of 5%. At higher ductilities, joint failure gradually developed, simultaneously with a confinement failure of the plastic hinge. In a real design, additional joint reinforcement would have been provided on the column axis on the sides of the cap beam, to satisfy longitudinal seismic response demands, and performance would have been even better than that shown in Fig. 5.68.

A further unit with alternate detailing based on prestress was also tested. This unit, which had details similar to those of Fig. 5.68 but with increased transverse reinforcement in the column plastic hinge region, was precast in two separate cap beam elements and a column element. Cap beam flexural

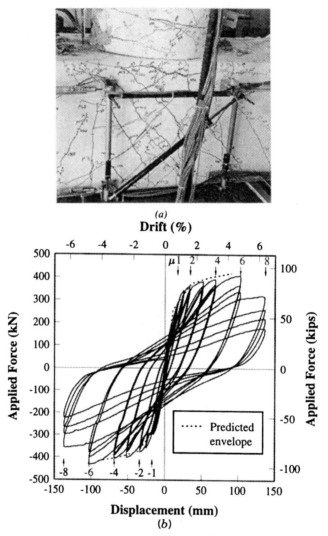

FIG. 5.68 Response of a tee joint with joint reinforcement satisfying Eqs. (5.100) to (5.104). (*Note:* Column details and cap beam size are scaled from the example of Fig. 5.65.) (*a*) Joint crack pattern at displacement ductility $\mu_\Delta = 6$; (*b*) Lateral force–displacement response.

strength was provided solely by the prestress connecting the three elements. Because of the level of prestress applied, Eq. (5.103) indicated that no horizontal joint hoops were needed, and only nominal hoops, in accordance with Eq. (5.96), were provided. Similarly, vertical stirrups outside the joint region were nominal, since predicted tension stresses from Eq. (5.89) were low. Figure 5.69(*a*) shows reinforcement details in the joint region. The unit performed

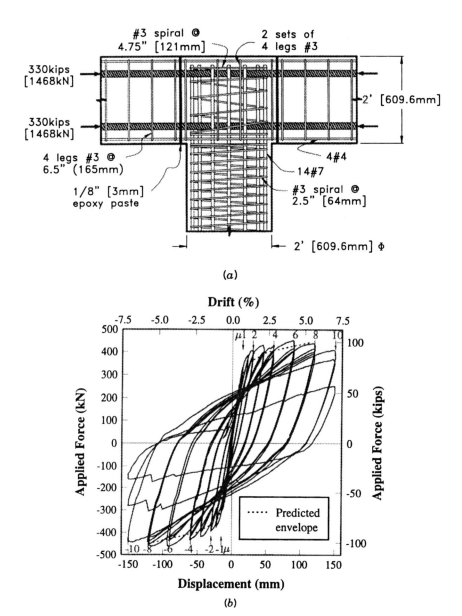

FIG. 5.69 Response of a precast tee joint with prestressed cap beam and minimal joint reinforcement. Dimensions and design strength are as for unit of Fig. 5.68. (*a*) Joint reinforcement details; (*b*) Lateral force–displacement response.

extremely well, attaining displacement ductilities of $\mu_\Delta = 10$ before the plastic hinge region failed due to excessive plastic rotation, as shown in Fig. 5.69(b). Behavior was significantly better than for the conventionally reinforced unit shown in Fig. 5.68, despite the greatly reduced amount of special joint shear reinforcement. Again, this illustrates the advantages of capbeam prestressing.

(c) Longitudinal Response. The general equilibrium equations and force transfer mechanisms governing longitudinal response of monolithic super-structure/column designs are essentially the same as for transverse design of tee joints. If the longitudinal girders are prestressed, however, Eqs. (5.85) and (5.86) must be modified to take account of the height of the resultant prestress force, which may not be at section midheight, as assumed in these equations.

It should be noted that longitudinal prestressing is unlikely to be as effective in joint shear force transfer as for transverse response with box girder construction because the prestress force will be provided to the cap beam from the deck and soffit slabs and through the webs. If the joint region is midway between webs, as shown in Fig. 5.70(b), it is unlikely that the prestress can provide any effective clamping force over the joint region. Until recently, little cover was provided on the sides of the cap beam over the column reinforcement, making satisfactory longitudinal response difficult to achieve. Extensive joint cracking and damage can be expected even if large amounts of hoop reinforcement are placed, since vertical force transfer is not achieved.

As with tee joints under transverse response, the three options shown in Fig. 5.66 can be considered for design. Of these, Fig. 5.66(c), also favored for transverse response, is the most constructable. For this mechanism to work for longitudinal response, the side cover of the cap beam must be sufficiently large to place the vertical reinforcement required by Eq. (5.100). Typically, this will mean a minimum of 12 in. (300 mm). As with transverse response, this mechanism will involve internal vertical reinforcement satisfying Eq. (5.102),

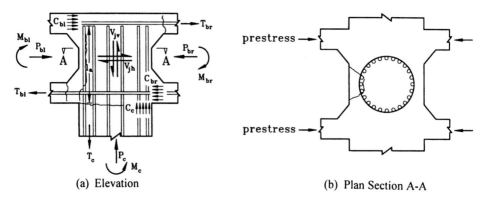

(a) Elevation   (b) Plan Section A-A

FIG. 5.70 Longitudinal response with monolithic superstructure–column design.

additional longitudinal soffit reinforcement satisfying Eq. (5.101), and hoop reinforcement in accordance with Eq. (5.104).

(d) Joint Reinforcement for Longitudinal and Transverse Seismic Response.
It has been established that the most practical joint reinforcement design involves utilization of the external reinforcement mechanism of Figs. 5.60(c) and 5.66(c). With monolithic superstructure–column details, joint reinforcement will be needed for both transverse and longitudinal response. Figure 5.71 defines, in plan view, the areas of cap beam within which vertical joint reinforcement should be placed, for internal and external columns of multicolumn bents, which require reinforcement for both longitudinal and transverse response, and for single-column bents, which normally require reinforcement only for longitudinal response. However, it should be noted that with highly curved continuous superstructures supported on single-column bents, hinging

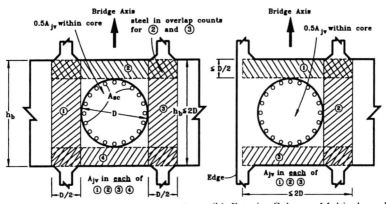

(a) Internal Column: Multicolumn Bent (b) Exterior Column: Multicolumn Bent

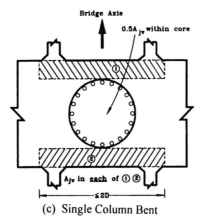

(c) Single Column Bent

FIG. 5.71 Locations for vertical joint reinforcement using external reinforcement. [*Note:* A_{jv} is given by Eq. (5.100).]

can occur at the column tops under transverse response, thus requiring joint reinforcement as for multicolumn bents.

In Fig. 5.71 the reinforcement for each direction of response is located within an area of width $2D$, for a circular column (or $b_c + h_c$ for a rectangular column). For an interior column of a multicolumn bent [Fig. 5.71(a)], four overlapping areas thus result. Within each area, vertical reinforcement satisfying Eq. (5.100) must be placed. The joint reinforcement within the overlap of areas can be counted as effective for both contributing areas, thus reducing the total quantity of reinforcement.

The extent of the additional beam longitudinal reinforcement to be placed near the bottom of the superstructure or cap beams in accordance with Eq. (5.101) is shown in Fig. 5.72. This reinforcement should extend past the column cage a distance not less than $l = 0.5D + l_d$, where l_d is the development length for the bar size, discussed further in Section 5.5. Note that for monolithic superstructure–column designs, as shown in Fig. 5.71(a) and (b), this additional reinforcement must be placed both longitudinally and transversally.

In all cases, transverse hoop reinforcement in accordance with Eqs. (5.103) or (5.104) [but not less than required by Eq. (5.96)] should be provided.

(e) ***Example of Joint Design for Combined Longitudinal and Transverse Response of an Internal Column of a Multicolumn Bent.*** The cap beam tee joint of Fig. 5.73 consists of a 5-ft (1524-mm)-diameter column reinforced with 32 No. 14 (44.5 mm) bars ($\rho_l = 0.025$) arranged in a single ring. The

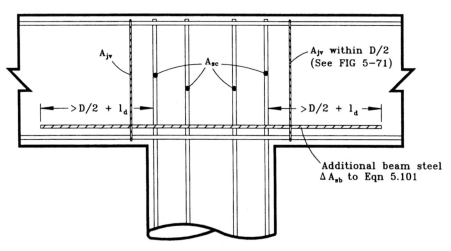

Note: l_d = development length
: ΔA_{sb} required longitudinally also

FIG. 5.72 Additional positive moment cap beam reinforcement for joint force transfer in accordance with Eq. (5.101).

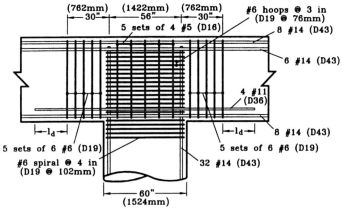

Cap Beam Elevation

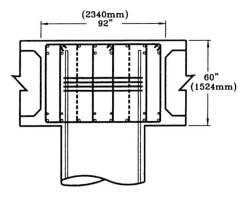

Cap Beam Section A–A
(Hoops Partly Omitted for Clarity)

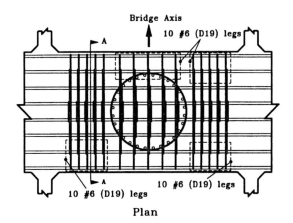

Plan

FIG. 5.73 Joint design for example.

column axial force of 700 kips (3115 kN) is not modified significantly by seismic loads. Grade 60 (f_y = 414 MPa) reinforcement is used throughout. Design of the joint reinforcement for longitudinal and transverse response is required. Concrete specified 28-day strength is f'_c = 4 ksi for column and cap beam.

From preliminary analysis, the column overstrength moment capacity is

$$M° = 13,000 \text{ kip-ft (17.6 MNm)}$$

CAP BEAM FLEXURAL REINFORCEMENT. Based on a capacity design approach, cap beam reinforcement is

Top steel	14 No. 14 bars in two layers
Bottom steel	8 No. 14 bars in one layer

This reinforcement is shown in Fig. 5.73, which also shows the final joint reinforcement details.

JOINT PRINCIPAL TENSION STRESS. Based on an assumed initial cap beam width of 72 in. (1830 mm), principal stresses, in accordance with Eq. (5.89), are calculated as 563 psi ($8.9\sqrt{f'_c}$ psi = $0.74\sqrt{f'_c}$ MPa). Thus full joint design is required—reductions of Section 5.4.4(a)(iii) do not apply.

DESIGN DETAILS. A design based on Figs. 5.66(c) and 5.71(a) is chosen. In accordance with Eq. (5.100), this requires external joint reinforcement of

$$A_{jv} = 0.125 \times (32 \times 2.25) \times 1.4 = 12.6 \text{ in}^2 \text{ (8127 mm}^2\text{)}$$

in each of the four areas of Fig. 5.71(a). This can be provided by 30 No. 6 (19.05 mm diameter) legs. To accommodate these, the cap beam width is increased to 92 in. (2340 mm). This will have the added benefit of reducing principal tension and compression stresses.

Internal joint vertical reinforcement of A_{vi} = 6.3 in^2 (4063 mm^2) is required within the column core, in accordance with Eq. (5.104). Transverse hoop reinforcement must satisfy Eq. (5.104), since the cap is not prestressed. Hence

$$\rho_s = \frac{0.3(32 \times 2.25)1.4}{54^2} = 0.0104$$

where the anchorage length of l_a = 54 in. (1372 mm) is taken as high as practicable. Note that Eq. (5.96) does not govern for this case. The value of ρ_s = 0.0104 is to be provided by No. 6 welded hoops. Hence the required spacing is

$$s = \frac{4A_h}{D'\rho_s} = \frac{4 \times 0.441}{56 \times 0.0104} = 3.03 \text{ in.}$$

Thus No. 6 hoops (19.05 mm diameter) (or spirals) at $s = 3$ in. (76 mm) are provided for the full height.

Details of the joint design are included in Fig. 5.73. Five sets of six No. 6 legs are provided in the cap beam adjacent to and on each side of the column core. These provide the full 30 No. 6 bars required for transverse response. The corner 10 No. 6 legs, shown within the dashed boxes, also contribute to longitudinal response. Hence, only 10 additional legs are required on each side of the cap beam for longitudinal response. These are provided by five sets of four-leg No. 6 stirrups, straddling the column. Five sets of four-leg No. 5 (15.9 mm diameter) stirrups provide the vertical internal reinforcement. These are hooked around the bottom beam steel. Along the cap beam, four No. 11 bars (35.8 mm) are placed in accordance with Eq. (5.101). These are not needed for longitudinal response, since the bottom horizontal legs of the external stirrups provide the necessary restraint to the outward-inclined struts of Fig. 5.60(c).

5.5 ANCHORAGE, DEVELOPMENT, AND SPLICING

5.5.1 Introduction

Anchorage, development, and splicing of reinforcing bars all rely on force transfer from the bars to surrounding concrete, and at least to some extent, on the tension strength of the concrete. The force transfer is improved by the deformations on the reinforcing bars and by 90° or 180° hooks, if used. Plain bars without hooks provide only limited force transfer, and this degrades under cyclic loading.

Two very different conditions may be identified for both lap splice and anchorage, depending on whether or not the splice or anchorage is effectively confined. Confinement may result from a large amount of transverse hoop reinforcement or by the influence of concrete and transverse reinforcement intended for flexure or shear strength in an adjacent member. In unconfined situations, anchorage or splice failure normally occurs as the result of splitting cracks developing in the concrete around the bar or bars considered. This is illustrated in Fig. 5.74(a). Where bars are close together, splitting cracks tend to form between the bars, reducing the surface area of the failure surface. Without the beneficial influence of confinement, these cracks dilate, and the ability of the bar or bars to maintain strength decreases, causing failure. If, however, the dilation of the splitting cracks is restrained by transverse reinforcement or concrete compression forces, a different form of failure develops, involving crushing of the concrete in front of the transverse deformations of the bar and development of a shear failure surface or "sleeve" of

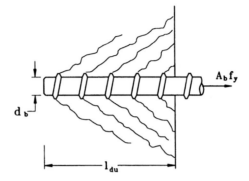

(a) Splitting failure in unconfined anchorage

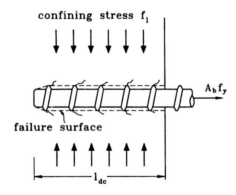

(b) Crushing and sleeving failure in
confined anchorage

FIG. 5.74 Anchorage failure of a single reinforcing bar.

diameter a little larger than that of the outer surface of the deformations, as shown in Fig. 5.74(b). Results of experiments on confined anchorage [E7], and from tests of elements with high moment gradients in a confined situation such as beam bars passing through interior beam–column joints under seismic loading [P4], indicate that much shorter development lengths are adequate for confined anchorage conditions than for the typical unconfined condition envisaged by most design codes.

Different situations, common in bridge design, involving anchorage and splicing are illustrated in Fig. 5.75. Anchorage conditions are represented by the lengths l_1 to l_4 and splicing by l_5 to l_7. At the column–cap beam connection shown in Fig. 5.75(c), the anchorage of bars close to the longitudinal axis is unlikely to benefit from clamping pressure other than that provided by transverse hoops. Bars close to the transverse axis may, however, have improved

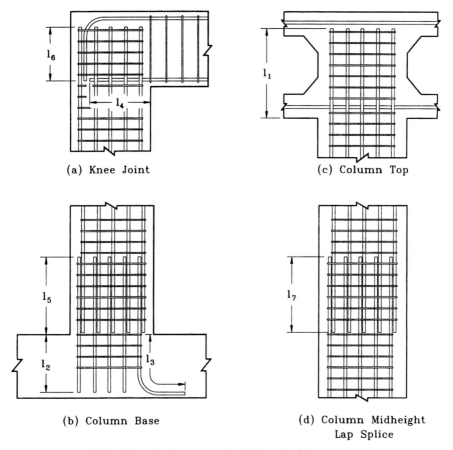

(a) Knee Joint   (c) Column Top

(b) Column Base   (d) Column Midheight
                                              Lap Splice

FIG. 5.75   Anchorage and splicing of reinforcement.

anchorage as a result of clamping action provided by transverse prestressing, or by diagonal compression struts set up by joint force transfer, as discussed in Section 5.4.4(a).

At the bottom of the column [Fig. 5.75(b)] with a moment-resisting column–footing connection, it is difficult to imagine a splitting failure of the sort represented by Fig. 5.74(a) occurring from either the straight (l_2) or bent (l_4) anchorage, provided that adequate footing–top horizontal reinforcement is present to clamp the crack. It should be noted, however, that requirements of Section 5.6.2(c) for amount and distribution of footing–top reinforcement must be satisfied to ensure that this clamping can occur, and joint shear reinforcement [discussed in Section 5.6.4(b)] would also be needed. Similarly, a splitting failure seems unlikely for the cap beam bottom reinforcement of Fig. 5.75(a), (l_4) provided that this is carried back into the compression zone of the column. Thus for these situations, the sleeving failure mode of Fig.

5.74(*b*) (but involving several parallel bars simultaneously) seems more likely, and reduced anchorage length could be adopted.

Lap splices in plastic hinge zones, such as the base of columns [Fig. 5.75(*b*), l_5], should not be used. If unconfined, these will break down under cyclic inelastic action, as discussed subsequently, even if very long splice lengths are used. If the splice is confined by large amounts of transverse reinforcement, splice failure may be inhibited but the consequence will be a shortening of the effective plastic hinge length since the spread of plasticity will be reduced because of the doubling of the effective longitudinal reinforcement ratio within the splice. This will result in higher-than-expected plastic curvatures and early onset of ductility failure. At the top of the column, with the typical knee joint detail of Fig. 5.75(*a*), the situation at the outside of the joint (length l_6) will be similar to that at the base of the column as a result of inclined shear cracking of the joint, as discussed in Section 5.4.4(*a*).

Finally, Fig. 5.74(*d*) shows a lap splice in the central region of a column. This is acceptable, provded that there is assurance that inelastic action cannot spread to the splice as a result of strain hardening in the plastic hinge region, tension shift effects, or higher-mode effects. These were discussed in relation to Fig. 5.37.

5.5.2 Codified Development Equations

Despite the differences in actions involved in anchorage, splicing, and flexural bond, and the considerable influence of confinement, it is common for codes to treat the conditions as essentially similar, with, perhaps, modifying factors applied to the basic equations for the different condition. Prior to about 1970, it was common to have a basic development length expressed simply as a multiple of bar diameter. For example,

$$l_d = \begin{cases} \dfrac{f_y d_b}{2} & (f_y \text{ in ksi}) & (5.108a) \\[2ex] \dfrac{f_y d_b}{13.75} & (f_y \text{ in MPa}) & (5.108b) \end{cases}$$

is implied by common provisions of the 1960s. More recently, the important role of the tension strength of concrete has been recognized, and current provisions include $\sqrt{f_c'}$ as a measure of tension strength in the development length equation. A common basic development length equation, incorporated in ACI 318-92 [A5] and other codes, is

$$l_d = \begin{cases} \dfrac{0.04 A_b f_y}{\sqrt{f_c'}} & (\text{in.}) \quad (f_y f_c' \text{ in psi}) & (5.109a) \\[3ex] \dfrac{0.0189 A_b f_y}{\sqrt{f_c'}} & (\text{mm}) \quad (f_y f_c' \text{ in MPa}) & (5.109b) \end{cases}$$

This is then modified by a number of factors representing the effects of bar spacing, cover, confinement, position in the concrete pour, ratio of amount of reinforcement provided to that required by analyses, and for lap splices, the severity of loading on the lap splice. Although suitably conservative for a general design situation, Eq. (5.109) is unfortunately dimensionally incorrect, since simple manipulation enables Eq. (5.109) to be expressed in the form

$$
\frac{l_d}{d_b} =
\begin{cases}
\dfrac{0.0314 f_y}{\sqrt{f_c'}}\, d_b & (f_y, f_c' \text{ in psi}) & (5.110a) \\[3mm]
\dfrac{0.0148 f_y}{\sqrt{f_c'}}\, d_b & (f_y, f_c' \text{ in MPa}) & (5.110b)
\end{cases}
$$

This implies that the development length, as a multiple of bar diameter d_b, increases as the bar size increases. For typical values of $f_y = 60,000$ psi (414 MPa), and $f_c' = 4000$ psi (27.6 MPa), Eq. (5.110) implies required basic development lengths of $l_d = 11.2d_b$ and $l_d = 67.0d_b$ for No. 3 (9.4 mm diameter) and No. 18 (57.2 mm diameter) bars, respectively. Dimensional analysis, supported by experimental research, indicates that the basic development length ratio should be independent of the bar size. To some extent the potential discrepancies implied by Eq. (5.110) are mitigated by the values of the correction factors for cover, spacing, and so on. At the time of writing this book, significant changes to the ACI-318 equations, removing this inconsistency, were being considered. Further information on the basic code approach, including numerical values for modification factors, are included elsewhere [A5,S2,P4].

5.5.3 Anchorage in Confined Conditions

For large-diameter reinforcing bars, the development lengths required by Eq. (5.110), in conjunction with the appropriate modification factors, can be excessive. It is not uncommon to find required l_d values for No. 14 (42.9 mm) and No. 18 (57.2 mm) bars as large as 110 to 180 in. (2790 to 4570 mm). It is almost impossible to accommodate such anchorage lengths within normal superstructures depths.

However, tests on large-scale column–cap beam connections indicate that the typical code requirements for confined anchorages are unnecessarily large. Figure 5.76 shows lateral force–displacement response for a full-size 60-in. (1524-mm) diameter column reinforced with 20 No. 18 bars connected to a prestressed superstructure and subjected to longitudinal seismic response [S6]. As evident in Fig. 5.76(a), the column was tested inverted, for convenience. The longitudinal No. 18 column bars were anchored by straight 63-in. (1600-mm) extension into the 72-in. (1829-mm)-deep cap beam. Since testing was carried out in the longitudinal direction, the column bars would not have

(a)

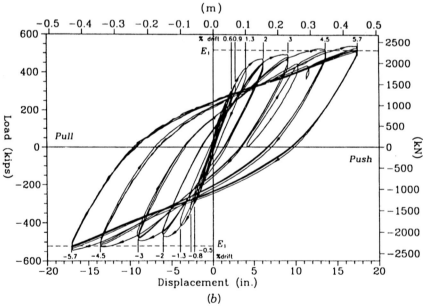

(b)

FIG. 5.76 Lateral force–displacement of a full-scale column–cap connection with No. 18 (57.1 mm) bars anchored 60 in. (1524 mm) (f_y = 77.5 ksi = 534 MPa). (a) Unit under test; (b) Longitudinal force–displacement response.

received any beneficial clamping from the superstructure longitudinal pre-stress. Joint reinforcement complied with the requirements of Section 5.4.4(c).

As is apparent from the longitudinal force–displacement response, excellent ductile response was achieved, with a well-controlled plastic hinge forming at the column–cap beam junction. Longitudinal column bars responded well into the strain-hardening range of response, despite the steel yield strength of 70 ksi (534 MPa), some 17% above the nominal yield. Strain gauge measurements within the joint region indicated that the column longitudinal reinforcement was yielding as close as 18 in. (450 mm) from the free end. This corresponds to a development length of about $l_d = 8d_b$. Data from this and other column–cap beam experiments [I5,M8] support ultimate bond stresses of about $30\sqrt{f_c'}$ psi ($2.5\sqrt{f_c'}$ MPa). Similar values are quoted elsewhere with reference to bond stresses in beam bars passing through beam–column joints [E7,P4]. Nevertheless, it would be unwise to rely on such high average values, since conditions close to the column–cap beam junction provide poor confinement. As a consequence it is recommended that column bars be anchored as close as possible to the far face of the cap beam (or footing, for a column–footing connection) but with an anchorage length of not less than

$$
l_{dc} = \begin{cases} \dfrac{0.025 d_b f_{ye}}{\sqrt{f_c'}} & \text{(in.)} \quad \text{(psi)} & (5.111a) \\[3mm] \dfrac{0.3 d_b f_{ye}}{\sqrt{f_c'}} & \text{(mm)} \quad \text{(MPa)} & (5.111b) \end{cases}
$$

To ensure conservative results, f_c' rather than f_{ce}' is used in Eq. (5.111) [see Section 5.3.1(d)(vii)].

Putting $A_b f_{ye} = \pi d_b u_u l_{dc}$, where u_u is the average ultimate bond stress, it can readily be shown that Eq. (5.111) corresponds to an average bond stress of only $14\sqrt{f_c'}$ psi ($1.17\sqrt{f_c'}$ MPa), for a reinforcement overstrength of $f_{yc}^o = 1.4 f_{ye}$.

Where confinement is not provided by adjacent members, or by joint reinforcement placed in accordance with Section 5.4.4, the anchorage should be confined by sufficient transverse reinforcement to transfer the column bar stress to the concrete by shear friction, using a coefficient of friction of $\mu = 1.4$. A situation where this would apply would be the knee joint of Fig. 5.77(b), which has already been discussed in Section 5.4.4(a). With respect to Fig. 5.77(a), the clamping force provided to each longitudinal bar by a hoop or spiral of area A_h at a stress of f_s is $A_h f_s \cdot 2\pi/n$. Hence to develop a stress of f_{yc}^o in the column bar, the following equality must hold:

$$
\mu \frac{A_h f_s \cdot 2\pi}{n} \frac{l_a}{s} = A_b f_{yc}^o
$$

which for $\mu = 1.4$ simplifies to

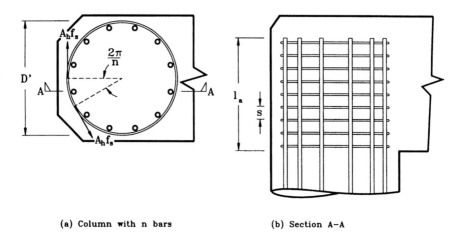

(a) Column with n bars   (b) Section A-A

FIG. 5.77 Anchorage by confining effect of transverse reinforcement.

$$\frac{A_h}{s} = \frac{0.36nA_b f^{\circ}_{yc}}{l_a \pi f_s} \tag{5.112}$$

Now since the volumetric ratio of transverse reinforcement is $\rho_s = 4A_h/D'S$, and $nA_b = A_{sc}$, this requires that

$$\rho_s = \frac{0.46A_{sc}}{D'l_a} \frac{f^{\circ}_{yc}}{f_s}$$

which has previously been designated Eq. (5.92). Expressing $A_{sc} = \rho_l \pi D^2/4$, approximating $D' = D$, and rounding up coefficients, this can be rewritten in a slightly more convenient form as

$$\rho_s = \frac{0.36\rho_l D}{l_a} \frac{f^{\circ}_{yc}}{f_s} \tag{5.113}$$

where, again, $f_s = 0.0015E_s$ and ρ_l is the column longitudinal steel ratio.

5.5.4 Splicing of Reinforcing Bars

As noted in Section 5.5.1, longitudinal reinforcement should not be spliced in plastic hinge regions, or within a distance from the plastic hinge regions equal to the column depth or diameter, to allow for tension shift effects. This restriction should apply whether the bars are spliced by lapping or welding, or by proprietary mechanical connectors. Mechanical or welded connections should be subjected to stringent quality control, including testing of representative samples prepared in the field, to ensure satisfactory strength and deformation characteristics. It should be noted that some mechanical connectors display significant slip before force transfer is effected. This can result in increased width of cracking and a consequent reduction in aggregate-interlock shear transfer. Conservative design for shear, including reduced spacing of transverse reinforcement, is needed in such cases. Welded longitudinal reinforcing bars should be connected by full penetration butt welds. They must never be connected by lap-splice welding, since the eccentricity of force transfer can result in bending moments developed in the bars, with a consequent reduction in axial strength.

The force-transfer mechanism in conventional lap splices is quite complex. However, insight into the strength of lap splices may be obtained from consideration of the mechanism of failure. As is indicated in Fig. 5.78 for both circular and rectangular columns, lap-splice failure, involving relative longitudinal movement of the spliced bars, requires the formation of a series of fracture surfaces perpendicular to the column surface to allow the bars to slide relative to each other and a further fracture surface parallel to the column surface to allow the radial cracks to dilate and to permit the bars to slide relative to the column core. There is thus a characteristic block of concrete of length l_s equal

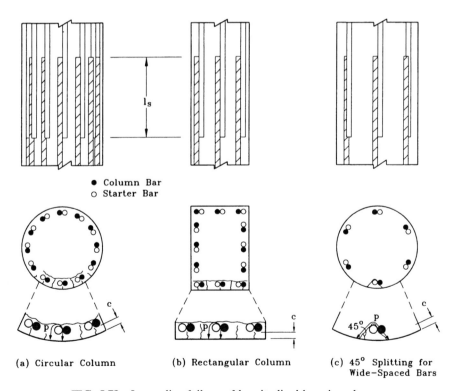

(a) Circular Column   (b) Rectangular Column   (c) 45° Splitting for
                                                        Wide-Spaced Bars

FIG. 5.78 Lap-splice failure of longitudinal bars in columns.

to the lap-splice length and perimeter p (see Fig. 5.78) associated with each bar. If we allow some plasticity in the tension strength of the concrete, the transverse force resisting formation of the fracture surfaces is thus $f_t p l_s$, where f_t is the tension strength of the concrete. Assuming that the resistance to sliding is provided by 45° diagonal struts between deformations of adjacent bars, or between bars and the column core, as shown in Fig. 5.79, the transverse tension force is equal to the longitudinal resistance. Thus the maximum bar force T_b that can be transferred without the assistance of special transverse reinforcement confining the splice is

$$T_b = A_b f_s = f_t p l_s \tag{5.114}$$

For the circular column of Fig. 5.78(a) and (c), the perimeter p of the characteristic block is

$$p = \frac{\pi D'}{2n} + 2(d_b + c) \le 2\sqrt{2}\,(c + d_b) \tag{5.115}$$

where n is the number of longitudinal bars of diameter d_b evenly spaced around the core, of diameter D', with cover c. For the rectangular column of Fig. 5.78(b),

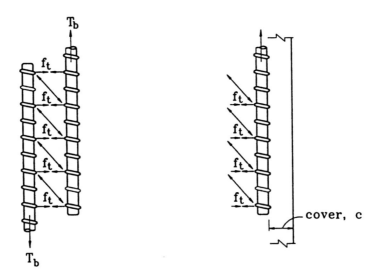

(a) Transfer between bars   (b) Transfer between bar and core

FIG. 5.79 Tension stresses induced by force transfer in lap splices.

$$p = \frac{s}{2} + 2(d_b + c) \leq 2\sqrt{2}\,(c + d_b) \tag{5.116}$$

where s is the average spacing between spliced pairs of bars along the critical column face. The upper limit to Eqs. (5.115) and (5.116) applies when bars are widely spaced, and as shown in Fig. 5.78(c), failure by the starter bars pulling off the core with a 45° wedge results in a lower effective perimeter.

Tests on column lap splices indicate that when the tension strength is taken as $f_t = 4\sqrt{f'_c}$ psi [$0.33\sqrt{f'_c}$ MPa], Eqs. (5.114) to (5.116) result in a conservative estimate of the tension force T_b that can be transferred across a lap splice. As discussed further in Chapter 7, this is particularly useful in assessing the strength of existing lap splices for seismic competence.

For circular columns it will be found that Eq. (5.115) is not very sensitive to the bar spacing for the range typical of columns. Assuming typical cover and making several conservative assumptions, Eqs. (5.114) and (5.115) can be rearranged to yield the following simplified equation for required splice length, in a form similar to that of Eq. (5.111):

$$l_s = \begin{cases} \dfrac{0.04 d_b f_s}{\sqrt{f'_c}} & \text{(in.)} \quad \text{(psi)} & \tag{5.117a} \\[2em] \dfrac{0.48 d_b f_s}{\sqrt{f'_c}} & \text{(mm)} \quad \text{(MPa)} & \tag{5.117b} \end{cases}$$

where f_s is the maximum bar stress to be transferred, which should be taken as f_{ye} unless a lesser value is supported by moment–curvature analysis of the section under a moment corresponding to flexural overstrength of the plastic hinges and with due recognition of tension shift effects.

It is advisable that the lap splice be adequately confined, to ensure that the required force can be transferred by shear friction in the event that the tension strength of the concrete is less than expected. In this case the resistance to sliding provided by concrete tension is likely to be completely lost by the stage that the cracks dilate sufficiently to induce significant tension stress in the hoop confinement. As a consequence, resistance of the original concrete tension mechanism and final shear friction mechanism should not be considered additive.

The confinement reinforcement can best be considered as providing a clamping stress across the fracture surfaces developing in accordance with the mechanisms of Fig. 5.78. Tests indicate that a coefficient of friction of $\mu = 1.4$ is appropriate, provided that the cracks do not dilate excessively, represented by an equivalent dilation strain of $\varepsilon_s = 0.0015$. Thus the maximum hoop tension stress relied on in the confining reinforcement should be $f_{sh} = 0.0015 E_s$. For circular columns the required volumetric ratio of transverse hoop reinforcement can then be shown to be

$$\rho_h = \frac{1.4 A_b f_s}{p l_s f_{sh}} \tag{5.118}$$

where p is given by Eq. (5.115).

For rectangular columns, the amount of transverse reinforcement will depend on the number of cross ties and the configuration of hoops adopted, and a first-principles approach, in accordance with the shear friction mechanism outlined above, is required.

5.5.5 Flexural Bond

In short columns, where plastic hinges of opposite sign develop simultaneously at the top and bottom of the column, bond conditions caused by the requirement to transfer force from bar to concrete as a result of the rapidly changing moment may be extreme. It is thus important to use smaller-diameter bars in such situations. It is recommended that requirements be similar to those for lap splices, and hence Eq. (5.117) may be taken to be applicable, where l_s is the distance from the critical section to the point of contraflexure. This may be inverted to obtain the following expression for the maximum bar diameter for a given column height:

$$d_b \leq \begin{cases} \dfrac{25\sqrt{f_c'}}{f_{ye}} \, l_b & \text{(in.)} \quad \text{(psi)} & (5.119a) \\[3mm] \dfrac{2.1\sqrt{f_c'}}{f_{ye}} \, l_b & \text{(mm)} \quad \text{(MPa)} & (5.119b) \end{cases}$$

where l_b is the distance from a critical section, where a plastic hinge may form, to the point of contraflexure. Equation (5.119) implies that for typical material properties of $f_c' = 4000$ psi (27.6 MPa), $f_{ye} = 66{,}000$ psi (455 MPa), a 10-ft (3048-mm)-tall column subjected to equal and opposite moments top and bottom, would have a maximum permissible bar diameter of 1.44 in. (36.9 mm). Thus a U.S. No. 11 bar, but not a No. 14 bar, would be permissible.

5.6 FOOTINGS AND PILE CAPS

5.6.1 Introduction

Different footing design options are discussed in Section 3.2.6. The choice between spread footings, pile-supported footings, or cylinder-supported footings will depend on ground conditions and local economic considerations. Analysis for soil–structure interaction effects was discussed in Section 4.4.2(d). Aspects of footing design to be considered in addition to the choice of footing support type include flexural strength, shear strength, column–footing and pile–footing connection details, and anchorage details.

5.6.2 Design for Flexure

(a) Stability. In normal seismic design of footings, the footing should be capable of resisting the input moment corresponding to flexural overstrength development in the plastic hinge mechanism. This requires that the footing system be stable under the overstrength response applied, as illustrated in Fig. 5.80, where three different footing types are subjected to axial load P (including seismic component), and overstrength moments and shears $M°$ and $V°$.

In stability calculations, ultimate capacities of the soil bearing pressure and pile compression and tension capacities should be used, together with a strength reduction factor ϕ. Thus for the spread footing of Fig. 5.80(a), the stability requirement is that

$$\phi(P + W_f)\left(\frac{L_f - a}{2}\right) \geq M° + V°h_f \tag{5.120}$$

where W_f is the total footing weight, L_f the footing length, h_f the footing depth, and

$$a = \frac{P + W_f}{p_u B_f} \tag{5.121}$$

is the depth of the ultimate compression block of soil pressure, p_u acting on the base of the footing, which has a width B_f. Equation (5.120) assumes that end bearing force H acting on the end of the footing is negligible and that all shear resistance is provided at the base of the footing. Where end bearing resistance is dependable and can be mobilized with small deformations, as

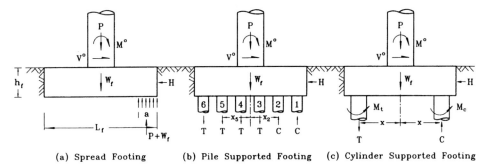

(a) Spread Footing   (b) Pile Supported Footing   (c) Cylinder Supported Footing

FIG. 5.80 Stability of footing systems.

might be the case when the footing is cast against an end-cut ground surface, modification of Eq. (5.120) to incorporate the small overturning resistance of the force H is obvious.

With the pile-supported footing of Fig. 5.80(b), ultimate pile compression and tension resistances may be used in the stability calculations. The former will typically be three to four times the safe bearing capacities given for gravity-load designs. For the details of Fig. 5.80(b), using the row numbering system defined therein, the stability requirement is

$$\phi[C(x_1 + x_2) + T(x_4 + x_5 + x_6 - x_3)]n \geq M° + V°h_f \qquad (5.122)$$

where it is assumed that each pile has a tension or compression capacity of T or C, and there are n piles in each row. Vertical equilibrium requires that

$$\Sigma C - \Sigma T = P + W_f \qquad (5.123)$$

It is further assumed in Eq. (5.122) that the moment resistance of the pile–footing connection is zero. For moment-resisting connections, a moment component should be added and the value of this found from analysis of the pile–soil interaction under the appropriate pile shear force (i.e., $V = V°/6n$, in this case, again assuming that $H = O$).

For the cylinder-supported footing of Fig. 5.80(c), the stability calculation requires that

$$\phi(Cx + Tx + M_c + M_t)n \geq M° + V°h_f \qquad (5.124)$$

where

$$n(C - T) = P + W_f \qquad (5.125)$$

and where, again, there are n cylinde:s in each row and M_c and M_t are the moments in the compression and tension cylinders under the appropriate proportion of the overstrength shear. If the axial loads on the tension and compression cylinders differ greatly, the moments M_c and M_t may also differ significantly and should correspond to the same curvature. This may require a detailed analysis of the soil–structure interaction, in accordance with the approach outlined in Section 4.4.2(d).

The value of the strength reduction factor ϕ included in the stability equations above depends, to some extent, on judgment. If the bearing strength of soil and pile axial load capacities are based on conservative strength estimates, as will usually be the case, a value of $\phi = 1$ will be satisfactory, since the probability of extreme overstrength moment capacity and low soil capacity occurring simultaneously are very low. In this context it should also be pointed

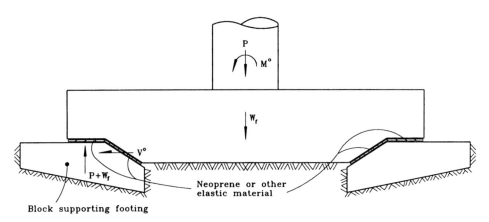

FIG. 5.81 Schematic details of a rocking footing.

out that there are very few, if any, reports of footing failures except where caused by liquefication or sliding of ground on sloping surfaces.

An alternative structural concept is outlined in Fig. 5.81. In this case the footing is deliberately undersized to *ensure* that it will rock under seismic response at *nominal* column capacity. Rocking acts as a form of seismic isolation, and is discussed further in Chapter 6. To ensure that plastic soil deformations do not reduce the effective length of contact between the footing and soil, it is advisable to support each end of the footing on special bearing blocks capable of transmitting the full axial force $P + W_f$ and shear force V, as shown. An elastic material such as neoprene should be placed between the footing and bearing blocks to reduce impact forces and spread the contact stresses.

(b) Design Moments. Determination of the actual moments and shears in the footing or pile cap under the overstrength plastic hinge input requires more precise analysis than that implied by the stability inequalities of Eqs. (5.120), (5.122), and (5.124). For the spread footing of Fig. 5.80(a), the pressure distribution could be elastic, with contact pressure over the full footing length; elastic, with partial length contact pressure; or plastic, with partial length contact pressure. The last two cases are shown in Fig. 5.82. In both cases, moment equilibrium at the base of the footing requires that the lever arm $\bar{y}$ between applied and resisting vertical forces is

$$\bar{y} = \frac{M° + V°h_f}{P + W_f} \tag{5.126}$$

For the elastic case of Fig. 5.82(a), the peak bearing stress p must then be

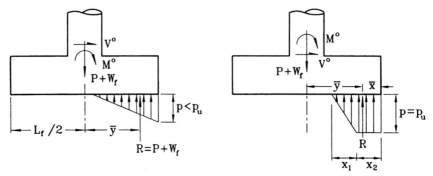

(a) Peak bearing stress $<$ p_u.   (b) Peak bearing stress $=$ p_u.

FIG. 5.82 Soil pressure distributions for footing.

$$p = \frac{2(P + W_f)}{3(0.5L_f - \bar{y})B_f} \leq p_u \qquad (5.127)$$

If Eq. (5.127) yields a peak stress $p > p_u$, the bearing stress distribution of Fig. 5.82(b) pertains, where bearing stress is constant and equal to p_u over the end distance x_2 and falls to zero over an additional length x_1. For equilibrium of axial force

$$B_f p_u(x_2 + 0.5x_1) = P + W_f \qquad (5.128)$$

For moment equilibrium,

$$\bar{x} = 0.5L_f - \bar{y} = \frac{0.5x_2 + 0.5x_1(x_2 + 0.33x_1)}{x_2 + 0.5x_1} \qquad (5.129)$$

Equations (5.128) and (5.129) can be solved simultaneously for x_1 and x_2, utilizing Eq. (5.12b) for $\bar{y}$, and the moment and shear distributions in the footing found accordingly from the final pressure distribution, noting that the footing weight provides bending moments in opposition to those of the pressure distribution.

For the pile-supported footing of Fig. 5.80(b), determination of the forces under the overstrength moment requires an incremental analysis if some of the piles reach the plastic tension or compression capacity. This is illustrated in Fig. 5.83. Assuming a rigid footing, the compression force per pile under axial forces is

$$D = \frac{P + W_f}{\sum_{i=1}^{m} n_i} \qquad (5.130)$$

where there are m rows of piles, with n_i piles per row.

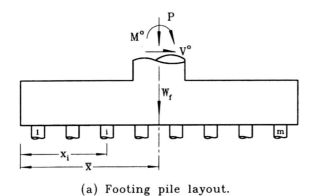

(a) Footing pile layout.

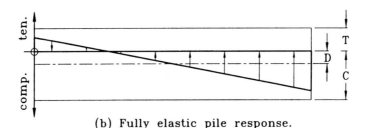

(b) Fully elastic pile response.

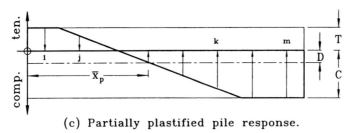

(c) Partially plastified pile response.

FIG. 5.83   Pile force distributions at column overstrength.

Under moments sufficiently small so that no pile reaches its tension capacity T or compression capacity C, as shown for example in Fig. 5.83(b), the pile forces are given by

$$T \leq \left[F_i = \frac{P + W_f}{\sum_1^m n_i} + \frac{M(x_i - \bar{x})}{\sum_1^m n_i(x_i - \bar{x})^2} \right] \leq C \qquad (5.131)$$

If substituting $M = M^\circ + V^\circ h_f$ in Eq. (5.131) results in extreme pile forces outside the limits of T and C, the moment required to plastify the first pile must be calculated and an incremental approach adopted for subsequent moment increments, where plastified piles are removed from the analysis, a new plastic centroid $\bar{x}_p$ is calculated, and a new moment of inertia of the remaining elastic piles is calculated. In the general case of Fig. 5.83(c), where $(j - 1)$ rows have plastified in tension and $(k + 1)$ rows have plastified in compression, the incremental forces induced in piles in rows j to k will be

$$\Delta F_i = \frac{\Delta M(x_i - \bar{x}_p)}{\sum_j^k n_i(x_i - \bar{x}_p)^2} \tag{5.132}$$

such that $T \le F_i + \Delta F_i \le C$, where F_i are the elastic forces in the piles prior to the increment ΔM of moment.

A more accurate simulation, including pile and footing flexibility, and pile top moments would require a detailed computer analysis and may be of unnecessary sophistication. Where cylinder-supported footings such as that shown in Fig. 5.80(c) are used, a detailed analysis using a distributed spring model to represent pile–soil interaction as discussed in Section 4.4.2(d) should be adopted.

Having found the force distribution at the base of the footing, moments and shears can be calculated directly by standard methods of structural analysis. Figure 5.84 shows typical results for a pile-supported footing, where pile-top moment capacity has conservatively been ignored. Bending moment and shear force envelopes are shown for the two opposite directions of response. Although not shown in Fig. 5.84, diagonal response should also be considered for stability and flexural design, where the corner piles receive maximum forces.

(c) Flexural Strength. For gravity-load design it is reasonable to expect the full footing width to be effective in resisting flexure, since the bending moments on opposite sides of the columns will be equal and have the same sign. As shown in Fig. 5.84(b), for seismic response, footing bending moments must change sign across the column dimension parallel to the direction of loading. This is effected solely by the column compression and tension resultants C and T [Fig. 5.84(a)]. It is unrealistic to expect reinforcement near the edges of the footing to be fully effective in flexure because of the rapid change in bending moment and the large distance from the column force resultants. This is essentially a shear lag effect which results in reinforcement tension stresses decreasing with distance from the side of the column. At ultimate footing capacity, the shear lag effect is reduced as yield strains eventually penetrate to the footing edge. However, by this stage, reinforcement strains and crack widths adjacent to the column will be very large, reducing the capacity of concrete shear-resisting mechanisms [see Section 5.3.4(b)]. Consequently, reinforcement close to the column face should remain elastic. Tests on column–footing connections [X1,X2] indicate that to ensure that footing reinforcement remains elastic under the design movement, corresponding to

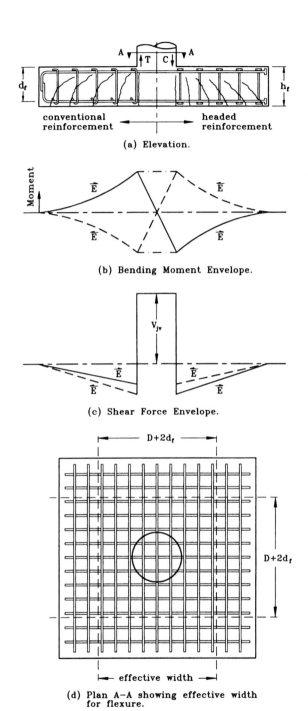

(a) Elevation.

(b) Bending Moment Envelope.

(c) Shear Force Envelope.

(d) Plan A–A showing effective width for flexure.

FIG. 5.84 Moments, shears, and reinforcement for footings (column and joint reinforcement omitted for clarity).

column flexural overstrength, only that reinforcement within an effective width of b_{eff} should be considered effective, where

$$b_{eff} = \begin{cases} D_c + 2d_f & \text{(circular columns)} & (5.133a) \\ B_c + 2d_f & \text{(rectangular columns)} & (5.133b) \end{cases}$$

where D_c and B_c are the diameter and width, respectively, of the circular or rectangular column and d_f is the effective depth of the footing. Thus the top and bottom footing reinforcement required to provide dependable footing flexural strength should be placed within an effective width of b_{eff}, as indicated in Fig. 5.84(d), with increased reinforcement spacing outside this width being permissible.

(d) Detailing. It is important that the footing reinforcement be anchored adequately since tension-shift effects and relatively short footing shear spans can result in high reinforcement tension stresses close to the ends of the footing. Two alternatives are shown in Fig. 5.84(a). To the left, conventional reinforcement is developed by 90° hooks bent parallel with the end face. The option shown to the right involves the use of headed reinforcement, which can develop its strength immediately adjacent to the enlarged head at the end. Note that depending on the mechanism chosen for joint shear resistance, additional footing-top reinforcement may be needed. This is discussed further in Section 5.6.4.

5.6.3 Design for Shear Strength

(a) Design Shear Force. Design shear forces corresponding to flexural overstrength of column plastic hinges is calculated from the same set of forces as that used to develop the design bending moments. A typical shear force envelope for a pile-supported footing is shown in Fig. 5.84(c). As is indicated in this figure, high vertical joint shear forces, V_{jv}, are expected within the column region as a result of the high vertical column force resultants C and T, shown in Fig. 5.84(a). Design for joint shear is discussed in the following section.

Because of the limited width of the compression support provided by the column, it is recommended that the footing be designed for the full shear force at the column face, not the reduced value a distance d_f from the column face. This is essential on the side of the footing adjacent to the tension face of the column, since a compression reaction is not available on this side to transfer shear input close to the column (Fig. 5.85).

(b) Shear Strength. Shear strength of footings should be assessed in accordance with Section 5.3.4(b)(iii). As with flexural design, it will be unrealistic to expect sections of the footing at large lateral distances from the column to be effective, and it is recommended that shear strength be assessed based on the effective width given by Eq. (5.133).

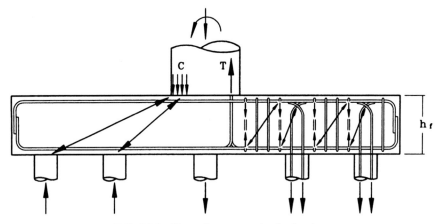

FIG. 5.85 Shear force transfer in footings.

On the compression side of the footing, shear force may often be transmitted by direct struts between the column flexural compression resultant and the compression piles, as indicated in Fig. 5.85. This mechanism can be expected to be effective for piles no further than $3h_f$ from the column centerline, and also for spread footings, where the center of soil pressure is no greater than $3h_f$ from the column centerline. In such cases nominal shear reinforcement should suffice.

As mentioned earlier, direct struts cannot be relied upon to transmit shears from the tension piles, which must use the conventional combination of concrete and transverse reinforcement shear-resisting mechanisms, as indicated on the right side of Fig. 5.85. Note that for these mechanisms to be effective, it is essential that the tension connection between the pile and the footing be anchored as high as possible in the footing, as discussed in Section 3.3.6(d).

(c) *Detailing.* Shear reinforcement in footings is normally provided by J bars, as shown on the left side of Fig. 5.84(a). It should, however, be recognized that the 90° hook provides only limited anchorage and support to the footing longitudinal reinforcement. An alternative, shown on the right of Fig. 5.82(b), uses headed reinforcing bars, which are easy to place and provide good support to both top and bottom longitudinal reinforcement layers.

It is recommended that regardless of the computed shear stress level, at least a minimum amount of transverse reinforcement satisfying

$$\rho v = \begin{cases} \dfrac{50}{f_y} & \text{(psi)} \qquad\qquad (5.134a) \\[2ex] \dfrac{0.35}{f_y} & \text{(MPa)} \qquad\quad (5.134b) \end{cases}$$

be provided, with spacing between the vertical legs of not more than $0.5h_f$. Equation (5.134) is equivalent to the familar minimum transverse reinforcement requirement for shear in ACI 318-92 [A5] and would require for a 36-in. (914-mm)-deep footing, transverse bars of grade 60 rebar ($f_y = 414$ MPa) of area $(50/60,000) \times 18 \times 18 = 0.27$ in^2 (174 mm^2) at 18-in. (457-mm) centers. This minimum requirement, also adopted by Caltrans [C1], has the secondary function of securing and locating the top mat of reinforcement relative to the bottom mat.

5.6.4   Design of Column–Footing Joints

(a)   Joint Cracking. Column–footing joints are essentially the same as inverted column–cap beam tee joints, with greater effective widths. If not designed for transfer of the forces from column into footing, shear failure may occur, as shown in the example of Fig. 5.86(*b*). As illustrated in Fig. 5.86(*a*), the vertical joint shear force can be assessed by subtracting the holddown force R_t due to the tension piles from the total reinforcement tension force T_c at the base of the column:

$$V_{jv} = T_c - R_t \tag{5.135}$$

Considering an effective joint width b_{jeff}, the average joint shear stress v_{jv} can be calculated as

$$v_{jv} = \frac{V_{jv}}{b_{\text{jeff}} h_f} \tag{5.136}$$

The effective width for joint shear will be less than for footing flexure and

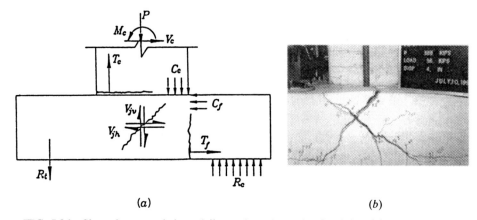

(a)   (b)

FIG. 5.86   Shear forces and shear failure of a column footing joint. (*a*) Forces acting on joint; (*b*) Shear failure of unreinforced joint.

shear, since the joint shear force results primarily from the column compression and tension stress resultants. As with cap beams, the effective width given by Eq. (5.88) should be assumed.

Principal tension stresses in the joint should be calculated in accordance with Eq. (5.89a). If these exceed $3.5\sqrt{f_c'}$ (psi) $[0.29\sqrt{f_c'}$ (MPa)] joint cracking is expected, although failure of the joint is unlikely to occur at principal tension stresses below $5\sqrt{f_c'}$ (psi) $[0.42\sqrt{f_c'}$ (MPa)] unless significant ductility levels are achieved within the plastic hinge region. Following recommendations of Section 5.4.4(a)(iii), reduced joint reinforcement may be placed when principal tension stresses are in the range $3.5\sqrt{f_c'} \le p_t \le 5\sqrt{f_c'}$ psi ($0.29\sqrt{f_c'} \le p_t \le 0.42\sqrt{f_c'}$ MPa.

(b) Mechanisms of Force Transfer. The mechanisms of force transfer discussed in Section 5.4.1(b) also apply to column–footing joint design. Contractors (and designers, as a matter of habit) prefer to provide anchorage for the column longitudinal reinforcement by bending the tails outward, thus making a stable platform for supporting the column cage. If this is done, any force transferred to the bend is directed away from the joint, increasing diagonal tension stress within the joint region. From a joint performance viewpoint, it is desirable to bend the column bars inward toward the joint. Provided that the bars are bent parallel to one of the two principal axes of the footing rather than bending along column radii and are arranged in two layers, this does not cause undue congestion. Both alternatives (i.e., inward and outward bending) are illustrated in Fig. 5.87(a), which also includes a third option, where the column bars are passed through the bottom mat of footing reinforcement into a drop cap.

When the column bars are bent inward, the tails should be restrained against straightening with vertical stirrups, providing a vertical restraint force of not less than $0.033A_bf_y$, centered $6d_b$ from the end of the bend, as discussed

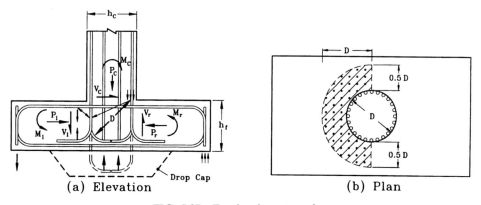

(a) Elevation   (b) Plan

FIG. 5.87 Footing force transfer.

in relation to Fig. 5.59(b). Transverse hoop reinforcement within the joint region should satisfy Eq. (5.104).

When column reinforcement is bent outward, it is best to rely on external joint reinforcement, using the mechanism developed for knee and tee joints, discussed in relation to Figs. 5.59(c) and 5.60(c). If all column reinforcement is bent outward, the amount of vertical reinforcement to be placed in each of the four quadrant areas outside the joint [see Fig. 5.71(a)] is given by Eq. (5.100). Again, vertical reinforcement in the overlapping corners can be taken to contribute to orthogonal directions of response. Where $X\%$ of the column bars are bent outward and the remainder are bent inward, the amount of external joint reinforcement to be placed in addition to that required for footing shear may be reduced proportionately.

As noted in relation to knee and tee joint designs, the external reinforcement mechanism for joint force transfer places extra demands on longitudinal reinforcement of the supporting member. In this case, extra top reinforcement in the footing is required. The area required will be 50% of the area of vertical external reinforcement placed and is given by Eq. (5.101) (or $X\%$ of this, where $X\%$ of column bars are bent outward and the remainder are bent inward). This should pass through the column or be placed as close as possible to the sides of the column, and extend a distance of not less than $l = 0.5D + l_d$, where l_d is the bar development length, beyond the column face on both sides of the column. In accordance with Eq. (5.102), internal vertical beam stirrup areas should be provided as a minimum within the column core.

Since the column inelastic action may develop in directions other than parallel to one of the principal axes of the footing, the requirements of Section 5.4.1(b) for external joint reinforcement can be interpreted as requiring a total vertical stirrup area of

$$A_{jv} = 0.50A_{sc} \frac{f_{yc}^\circ}{f_{yv}} \tag{5.137}$$

to be uniformly placed around the column. Figure 5.87(b) shows the region over which 50% of this area would be placed. Again, if only $X\%$ of the column bars are bent outward, with the remainder bent into the core, the reinforcement area given by Eq. (5.137) may be reduced proportionately.

(c) Summary of Footing Joint Design

1. (a) For column bars bent inward, provide a vertical restraint force for the tail of each bar of at least $0.033A_b f_y$. If the column bar and restraining bar have equal yield strengths, this implies a restraining bar diameter of $0.18d_b$, where d_b is the column bar diameter. To allow for possible material overstrength of the column bar, it is recommended that the minimum restraining bar diameter be one-fourth that of the column bar.

(b) Transverse hoop reinforcement around the column bars should satisfy Eq. (5.104).

2. (a) Where some or all of the column bars are bent outward or anchored by straight bar extensions into the joint, vertical external stirrups should be placed in the region defined in Fig. 5.87(b) in a total amount satisfying Eq. (5.137), reduced in proportion to the fraction of column bars bent inward. These vertical stirrups should be anchored by at least 135° hooks at each end, or headed stirrups should be used.

(b) Internal vertical stirrups in the joint core should not be less than required by Eq. (5.102).

(c) Additional top reinforcement satisfying Eq. (5.101) (reduced in proportion to the amount of column rebar bent inward) should be placed in the upper mat of footing reinforcement in the two orthogonal directions.

(d) Transverse hoop reinforcement round the column bars should satisfy Eq. (5.104) but not be less than required by Eq. (5.96).

(d) Footing Joint Design Example. The example used to illustrate tee joint design and shown in Fig. 5.73 is now considered for design of the footing joint, where the column is supported by a 4-ft (1219-mm)-deep pile-supported foundation. Under the column plastic moment capacity of 13,000 kip-ft (17,600 kNm), an uplift shear force of 350 kips (1560 kN) (footing weight plus pile tension) is developed. Preliminary calculations for principal tension stress result in a value exceeding $5\sqrt{f_c'}$ psi ($0.42\sqrt{f_c'}$ MPa), indicating that full joint design is necessary.

DESIGN OF JOINT REINFORCEMENT. Two alternatives are considered, as shown in Fig. 5.88. The first has all column steel bent inward in two orthogonal layers below the footing bottom reinforcing mat, in a drop cap. The horizontal end leg of each bend must be restrained vertically by a bar of diameter no less than 0.18×1.69 in. (No. 14) $= 0.304$ in. (7.73 mm). A No. 3 bar (0.375 in. $= 9.52$ mm diameter) provides adequate excess to cope with possible overstrength of the No. 14 bar. Since the column contains 32 No. 14 bars, 32 No. 3 support legs arranged in 16 closed stirrups are provided. (*Note:* The No. 3 bar is larger than $0.18d_b$, but slightly smaller than the recommended $0.25d_b$.) Horizontal hoop reinforcement is required in accordance with Eq. (5.104). Hence conservatively taking $l_a = 54$ in. (1219 mm) and $f_{yc}^o = 1.4 f_{yh}$ yields

$$\rho_s = \frac{4A_h}{D_s'} = \frac{0.3 \times (32 \times 2.25) \times 1.4}{54^2}$$

$$= 0.0104$$

As with the tee-joint example of Fig. 5.72, this is provided by No. 6 (19.05 mm diameter) hoops or spirals at 3.0-in (76-mm) centers.

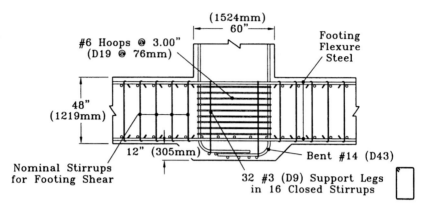

(a) 100% Column bars bent inward in drop cap

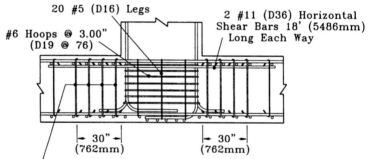

Extra External Stirrups: #5 @ 10" (D16 @254mm) Each Way
for 30" (762mm) Ring Around Column

(b) 50% Column bars bent out

FIG. 5.88 Alternative joint designs for footing example.

The second alternative supports the column steel on the bottom footing reinforcement mat, with alternate bars bent inward and outward. The outward-bent bars need to be supported by vertical stirrups within a circle of diameter $2D$ but outside the column. Since 50% of the bars are bent outward, the total vertical joint rebar required is 50% of that required by Eq. (5.137). Thus, again with $f_{yc}^o = 1.4 f_{yv}$,

$$A_{jv} = 0.5 \times 0.5 \times 32 \times 2.25 \times 1.4 = 25.2 \text{ in}^2 \quad (16,250 \text{ mm}^2)$$

using No. 5 (15.9 mm diameter) stirrups, $n = 25.2/0.31 = 81$ legs. Area over which these are distributed = $(\pi/2) (10^2 - 5^2) = 58.9$ ft^2 (5.5 m^2). Therefore, we require $81/58.9 = 1.38$ legs/ft^2, or a spacing of 10 in. (305 mm) both ways. It must be emphasized that the No. 5 stirrups at 10-in. centers are additional to any required for footing shear force but may include any stirrups placed for purely nominal requirements.

Within the joint, provide stirrups in accordance with Eq. (5.102). Thus $A_{vi} = 0.625 \times (32 \times 2.25) \times 1.4 = 6.3$ in^2 (4064 mm^2). This can be provided by 20 No. 5 (15.9 mm diameter) stirrups, which can also be used to restrain the hook ends of the 16 bars bent inward. Transverse hoop reinforcement, again satisfying Eq. (5.104) and hence supplied by No. 6 hoops or spirals at 3-in (76-mm) pitch, is required.

Additional footing-top reinforcement must satisfy Eq. (5.101). However, since only 50% of the bars are bent out, the required area can be reduced by 50%, requiring $\Delta A_{sb} = 0.5 \times 0.0625 \times 72 \times 1.4 = 3.15$ in^2 (2032 mm^2). This is provided by two additional No. 11 (35.8 mm) bars 18 ft (5.5 m) long each way.

5.6.5 Design of Piles

Because of difficulties in investigating pile conditions after an earthquake and because of expense in repairing pile damage, it will be normal to design piles and foundation cylinders to remain elastic under the design-level seismic response. A capacity design approach, with ductile elements responding at overstrength flexural capacity in accordance with Section 5.3.3, should be used to determine required flexural and shear strength of the piles. Generally, this can be achieved with adequate accuracy and conservatism by assuming all piles or cylinders to have equal lateral stiffness and hence dividing the overstrength shear input evenly between the piles. However, it should be recognized that passive pressure on the vertical face of the footing may result in a significant horizontal force H (see Fig. 5.80) which may greatly reduce pile shear forces. The critical pile for shear strength will then be the pile with lowest axial compression (or highest axial tension).

Design moments for the piles should be calculated with regard for the fixity of the pile–pile cap connection and the relative stiffness of soil and pile, as discussed in Section 4.4.2. Since pile moments will be greatest for low values of soil stiffness, conservatively low values for soil modulus should be assumed. If partial liquefaction of upper levels of the soil is possible during seismic response, the stiffness of the soil at these levels should be discounted when calculating pile moments. This will, of course, result in increased pile design moments for the constant shear input of $V°/n$, where $V°$ is the overstrength shear to be transmitted to ground by the piles (i.e., total overstrength minus dependable passive pressure) and n is the total number of piles.

In some cases the averaging procedure implied above, where all piles receive equal shear load, may be unnecessarily conservative, since piles with increased axial compression will have both increased stiffness, hence attracting a greater proportion of the total shear force and higher shear strength. This is examined further in Section 5.10, Example 5.1. More detailed analyses, as described further in Section 7.4.10(g), may be made to refine the estimate of pile design forces.

5.7 SUPERSTRUCTURE LONGITUDINAL DESIGN

Rather surprisingly, designers often seem to forget that with moment-resisting superstructure–column details, seismic forces of considerable magnitude are induced in the superstructure. Since the preferred design philosophy involves development of column plastic hinges, a capacity design approach in accordance with Section 5.3.3 must be adopted to ensure that inelastic action is not developed in the superstructure.

At all times moment equilibrium must exist at the center of the column–superstructure joint. Hence with reference to Fig. 5.89(b), at the top of column C, the sum of the beam moments, $M_{lc} + M_{rc}$, must equal the column moment M_{cc}, all moments being extrapolated to the joint center. Typically, as a result of negative dead-load moments in the superstructure over the column [Fig. 5.89(a)], $M_{lc} > M_{rc}$ for the direction of seismic response represented by the solid-line moment profile. However, at column B, the moment profile is distorted by proximity to a movement joint, at which location the moment is always zero. Although this provides little influence on the dead-load moment profile since the hinge location will typically be chosen to correspond to the point of inflection for a continuous structure, the influence on seismic moments can be severe. As shown in Fig. 5.89(b), the negative moment M_{lb} will be much less than M_{lc} because the point of inflection is constrained to be at the hinge. Thus, for equilibrium, the moment M_{rb} to the right of the column must be correspondingly larger. If this effect is not designed for, the plastic hinge will probably form in the superstructure instead of the column. As shown by the dashed line in Fig. 5.89(b), corresponding to the reversed direction of

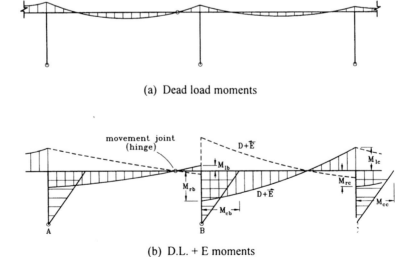

(a) Dead load moments

(b) D.L. + E moments

FIG. 5.89 Influence of movement joint on superstructure longitudinal moments.

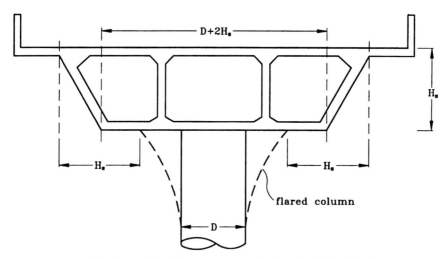

FIG. 5.90 Effective width of superstructure for longitudinal seismic response.

seismic response, the superstructure negative moment required to the right of the column will also be increased compared with other columns.

When computing the flexural resistance of the superstructure to support the moments implied by Fig. 5.89, the effective superstructure width will often be less than the full section width, particularly for single-column bents (Fig. 5.90). Since the superstructure will be required to develop bending moments of opposite sign at opposite faces of the cap beam, girders at some distance from the column centerline will be less effective than the closer girders, since their moment input will tend to twist the cap beam, reducing the stresses in the outer sections of the superstructure.

With solid-section or box-girder sections, an effective tributary width equal to the superstructure depth H_s may be taken on either side of the column. However, with open-soffit sections such as those formed by I girders or multiple spine boxes, less resistance to the torsional rotation of the cap beam is provided and the effective tributary width should be taken as $0.5H_s$. Note that providing a structural flare at the top of the column, as suggested by the dashed profile in Fig. 5.90, will increase the effective width of the superstructure proportionately. An architectural flare with little longitudinal reinforcement is likely to be less effective.

5.8 MOVEMENT JOINTS

Movement joints, placed in bridge superstructures to accommodate length changes due to thermal, creep, and shrinkage effects, have a tendency to become locations of minor spalling, corrosion, and weather penetration, requiring continual maintenance. As a consequence, there has been a move toward minimizing the number of movement joints in long-span bridges. In the 1950s and 1960s, maximum distances between movement joints were on the order of 500 ft (150 m). In recent years this has increased to more than 1000 ft (300 m).

The reduction in usage of movement joints is also a sound choice based on seismic considerations. Differential movements between bridge segments separated by movement joints can result in impact, with concrete spalling under differential response to seismic excitation. Estimation of the maximum expected differential displacement across movement joints is one of the most difficult analytical problems facing a designer, as discussed subsequently. If seating at the movement joint is inadequate, superstructure failure can occur during seismic response as a consequence of loss of support. Examples were discussed in Section 1.2.1(*a*). To ensure satisfactory performance at movement joints, restrainers and shear keys are commonly placed, designed according to simplistic analyses which typically bear little resemblance to probable response.

Because of these complexities, it is advisable to avoid movement joints wherever possible. Where absolutely necessary, they should be designed very conservatively. Typically, this results in only very marginal cost increase. Three aspects of design will be discussed here briefly: seating length, shear key design, and restrainer design.

5.8.1 Seating

Current policy in California requires a seating length of 24 in. (610 mm) at internal movement joints, and the value of abutments found from elastic analysis or the value of N_A resulting from the following equation:

$$
N_A = \begin{cases} (1 + 0.0025L + 0.01H)\left(1 + \dfrac{S^2}{8000}\right) & \text{(ft)} \quad (5.138a) \\[4mm] (0.30 + 0.0025L + 0.01H)\left(1 + \dfrac{S^2}{8000}\right) & \text{(m)} \quad (5.138b) \end{cases}
$$

where L is the length of the bridge deck to the next expansion joint, H the average height of columns supporting the bridge deck to the next expansion joint, and S the abutment skew in degrees. The dependence on length is intended to provide some allowance for thermal, creep, and shrinkage effects and for the effects of seismic traveling waves.

Equation (5.138) would appear to be too strongly dependent on the span length and too weakly dependent on the average column height. At maximum ductility, bridge pier displacements are likely to correspond to drift angles in the range 0.015 to 0.035 compared with the angle of 0.01 implied by Eq. (5.138), but bridge length effects of 2.5 ft per 1000 ft (2.5 m per 1000 m) of span length seem excessive and would result in excessive seating lengths, particularly when applied to internal expansion joints. It is consequently recommended that the coefficients in Eq. (5.138) be changed to 0.0015L and 0.03H, for abutment seating length.

Relative displacements of internal movement joints are affected by the stiffness of the two frames separated by the movement joint, the yield strengths of the frames, the frictional restraint of sliding, the impact on closing the joints, and the characteristics of restrainers connecting the frames, as discussed in Section 4.4.2(f). Comparison of results from dynamic time-history analyses with those from elastic response spectrum analyses indicates that although the elastic response spectrum approaches produce reasonable values for maximum absolute displacement response of each frame, the relative displacement (i.e., the movement joint opening) is poorly predicted [Y1]. This is illustrated in Fig. 5.91(a), where results from inelastic time-history analyses are compared with results from elastic time history with full modeling of the movement joint complexities, with simplified elastic time-history analysis where the stiffness of connection between the frames was based on restrainer stiffness alone, and with elastic response spectrum analysis using restrainer tension stiffness between the frames.

Analyses were carried out for a range of relative stiffnesses between the two frames [stiffness ratio, in Fig. 5.91(a)] and different levels of seismic intensity, sliding friction, and so on. Typically results for moderate seismic intensity are shown in Fig. 5.91(a). It is evident that the simplified methods greatly overpredict the relative displacements across the movement joint. It was found that a simple method for predicting the relative longitudinal displacement at the movement joint with reasonable accuracy could be found from the difference Δ_L between the absolute magnitude of peak longitudinal displacements calculated for the two frames separated by the joint, where each frame was considered as a stand-alone element and the absolute magnitudes of peak displacement have the same sign. A comparison between results from this approach and those from the full inelastic time-history analysis is presented in Fig. 5.91(b). It will be seen that the agreement is reasonable.

To estimate design relative displacements at internal movement joints, the displacement Δ_L calculated above should have components added for transverse response and for traveling-wave effects. The following equation is suggested as a conservative option for regions of high seismicity:

$$N_E = \Delta_L + 0.015W + 0.001L \qquad (5.139)$$

where Δ_L is the longitudinal relative displacement calculated as described above, W the width of the seating in the bridge transverse direction at the movement joint, and L the average distance to the adjacent movement joints.

The displacements implied by the seat lengths of Eqs. (5.138) and (5.139), or the default option of 24 in. (610 mm), will be seen to be impossibly large for short bridges of one or two frames, where the cumulative gap width at

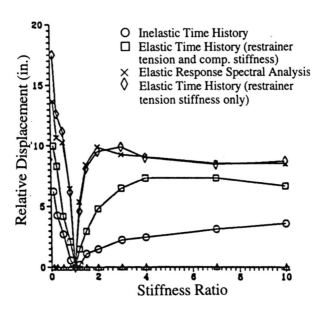

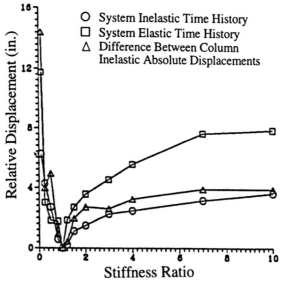

FIG. 5.91 Relative longitudinal displacements between two frames connected across a movement joint. (*a*) Relative movements predicted by different analytical approaches; (*b*) Comparison between relative movements predicted from nonlinear time-history analysis and difference between absolute displacement of independent adjacent frames.

abutments and internal movement joints will be on the order of 3 to 6 in. (75 to 150 mm). In such cases the seating lengths are intended to provide protection when abutment fill material and backwalls fail, permitting relatively unrestrained longitudinal movements. If the abutment detail involves a knock-off backwall [see Fig. 3.1.3(d)], the seating lengths are more directly applicable.

5.8.2 Shear Keys

Accurate determination of design forces on abutment or movement joint shear keys is difficult, particularly for bridges with ductile columns. In such cases the results predicted from elastic analyses are likely to differ considerably from the real forces. Section 4.4.2(e) provides some guidance in determining design forces.

The most realistic method for obtaining the expected force levels will be by dynamic inelastic analysis, but a reasonable estimate may be obtained by considering a plastic collapse analysis, where the movement joints and shear keys are included, as illustrated for a two-frame bridge in Fig. 5.92. In this example it is assumed that there are shear keys at abutments 1 and 8 and at the movement joint, as in span 4–5. Figure 5.92(b) shows resultant inertia forces F_L and F_R acting on the left and right frames, and the ultimate resisting forces acting in opposition to the inertia forces. Assuming the bending strength of the movement joints at 1, 8, and m about the vertical axis to be zero, the system is statically determinate and can be solved for the five unknowns F_L, F_R, F_{sk1}, F_{sk8} (the abutment shear key forces), and V_m (the shear force across the movement joint at m). This requires a knowledge of the displaced shape, as shown in Fig. 5.92(c). Thus, taking moments about m yields

$$F_L(x_m - x_l) = F_{sk1}x_m + \sum_2^4 F_{ui}(x_m - x_i) \tag{5.140}$$

$$F_R(x_r - x_m) = F_{sk8}x_m + \sum_5^7 F_{ui}(x_i - x_m) \tag{5.141}$$

Lateral equilibrium requires that

$$F_L = F_{sk1} + \sum_2^4 F_{ui} + V_m \tag{5.142}$$

$$F_R = F_{sk8} + \sum_5^8 F_{ui} - V_m \tag{5.143}$$

and assuming the tributary mass associated with each bent is m_i, the relationship between F_L and F_R is

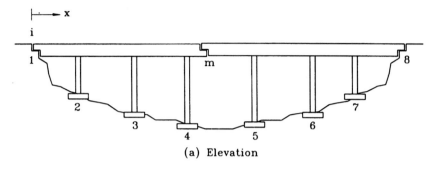

(a) Elevation

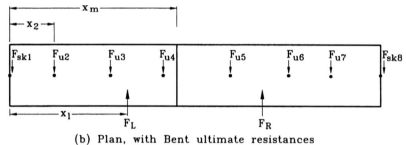

(b) Plan, with Bent ultimate resistances

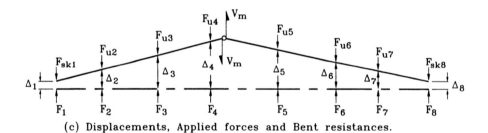

(c) Displacements, Applied forces and Bent resistances.

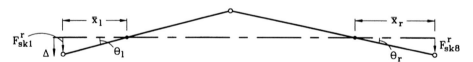

(d) Rotational inertia component for plastic Bents.

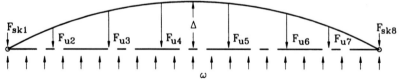

(e) Forces on Bridge without internal movement joint.

FIG. 5.92 Forces developed in shear keys.

$$F_R = F_L \frac{\sum_5^8 m_i \Delta_i}{\sum_1^4 m_i \Delta_i} \tag{5.144}$$

To obtain the displaced shape shown in Fig. 5.92(c) may involve trial and error, particularly if the abutment deformations Δ_1 and Δ_8 are significant and if superstructure flexibility must be considered. In the latter case, F_L and F_R should be distributed between the bents as forces F_1 to F_8, shown in Fig. 5.92(c) and given by

$$F_i = F_L \frac{m_i \Delta_i}{\sum_1^4 m_i \Delta_i} \tag{5.145}$$

Note that the forces F_L and F_R do not necessarily act through the center of mass of the left and right frames, respectively, but at distances given by

$$x_l = \frac{\sum_1^4 m_i \Delta_i x_i}{\sum_1^4 m_i \Delta_i} \tag{5.146a}$$

$$x_r = \frac{\sum_5^8 m_i \Delta_i x_i}{\sum_5^8 m_i \Delta_i} \tag{5.146b}$$

Solution of Eqs. (5.140) to (5.146) enables the maximum forces capable of being generated in the shear keys at inelastic response of the bridge. In accordance with the principles of capacity design, high estimates of the bent resistance F_{u1} to F_{u8} should be adopted in the analysis. Results should be checked to ensure that all bents assumed to respond inelastically actually develop displacements larger than the nominal yield displacement and appropriate adjustments made to the resisting bent forces where this is not the case.

The approach developed above is incomplete since it ignores possible increased abutment forces due to superstructure rotational inertia with the piers in the fully plastic state. This is represented in Fig. 5.92(d), where the stiffness of all bents has been removed. Ground displacement Δ_g and acceleration $\ddot{\Delta}_g$ will cause no displacement at the center of mass of the left and right frames, as shown in Fig. 5.92(d), but will cause rotations and rotational accelerations of $\theta_l = \Delta_g/\bar{x}_l$, $\ddot{\theta}_l = \ddot{\Delta}_g/\bar{x}_l$ of the left frame, and $\theta_r = \Delta_g/\bar{x}_r$, $\ddot{\theta}_r = \ddot{\Delta}_g/\bar{x}_r$, of the right frame. It can easily be shown that for rigid frames, the maximum abutment forces from rotational inertia will be

$$F'_{sk1} = \frac{W_L}{3} a_{g,max} \tag{5.147a}$$

$$F'_{sk2} = \frac{W_r}{3} a_{g,max} \tag{5.147b}$$

where W_L and W_r are the weights of the left and right frames, respectively, assumed uniformly distributed along the frames, and $a_{g,\max}$ is the peak ground acceleration, expressed as a fraction of g.

It would thus seem that the maximum possible shear key force would be the sum of that given by solution of Eqs. (5.140) to (5.146), and by Eq. (5.147). Inelastic time history analyses show this approach to be reasonable, but increasingly conservative as the abutment lateral flexibility increases. Further details are available in [B8].

When the bridge does not have an internal movement joint (e.g., the bridge in Fig. 5.92 is continuous through m), the approach developed above is inappropriate, since abutment shear key forces are dictated largely by displacements and stiffness of the superstructure, as shown in Fig. 5.92(e). Although the superstructure will be designed to remain elastic, the shear key forces predicted by elastic dynamic modal analysis will generally *underpredict* the true forces. This is because displacements from the elastic analysis can be expected to be reasonable estimates of actual response, but restraining forces at the internal bents will be overpredicted, due to bent ductility. In this case a conservative estimate of the shear key forces may be found by analyzing the bridge under the ultimate restraint forces F_{u2} to F_{u7}, as before, opposing a uniformly distributed load w [see Fig. 5.92(e)], where

$$w = \frac{384\ EI_s\Delta_m}{5L^4} \tag{5.148}$$

where L is the total bridge length, Δ_m the maximum lateral displacement predicted at midspan, and I_s the superstructure moment of inertia about the vertical axis. Equation (5.148) is based on the conservative assumptions of uniformly distributed inertial force and zero abutment displacement.

Design of shear keys is normally based on a shear friction approach. With reference to Fig. 5.93, the design shear strength of the key is

$$V_{sk} = \phi_s\mu A_s f_y \tag{5.149}$$

where the coefficient of friction may be taken as $\mu = 1.4$ for the naturally occurring shear crack and A_s is the total area of reinforcement crossing the critical interface. A shear strength reduction factor of $\phi_s = 0.85$ should be adopted.

Note that the actual strength may be less than expected if the lateral force P is transmitted too high on the shear key, causing the shear key to behave as a squat cantilever with a flexural failure mode. The maximum lateral force that can be transmitted by flexure will be

$$P \approx \phi_f \frac{0.9}{y}\left(A_s f_y \frac{L_{sk}}{2}\right) \tag{5.150}$$

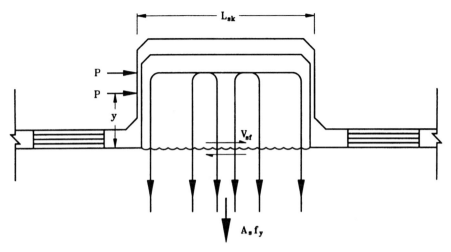

FIG. 5.93   Shear key design.

Hence, equating Eqs. (5.149) and (5.150), the height y of load application should be less than

$$y = 0.45 \frac{\phi_f}{\phi_s} \frac{L_{sk}}{\mu} \approx 0.3 L_{sk} \qquad (5.151)$$

To ensure this, the shear key length should be about three times its height, or a special detail incorporated at the sides to positively locate the point of lateral force application.

5.8.3   Restrainer Design

No satisfactory methods have been developed for designing the strength and stiffness of restrainers to be placed across movement joints. The approach taken in this section has been to provide sufficient seating at movement joints so that the maximum displacements developed on the assumption of no restrainers across the joint can be safety accommodated without unseating. With this philosophy it can be argued that there is no need for restrainers.

However, even with conservative assumptions about relative displacement, the general uncertainties about seismic input characteristics will often lead designers to place restrainers as a "belt and braces" approach. A method generally utilized in California for design of restrainers strength and stiffness [C1] has been found to yield results that differ very significantly from results from inelastic time-history analyses [Y1] and should not be used.

Two rational methods can be predicted. The first involves dynamic time-history analysis, where the strength and stiffness of the restrainers can be varied until acceptable results are obtained. In this context it should be noted that the parameter studies [Y1] for influence of restrainer stiffness on opening of the movement joint indicate that restrainers are relatively ineffective unless the restrainer stiffness is at least as large as the stiffness of the more flexible of the two frames connected across the joint. This is illustrated in Fig. 5.94, where results for restrainer stiffnesses varying from zero to six times a typical "standard" value are shown. Discounting the zero-stiffness case, a change in restrainer stiffness of more than 100 times reduc'd displacements by less than 50%.

The second approach is based on simple observations from the dynamic analysis. The analyses indicated that the connected frames responded essentially in-phase during maximum response. As a consequence, the maximum tensile force transfer between the frames should be equal to the *difference* between the frame overstrength longitudinal shear capacities. Thus the maximum restrainer design force should be

$$F_R = V^\circ_{F1} - V^\circ_{F2} \tag{5.152}$$

where V°_{F1} and V°_{F2} are the overstrength capacities, found from summing the overstrength capacities of all columns in each frame.

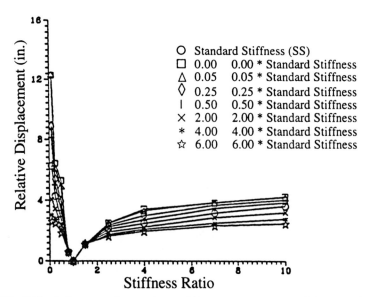

FIG. 5.94 Influence of restrainer stiffness on movement joint opening (1 in. = 25.4 mm).

If the stiffness is adjusted so that the yield strength of the restrainers is not attained until a relative displacement based on the difference between absolute displacements of the frames, as above, that is,

$$\Delta L = |\Delta_{F2}| - |\Delta_{F1}| \qquad (5.153)$$

the required restrainer stiffness will be

$$K_R = \frac{F_R}{\Delta L} = \frac{V_{F1}^{\circ} - V_{F2}^{\circ}}{|\Delta_{F2}| - |\Delta_{F1}|} \qquad (5.154)$$

The strengths and stiffnesses of Eqs. (5.152) and (5.154) are those for the total restrainers crossing the joint. Note that Eqs. (5.152) and (5.153) imply that frame 1 is stiffer and stronger than frame 2.

Designing a restraining system based on the simple principles noted above will ensure that relative displacements are less than those predicted by Eq. (5.153), though the degree of reduction will not be clear unless determined by an inelastic time-history analysis with full modeling of the movement joint as described in Section 4.4.2(*f*).

5.9 *P*–Δ EFFECTS

As a bridge displaces laterally, as shown, for example in Fig. 5.95(*a*), gravity loads induce column moments in addition to those resulting from lateral inertia forces. Using the nomenclature of Fig. 5.95, the base moment *M* for a simple vertical cantilever is

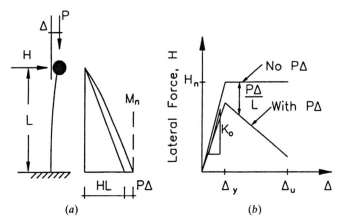

(a)   *(b)*

FIG. 5.95 Effect of *P*–Δ moments on lateral response characteristics of a bridge pier. (*a*) Moment diagram; (*b*) Force–displacement response.

$$M = FL + P\Delta \tag{5.155}$$

If the base moment capacity is equal to the moment capacity M_n, then the lateral inertia force which may be resisted reduces as the displacement increases, according to the relationship

$$F = \frac{M_n - P\Delta}{L} \tag{5.156}$$

This effect is illustrated in Fig. 5.95(b), where it is shown that the $P-\Delta$ effect not only reduces the maximum lateral strength but also modifies the lateral force–deformation characteristic. The effective initial stiffness is reduced and the postyield stiffness may become negative.

The importance of $P-\Delta$ effects in modifying structural response under seismic action has been the subject of considerable research activity over recent years [B5,M9,M10]. However, there is still controversy about when and how it should be considered in routine design.

Although dynamic inelastic time-history analyses of simple bridge bents including $P-\Delta$ effects indicate that displacements are generally higher than when $P-\Delta$ effects are ignored, the increase is often very small and dependent on the shape of the hysteretic lateral force–displacement characteristic used to describe the response. With the adoption of an elastoplastic characteristic, it can be shown that if the earthquake record is long enough, instability (i.e., increase in displacements until $P\Delta = M_n$) will eventually occur when $P-\Delta$ effects are considered. This can be explained by reference to Fig. 5.96(a). Consider an initial inelastic pulse that results in maximum displacement reaching point A. The structure unloads down a line of stiffness

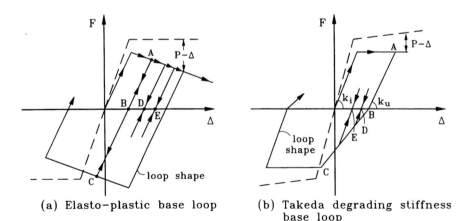

(a) Elasto–plastic base loop   (b) Takeda degrading stiffness base loop

FIG. 5.96 Influence of hysteresis characteristic on $P-\Delta$ effects.

equal to the initial stiffness to point B. If further elastic cycles cause oscillation about the residual displacement (B), it is probable that the yield strength at A rather than at C will first be attained since a much higher elastic response level is required to get to C than to A. Consequently, once the first inelastic pulse occurs, it creates a tendency for continued displacement in the same sense, and response continues incrementally to D and E and eventually to failure.

The behavior just described depends on the special characteristics of elastoplastic response, which, as discussed in Section 4.5.4, is not representative of column inelastic response. This is better represented by the Takeda degrading stiffness model [Fig. 4.44(*d*)], which has a positive postelastic stiffness, an unloading stiffness less than the initial stiffness, and reduced reloading stiffness, as suggested in Fig. 5.96(*b*). Even if *P*–Δ effects result in negative stiffness for the postelastic envelope, behavior tends to be stable. Since the unloading stiffness is less than the initial loading stiffness, unloading from point A results in a less residual displacement at B than for the elastoplastic case. Subsequent elastic cycles result in gradual reduction to the residual displacement due to the reduced stiffness in the reverse direction. Consequently, with this hysteretic characteristic, there is a tendency for residual displacements to decrease rather than increase on further cycling, and no preferential direction for cumulative displacement develops.

Analyses have shown [M10] that provided the second slope stiffness $r_p K_e$ of the stabilized loop shape including *P*–Δ effects is positive, structural response is stable, without significant increase in displacement. This is illustrated in Fig. 5.97, where hysteretic response with and without *P*–Δ effects, based on the Takeda degrading stiffness model, are shown.

Sensitivity to *P*–Δ effects is generally related to a stability index, θ, measured at first yield, and defined as

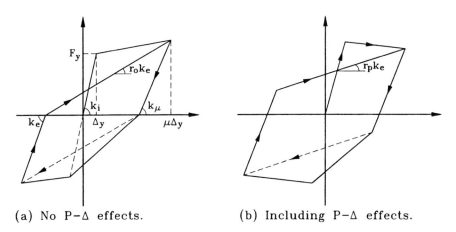

(a) No P–Δ effects.   (b) Including P–Δ effects.

FIG. 5.97 Parameters for determining stability under *P*–Δ effects.

$$\theta = \frac{P\Delta_y}{F_y L} \qquad (5.157)$$

where F_y and Δ_y are yield force and displacement and L is the column height. Thus θ is the ratio of P–Δ moment at yield to the base moment capacity. Equation (5.157) may be expressed as

$$\theta = \frac{P}{K_e L} \qquad (5.158)$$

where K_e is the elastic stiffness. In terms of the stabilized loop shape of Fig. 5.97(a), it is the unloading stiffness K_μ that should be used for the elastic stiffness. Using the Takeda stiffness model, an unloading stiffness of $K_\mu = K_i \mu^{-0.5}$ is appropriate for typical column response.

The postelastic stiffness ratio r_p including P–Δ effects can be related to the stiffness r_0 ignoring P–Δ effects by the relationship

$$r_p = \frac{r_0 - \theta}{1 - \theta} \qquad (5.159)$$

If we require a minimum value for $r_p = 0.05$ to allow for uncertainty of hysteretic characteristics and noting that $r_0 = 0.19$ seems a lower limit for stabilized loop stiffness ratio at $\mu_\Delta = 6$ from column tests [M10], then, from Eq. (5.159), the maximum permissible value for θ would be

$$\theta_{\max} = \frac{P}{K_e L} = \frac{P\mu^{0.5}}{k_i L} = \frac{r_0 - r_{p,\min}}{1 - r_{p,\min}} = \frac{0.19 - 0.05}{0.95} = 0.147$$

or for $\mu_\Delta = 6$,

$$\frac{P}{k_i L} \leq 0.06 \qquad (5.160)$$

This can be rearranged, with conservative rounding up, to require that

$$\frac{F_y}{P} \geq 20 \frac{\Delta_y}{L} \qquad (5.161)$$

Thus to ensure relative insensitivity to P–Δ effects for displacement ductilities up to $\mu_\Delta = 6$ would require, for a yield drift ratio of $\Delta_y = 0.005L$, that $F_y > 0.1P$. For more flexible structures, minimum strength should be increased proportionately. However, with more flexible structures it is unlikely that a ductility of $\mu_\Delta = 6$ would be required, and hence $r_0 > 0.19$, $k_e > 0.41 k_i$, and $P/k_i > 0.06$ would result.

Equation (5.161) was based on $\mu_\Delta = 6$ and hence may be expressed in terms of maximum expected displacement as

$$\frac{F_y}{P} \geq 3.3 \frac{\Delta_u}{L} \tag{5.162}$$

Since r_p increases with decreasing ductility, Eq. (5.162) will be conservative for $\mu_\Delta < 6$.

5.10 DESIGN EXAMPLES:

5.10.1 Example 5.1: Cantilever Column with Curved Continuous Box-Section Superstructure

The bridge shown in Fig. 5.98 was set by the authors as an example for comparative analysis for the 2nd International Workshop on the Seismic Design of Bridges [P26]. Results from one of the analytical solutions will now be used to design the critical pier 5. Note that the superstructure has high horizontal curvature and is supported on two bearings, 4 m apart, which allow longitudinal sliding but are restrained against sliding in the radial direction. Preliminary *elastic* analyses indicate that the maximum radial displacement at the superstructure midheight is 160 mm (6.3 in.), and that of this, 95 mm (3.74 in.) was due to footing displacement [75 mm (2.95 in.)] and footing rotation [20 mm (0.79 in.)]. Moments in the column resulting from the elastic modal analysis are indicated in Fig. 5.99(a) by the line marked "elastic." It is noted that this profile implies a moment of 20 MNm at the level of the bearings, which exceeds the moment of 16 MNm required to cause uplift. The high reversed moment of the top of the column is due to the torsional stiffness of the superstructure, commented on previously in relation to Fig. 3.3. The structure first-mode period, at $T = 2.2$ s, exceeds the period at peak elastic spectral response of $T = 0.6$ s sufficiently that the force-reduction factor $Z = \mu_\Delta$, from Eq. (5.22). Figure 5.15 would indicate, with $L/D = 2$, that a column displacement ductility of $\mu_\Delta = 6$ might be permissible. However, the elastic analysis indicated that 59% of the elastic displacement resulted from soil displacement, implying an additional flexibility coefficient of $f_a = 95/(160 - 95) = 1.46$. In accordance with Eq. (5.21), this would imply a structure ductility of

$$\mu_{af} = 1 + \frac{6 - 1}{1 + 1.46} = 3.03$$

It is decided to design for $\mu_\Delta = Z = 4$ and check ductility capacity, increasing transverse reinforcement above the design level given by Eq.

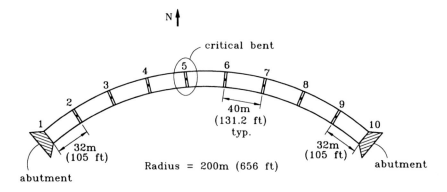

a) Plan View

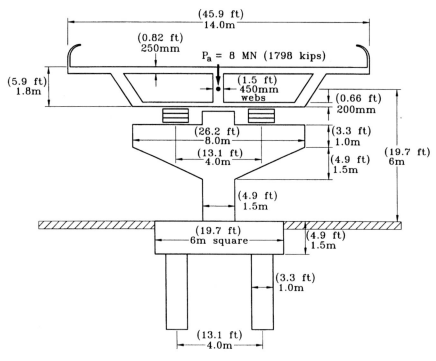

b) Bent 5 dimensions

FIG. 5.98 Bridge dimensions for Example 5.1.

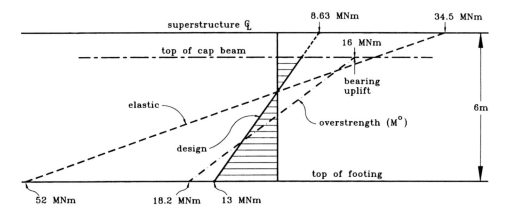

(a) Column design moments for Example 5.1 (1 MNm=737 kip-ft)

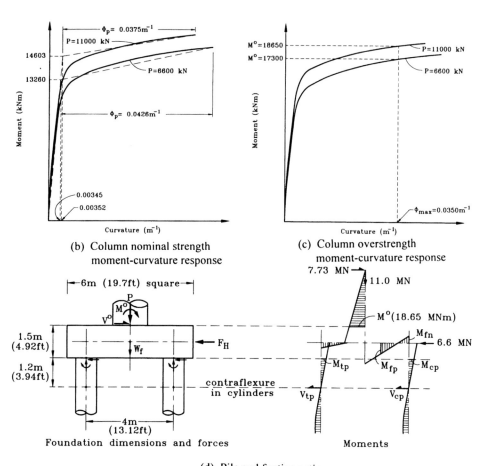

(b) Column nominal strength
moment-curvature response

(c) Column overstrength
moment-curvature response

(d) Pile and footing actions

FIG. 5.99 Design details for Example 5.1.

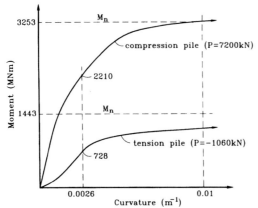

(e) Partial moment-curvature response for piles

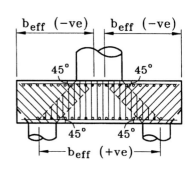

(f) Footing effective widths for flexure

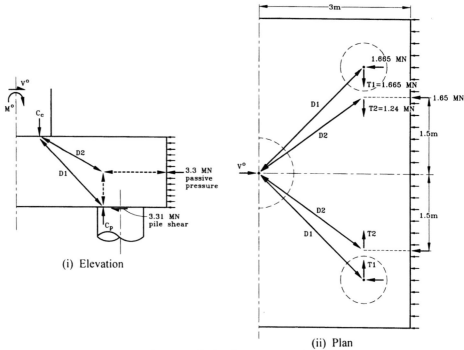

(i) Elevation

(ii) Plan

(g) Footing shear transfer by direct compression struts

FIG. 5.99 *(Continued).*

(5.47) if necessary. Consequently, design moments are found by dividing the elastic profile in Fig. 5.99 by 4, giving the profile marked "design." The base design moment is thus 13 MNm (9586 kip-ft), and yield displacement is nominally 40 mm (1.57 in.), of which 23.8 mm (0.94 in.) results from foundation flexibility.

Column Flexural Design. Specified material strengths are $f'_c = 30$ MPa (4.35 ksi), $f_y = 400$ MPa (58 ksi). In accordance with recommendations of Section 5.3.1(d)(vii), we design for

$$f'_{ce} = 1.3 f'_c = 1.3 \times 30 = 39 \text{ MPa (5.66 ksi)}$$
$$f_{ye} = 1.1 f_y = 440 \text{ MPa (63.8 ksi)}$$
$$\varepsilon_c = 0.004 \text{ (at design strength)}$$

Including the weight of the column, axial load due to dead load at the column base is 8800 kN. We design for a cover of 50 mm, and in accordance with Section 5.3.1(d)(x), the design axial load, including vertical acceleration effects, is

$$P = (1 \pm 0.5\text{PGA})8800 = (1 - 0.5 \times 0.5)8800 = 6600 \text{ kN}$$

since reduced axial load represents the critical case. Section design using the program $SE-M\phi$ [S12] indicates that the required moment capacity of 13,000 kNm (9586 kip-ft) can be provided by 64 D28 (1.10 in. in diameter), 49 H32 (1.26 in. in diameter), or 32 H40 (1.57 in. in diameter). The latter option is chosen:

Use column longitudinal steel 32 H40.

This corresponds to a longitudinal steel ratio of 2.28%.

Column Confinement Design. Check confinement provided by Eq. (5.47): maximum axial load $= 1.25 \times 8800 = 11,000$ kN (2473 kips):

$$\rho_s = 0.16 \times \frac{39}{440} \left(0.5 + 1.25 \times \frac{11}{39 \times 1.766} \right) + 0.13(0.0228 - 0.010) = 0.0125$$

Using a D20 spiral yields

$$s = \frac{4A_b}{D'\rho_s} = \frac{4 \times 314}{1420 \times 0.0125} = 71 \text{ mm (2.8 in.)}$$

The confinement is comparatively high because of the high axial load on the column. Note that Eq. (5.56) for antibuckling requirements does not govern, since $\rho_s = 0.002 \times 32 = 0.0064$ and $M/VD < 4$. Because of conservatism in Eq. (5.47), it is elected to try D20 at 80 mm (3.15 in.) for confinement. The plastic displacement capacity is checked by carrying out a moment–curvature analysis.

Ultimate concrete strain is given by Eq. (5.14). This requires an estimate of the confined concrete strength f'_{cc}. From Eq. (5.4),

$$f_l = 2 \times 440 \times \frac{314}{1420} \times 80 = 2.44 \text{ MPa (354 psi)}$$

From Eq. (5.5),

$$f'_{cc} = 39 \left(2.254 \sqrt{1 + 7.92 \times \frac{2.44}{39}} - 2 \times \frac{2.44}{39} - 1.254 \right)$$
$$= 53.7 \text{ MPa (7787 psi)}$$

From Eq. (5.14), with $\varepsilon_{su} = 0.12$,

$$\varepsilon_{cu} = 0.004 + 1.4 \times 0.0108 \times 440 \times \frac{0.12}{53.7} = 0.0189$$

Results of moment–curvature analysis calculated using SEQ–Mϕ [S12] for maximum and minimum axial loads are shown in Fig. 5.99(b). The case with high axial load is critical and yields a plastic curvature capacity of

$$\phi_p = 0.0410 - 0.0035 = 0.0375 \text{ m}^{-1} \ (947 \times 10^{-6} \text{ in}^{-1})$$

From Eq. (5.39), the plastic hinge length is

$$L_p = 0.08L + 0.022 f_{ye} d_{bl} \geq 0.044 f_{ye} d_{bl}$$

To determine the appropriate value for L we must consider the moment profile at overstrength. Assuming, at this stage, an overstrength factor of $\phi^\circ = 1.4$ for the column base plastic hinge, the overstrength moment profile is shown in Fig. 5.99(a) as "overstrength." Note that the moment at the level of the cap beam will rise with rotation of the top of the cap beam to the limit value of 16 MNm (11,790 kip-ft), corresponding to lift-off of one bearing. Thus the point of contraflexure drops to 2.7 m (8.86 ft) above the base. Consequently, from Eq. (5.39), the plastic hinge length is

$$L_p = 0.08 \times 2700 + 0.022 \times 440 \times 40 \geq 0.044 \times 440 \times 40 = 774 \text{ mm}$$

Therefore,

$$\theta_p = L_p \phi_p = 0.774 \times 0.0373 = 0.0289 \text{ rad}$$

The plastic displacement capacity measured at superstructure midheight is thus

$$\Delta_p = 0.0289 \times 5000 = 144 \text{ mm (5.68 in.)}$$

Note that (1) the plastic hinge has been assumed to be centered at the column base because of equal strain penetration into the footing and up into the column, and (2) a length of 5 m, not 6 m, has been used to calculate the displacement, since the superstructure torsional rigidity means that it will not rotate significantly.

Ultimate displacement capacity is thus $\Delta_u = \Delta_y + \Delta_p = 40 + 144 = 184$ mm (7.24 in.). This exceeds the expected displacement under the design earthquake of 160 mm, and hence the confinement is satisfactory.

Confinement reinforcement: D20 at 80 centers (0.79 in. at 3.15-in. centers)

In accordance with Section 5.3.2(c)(iii), this confinement must extend at least 1.5 m up the column (the 80% rule does not govern).

Column Shear Design. Two cases are considered, corresponding to maximum and minimum axial load. In accordance with the recommendations of Section 5.3.3, the maximum column moments are checked at overstrength using a moment–curvature analysis with $f'_{co} = 1.7 \times 30 = 51$ MPa (7400 psi). A value of $f_{yo} = 500$ MPa (72.5 ksi), the maximum permitted for the grade is adopted. From the previous moment–curvature analyses, the ultimate curvature corresponding to the design displacement of 160 mm will be

$$\phi_u = 0.0035 + \frac{160 - 40}{184 - 40} \times 0.0375 = 0.0348 \text{ m}^{-1} \ (883 \times 10^{-6} \text{ in}^{-1})$$

Results of the overstrength calculations are plotted in Fig. 5.99(c). Maximum overstrength moments at the design displacement are:

(a) $P = 6600$ kN: $M° = 17,300$ kNm (12,750 kip-ft)
(b) $P = 11,000$ kN: $M° = 18,650$ kNm (13,750 kip-ft)

Bearing uplift moments for the two cases are 12,000 kNm and 20,000 kNm, respectively, so the overstrength shears are

(a) $P = 6600$ kN: $V° = (17.3 + 12.0)/5 = 5860$ kN (1317 kips)
(b) $P = 11,000$ kN: $V° = (18.65 + 20)/5 = 7730$ kN (1737 kips)

Concrete Component V_c. Curvature ductility demand $\mu_\phi = 0.0348/0.0037 = 9.4$. From Fig. 5.46, using the curve for uniaxial ductility, since the sliding bearings protect the column against yield in the longitudinal direction,

$$v_c = 0.0705\sqrt{f'_c} \text{ MPa} = 0.0705\sqrt{30} = 0.386 \text{ MPa (56 psi)}$$

Therefore, from Eq. (5.75),

$$V_c = 0.386 \times 0.8 \times 1.766 = 545 \text{ kN (122.5 kips)—plastic end region}$$

Outside the plastic end region, $k = 0.25$ and hence

$$V_c = 0.25\sqrt{30} \times 0.8 \times 1.766 = 1935 \text{ kN (435 kips)—nonend regions}$$

Axial Force Component V_p. The point of contraflexure is 2700 mm up the column. Hence, with $c/2 = 230$ mm, from Eq. (5.77b):

(a) $P = 6600$ kN: $V_p = 0.85 \times 6600(750 - 230)/2700 = 1080$ kN (243 kips)
(b) $P = 11,000$ kN: $V_p = 0.85 \times 11,000(750 - 230)/2700 = 1802$ kN (405 kips)

Truss Mechanism Component V_s. Required shear strength of transverse reinforcement:

$$V_s = \frac{V^\circ}{\phi_s} - V_c - V_p$$

(a) $P = 6600$ kN: V_s

$$= \begin{cases} 5860/0.85 - 545 - 1080 = 5270 \text{ kN (1184 kips) (plastic end regions)} \\ 5860/0.85 - 1935 - 1080 = 3879 \text{ kN (872 kips) (nonend regions)} \end{cases}$$

(b) $P = 11,000$ kN: V_s

$$= \begin{cases} 7730/0.85 - 545 - 1802 = 6747 \text{ kN (1516 kips) (plastic end regions)} \\ 7730/0.85 - 1935 - 1802 = 5357 \text{ kN (1204 kips) (nonend regions)} \end{cases}$$

It is seen that $P = 11,000$ kN governs the shear design. As noted earlier, some small conservatism has resulted from using nominal compression strength f'_c to determine V_c while using f'_{co} to determine V°.

Using D20 bars with $f_y = 400$ MPa (58 ksi), Eq. (5.76a) with $\theta = 35°$ results in the following required spacings:

$$s \leq \begin{cases} \dfrac{\pi}{2} \times \dfrac{314 \times 400 \times 1420 \times 1.43}{6747} = 59.4 \text{ mm (2.34 in.) (plastic end regions)} \\[3mm] \dfrac{\pi}{2} \times \dfrac{314 \times 400 \times 1420 \times 1.43}{5357} = 74.8 \text{ mm (2.95 in.) (nonend region)} \end{cases}$$

Thus shear governs spiral spacing for the full column height. Since the spacing is small, it is decided to use double (i.e., bundled) spirals as follows, where the end region is in accordance with Section 5.3.4(*b*)(v).

End region [lower 3.0 m (9.84 ft)]:   double D20 at 115 mm (0.79 in. diameter at 4.53 in.)

Remainder of column: double D20 at 150 mm (0.79 in. diameter at 5.9 in.)

Finally, the shear stress is checked:

$$v_{max} = \frac{7730}{0.8 \times 1.766} = 5.47 \text{ MPa} = 0.182 f'_c$$

This is (just) less than the limit of $0.2 f'_c$ recommended in Section 5.3.4(*b*)(iv).

It is noted that although feasible, the design is very close to limits for ductility and shear capacity. This has been largely a result of the unsatisfactory structural configuration, with the very short column at bent 5 attracting a disproportionately high lateral force and the use of a force-based design approach. It is of interest to note that if the amount of longitudinal reinforcement was decreased to 1%, the design moment capacity would decrease from 13,000 kNm to 8400 kNm (6190 kip-ft) and the overstrength moment at $P = 11,000$ kN from 18,650 kNm to 12,000 kNm. Thus the overstrength shear would be 6400 kN (1438 kips), and spiral spacing could be increased to 77 mm, or 154 mm (6.06 in.) for a double D20 spiral, which would be more manageable. Although the reduced strength implies a column force reduction factor of $Z = 6.19$, it is found that the ultimate displacement capacity has increased to 193 mm (7.62 in.). This emphasizes the illogicality of a force-based design.

Foundation Design. Although it is clear that displacement-based design would result in lower design forces than developed above, we continue with the original example, for completeness. The next stage in the design is to determine actions on the four 1-m (39.4-in)-diameter foundation cylinders. The overstrength base shear force $V°$ will be partly resisted by passive pressure F_H on the face of the footing and partly by shear force in the cylinders, as shown in Fig. 5.99(*d*). Using the ultimate passive pressure recommended in Section 4.4.2(*e*) for backfilled and compacted material of 370 kN/m² (7.7 kips/ft²),

$$F_H = 6 \times 1.5 \times 370 = 3330 \text{ kN (748 kips)}$$

Therefore, average pile shear force $V_p = (7.73 - 3.33)/4$ MN $= 1100$ kN (247 kips).

Initial elastic analyses indicate a point of contraflexure 1.2 m (3.94 ft) below the base of the pile cap, and approximately equal moments, but of opposite sign at the pile cap and at a depth of 3 m (9.84 ft). Thus the *average* pile moment is $M_p = 1.2 \times 1100 = 1320$ kNm (973 kip-ft).

Axial Forces on Piles. Assuming that all vertical forces are resisted by piles, and noting that the footing weight is $W_f = 6^2 \times 1.5 \times 23.5 = 1270$ kN (285 kips), the pile force from vertical loads is,

$$P_v = \frac{11 + 1.27}{4} = 3.07 \text{ MN (690 kN)}$$

Taking moments about pile contraflexure point for overturning forces gives

$$T = C = \frac{V°(1.5 + 1.2) + M° - F_H(0.75 + 1.2)}{2 \times 4}$$

$$= \frac{7.73 \times 2.7 + 18.65 - 3.33 \times 1.95}{8} = 4.13 \text{ MN/pile (928 kips)}$$

Therefore, maximum pile compression $C_p = 3.07 + 4.13 = 7.20$ MN (1618 kips) and maximum pile tension $T_p = 4.13 - 3.07 = 1.06$ MN (238 kips) tension.

Pile Moment Capacity. Initial trials indicate that pile moment capacity will be no problem. 16 D28 bars, giving a longitudinal steel ratio of 1.25%, are chosen, although it would be possible to decrease this. Based on moment–curvature analyses for the piles [see Fig. 5.99(e)] nominal moment capacities are 1443 kNm (1063 kip-ft) and 3253 kNm (2397 kip-ft) for tension and compression piles. Required nominal capacity is $M_r/\phi_f = 1320/0.9 = 1467$ kNm (1081 kip-ft), which is slightly higher than the nominal strength of the tension column. However, we note that curvatures in the tension and compression piles must be essentially equal because of equal displacements, and that the tension pile is much more flexible than the compression pile. At curvatures of 0.0026 m^{-1} (66×10^{-6} in^{-1}), moments in the tension and compression piles are 728 kNm (536 kip-ft) and 2210 kNm (1629 kip-ft), giving an *average* capacity of 1469 kNm (1082 kip-ft), which equals the required demand. At this curvature, tension and compression piles are at 50% and 68% of nominal capacity. Thus both piles are satisfactory, and it is the *compression* pile that is more critical. Again displacement considerations lead us to a more logical conclusion than that based on force and strength considerations.

Pile Confinement. The piles have adequate protection against formation of plastic hinges even if the footing passive pressure is discounted completely. Therefore, confinement is not strictly necessary. However, the axial load ratio on the compression pile is $P/f'_c A_g = 7.2/(30 \times 0.785) = 0.306$, which is moderately high. Consequently, it is conservatively elected to satisfy Eq. (5.49) in case of unexpectedly large curvatures, perhaps due to differential soil displacement at different levels. Hence

$$\rho_s \geq 0.45 \left(\frac{1.0^2}{0.91^2} - 1 \right) \frac{30}{400} = 0.007$$

This is provided by D12 (0.47 in. diameter) spirals at 70 mm (2.75 in.) spacing.

Pile Shear Force. From the analysis described above, the pile shears are distributed between tension and compression piles in proportion to their stiffnesses as

$$V = \begin{cases} \dfrac{728}{728 + 2210} \times 2 \times 1100 = 545 \text{ kN (122.5 kips) (tension pile)} \\[4mm] \dfrac{2210}{728 + 2210} \times 2200 = 1655 \text{ kN (372 kips) (compression pile)} \end{cases}$$

Shear Strength	Tension Pile		Compression Pile
V_c [Eq. (5.75), $k = 0.25$]	$0.25\sqrt{30} \times 0.8 \times 0.785 =$	860 kN	= 860 kN
V_s [Eq. (5.76a), $\theta = 35°$]	$\dfrac{\pi}{2} \times \dfrac{113 \times 400 \times 912 \times 1.43}{70} =$	1323 kN	= 1323 kN
V_p [Eq. (5.77b)]	$-1060 \times \dfrac{(0.5 - 0.08)}{1.2} =$ -371 kN		$7200 \times \dfrac{0.5 - 0.3}{1.2} = 1200$ kN
	$V_n =$ 1812 kN (407 kips)		$V_n =$ 3383 kN (760 kips)

With a shear strength reduction factor of $\phi_s = 0.85$, these still greatly exceed demand.

Footing Design Moments. Footing moments and shears can now be calculated directly. Only those on the compression side [i.e., M_{fn} and M_{fp}, Fig. 5.99(d)] are significant. Extrapolating up the compression pile moments to the footing centerline, we have

$$M_{fn} = 2 \times (-2210) \times \frac{1.2 + 0.75}{1.2} = -7180 \text{ kNm (5292 kip-ft).}$$

Noting that the footing shear force is $2 \times 7.2 = 14.4$ MN (3.24 kips), the maximum positive moment, at the column face is

$$M_{fp} = -7180 + 14{,}400 \times 1.25 = 10{,}820 \text{ kNm (7974 kip-ft)}$$

Positive Moment. From Eq. (5.133), with $d_f \approx 1.4$ m (4.59 ft),

$$b_{\text{eff}} = 1.5 + 2 \times 1.4 = 4.3 \text{ m}$$

$$M_n \geq \frac{M_{fp}}{\phi_f} = \frac{10{,}820}{0.9} = 12{,}020 \text{ kNm (8860 kip-ft)}$$

Part of this moment $[6600(0.75 - 0.1) = 4290 \text{ kNm}]$ is carried by the axial compression provided by passive pressure and pile shear. The remainder, 7730 kNm, requires over the effective width of 4.3 m shown in Fig. 5.99(f), $A_s = 14{,}865$ mm^2, which can be provided by D32 at 225 mm (No. 10 at 9-in. centers). Although the spacing could be increased in the outer region, extra steel may be needed for shear resistance, as discussed subsequently, and the spacing is maintained over the full width.

Negative Moment. The moment input is provided by the piles, and hence reinforcement should be concentrated over the top of the piles. However, as shown in Fig. 5.99(f), the effective width for negative moments, assuming a 45° spread, extends virtually to the column center.
   Therefore, use $b_{\text{eff}} = 6$ m:

$$M_n \geq \frac{7180}{0.9} = 7978 \text{ kNm (5880 kip-ft)}$$

This can be provided by D32 at 325-mm centers (No. 10 at 12 in.).

Footing Shear. The footing shear force on the compression side is 14.4 MN (3.24 kips). With an effective width of 4.3 m, this corresponds to an average shear stress of

$$v_f = \frac{14.4}{1.4 \times 4.3} = 2.4 \text{ MPa (347 psi)}$$

This corresponds to a stress ratio of $0.44\sqrt{f_c'}$ MPa($5.26\sqrt{f_c'}$ psi), which implies that full design for shear is necessary, with considerable amounts of shear reinforcement. However, it is evident that the shear can be transferred by direct compression struts, as shown in Fig. 5.99(g). Two struts exist, D1 and D2, each carrying part of the shear. In the vertical plane [Fig. 5.99(g)(i)], D1 is equilibrated by the compression pile shear forces (3.31 MN) and part of the pile compression force C_p, while the strut D2 is equilibrated by passive pressure (3.3 MN) and part of the pile compression force C_p. In plan [Fig. 5.99(g)(ii)] it is seen that strut D1 consists of two 45° components, spreading from the column to the piles.

It is clear that a horizontal force T_1 = 1.665 MN (374 kips) is needed perpendicular to the shear direction to balance this. Strut D2 is seen to be a fan spreading to the full width of the footing. Resolving the passive pressure into two equal resultants of 1.65 MN, it is seen that a second tension force T_2 = (1.5/2.0)1.65 = 1.24 MN (278 kips) is needed.

Thus a total perpendicular tension capacity of 2.91 MN (560 kips) is needed over the outer 2 m (6.56 ft) of the footing. Steel provided for positive moment flexure is not utilized for flexure in this direction of applied shear and hence can be utilized. The available capacity (D32 at 225 cm) is

$$804 \times 400 \times \frac{2000}{225} \text{ kN} = 2.86 \text{ MN}$$

This is essentially identical to the required strength. Since some capacity will also be available from concrete mechanisms, the design is satisfactory for shear as is.

Nominal stirrups in accordance with Eq. (5.134b) are required, at a spacing not exceeding 0.5 × 1500 = 750 mm (29.5 in). Since ρ_v = 0.35/400 = 0.000875, this is satisfied by D20 vertical legs at 600-mm centers (No. 6 at 24 in.).

Column–Footing Force Transfer. The procedure has already been covered by an example in Section 5.6.4(d) and is therefore not repeated here. Half of the column bars are anchored with inward 90° hooks at the base of the footing, and half by outward hooks. External joint reinforcement is provided by reducing the spacing of the D20 stirrups from 600 to 300 mm (12 in.) for a distance of 750 mm from the column. Existing column hoop reinforcement is adequate for joint force transfer and is carried down into the core. Sixteen D10 legs in the form of eight closed stirrups support the hooks of the column bars bent into the joint.

5.10.2 Example 5.2: Two-Column Bent Designed to Displacement-Based Considerations

The two-column bent in Fig. 5.100 supports a total dead load, including the cap beam weight, of 1800 kips (8006 kN). The columns are to be circular, with monolithic connections to the cap beam and spread footings, which are supported on firm ground with an ultimate bearing capacity of 50 kips/ft^2 (2.4 MPa). For displacement calculations the footings may be considered effectively rigid for both translational and rotational degrees of freedom. The bent is part of a regular elevated viaduct composed of similar bents at 120-ft (36.6-m) spacing. Seismic design of the bridge is to be for a peak ground accelertion of 0.7g, with the Eurocode spectral shape (see Fig. 2.8), using a displacement-based approach for a maximum permissible drift angle of 0.0375. Total design live load on the cap beam is 600 kips,

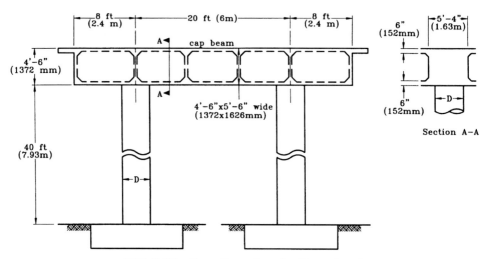

FIG. 5.100   Bent dimensions for Example 5.2.

uniformly distributed, and a seismic reduction factor of $r = 0.25$ is appropriate. Nominal material strengths are $f'_c = 4$ ksi (27.6 MPa) at 28 days
and $f_y = 60$ ksi (414 MPa).

Column Design Strength. The displacement-based procedure of Section
5.3.1(c) is followed. Figure 5.101(a) shows the design displacement spectra,
and Fig. 5.101(b) shows the relationship between displacement ductility factor
μ_Δ and equivalent viscous damping ratio ξ. An initial estimate of yield displacement of $\Delta_y = 4$ in. (101.6 mm) is made. [*Note*: This has deliberately been
chosen to be rather low, to illustrate how rapidly the approach converges. A
more realistic estimate could be obtained from estimating the yield curvature
in accordance with Eq. (7.15).] For a drift angle of 0.0375, the design maximum
displacement is $\Delta_d = 0.0375 \times 480 = 18$ in. (457 mm). The implied displacement
ductility is hence

$$\mu_\Delta = \frac{18}{4} = 4.5$$

From Fig. 5.101(b), $\xi = 19.2\%$, and from Fig. 5.101(a), with $\Delta_u = 18$ in., the
effective period at maximum displacement is $T_d = 3.15$ s. From Eq. (5.23a),
the effective stiffness per column at maximum response, with $P = 900$ kips/
column (4003 kN), is

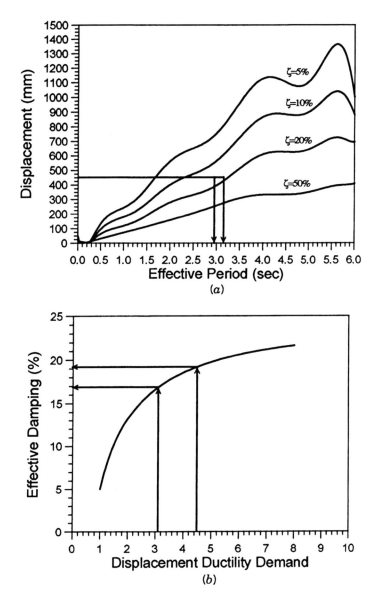

FIG. 5.101 Data for displacement-based design of Example 5.2. (*a*) Design displacement spectra; (*b*) Damping–ductility ratio.

$$K_e = 4\pi^2 \times \frac{900/32.2}{3.15^2} = 111.2 \text{ kips/ft (1619 kN/m)}$$

Hence the design lateral force at maximum displacement is, from Eq. (5.23b),

$$F_u = K_e\Delta_d = 111.2 \times \frac{18}{12} = 166.8 \text{ kips (729 kN) per column}$$

Assuming a typical bilinear stiffness ratio of $r = 0.05$, the design lateral force per column at nominal strength is, from Eq. (5.24),

$$F_n = \frac{F_u}{r\mu_\Delta - r + 1} = \frac{166.8}{0.05 \times 4.5 - 0.05 + 1} = 139 \text{ kips (620 kN)}$$

The design nominal moment capacity is thus

$$M_d = 139 \times \frac{40}{2} = 2780 \text{ kip-ft (3770 kNm)}$$

Preliminary analysis indicates that a 40-in (1016-mm)-diameter column with $\rho_l = 2\%$ would be adequate.

In accordance with the recommendations of Section 5.3.1(d)(vii), characteristic material strengths of $f'_{ce} = 1.3 \times 4000 = 5200$ psi (35.9 MPa) and $f_{ye} = 1.1 \times 60 = 66$ ksi (455 MPa) are used in the design. With $D = 40$ in., the average axial load ratio is

$$\frac{P}{f'_{ce}A_g} = \frac{900}{5.2 \times 1256} = 0.139$$

With $\rho_l = 0.02$, the elastic stiffness ratio, from Fig. 4.9(a), is $I_{cr}/I_g = 0.47$. With $I_g = \pi \cdot 40^4/64 = 125,700$ in^4, $I_{cr} = 59,100$ in^4 (0.0245 m^4). The effective column elastic stiffness is thus

$$K_{cr} = \frac{12EI}{L^3} = \frac{12 \times 4100 \times 59,100}{480^3} = 26.2 \text{ kips/in. (4590 kN/m)}$$

The revised estimate for yield displacement is thus

$$\Delta_y = \frac{139}{26.2} = 5.31 \text{ in. (135 mm)}$$

On past experience, convergence will be improved by choosing a value slightly larger than this—say, $\Delta_y = 5.7$ in. (145 mm). This results in $\mu_\Delta = 3.16$ and

$\xi = 16.9\%$, which results in $T_d = 2.95$ s. Proceeding through the steps above yields

$$K_e = 126.8 \text{ kips/ft } (1850 \text{ kN/m})$$

$$F_u = 190.2 \text{ kips } (846 \text{ kN})$$

$$F_n = 171.6 \text{ kips } (762 \text{ kN})$$

$$M_d = 3430 \text{ kip-ft } (4650 \text{ kNm})$$

This can be provided by a 40-in. (1016-mm)-diameter column with $\rho_l = 0.0272$. With this steel ratio, $I_{cr}/I_g = 0.55$, the elastic cracked-section stiffness is 30.6 kip/in. (5360 kN/m), and the predicted yield displacement is

$$\Delta_y = \frac{171.6}{30.6} = 5.61 \text{ in. } (142.5 \text{ mm})$$

which is very close to the assumed value of 5.7 in. (145 mm). The design has converged in two iterations.

Note: At this stage of design we have considered only the average column load of 900 kips, on the assumption that the flexural strength of the column with seismic compression will be enhanced by the same amount as that of the column with seismic tension will be reduced. This will be checked shortly. When calculating the elastic stiffness, we have also ignored column moments due to dead load (which are very small) and have assumed that the cap beam is rigid. Simple calculations will show this to be an acceptable approximation.

At this stage, since the columns are quite slender and the maximum displacement is large, it is advisable to check for $P–\Delta$ effects, in accordance with Eq. (5.162). Thus

$$\frac{F_y}{P} = \frac{171.6}{900} = 0.191$$

$$\frac{3.3\Delta_u}{L} = \frac{3.3 \times 18}{480} = 0.124 < 0.191 \quad (\text{OK})$$

Thus $P–\Delta$ effects are not expected to increase maximum displacements.

Column Flexural Design. The required column longitudinal steel area is

$$A_{sc} = 0.0272 \times 1256 = 34.2 \text{ in}^2 (22,060 \text{ mm}^2)$$

We choose 16 No. 14 (D43) bars ($A_{sc} = 36.0 \text{ in}^2$). For an initial estimate of confinement requirements, we need an estimate of the maximum column axial load. Taking $M_p = 1.4M_d = 4900$ kip-ft (6650 kNm), the maximum seismic axial load is thus

$$P_E = 4900 \times \frac{22.25}{20} \times \frac{2}{20} = 545 \text{ kips (2425 kN)}$$

Therefore, the maximum axial compression is $900 + 545 = 1445$ (6430 kN). Preliminary analysis indicates a neutral-axis depth of $c = 15.3$ in. (386 mm) at maximum displacement. The plastic hinge length, from Eq. (5.39), is $L_p = 0.08 \times 240 + 0.15 \times 66 \times 1.69 = 35.9$ in. (913 mm).

The required plastic rotation is

$$\theta_p = \frac{\Delta_d - \Delta_y}{480 - L_p} = \frac{18 - 5.6}{444} = 0.0279 \text{ rad}$$

$$\phi_p = \frac{\theta_p}{L_p} = \frac{0.0279}{35.9} = 780 \times 10^{-6} \text{ in}^{-1} \text{ (0.307 m}^{-1})$$

The preliminary analysis indicates a yield curvature of

$$\theta_y = 147 \times 10^{-6} \text{ in}^{-1} \text{ (0.0058 m}^{-1})$$

Therefore,

$$\phi_d = \phi_y + \phi_p = (780 + 147) \times 10^{-6} = 927 \times 10^{-6} \text{ in}^{-1} \text{ (0.0365 m}^{-1})$$

To estimate the confined compression strength f'_{cc}, we assume a typical transverse confinement ratio of $\rho_s = 0.0075$. Lateral confining pressure is

$$f_l = 0.5\rho_s f_{yh} = 0.5 \times 0.0075 \times 60 = 0.225 \text{ ksi}$$

Therefore,

$$\frac{f_l}{f'_{ce}} = 0.0433$$

From Eq. (5.6),

$$f'_{cc} = 5.2 \, (2.254\sqrt{1 + 7.94 \times 0.0433} - 0.0866 - 1.254)$$
$$= 6.62 \text{ ksi (47.7 MPa)}$$

With $c = 15.3$ in. (see above), from Eq. (5.42),

$$\varepsilon_{cu} = \phi_u c = 927 \times 10^{-6} \times 15.3 = 0.0142$$

The required transverse volumetric ratio is thus, from Eq. (5.14)

$$\rho_s = (\varepsilon_{cu} - 0.004) \frac{f'_{cc}}{1.4 f_{yh} \varepsilon_{su}} = \frac{0.0102 \times 6.62}{1.4 \times 60 \times 0.12} = 0.0067$$

Note that this is significantly less than would be required by the prescriptive Eq. (5.47), which would require that $\rho_s = 0.00963$. Note further that Eq. (5.56) requires $\rho_s > 0.0002n = 0.0002 \times 16 = 0.00032$ for antibuckling restraint, which thus does not govern.

The design confinement of $\phi_s = 0.0067$ can be provided by a No. 5 spiral at $s = 5$-in. pitch (D16 at 125 mm). This satisfies requirements of Section 5.3.2(c)(v) that

$$s < 6d_{bl}[6 \times 1.69 = 10.1 \text{ in. } (258 \text{ mm})] \quad \text{and} \quad s < D/5 = 8 \text{ in. } (203 \text{ mm})$$

The plastic end region over which this confinement must be provided is, in accordance with Fig. 5.34, $h = D$ but not less than the region over which $M > 80\%$ of M_{max}. Thus $h = 3.3$ ft but not less than $0.2 \times 20 = 4$ ft. Thus special confinement is required for 48 in. (1220 mm) at the top and bottom of the column. If not required for shear, the spacing could be increased to 8 in. (203 mm), as above, outside the plastic end region.

Results of moment–curvature analyses run at characteristic material strength [$f'_{ce} = 5.2$ ksi (35.9 MPa), $f_{ye} = 66$ ksi (455 MPa)] and at material overstrengths [$f'_{co} = 1.7 f'_c = 6.80$ ksi (46.9 MPa), $f_{yo} = 78$ ksi (540 MPa)], calculated using SEQ $-$ Mϕ [S12] are shown in Fig. 5.102(a) and (b), respectively, for maximum [$P = 1445$ kips (6427 kN)] and minimum [$P = 355$ kips (1597 kN)] axial force levels. At $\varepsilon_c = 0.004$, design flexural strengths are 3835 kip-ft (5204 kNm) and 3196 kip-ft (4337 kNm). The average of these [3516 kip-ft (4770 kNm)] exceeds the required design moment of $M_d = 3430$ kip-ft (4650 (kNm)) by 2.5%.

Column Shear Design. At the design ultimate curvature of $\phi_d = 927 \times 10^{-6}$ in^{-1} (0.0365 m^{-1}) (see above), overstrength column moments are $M_p = 4640$ kip-ft (6300 kNm) and 4090 kip-ft (5552 kNm) for compression and tension columns, respectively. Thus the average ratio of overstrength to design flexural strength is $(4640 + 4090)/(3835 + 3196) = 1.242$, considerably less than the value of 1.4 recommended for prescriptive design. Thus, using moment–curvature analysis allows significant economics in reduced overstrength forces for shear and cap beam design.

Revised seismic axial loads on the columns are

$$\pm(4640 + 4090) \frac{22.25}{20} \times \frac{1}{20} = \pm486 \text{ kips (2160 kN)}$$

Thus, considering axial loads including vertical acceleration effects in accordance with Eq. (5.58),

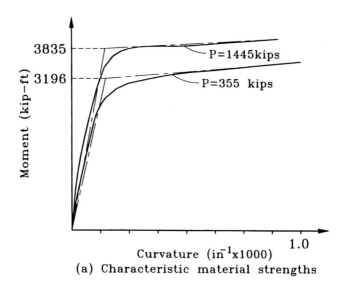

(a) Characteristic material strengths

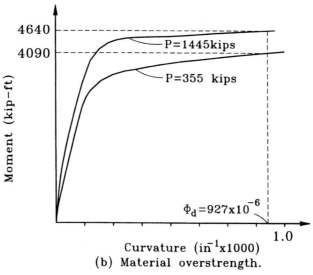

(b) Material overstrength.

FIG. 5.102 Column moment–curvature relationship for Example 5.2.

$$P_{e,max} = 486 + 900(1 + 0.5 \times 0.7) + 0.25 \times \frac{600}{2} = 1776 \text{ kips (7903 kN)}$$

$$P_{e,min} = -486 + 900(1 - 0.35) = 99 \text{ kips (440 kN)}$$

It would be appropriate to rerun moment–curvature analyses for these loads, but the difference from the two curves in Fig. 5.102(b) is less than 3%. The minimum load condition is expected to govern, and it will thus be conservative to check for the slightly high moment of Fig. 5.102(b). Maximum column shear forces are thus

$$V° = \begin{cases} 2 \times \dfrac{4640}{40} = 232 \text{ kips (1032 kN) (compression column)} \\[2mm] 2 \times \dfrac{4090}{40} = 204.5 \text{ kips (910 kN) (tension column)} \end{cases}$$

In accordance with recommendations of Section 5.3.4(b)(iii), the shear strength of the tension column is found as follows:

V_c: Curvature ductility $\mu_\phi = 927/147 = 6.3$. Hence, from Fig. 5.46, assuming biaxial ductility, $v_c = 0.9\sqrt{f'_c}$ psi$(0.075)\sqrt{f'_c}$ MPa,

$$V_c = 0.9\sqrt{4000} \times 0.8 \times 1256 = 58.5 \text{ kips (260 kN)}$$

V_p: Compression zone depth = 11 in.:

$$V_p = \frac{99(40 - 11)}{480} = 6.0 \text{ kips (27 kN)}$$

V_s: Using an angle of 35° for the truss mechanism yields

$$V_s = \frac{\pi}{2} \times 0.306 \times 60 \times \frac{36.6}{5} \times 1.43 = \underline{301.9 \text{ kips (1343 kN)}}$$

Therefore

$$V_n = 366.4 \text{ kips (1630 kN)}$$

Dependable shear strength is thus $\phi_s V_n = 0.85 \times 366.4$ (311.4 kips > 204.5). Thus shear in the plastic end region is satisfactory. Since $\phi_s V_n$ exceeds 232 kips (1032 kN) it is clear that the compression column is also satisfactory. Outside the plastic end region, where spiral spacing increases to 8 in. (302 mm), V_s decreases to 188.7 kips (840 kN) but V_c increases to 195 kips (868 kN), and again, shear is satisfactory. However, since V_c is reduced over the end 2D length, it is elected to carry the 5-in. (126-mm) spacing of the spiral over 80 in. (2.0 m) at the top and bottom of the column. Final column rebar details are thus:

Column longitudinal steel: 16 No. 14 (16D43)
Column traverse steel: End 80 in.:   No. 5 spiral at 5-in. pitch (D16 at 125)
Central region: No. 5 spiral at 8-in. pitch (D16 at 205)

Consideration could be given to reducing the longitudinal steel quantity in the central part of the column in accordance with Fig. 5.37.

Cap Beam Design. Cap beam design forces are given in Fig. 5.103. Moments [Fig. 5.103(a)] and shears [Fig. 5.103(b)] are given for the two load combinations defined by Eq. (5.58), based on maximum column plastic moments at $\phi_d = 927 \times 10^{-6}$ in^{-1}, from Fig. 5.102. The two load conditions of Eq. (5.58) correspond to distributed cap beam loads of 71.7 kips/ft (1047 kN/m) and 32.5 kips/ft (477 kN/m), respectively.

Flexural Design. From Fig. 5.103(a), maximum cap beam design moments at the column faces are -5450 kip-ft (7395 kNm) and $+3200$ kip ft (4342 kNm) for negative and positive moments, respectively. These occur in conjunction with cap beam axial forces of 115 kips (512 kN) compression and 87 kips (387 kN) tension, respectively, from Fig. 5.103(c). Note that the cap beam axial forces were found by distributing the total cap inertia force, $232 + 204.5$ kips, uniformly along the length of the cap beam. In this example the cap beam axial forces are not particularly significant, contributing about 5 to 8% of the moment demand, but should still be considered.

Adopting a strength reduction factor of $\phi_t = 0.9$ for flexure, the design nominal strengths are thus M_n (negative) $= -6056$ kip-ft (8216 kNm) and M_n (positive) $= 3556$ kip-ft (4824 kNm). Figure 5.104 shows the reinforcement selected, consisting of eight No. 14 top bars, four No. 14 bottom bars, and six No. 14 side bars (No. 14 = D43). Analysis using SEQ$-$Mϕ [S12] results in nominal flexural strengths of -5856 kip-ft (7946 kNm) and 3699 kip-ft (5019 kNm) for negative and positive moments, respectively. The 3.3% shortfall for negative moment capacity will easily be provided by deckslab transverse reinforcement, which has not been considered in estimating strength. Tension shift effects indicate that the cap beam reinforcement cannot be reduced before a distance of $d_{cb} = 50$ in. (1270 mm) from the column force. Since this is only 50 in. (1270 mm) from the cap beam centerline, there is no point considering termination of top or bottom cap beam flexural reinforcement.

Cap Beam Shear Design. From Fig. 5.103(b), the critical shear force distance d_{cb} from the column on the negative moment side is 740 kips (3293 kN). This occurs in conjunction with an axial compression force of 64 kips (283 kN). Design for shear uses the recommendations for beams in Section 5.3.4(b)(ii).

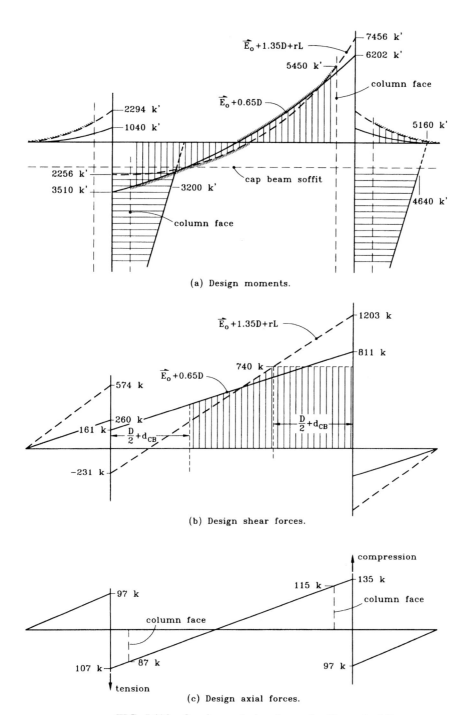

FIG. 5.103 Cap beam design forces for Example 5.2.

V_c: $\rho_l = (10 \times 2.25)/(64 \times 50) = 0.00703$. Then

$$v_b = (0.8 + 120 \times 0.00703)\sqrt{4000} = 104 \text{ psi } (0.717 \text{ MPa})$$
$$V_c = 0.104 \times 0.8 \times 64 \times 54 = 288 \text{ kips } (1282 \text{ kN})$$

V_p: $64(54 - 5)/240 = 13$ kips (58 kN)
V_s: $V_s = V_u/\phi_s - (V_c + V_p) = 740/0.85 - (288 + 13) = 570$ kips (2535 kN). Then

$$\frac{A_v f_y d}{s} \cot 35° = 570$$

Therefore,

$$\frac{A_v}{s} = \frac{570}{60 \times 50 \times 1.43} = 0.133 \text{ in}^2/\text{in. } (3.38 \text{ mm}^2/\text{mm})$$

This can be provided by four No. 4 (D13) legs at 6 in. (152 mm) spacing. This spacing is required from the column face for a distance of 70 in. (1778 mm). There is little point in reducing the spacing in the central part of the beam. In the cantilever portions, gravity-load design will govern. Note that in the positive moment region [left column in Fig. 5.103(b)] the shear is less than V_c, and all shear reinforcement will be available for resisting joint shear forces.

Joint Force Transfer: Joint Shear Force. From Eq. (5.83),

$$V_{jh} = \frac{M°}{h_b} = \frac{4640}{4.5} = 1031 \text{ kips } (4588 \text{ kN})$$

Therefore,

$$v_{jh} = \frac{1031}{40 \times 64} = 403 \text{ psi } (2.78 \text{ MPa})$$

Axial stress $f_h \approx 0$.

$$f_v = \frac{1400}{64 \times 114} = 192 \text{ psi } (1.32 \text{ MPa})$$

Then from Eq. (5.89),

$$p_t = 96 - \sqrt{96^2 + 403^2} = 318 \text{ psi } (2.20 \text{ MPa})$$

This principal tension stress corresponds to $5.03\sqrt{f'_c}$ psi, so full joint shear design is required. Joint reinforcement is provided in accordance with the suggestions of Figs. 5.71 and 5.72. For external joint reinforcement, Fig. 5.71(b) pertains. In each of the three overlapping areas an amount of $A_{jv} = 0.125 \times 36 \times 1.3 = 5.85$ in^2 (3773 mm^2) is required [Eq. (5.100)]. In area 2 [see Fig. 5.71(b)], the amount of shear steel provided over a distance $h_b = 27$ in. from the column face is five sets $\times 4 \times 0.2 = 4$ in^2. We replace these sets by sets of No. 5 ties at the same spacing, providing $A_{jv} = 6.12$ in^2 (3947 mm^2). Closing moment force transfer can be effected by a similar quantity of No. 5 stirrups outside the joint in the cantilever. In areas 1 and 3, on the sides of the cap beam we provide five sets of two-leg No. 5 ties. Together with the overlap with 50% of the steel in area 1 and the cantilever, this is sufficient. The bottom horizontal leg satisfies the requirement for additional soffit steel.

Within the column core, an extra 2.93 in^2 (1887 mm^2) of vertical rebar is required to clamp the top cap beam reinforcement. This is provided by 10 No. 5 inverted J bars, 40 in. (1016 mm) long.

In the direction of the cap beam axis, in accordance with Eq. (5.101) and Fig. 5.72, additional beam steel is required: $A_{sb} = 0.0625 \times 36 \times 1.3 = 2.93$ in^2 (1887 mm^2). This is provided by two No. 11 (D36) bars close to the soffit, extending 80 in. (2030 mm) past the column face, as shown in Fig. 5.104.

Hoop or spiral joint reinforcement is required in accordance with Eq. (5.104). With an average length of $l_a = 48$ in. and $f^o_{yc} = 1.4 \times 60$ ksi:

$$\rho_{sj} = \frac{0.3 \times 36 \times 1.4}{48^2} = 0.00656$$

This is slightly less than that provided in the column hinge region. Thus use No. 5 hoops or spirals at 5-in. centers (D16 at 125) in the joint region.

Footing Design. Footing design is comparatively straightforward and will not be considered in detail in this example. However, it will be noted that the low axial compression on the tension column causes problems for the spread footing, and uplift resistance is needed. A footing size of 12 ft $\times$ 12 ft $\times$ 4.5 ft (3.66 m $\times$ 3.66 m $\times$ 1.37 m) is chosen, and soil anchors are used to provide an elastic uplift resistance of 5000 kips (2225 kN) on either side of the column. The footing size is such that the full width can be considered effective for flexure and shear. Flexural reinforcement is provided by No. 8 (D25.4) headed bars at 9-in. (229-mm) centers in the top layer and No. 11 headed bars (D35.8) at 9-in. (229-mm) centers in the bottom layer (see Fig. 5.84a). Shear can be carried by direct compression struts, and only nominal shear reinforcement in accordance with Eq. (5.134) is provided. Thirty No. 6 headed vertical bars are placed to resist vertical joint shear, in accordance with Eq. (5.137), and two additional No. 11 bars (D35.8) are placed in the top mat through the column as required by Eq. (5.101). Finally, we check anchorage length for the No. 14 bars in the joint. Taking these to within 6 in. (152 mm) of the top of the cap beam, the anchorage length is 48 in. (1220 mm). The length required to satisfy Eq. (5.111a) is $l_{dc} = 0.025 \times 1.41 \times 66000/\sqrt{4000} = 36.8$ in. < 48 in. (OK). Final reinforcement details for the bent are shown in Fig. 5.104.

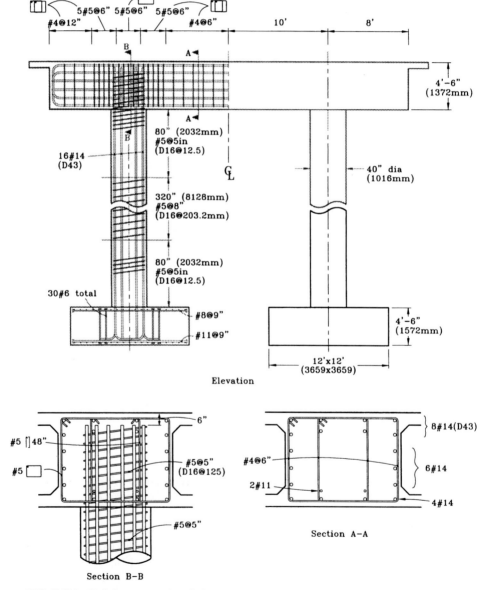

FIG. 5.104 Reinforcement details for Example 5.2 (footing uplift restraint not shown).

6

DESIGN OF BRIDGES
USING ISOLATION AND
DISSIPATION DEVICES

6.1 INTRODUCTION

6.1.1 Concepts of Seismic Isolation and Energy Dissipation

Since the beginnings of seismic engineering it has been recognized that the damage potential of an earthquake is due essentially to the unfortunate correspondence between the fundamental periods of vibration of the majority of structures and the frequency content of the seismic input. It is also well known that despite the correspondence of dynamic characteristics of structures and earthquakes, many structures are able to find ways to survive by escaping the frequency range where the earthquake has greatest power, as a result of the period elongation due to accumulated damage.

A second fundamental concept in earthquake-resistant design of structures is evident from an examination of the equation of motion of a linear system [see Eq. (4.2)]: the higher the viscous damping, the lower the forces to be resisted by the structure. Viscous damping is actually a way to dissipate energy, which can also be dissipated through hysteresis, friction, and in general any inelastic response. For a numerical simulation of the seismic response, it has often been found convenient to represent the effects of any kind of inelastic behavior by means of an equivalent viscous damping, creating an equivalent model (or substitute structure), as discussed briefly in Section 4.4.3.

It is clear from the discussion above that the possibility of artificially increasing both the period of vibration and the energy dissipation capacity of a structure has to be regarded as a very attractive way of improving its seismic resistance. This objective can be pursued as a specific design intention, taking advantage of natural characteristics of the soil–foundation–structure system or making use of specific artificial elements designed to isolate part of the structure from the full intensity of the seismic input and to dissipate a large amount of energy. These elements are often called *isolators*, *dampers*, or *isolation/dissipation* (I/D) *devices*, in most cases overstating the true case or neglecting important aspects of their response; all these terms will be used alternatively in the following.

The concepts of isolation and dissipation are particularly interesting for bridges, because of a series of potential advantages related to their specific structural characteristics. In most cases bridges are strategic structures that require a higher degree of protection to ensure their functionality after a seismic event. It can therefore be convenient to concentrate the damage potential into a few mechanical elements that may be easily checked and replaced, if need be. In addition, most of the mass of a bridge is concentrated at the deck level, and decks are usually designed to remain elastic under seismic action. A common structural configuration (particularly in Europe) is made by a continuous deck supported on bearings at the tops of piers. Advantages of this design choice were discussed in Section 3.3.1. In this case the bearings themselves can be designed as I/D devices, selecting their stiffness, yield strength, and ultimate elongation capacity as a function of the desired protection and of the seismic intensity expected. This option is particularly common in the case of seismic upgrading of existing structures.

Since bridges are usually simple structures from the point of view of the expected structural response, it is easier to conceive an appropriate correction of the stiffness distribution than with more complex structures. I/D devices can therefore be used to correct or regularize the expected response, adding flexibility to stiffer piers, thus avoiding possible undesirable concentration of ductility demand. Also, bridges often have long natural periods, particularly if some nonlinear response is accepted and considered. Large displacements are then already expected and accepted. The addition of I/D devices may thus have little effect on the maximum displacements, nonetheless assuring a higher level of protection to selected structural elements and larger energy-dissipation capacity.

6.1.2   Period Shift

The main aspects of ground motion and response spectra have been discussed in Section 2.4. As an example to be used in this section, the acceleration and displacement response spectra suggested in EC8 [E5] for the case of medium-dense soil are plotted in Figs. 6.1 and 6.2 for a peak ground acceleration of 0.8*g* using the equivalent viscous damping as a parameter. It is immediately clear that for such spectra an elongation of the period of vibration (above 0.6 s) results in lower acceleration, but it is also evident that this corresponds to a linear increase in the structure displacement (up to a period of 3 s). For longer periods of vibration the structure displacement no longer increases but remains constant, while in a real case it should tend to reduce and converge to the ground displacement for a long period of vibration. Both spectra are

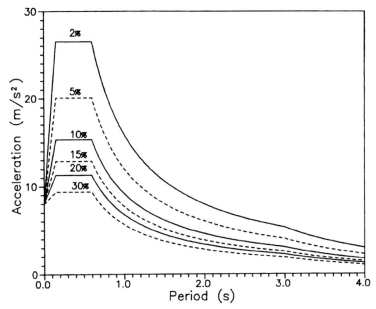

FIG. 6.1 EC8 acceleration response spectrum for medium-dense soil as a function of viscous damping (ground acceleration = 0.8g).

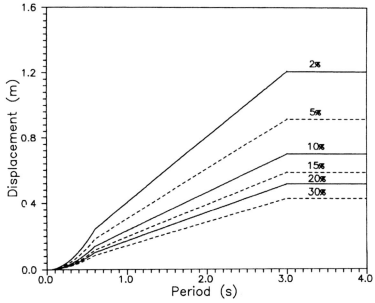

FIG. 6.2 EC8 displacement response spectrum for medium-dense soil as a function of viscous damping (ground acceleration = 0.8g).

459

combined in Fig. 6.3, in the form discussed in Section 2.4.1, which is well suited for design of isolated bridges.

For this acceleration spectrum a period shift is therefore particularly useful when the structure has a fundamental period of vibration around 0.5 to 0.6 s because in this case a small period elongation produces a significant force reduction, while the linearly increasing displacement may still be accepted. For example, a period elongation from 0.6 s to 1 s implies a force reduction of 40% while the displacement increases 67%. Bridges often have comparatively long fundamental periods of vibration, and in this case do not fall in the category discussed above. If the period of a rather flexible structure is elongated further, the displacement still increases linearly, while the benefit in terms of force reduction becomes less significant. For extremely flexible structures (period larger than 3 s) a further increase in the period continues to reduce the acceleration, while the structure displacement no longer increases (and possibly decreases in a real case). The extension of the concepts expressed above to other response spectra is rather obvious.

6.1.3 Damping

From Figs. 6.1 and 6.2 it is also clear that an increase in the equivalent viscous damping is always favorable, inducing significant reduction of spectral

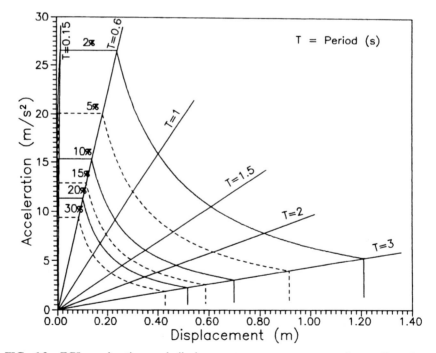

FIG. 6.3 EC8 acceleration and displacement response spectrum for medium-dense soils as a function of equivalent viscous damping (ground acceleration = 0.8g).

acceleration and displacement. The potential benefit can be particularly important for low equivalent damping values, while the response tends to be more uniform for high damping ratios. Damping ratios of 20 to 30% are usually considered of interest in designing isolated bridges. Note that the term *equivalent damping* is the viscous damping equivalent to the global nonlinear response of the structure with the I/D devices. It therefore results from a complex combination of different sources of energy dissipation. It is therefore not easy to select an appropriate hysteresis curve for each type of I/D devices. Damping can be added in at least four different forms: viscosity, friction, material hysteresis (internal work), or impact and radiation. This last kind of energy dissipation is typical of soil–foundation interaction, where a combination of all of the forms of dissipation mentioned is often present.

The four ways of dissipating energy can correspond to various I/D devices described in Section 6.2. The most widely used are based on material (steel or lead) hysteresis, because of the reliability of the response expected and because the basic concept is more traditional for engineers, but interesting applications have also been based on friction-based and pendulum devices. It is also important to observe that for most I/D devices a large energy-dissipation capacity implies large potential residual displacements in the absence of an appropriate restoring force. Whereas residual displacement in conventional structural response involving damage to structural members may be difficult to repair, residual deformation with I/D may generally be repaired by pulling back the deck at the original position and replacing the damaged I/D devices. This kind of operation should be planned for in the design phase to make later implementation easier.

6.1.4 Seismicity Aspects

Aspects related to ground motion, from potential earthquake sources to local soil effects, have been discussed in Chapter 2. It is important here to observe that the frequency content of the expected ground motion can be of the utmost importance in the case of isolated bridges, particularly if a period shift rather than added damping is considered as the important design parameter. If the possibility of different spectra, characterized by high response at longer periods (see, e.g., Fig. 2.8 or 2.23), cannot be excluded, an artificial period elongation could result in a catastrophic situation. A detailed analysis of the expected ground motion is therefore of fundamental importance if some possible isolation is taken into consideration [P24].

6.1.5 Vertical Response

The insertion of flexible elements in the horizontal direction in most cases implies an increment of the vertical flexibility as well. This aspect has to be considered carefully, particularly in case of near-field earthquakes, because of the potential amplification of the vertical motion and of possible deck unseating. Even if a possible tensile stress in the bearing is permissible, it must be recognized that many bearing isolators have different vertical flexibilities when compressed and when stretched, usually being much stiffer in compression: see, for example, Fig. 6.4, where the stress–strain curve for a laminated-rubber bearing axially loaded is shown. It cannot be excluded that fracture of the bearing may result as a consequence of resonance in vertical response. Also, high variation in the vertical load could cause significant variation in the horizontal response of the bearing, not taken into account to the appropriate extent in the design phase.

6.1.6   Design Philosophy and Structural Aspects

As usual in seismic engineering, design of an isolated bridge cannot be a simple application of mathematical equations and safety factors, but on the contrary, requires an understanding and acceptance of specific design strategies. The design strategy for nonisolated bridges, discussed in detail in Chapters 1, 3, and 5, assumes that there are specific parts of the structure, typically the piers ends, where significant nonlinear behavior is accepted and used as a fundamental means to resist the expected earthquake motion. Ductile behav-

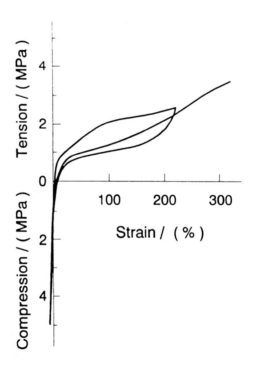

FIG. 6.4   Stress–strain curve for a laminated-rubber bearing axially loaded [T8].

ior is obtained by careful detailing, but significant damage is necessarily accepted as accompanying force reduction; the emphasis is therefore placed on the low sensitivity to unexpected earthquake motions, on the acceptable remaining capacity after the seismic event, and on the repairability of the damage.

In the case of isolated bridges, the critical role is assumed by the isolating and dissipating devices introduced between piers, abutments, and deck, or in fewer cases at the base of the piers. Most of the damage is then concentrated in these devices, with significant plasticization and possibly large residual displacements. The response of the remaining part of the structure is therefore in principle linear elastic, with some possible damage localized at movements joints due to large structural displacements. Some fundamental questions then arise naturally: What happens in the case of an earthquake of unexpected characteristics (either in terms of frequency content, ground acceleration, or duration)? Can some structural (e.g., pier) damage be taken into account provided that catastrophic failure is excluded? In this case how should the nonlinear response of the structure be considered in the design process? These fundamental questions will be discussed in operative terms later in the chapter, showing that the same capacity design principles and displacement-based design concepts as those used for standard bridges can also be rigorously applied to the case of isolated bridges.

When designing an isolated bridge it is fundamental to have clear ideas on the global structural response in both directions, as resulting from the structural concepts. The location of the I/D devices at the tops of piers is due primarily to practical and economical concerns, but in most cases it suits well the structure layout, with dominant mass at the deck and relatively short piers. With superstructure isolation the piers tend to respond as independent structures, fixed at the bottom and elastically restrained at the top. In the case of tall and massive piers their response can dominate the seismic loads on piers and foundation. It can be therefore appropriate to consider the possibility of a base isolation of the piers, adopting some moment-limiting detail close to the pier base or at the foundation level. Rocking as an isolation system is discussed in Section 6.4.

At the abutments, isolation is always at the deck level. Care has to be taken to select the appropriate stiffness and strength of the devices, since strength and stiffness of the abutment–soil system are usually less certain than those of piers. If the bridge has long spans and a flexible deck in the horizontal plane and the abutments are overdesigned, as sometimes happens in Europe, the possibility of keeping a hinged connection between deck and abutments could also be considered.

6.1.7 Displacement Problems

Displacements are often a major concern when designing an isolated bridge, since the increased flexibility generally results in increased displacements, which in turn may result in a series of potential problems, discussed briefly below. P–Δ effects can become more important. In particular, it must be kept in mind that the displaced position of the vertical load increases the second-order moment at the base of the pier, so reducing the available first-order strength. Since the yielding strength of the I/D devices is usually designed to avoid yielding of the piers, this effect could impair the global design philosophy, inducing premature yielding in the pier. The pier strength reduction due to P–Δ effects should therefore always be considered. $P - \Delta$ effects are considered in detail in Section 5.9.

Isolated bridges often have a long continuous superstructure, which implies extremely large joints and corresponding road surface links. Sacrificial joints, knock-off details, and in general limiting displacement details have often been used in the past in conjunction with isolation systems. Their use has to be considered in combination with the force–displacement response of I/D devices, which in some cases have built-in features limiting their displacement. Reaching the displacement limit obviously implies structural damage, which could be accepted because of the extreme nature of the seismic event. The main concern is therefore careful consideration of the structural response to the impact resulting in higher-than-expected seismic forces.

An isolated bridge is in general more prone to suffer permanent deformations due to temperature cycles and creep: one-way bearing slip (crawling) connected to seasonal or diurnal temperature variations have been reported. To avoid problems related to seismic gaps becoming blocked or too small, or limit displacement becoming significantly unsymmetric, periodic inspections should be programmed and ways to reposition the deck at the right place considered in the design phase (the same system could be needed in a post-earthquake situation, due to permanent deformations of the isolators).

6.1.8 Past Applications and Response in Past Earthquakes

A rather complete and accurate list of the structures endowed with seismic isolation devices is presented in [S20]; from this source it is found that a total of 255 isolated bridges—5 in Iceland, 49 in New Zealand, 12 in Japan, 21 in the United States, and 168 in Italy—had been built by 1993. Lead–rubber bearings were used in three Italian bridges and in 66 of the other 87 bridges, showing clearly that this is the preferred choice except in Italy, where more frequently, some damper (in most cases a steel damper) is coupled with traditional sliding support. Economy, simplicity, reliability, and low maintenance are the reasons why it appears that lead–rubber bearings and steel dampers, coupled with polytetrafluoroethylene (PTFE) bearings, dominate the market.

Unfortunately, very little information is available on real response to earthquakes of isolated bridges. The Sierra Point overhead bridge, San Francisco, built in 1956 and later retrofitted using lead–rubber bearings, experienced a ground acceleration of 0.09g during the Loma Prieta earthquake. The test was therefore not significant. The Te Teko bridge (New Zealand), also isolated with lead–rubber bearings, was hit by the Edgecumbe earthquake (1987), with a ground acceleration probably in the range 0.3 to 0.35g. The bridge has five 20-m (66-ft) spans, precast U beams, and cast-in-place deck and single-column cylindrical piers. Each end of each span is supported by a pair of lead–rubber bearings. Due to a construction deficiency [the isolators were designed to have retaining rings to prevent horizontal sliding with 20 mm (0.79 in.) effective height, whereas the effective height of the ring of one of the isolators at the western abutment was only 5 mm (0.20 in.)], the action of one abutment bearing was lost, resulting in moderate structural damage [including some loss of cover concrete at the base of the western column and pushing the abutment knockout fuse about 75 to 100 mm (3 to 4 in.)]. It was reported that if the action of one abutment bearing had not been lost, it was unlikely that the columns would have suffered visible cracking. The bridge was rendered serviceable by jacking up the western end of the superstructure and replacing the bearings, repositioning the knockout beams, and filling and compacting the approach material displaced by these beams [P28].

6.1.9 Viewpoints on the Applicability of Seismic Isolation to Bridges

Designers and those responsible for seismic protection of bridges are presently sharply divided into those who consider isolation as a panacea that will solve all seismic problems and those who believe that isolation is not reliable and should never be used. The limited experience on response during severe earthquakes is certainly contributing to this dichotomy, even if there is a good capacity for simulating the nonlinear response of isolated structures and a high and reliable technological production level. On the other hand, broad probabilistic studies to estimate the safety level associated with various protection factors and response parameters have not yet been completed and are an important future research field.

For the time being, possibly larger safety factors should be used when designing an isolated bridge, nonlinear time history analysis should be performed in most cases, the I/D devices should always be tested, and a maintenance program with periodic inspection and possible retesting of devices should be scrupulously followed. Provided that this general philosophy is accepted and the specific requirements described in this chapter are met, dynamic isolation constitutes a powerful, flexible, and economical tool for designing new bridges, and to an even larger extent, for retrofitting existing bridges. The extreme variability of the potential design variables encourages innovative solutions to difficult problems and specific protection of elements without recourse to ductile details. The same possible wide selection of design parameters makes the preparation of simple design rules difficult and confirms from another point of view the need for detailed analysis. Design of isolated bridges is therefore still more an art than a combination of rules: For this reason the final design product has to be checked more reliably and its effectiveness proved with models capable of capturing the real structural response.

6.2 ISOLATION AND DISSIPATION DEVICES

6.2.1 Main Properties of the Most Common Devices

It has been clarified in Section 6.1 that an isolation system should be able to support a structure while providing additional horizontal flexibility and energy dissipation. The three functions could be concentrated into a single device or could be provided by means of different components; consider, for example, a traditional bridge whose continuous deck lies on polytetrafluoroethylene (PTFE) supports to allow thermal deformations: Its displacements could be restrained by means of hydraulic or hysteretic dampers, which will provide additional energy dissipation (i.e., equivalent damping).

A few parameters have to be considered carefully in the choice of an isolation system, in addition to its general ability of shifting the vibration period and adding damping to the structure, such as:

- Deformability under frequent quasistatic load (i.e., initial stiffness)
- Yielding force and displacement
- Ultimate displacement and postultimate behaviour
- Capacity for self-centering after deformation (i.e., restoring force)
- Vertical stiffness

A summary of the devices most commonly used and readily available in the world is presented below, together with the main characteristics of these devices.

(a) Laminated-Rubber Bearings. A cylindrical or rectangular block of rubber constitutes the simplest isolator for a bridge superstructure but presents a number of inconveniences, essentially related to its high deformability under vertical loads. The insertion of a number of horizontal steel plates, as with elastomeric bearing pads, solves most problems (see Fig. 6.5) by increasing the vertical stiffness and improving the stability of the behavior under horizontal load. This kind of bearing shows a substantially linear response, governed essentially by the properties of the rubber. It is therefore unusual to utilize it without some other element able to provide increased damping and stability under nonseismic loads, except where the rubber exhibits high natural inherent damping.

The shape, plan, and number of bearings are governed by the vertical load to be transmitted: If the maximum horizontal displacement has been assumed, the strength of the bearings is proportional to the bonded area and inversely proportional to the rubber layer thickness, which in turns governs the vertical

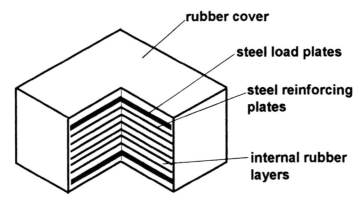

steel load plates

steel reinforcing
plates

internal rubber
layers

rubber cover

FIG. 6.5 Laminated-rubber bearing.

and torsional stiffness. The total rubber thickness (i.e., the number of layers) influences essentially the maximum allowable lateral displacement and the period of vibration. An approximate summary of the complex response of such bearings is given in the following [S20]:

- Vertical load capacity:

$$W < A'GS\gamma_{xz} \tag{6.1}$$

where W is the allowable weight, γ_{xz} the allowable shear strain (when the rubber is assumed incompressible, a vertical compressive strain ε_z causes the rubber to bulge by an amount proportional to its distance from the center of the disk; when the bulge profile is approximated by a parabola, constant rubber volume gives the maximum shear strain as $\gamma_{xz} = 6S\varepsilon_z$, with S as defined below), A' the overlap of top and bottom area of a bearing at maximum displacement, G the shear modulus of rubber, and S a shape factor (loaded area/force-free area, e.g., for a circular disk of diameter D and layer thickness t, $S = D/4t$).
- Bearing horizontal stiffness:

$$K_b = \frac{GA}{h} \tag{6.2}$$

where h is the total rubber height and A is the gross rubber area.
- Bearing lateral period:

$$T_b = 2\pi \left(\frac{M}{K}\right)^{1/2} = 2\pi \left(\frac{Sh\gamma_{xz}A'}{Ag}\right)^{1/2} \tag{6.3}$$

where g is the acceleration due to gravity.

- Bearing vertical stiffness:

$$K_z = \frac{6GS^2Ak}{(6GS^2 + k)h} \qquad (6.4)$$

where k is the rubber bulk modulus.
- Allowable seismic displacement:

$$\Delta_b = B\left(1 - \frac{A'}{A}\right) \qquad \text{(rectangular bearing)} \qquad (6.5)$$

where B is the side dimension of the bearing in the direction considered.

It should be noted that the bearing lateral period of vibration has very limited practical applicability, since the bearing stiffness has to be combined with the structural (pier–foundation system) stiffness to allow the calculation of an effective period of vibration. Typical values for bridge elastomeric bearings are as follows: $G = 1$ MPa (145 psi), $k = 2000$ MPa (290 ksi), $\gamma_{xz} = 0.2\,\varepsilon_{tu}$ (with ε_{tu} the failure tensile strain, typically equal to 4.5 to 7), $S = 3$ to 40, and $A'/A = 0.4$ to 0.7. As expected, the major variability lies in S, which is a function of plan dimensions and rubber layer thickness.

It is then easy to check that for average values of the variable parameters, the allowable vertical stress on the gross area is on the order of 5 to 10 MPa (725 to 1450 psi), the horizontal stiffness is on the order of 1 to 2 MN/m (70 to 140 kips/ft), the period of vibration is of the order of 2 to 3 s, the vertical stiffness is of the order of 1000 to 2000 MN/m (70,000 to 140,000 kips/ft), and the allowable seismic displacement is on the order of one-half of the bearing dimension in plan. The allowable displacement can be increased by segmenting the bearing and introducing stabilising plates, as shown in Fig. 6.6, with obvious corrections to the mathematical relations given above.

The bearing damping has not been mentioned because it is provided only by the viscous behavior of the rubber. It is therefore velocity dependent and usually quite low, up to approximately 5% of the critical damping. Specially formulated high-damping rubbers have been manufactured, obtaining viscous damping ratios of about 15%. Unfortunately, the response of high-damping rubbers is strongly amplitude and history dependent, with stiffness variation

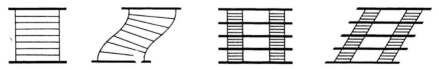

FIG. 6.6 Deformation of standard and segmented rubber bearings.

on the order of 50%. The effects of rate of loading and temperature are also not predictable with sufficient accuracy. It is therefore felt that these kind of devices are not yet mature for extensive practical use [S20].

(b) Lead–Rubber Bearings. The laminated-rubber bearings described in the preceding section present several convenient features, with fundamental drawbacks related to the negligible increase in damping and the high deformability for low static loads. With the insertion of a lead plug (Fig. 6.7), which provides energy dissipation for seismic response and stiffness for static loads, a single compact device is obtained, able to satisfy most of the requirements for a good isolation system. For these reasons, lead–rubber bearings have been used extensively in practical application to bridge structures.

The reasons why lead is an appropriate material are related to its mechanical properties, which allow a good combination with the characteristics of laminated bearings: low yield shear strength [about 10 MPa (1450 psi)], sufficiently high initial shear stiffness [G approximately equal to 130 MPa (18.8 ksi)], behavior essentially elastic–plastic and good fatigue properties for plastic cycles. Considering the characteristics of rubber it is easy to check that for a lead plug with diameter equal to one-fourth of the diameter of a circular bearing, the initial horizontal stiffness is increased by about 10 times, with obvious advantages under wind and braking loads. The lead responds essentially with elastic–perfectly plastic loops. After yielding, the stiffness is therefore equal to the stiffness of the rubber bearing alone. The global hysteresis loops are therefore almost bilinear, as shown in Fig. 6.8.

In the case of laminated bearings the maximum lateral force can be calculated as a function of stiffness and allowable displacement [from Eqs. (6.2) and (6.5)], resulting approximately in

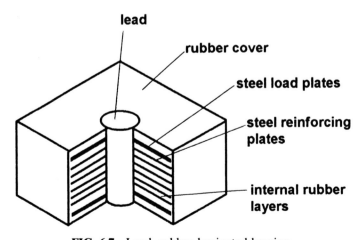

lead

rubber cover

steel load plates

steel reinforcing plates

internal rubber layers

FIG. 6.7 Lead–rubber laminated bearing.

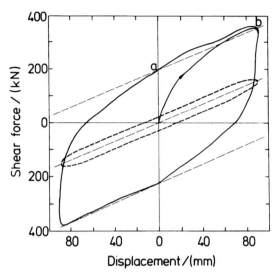

FIG. 6.8 Idealized hysteretic loop for a lead–rubber bearing with a lead plug contributing to one-half of the total strength and assuring an initial stiffness equal to 10 times the stiffness of the rubber bearing alone [rubber diameter 650 mm (25.6 in.), lead plug diameter 170 mm (6.7 in.), vertical compression 3150 kN (709 kips), horizontal force frequency 0.9 Hz] [R6].

$$V_r = \frac{B^3}{2h} \tag{6.6}$$

where it has been assumed that $G_r = 1$ Mpa (145 psi), $A'/A = 0.5$, and the bearing has a square plan with side B (mm), height h (mm), and V_r is in (N). Again assuming a lead plug insert with diameter $B/4$ and shear strength equal to 10 MPa (1450 psi), the maximum horizontal load for the lead plug is found to be

$$V_l = \frac{10\pi(B/4)^2}{4} \tag{6.7}$$

or, approximately, $V_l = B^2/2$.

It is possible to conclude that the contributions of lead and rubber to the total strength of the bearing are approximately equal for a plug with diameter one-fourth of the bearing side and for a total height $h = B$. The idealized resulting force–displacement loop is shown in Fig. 6.8, where the response of a lead–rubber bearing with the following characteristics is represented: rubber diameter 650 mm (25.6 in.), lead plug diameter 170 mm (6.7 in.), vertical compression 3150 kN (709 kips), horizontal force frequency 0.9 Hz, and stroke 90 mm (3.54 in.). The dashed curve is for the rubber bearing alone, the solid line for the lead–rubber bearing [R6].

It is clear that the size of the lead plug can be used as an additional design variable to obtain the desired characteristics from the isolation system. The conceptual effects of geometrical variations of both lead plug and rubber bearing are summarized in Fig. 6.9. The size of the lead plug is proportional to the yield strength of the isolator, while the postyielding stiffness is proportional to the rubber bearing stiffness, therefore increasing when the plan size of the rubber bearing increases and when its height decreases.

It has been shown experimentally [S20] that lead–rubber bearings have little strain-rate dependence for a wide frequency range which contains typical earthquake frequencies, have a stable behavior under repeated loads (the reduction of the dissipated energy during an hysteresis loop after 20 cycles at maximum displacement is approximately 20%), and are not strongly temperature dependent [approximately 30% force variation for 60°C (108°F) temperature variation]. Also, lead–rubber bearings respond well to creep loading conditions: At typical thermally induced velocities the lead shear stress is reduced to approximately 30% of the stress at higher (seismic) rates of loading and the elastic restoring force due to rubber stiffness is large enough to drive the structure back to its original position.

Lead–rubber bearings therefore provide an economical and effective solution for bridge isolation, incorporating period shifting and increased damping in a single device for vertical support. It has to be noted, however, that most of the self-centering property of the laminated rubber bearing, under seismic forces, is lost after insertion of the lead plug. Finally, it has to be noted that from a manufacturing point of view, it is imperative that the lead plug be well confined by the steel plates, to assure a pure shear behavior of the lead plug. This condition will assure that the hysteresis loops will be rather insensitive to a wide range of vertical loads. For this reason the lead-plug volume should be slightly greater than the hole volume.

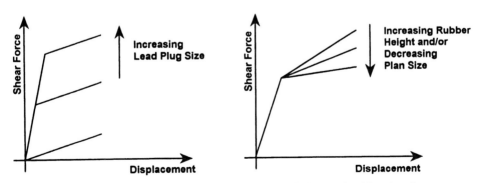

FIG. 6.9 Effects of geometrical variations of the lead plug and rubber bearing on the overall response.

(c)   Sliding Bearings. Stainless steel-PTFE bearings have been used as sliding supports for bridge superstructure for about 30 years, to allow thermal movements. The friction coefficient of PTFE on stainless steel is about 0.02 to 0.03 (depending on possible lubrication) for very low slip rates, such as in the case of temperature or creep movements, but it is much higher and dependent on pressure and sliding velocities at typical earthquake velocities. For typical earthquakes velocities and typical pressure for bridge bearings, the friction coefficient ranges from about 0.10 to 0.15 or more for unlubricated bearings [T7]. The friction coefficient remains very small (typically less than 0.02) under seismic conditions in the case of lubricated PTFE bearings, which require regular inspection and service. PTFE bearings respond with almost rigid-plastic hysteresis loops, thus being able to dissipate large amounts of energy. Because the equivalent yield strength depends on the friction coefficient, it is somewhat unreliable. Any centering force is absent from the system, with a significant potential for large drifts, crawling, and ultimately, support unseating.

The unsatisfactory predictability and reliability of the response and the absence of any centering force suggest that PTFE bearings should be used as seismic isolators only in combination with other centering devices. Effective combinations could result when PTFE bearings are used in conjunction with steel dampers or rubber bearings. In the first case, where all the vertical load is taken by the PTFE bearings, the friction coefficient should be kept as low as possible, with the steel dampers providing some self-centering force and additional damping. In the second case, where the vertical load is shared by the two systems, the axial load taken by the PTFE bearings will be reduced and the rubber bearings will provide a self-centering force. Rubber bearings and PTFE could also be mounted in series (i.e., one above the other) to provide flexibility at forces lower than the bearing sliding force.

The concept of sliding bearings has also been combined with the concept of a pendulum-type response, obtaining a conceptually interesting seismic isolation system known as a friction pendulum system (FPS, Fig. 6.10) [Z4]. The concept is simple and can be illustrated with reference to Fig. 6.11. The deck weight is supported on rollers sliding on a spherical surface; any horizontal movement would therefore imply a vertical uplift of the weight. If the friction is neglected, the equation of motion of the system is similar to the equation of motion of a pendulum, with equal mass and length equal to the radius of curvature of the spherical surface. For a pendulum with weight W, mass M, and radius of curvature r, it is well known that the period of vibration (T_p) and the associated stiffness (K_p) are as follows (g is the acceleration due to gravity):

$$T_p = 2\pi \left(\frac{r}{g}\right)^{1/2} \tag{6.8}$$

$$K_p = \frac{W}{r} \tag{6.9}$$

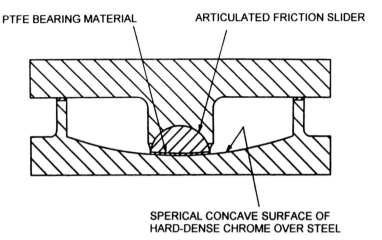

PTFE BEARING MATERIAL   ARTICULATED FRICTION SLIDER

SPERICAL CONCAVE SURFACE OF
HARD-DENSE CHROME OVER STEEL

FIG. 6.10 Section view of a friction pendulum device.

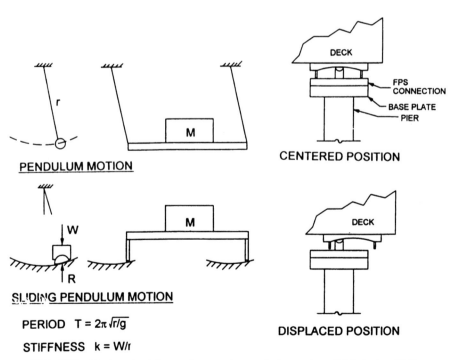

r

M

PENDULUM MOTION

W

R

SLIDING PENDULUM MOTION

PERIOD $T = 2\pi \sqrt{r/g}$

STIFFNESS $k = W/r$

DECK

FPS
CONNECTION
BASE PLATE
PIER

CENTERED POSITION

DECK

DISPLACED POSITION

FIG. 6.11 Basic principles and operating technique of an FPS device [Z4].

473

The expected force–displacement response of an FPS bearing is therefore rigid for a horizontal load lower than the friction force level of the bearing material and proportional to K_p for higher loads. In other words, if the horizontal force exceeds the friction force level, the structure should oscillate with period T_p.

The fundamental design variables are in this case the radius of curvature r and the material friction coefficient. Any of the shapes shown qualitatively in Fig. 6.12 are in principle obtainable, but the value of the friction force is less reliable than the pendulum stiffness. The system is not necessarily self-centering, since the friction force could be in equilibrium with the horizontal component of the weight, but it is very easily recentered after a seismic event.

The range of vertical load capacity, stiffness to lateral force, and period of vibration are of the same order of magnitude as those of lead–rubber bearings

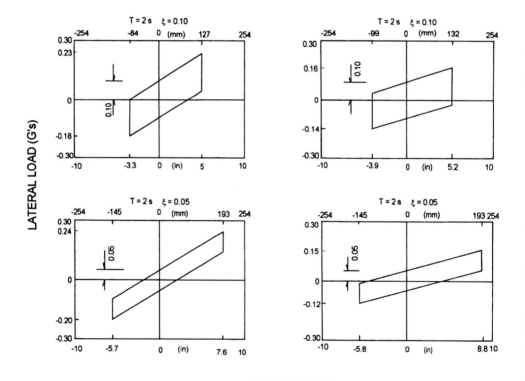

DISPLACEMENT

FIG. 6.12 Qualitative response of FPS bearings as a function of friction coefficient and radius of curvature of the pendulum [Z4].

of similar size. For example, a required period of vibration of 2.5 s results in a required radius of curvature of approximately 1250 mm (49.2 in.), from Eq. (6.8). Typical stiffness values for rubber bearings (i.e., the postyielding stiffness of lead–rubber bearings) were in the range 1000 to 2000 kN/m (70 to 140 kips/ft), which corresponds [Eq. (6.9)] to a supported weight of 1250 to 2500 kN (281 kips) for an FPS with a 1250-mm (49.2-in.) radius of curvature. These values would require rubber bearings with a diameter in the range 400 to 800 mm (16 to 32 in.). The required displacement would govern the height of the rubber bearing and the plan dimension of the FPS device. In the case of the rubber bearing it has been shown that a limit for the maximum displacement is approximately one-half of the plan dimension. In the case of the FPS bearing, the horizontal displacement should be limited to one-fifth of the radius of the spherical surface. The values resulting from the example above are again comparable.

Despite all the favorable features of FPS bearings, the authors are not aware of any current application for bridge structures. This could be due to the more recent development of this technology as well as to some concern related to the long-term stability of the response characteristics in connection to maintenance problems. A model bridge based on a true pendulum concept where the bridge is suspended from the abutments by hangers is shown under test at the University of California–San Diego in Fig. 6.13.

(d) Steel Hysteretic Dampers. The idea of increasing the seismic resistance of a structure by adding a structural element able to dissipate a large amount of energy and therefore to increase the equivalent damping in an elastic-equivalent structure is obviously tempting since it presents only advantages for bearing-supported structures. For this reason, steel dampers of various

FIG. 6.13 Bridge based on a true pendulum concept under test at the University of California–San Diego.

configuration have been designed and produced for many years, using innovative structural shapes and materials. The optimum objectives have been pursued: a stable elastic–plastic behavior (see, e.g., Fig. 6.14) and a long fatigue life under high-intensity plastic cycles (see, e.g., Fig. 6.15). The basic principles to pursue these objectives are to:

- Avoid possible buckling using plastic beams of compact section.
- Limit stress concentration at the connections between dampers and structures.
- Avoid possible weld failure due to fatigue and stress concentration using principles of capacity design.
- Design dampers with nominal equal strain ranges over a large volume of damper material.
- Limit the maximum strain range during earthquakes according to the design phylosophy and limit states adopted. Typically, several design earthquakes and at least one extreme earthquake should be resisted without problems. Typical values of maximum strain for mild steel dampers are in the range of 3% for the design earthquake and 5% for the extreme earthquake (it can be seen from Fig. 6.15 that these values allow a sufficiently large number of cycles).

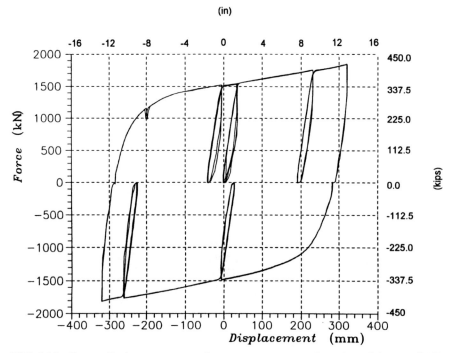

FIG. 6.14   Force–displacement curve for a crescent moon–shaped steel damper [U2].

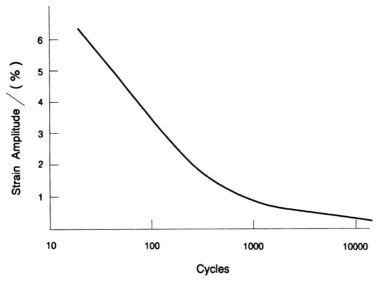

FIG. 6.15 Fatigue life for a typical steel damper [T6].

Three typical kinds of steel dampers are shown in Fig. 6.16: a uniform-moment bending beam with transverse loading arms [Fig. 6.16(a)], a tapered-cantilever bending beam [Fig. 6.16(b)], and a torsional beam with transverse loading arms [Fig. 6.16(c)]. The principle of keeping a constant strain range

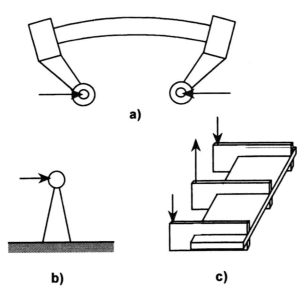

FIG. 6.16 Working principles for bending-beam (a), cantilever (b), and torsional-beam (c) dampers.

for each cross section clearly accounts for the constant section of the bending beam and for the tapered shape of the cantilever beam. Scaling factors for a preliminary design of these kinds of mild steel dampers are given in [S20], but the final design requires extensive technological knowledge. The conceptual design of the device to be used should always be verified by testing.

The choice between different dampers usually depends on location, available space, connection with the structure, and force and displacement level. In the case of superstructure isolation in bridges, it is particularly important that the device permit large displacement and generally allow damping of response in all directions. For example, this is possible using several cantilever beams, as shown in Fig. 6.17. A complex and efficient damper has recently been developed specifically for bridges, to assure a stable multidirectional response for very large-amplitude cycles (Fig. 6.18). The dissipation capacity of a crescent moon–shaped steel element is very efficient and the in-plane deflection allows relatively thin devices even for very large design displacements. Devices with a design displacement of 480 mm (19 in.) have been constructed and tested (U2).

Steel dampers are usually employed in conjunction with sliding bearings (e.g., PTFE, as shown in Fig. 6.17), which take care of transmitting the vertical load and are clearly well suited for application in bridge design and strengthening. The frictional response of PTFE bearings should always be considered in the response simulation, because increased yielding strength of the isolation system could consume the designed capacity protection of the supporting pier.

(e) Hydraulic Dampers. Hydraulic dampers have been used in standard bridge design since the early 1970s, with the objective of permitting slowly

FIG. 6.17 PTFE bearing coupled with a series of tapered cantilever steel dampers under test. (Courtesy of A. Parducci.)

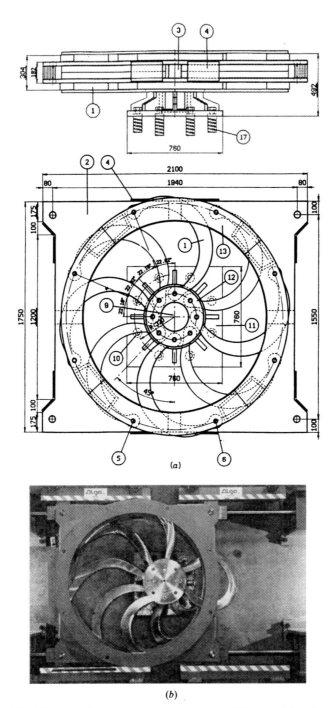

(a)

(b)

FIG. 6.18 Multidirectional crescent moon–shaped steel damper (a) and the deformed shape of a device under test (b). (Courtesy of ALGA.)

479

developing displacements such as those due to thermal and creep effects but limiting the response under dynamic actions, not necessarily due to earthquakes (e.g., braking loads). The jacks were therefore mounted in the longitudinal direction, usually at locations of thermal gaps. Typical devices were characterized by low forces for velocities slower than about 1 mm/s and by a substantially rigid response for higher velocities (Fig. 6.19).

It is obviously possible, and it has been attempted, to design hydraulic devices with damping forces proportional to the velocity of isolator deformation for the velocity range typical of earthquake actions. High-viscosity silicone liquids have been tried, without great success because of a number of difficulties, such as the increase in silicone volume due to temperature and to the tendency of the silicone liquid to cavitate under negative pressure [S20].

(f) Lead-Extrusion Dampers. At a superficial look, lead-extrusion dampers are very similar to hydraulic dampers, but this is a false impression. In fact, the energy dissipation is in this case provided by the physical processes that take place in a metal when it is forced through an orifice. The application of an extrusion process to create a seismic damper was suggested in 1976 [R5]. Following this basic research, lead-extrusion dampers have been used in bridge application in New Zealand and Japan. Details of the working principles and technical features are given elsewhere [S20].

The longitudinal sections of two typical dampers are shown in Fig. 6.20, where the basic concept is evident: The reduction of the cross section of the

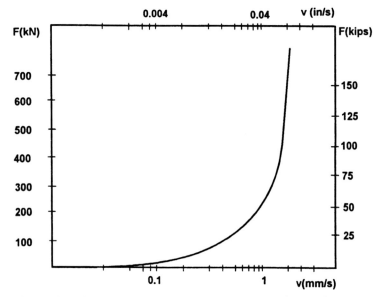

FIG. 6.19 Typical characteristic force versus velocity curve for a hydraulic damper [S20].

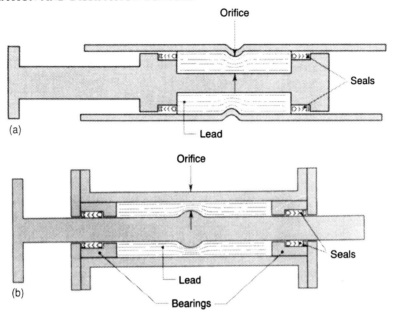

FIG. 6.20   Longitudinal sections of typical lead-extrusion dampers [S20].

lead, forced to pass through the orifice, involves plastic deformation, with significant surface friction and heat production. If the temperature increases, the extrusion force and heat generated decrease. At the end of the dynamic process the lead goes through the physical processes of recovery, recrystallization, and grain growth, returning to the previous situation. Typical load–displacement hysteresis loops are shown in Fig. 6.21.

Lead-extrusion dampers are almost pure Coulomb dampers, with force–displacement loops approximately rectangular and practically rate independent at typical earthquake frequencies. Obviously, the dampers have no centering effect. Since the lead recrystallizes at ambient temperature after each extrusion process, these devices are not affected by problems of hardening or fatigue and have a virtually unlimited operating life. It is possible to design dampers for large displacements [e.g., a damper with a maximum total stroke of 800 mm (31.5 in.) has been produced], but it can be difficult simultaneously to specify large forces because of the possibility of buckling of the internal shaft during compression.

(g) Initiating and Limiting Devices. Depending on the properties of the isolating system it may be necessary to design initiating or limiting devices. The first case applies to systems that would be too flexible under nonseismic loads (e.g., wind): Any of various types of knockoff shear keys will solve the problem, obviously implying some local damage under earthquake forces. Limiting devices are required to avoid excessive displacement in the isolators

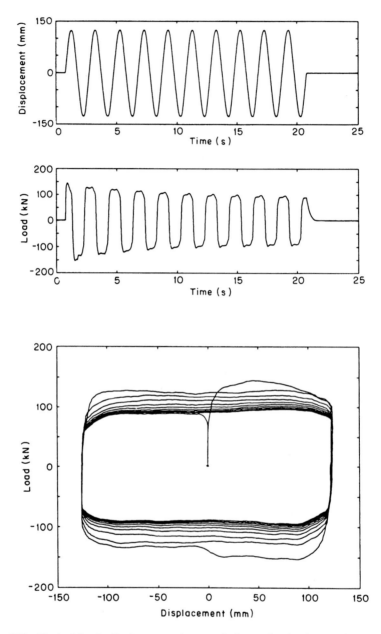

FIG. 6.21 Typical load–displacement hysteresis loops for lead-extrusion dampers [S20].

in case of a low-probability, extreme seismic event. Some kinds of I/D devices (e.g., some steel dampers) show significant strain hardening when the displacement increases beyond a certain level and generally do not need limiting devices. In other cases, such as lead/rubber bearings that might become unstable under excessive deformation, a limit to the displacement could be obtained with rigid stoppers or with deformable buffers, in case there are concerns on the response of the structure under an impact force. Steel tapered beams or stiff rubber buffers could be used to this purpose. In all cases the structure will be subjected to higher-than-expected forces, and there will be some ductility demand in the piers. Since for extreme-level earthquakes only the survival limit state is of interest, this is not undesirable per se, as discussed in more detail in Section 6.3.

6.2.2 Performance Requirements and Testing of Isolating Devices

The choice of an isolation system for a bridge will be based on a number of factors, such as availability and economy, which cannot be discussed in this book, and on other factors, of more technical nature, discussed below. Increasingly, damper design and verification are governed by code guidelines. The subject could be discussed from three different points of view, equally important: what requirements should always be guaranteed, how to compare the performance of different systems, and how the properties of a system should be tested and certified.

(a) Fundamental Features. The basic aspects of seismic isolation were discussed at the beginning of this chapter, showing that a high flexibility can reduce the seismic forces but that only in conjunction with good energy-dissipation properties (i.e., equivalent damping) can the structural displacements be adequately controlled. Apart from these two basic aspects, there are a number of features that have to be provided by a good isolating system.

The bridge should not be too flexible under wind or braking loads. If this is not assured by the isolating system itself, it may be necessary to add some rigid or elastic restraint system to control service load movements. After even moderate earthquakes it is very likely that some element of this system would need replacement.

Good energy dissipation properties are often associated with large residual displacements, which could constitute a problem for the post-earthquake utility of the structure. If a self-centering force is not incorporated into the isolating system, consideration must be given to procedures required after a seismic event to return the bridge superstructure to the correct position.

Stability under vertical loads is a crucial feature for good isolating devices. Two situations should be carefully considered: a possible increment of the vertical load in the undeformed state, and lateral displacements considerably larger than the design displacement. For the first case it is enough to assure that the sliding properties will not be strongly influenced by a variation of the normal force; for the second case it is essential to determine that some buckling failure mode of the bearing will not be triggered. This requirement is obviously more important for systems without a self-centering restoring force.

The possibility of larger-than-expected displacements should be also considered for other possible failure modes, avoiding out-of-stroke or locking problems, which may result in sudden catastrophic failure. In other words, for longer-period events, more damage has clearly to be accepted, but the response should deteriorate smoothly. For example, a good way of ensuring this kind of behavior is offered by progressive hardening at large displacements shown by some steel dampers. In this way, when the required displacement tends to be larger than expected, the response of the piers tends to penetrate the nonlinear range since a force cutoff lower than the yielding force is no longer provided.

(b)   Performance Comparison. It has been suggested [M3] that a comparison of the relative performances of different isolating systems is possible by imposing one of the three following conditions and checking for the two remaining parameters:

- The isolated structures must have the same base shear.
- The isolated structures must have the same displacements.
- The isolated structures must have the same isolated period.

This criterion can be applied very simply if a number of simplifications are accepted, [an elastic response of the structure is assumed, defining an equivalent device characterized by its effective stiffness and by a viscous damping factor equivalent to its hysteretic behavior, as discussed in Section 4.4.3 (Figs. 4.35 and 4.36)], but in this case it only leads to the obvious conclusion that a higher damping is beneficial. Given a specific damping value and imposing one of the conditions above, the other two parameters are uniquely determined; to a higher damping correspond in an obvious way shorter periods and smaller displacements for the same force, shorter periods and smaller forces for the same displacement, and smaller forces and smaller displacement for the same period—all provided that the period of vibration falls in the descending branch of the response spectrum. This is illustrated in Fig. 6.22 where specification of displacement (0.25 m) and damping (4%) uniquely define the acceleration ($6 \, \text{m s}^{-2}$). Unfortunately, a more valid performance comparison can only result from complex nonlinear analyses of the different isolated structures, after having completed alternative design options. Therefore, it will be more appropriate to discuss this topic in Section 6.3 in relation to design principles and methods.

(c)   Testing Specifications. Typically, these should be defined rigorously in the form of codified requirements, which should specify at least the following items:

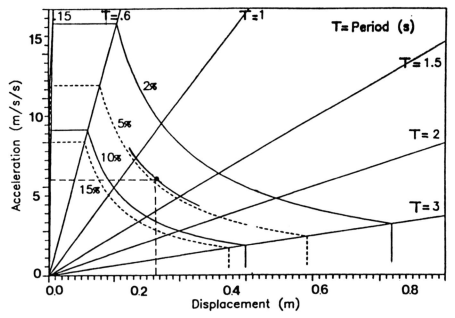

FIG. 6.22   Relations between force, displacement, and period as a function of equivalent damping (ground acceleration = 0.5g).

- How many devices have to be tested
- Whether tests on scaled specimens are allowed and under what conditions
- Which loading history should be applied
- What loading rate should be used and under what conditions quasistatic loads should be accepted
- What vertical load should be applied for devices with bearing functions
- Which durability tests should be performed
- Acceptable levels of strength and stiffness deterioration at the end of the test

In the absence of a code of standards, the list above could be used to check appropriate certification of the characteristics of the I/D devices. At least two tests on full-scale devices at a velocity comparable to the expected seismic velocity should be required. Scaled tests and lower velocity levels could be accepted in conjunction with numerical studies and tests on subelements. The loading history should depend on the seismic input expected, but considering the related uncertainties and the potentially catastrophic consequences of a misfunctioning of the isolation devices, appropriate safety factors should be used. The vertical load should also be factorized in the case of a bearing support, since the load could be increased or decreased by the structural response during a seismic event and could cause instability to the bearing or modify the frictional forces. The isolating devices should not show deterioration under such realistic testing conditions.

Considerable attention is paid to the properties of the isolators in a Japanese document (M17), where it is also prescribed that:

- The devices should be stable when tested at 50 cycles of harmonic loading to the design displacement.
- The residual displacement of a device smoothly released from the design displacement should be less than 10% of the design displacement (note that many common I/D systems would not satisfy this requirement, which is incompatible with high damping)
- The equivalent stiffness and damping should be stable against elongation and shrinkage of the deck due to temperature effects, loading hysteresis, variation of vertical load, loading rate, deformation due to creep and shrinkage, direction of excitation, and variation of temperature [a maximum of 50% variation of stiffness is accepted for a 50°C (90°F) temperature variation].

Specific tests to check the dynamic and static response of the devices are described in the document, requiring a minimum of seven different tests, four for dynamic loading and three for static loading. Each test corresponds to one of the following properties to be confirmed:

- Equivalent stiffness and damping
- Stability against 50 cycles of harmonic loading at the design displacement
- Residual displacement requirements
- Stability against environmental variation such as temperature and rate of loading
- Durability and stability against cyclic loading associated with daily and yearly elongation and shrinkage
- Stiffness for extremely low-rate loading such as yearly temperature change and concrete creep
- Stability of stiffness for large temperature variation

Another document that pays some attention to the properties of I/D devices has been published in Italy [A11]. For the case of dissipating devices a stable response is required after energy dissipation twice that of the energy resulting from an analysis performed using the survival-limit-state earthquake, a dissipated energy per cycle at least equal to 70% of the energy dissipated by a rigid–perfectly plastic system, and a yield strength greater than 60% of the force corresponding to ultimate displacement for the extreme earthquake.

Two tests are always required, the first characterized by eight cycles at the displacement corresponding to one-quarter of the design earthquake, in which case no residual displacement should result; in the second one a number of cycles at the maximum displacement calculated for the extreme earthquake are performed, until the dissipated energy is at least equal to twice the energy dissipated during the numerical analysis. Stability against temperature, rate of loading, and time is required, but specific figures are not given and tests are not described.

6.3 MODELING, ANALYSIS, AND DESIGN

6.3.1 Modeling

Modeling an isolated bridge requires a somewhat different attitude compared with that for a nonisolated structure. In particular, a number of connection elements need to be considered more carefully because of their increased influence on the structural response. These comprise not only the isolating devices but also the bearing support (if any) with the appropriate friction forces, any displacement limiting device to protect possible failure due to I/D displacement limitation, and any connection element between different adjacent decks at movement joints. On the other hand, if the structure is assumed to be protected through the isolating system, there is no need to take into account nonlinear response of the pier elements, where the reinforcement should not yield (and, of course, this has to be checked at the end of the analysis). It can be more important, however, to take into account the pier masses and their own modes of vibration, since isolation of the lower-frequency modes involving the deck mass may increase the importance of higher-frequency modes, due to the response of the piers alone, fixed at the base and restrained at the top. The deck can obviously be simulated with linear elements, but again a more detailed mass distribution and a more careful consideration of higher modes of vibration could be appropriate. Among the several combinations of choices to model the whole soil–foundation–pier–I/D device–deck–abutment system, it is felt appropriate to recommend the following.

(a) Preliminary Design. The effects of soil–structure interaction could generally be neglected. If the soil flexibility is of some concern, a linear spring simulating the combined effect of horizontal and rotational flexibility should be used: in this case 2 to 5% total damping could be considered. The pier can be modeled by means of a linear spring, with stiffness equal to the secant stiffness to the yielding point. A 2% damping will generally be appropriate. The I/D device will be modeled using the substitute structure approach (Section 4.4.3), with a linear spring, with stiffness equal to the secant stiffness to the equivalent displacement. The equivalent displacement would correspond to the expected maximum displacement if all the cycles during the response should go to the same displacement. During the response to a real accelero-

gram only one or two cycles will go to the extreme displacement; several other cycles will typically remain in a range around 50% of the maximum displacement. An equivalent displacement corresponding to the maximum expected displacement will be used to define the equivalent stiffness, while it will be appropriate to compute the damping ratio of the substitute structure as equivalent to the hysteretic energy dissipated in a full cycle to a displacement smaller than the maximum displacement. The coupling effect of the deck could generally be neglected, provided that the design process allows the definition of a rather regular structure (i.e., the global stiffness of each soil–pier–isolator system is similar and the expected displacement at the tops of different piers is similar, which means that the deck will remain substantially undeformed). In this case the design of each bent will be independent of the others. If it is felt that the coupling effect of the deck could have significant effects (e.g., a stiff deck is supported on piers of different flexibility), a linear multidegree-of-freedom (MDOF) system should also be used in the preliminary design phase.

(b)   Models for the Design Earthquake.   After having completed the preliminary design of the bridge, a more refined analysis will usually be performed to check that the simplified model adopted in the preliminary phase has not distorted the response, ending with an inappropriate structure. At the design earthquake value it is expected that the entire structure will still respond almost linearly, the only exception being the isolation system. This expected behavior will be reflected in the modeling choices and will obviously be checked. A MDOF model will generally be used, to take into account the coupling effect of the deck, modeled with linear beams with mass. The deck stiffness should take into account an appropriate level of cracking. The soil–foundation system will be modeled with linear springs, and the piers will be modeled using linear beam elements with mass, considering the shear flexibility in case of very squat piers. The I/D devices will be modeled with linear equivalent highly damped elements, as in the previous case, or, more appropriately, with bilinear springs.

(c)   Models for the Extreme Earthquake.   The response of the structures in case of a larger-than-expected or maximum credible earthquake could be of interest if there are concerns about a possible collapse due to the sensitivity of the bridge response to the level of the seismic action. For example, some significant strain hardening in the isolating system could result in yielding in the pier, where the ductility demand could soon become excessive because of the high stiffness of the pier with respect to the whole system [see Eq. (5.3.2)]. An example of this kind of response is represented in Fig. 6.23, where assuming that the pier yielding displacement (Δ_{yp}) is equal to 5% of the bearing displacement at which the pier yielding starts (Δ_{ub}), it is shown that the additional displacement due to a pier plastic deformation corresponding to a pier ductility equal to 4 is approximately equal to $0.15\Delta_{ub}$ (i.e., 15% displace-

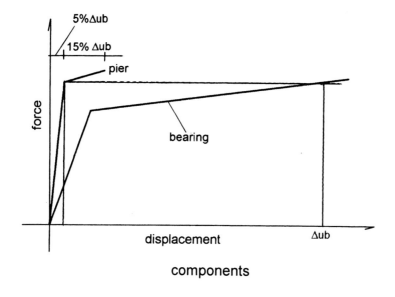

components

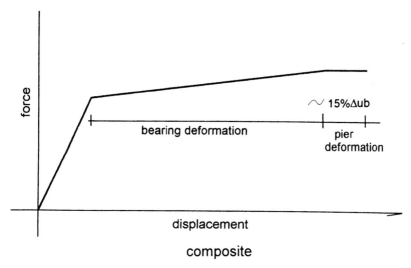

composite

FIG. 6.23 Response of a stiff pier coupled with a flexible isolating system to an extreme earthquake. In this example a pier ductility equal to 4 increases the global displacement capacity of approximately 15%.

ment overrun could induce failure in the pier). Clearly, an appropriate model for this extreme case has to be much more refined, at least in the elements where a highly nonlinear response has to be expected. A nonlinear simulation of the soil response will be required only rarely. The pier should be modeled with nonlinear member models or with fiber elements (Sections 4.4.1 and 4.4.2). The spring element representing the isolating system will generally require a third linear branch to simulate the increasing strain hardening or the simulation of displacement-limiting devices. The deck could still respond almost linearly, but the cracked stiffness should probably be used.

6.3.2 Analysis

As discussed in Section 4.5, several choices are in principle possible to analyze a bridge, resulting from the combination of the options: linear/nonlinear, static/dynamic, and single/multidegree of freedom. The appropriate choice will depend on the design phase, as discussed in the preceding section, on the level of the expected nonlinear response and on the complexity of the bridge to be designed. A summary of the more useful combinations for isolated bridges follows.

(a)   Static, Linear, Single-DOF Analysis. This procedure should in general be applied to very simple bridges and to a preliminary design phase of bridges for which the coupling effects of the deck can be neglected (as discussed in the preceding section), in which case each bent will be considered a single-DOF system. The seismic force is represented by an acceleration/displacement spectrum. The bridge model results from the appropriate combination of stiffness, mass, and damping of the structural elements. The period of vibration and equivalent global damping of the substitute structure will allow acceleration and displacement to be read directly from the spectra. The relevant parameters of the substitute structure are therefore its tributary mass, its effective global stiffness, and its effective global damping, which shall be calculated considering a model of the type shown in Fig. 6.24, as discussed briefly in the following.

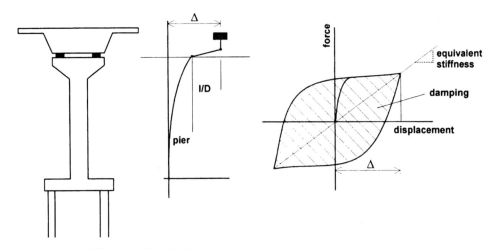

FIG. 6.24   Qualitative response of a bent of an isolated bridge.

(i) Stiffness. The effective stiffness of each pier can be taken as equal to the secant stiffness to yielding (K_{Py}), the effective stiffness of the isolating system located at its top as the secant stiffness to the expected maximum displacement (K_{DE}). The global effective stiffness of the system is therefore

$$K_G = \frac{1}{1/K_{Py} + 1/K_{DE}} \tag{6.10}$$

In some design, as discussed subsequently, the global pier-bearing stiffness will be established to optimize response. The required effective stiffness of the damper can then be computed as a function of global stiffness and pier stiffness by inverting Eq. (6.10), obtaining

$$K_{DE} = \frac{K_{Py} K_G}{K_{Py} - K_G} \tag{6.11}$$

It should be noted that the effective stiffness of the isolator is not taken at the maximum displacement capacity of the isolator but at the expected displacement demand. The ratio between the two values, to be assumed in the design phase, will depend on the type of response of the isolator and on the desired protection, as discussed in Section 6.3.3.

(ii) Damping. The soil–foundation–pier system is assumed to respond elastically, with a standard viscous damping (ξ_P) equal to, say, 5%. As discussed in Section 6.2.1, the isolating system could be characterized by a high viscous damping (ξ_{DV}) and by an essentially linear response, or by an effective damping (ξ_{DE}) equivalent to the dissipated hysteretic energy. In the first case the global damping (ξ_G) of the pier–isolator system can be calculated according to the following relation, which assumes a stiffness-proportional damping (Fig. 6.25):

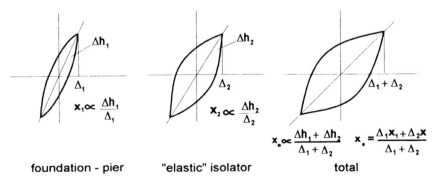

foundation - pier	"elastic" isolator	total

FIG. 6.25 Equivalent global damping of two quasielastic viscous systems in series, applying a stiffness-proportional damping concept.

$$\xi_G = \frac{\Delta_P \xi_P + \Delta_D \xi_{DV}}{\Delta_P + \Delta_D} \qquad (6.12)$$

In the second case, the effective damping of the isolator should be evaluated as a function of its hysteretic response and of the expected ductility demand at the equivalent displacement, as discussed in Section 4.3.1. For example, in the case of a response essentially elastic–perfectly plastic, the following relation applies:

$$\xi_{DE} = \frac{2(1 - 1/\mu_D)}{\pi} \qquad (6.13)$$

while a smaller damping corresponds to thinner cycles [Z2]. It is often convenient to express the hysteretic dissipation capacity of a damper giving the ratio of the area of its typical cycle to the area of a corresponding elastic–perfectly plastic cycle (i.e., an efficiency factor), in which case expression (6.13) can simply be factored by the calculated efficiency ratio.

With a damper whose energy dissipating characteristics depend on ductility level, as above, the global ductility is reduced from the damper ductility as a consequence of additional flexibility of the pier, in accordance with principles discussed in Section 5.3.1(a). Thus, using the nomenclature of Fig. 6.26, the effective global ductility μ_G is related to the damper ductility by the expression

$$\mu_G = 1 + (\mu_D - 1)\frac{\Delta_{DE}}{\Delta_S + \Delta_{DE}} \qquad (6.14)$$

The effective damping provided by the dissipator can then be found from Eq. (6.13) using μ_G instead of μ_D, and reducing by the appropriate efficiency factor, as above. Global damping, incorporating structural damping provided by the pier is then found from Eq. (6.12). In most cases the contribution of

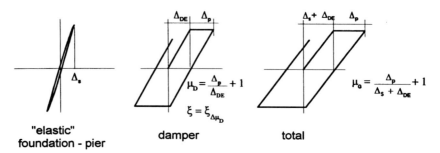

FIG. 6.26 Equivalent global damping of a quasielastic system in series with a hysteretic system, applying a stiffness-proportional damping concept.

the viscous damping of the pier to the global damping is negligible, and can be conservatively ignored in the preliminary design phase.

(b) Dynamic Modal Analysis. By definition a modal analysis can be applied to linear MDOF systems. A substitute structure model will again be used, considering equivalent linear stiffness to the expected displacement and viscous damping equivalent to the energy dissipated by hysteresis.

As discussed in the preceding section, the structure will essentially be modeled using the following elements:

- Linear springs to simulate the soil–foundation system
- Beam elements to simulate the piers, with an effective stiffness that takes cracking into account (i.e., a secant stiffness to the yield value)
- Spring elements to simulate the isolation system, with an effective stiffness that takes yielding into account (i.e., a secant stiffness to the design displacement); it may be necessary to perform iterative analyses to obtain the correct value of the design displacement
- Beam elements to simulate the deck, with a stiffness corresponding to the uncracked value

A simplified modal structure could also be formulated coupling the equivalent global springs simulating each soil–foundation–pier–isolator system (as discussed for the case of single-DOF systems) with beam elements simulating the deck. If a complete model is used, the concept of equivalent damping would imply a different equivalent viscous damping for several structural elements, such as soil, foundation, cracked piers, yielded dampers, and undamaged deck. In general, a common value (2 or 5%) is adopted for all elements, with the exception of the dampers, which will be characterized by a much higher damping equivalent to their hysteresis cycles.

Unfortunately, commercial modal analysis programs, available to most engineering offices, do not allow the selection of different damping ratios for different elements, and therefore an appropriate correction of the response spectrum has been proposed as the only feasible solution to the problem [M3]. This correction is based on the observation that the equivalent damping of the isolation system is effective only for cycles that involve significant yielding of the damper, and this corresponds to longer periods of vibration. It has therefore been proposed that the response spectrum be modified according to Fig. 6.27: The result is a composite spectrum with a step between two spectra with different damping. The step corresponds to a period value close to, but less than, the period of vibration of the isolated structure (considering its equivalent stiffness), so that only modes effectively isolated fall into the reduced acceleration spectrum, while higher modes of vibration that do not involve isolator deformation are damped only by the structural viscous damping.

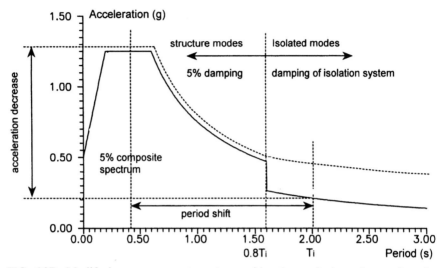

FIG. 6.27 Modified response spectrum to consider the equivalent viscous damping of the isolation system.

The concept is clear and ingenious, but certainly not rigorous, and it is reasonable to expect problems when a clear distinction between isolated and not-isolated modes is not possible or when different periods of vibrations, involving different ductility demand from the dampers, are close to each other. These aspects should be considered before relying on results obtained from this kind of analysis. Extensive parametric studies and comparison with results obtained from more accurate methods are not presently available. It is therefore not possible to give quantitative indications on cases for which the method is not applicable and on the most appropriate length and location of the step between the different spectra.

Example 6.1. Consider the simple bridge shown in Fig. 6.28: It is a simple four-span bridge with three piers of rather irregular height and two abutments. Five isolators have been designed and located at the top of the piers and between deck and abutments. It is assumed that the soil flexibility can be neglected and that the abutments can be taken as infinitely stiff and strong in this phase of analysis. It is also assumed that the tributary weight at each pier resulting from the deck is 10,000 kN (2250 kips), while the tributary weight at the abutments is 5000 kN (1125 kips).

All the piers have identical cross sections but different reinforcement, with yield bending moments (M_{py}), yield shear forces at the top of the piers (V_{py}), and secant stiffnesses to yield (K_{py}) equal to

$$M_{p1y} = 34{,}790 \text{ kNm (25,700 kip-ft)}, \qquad V_{p1y} = 2485 \text{ kN (560 kips)},$$
$$K_{p1y} = 30{,}700 \text{ kN/m (2108 kips/ft)}$$

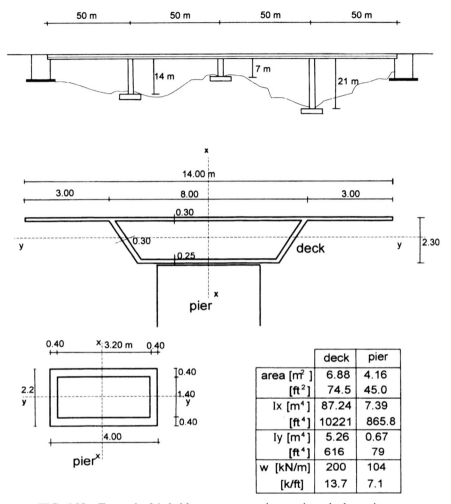

FIG. 6.28 Example 6.1: bridge geometry, pier section, deck section.

M_{p2y} = 17,395 kNm (12,855 kip-ft), V_{p2y} = 2485 kN (560 kips),
K_{p2y} = 124,400 kN/m (8537 kips/ft)

M_{p3y} = 52,185 kNm (38,565 kip-ft), V_{p3y} = 2485 kN (560 kips),
K_{p3y} = 13,650 kN/m (937 kips/ft)

The piers thus have equal strength in terms of lateral force resistance. The corresponding dampers have the bilinear response shown in Fig. 6.29, characterized by a yield strength equal to 2100 kN (473 kips), by the absence of any strain hardening, and by secant stiffnesses to expected equivalent displacements equal to 16,100, 11,500, and 47,000 kN/m (1105, 789, and 3225

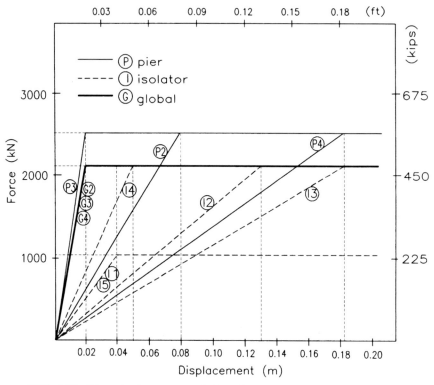

FIG. 6.29 Linearized response of piers–isolator models of Example 6.1.

kips/ft). It is assumed that the hysteretic cycles of the dampers dissipate 70% of the energy dissipated in an elastic–perfectly plastic cycles. The secant stiffness of the dampers at the abutments is equal to 5300 kN/m (364 kips/ft). These values had obviously been selected after a preliminary design phase (see Section 6.3.3), where the superstructure displacement to be expected at the pier top had been estimated.

The global stiffness of each pier–isolator system can be evaluated according to Eq. (6.10), obtaining the following values:

$$K_1 = K_5 = 5300 \text{ kN/m (364 kips/ft)},$$

$$K_{p2} = \frac{1}{1/30,700 + 1/16,100} = 10,570 \text{ kN/m (725 kips/ft)},$$

$$K_{p3} = \frac{1}{1/12,4400 + 1/11,500} = 10,547 \text{ kN/m (724 kips/ft)},$$

$$K_{p4} = \frac{1}{1/13,650 + 1/47,000} = 10,593 \text{ kN/m (727 kips/ft)},$$

Since the tributary masses are 5000 kN (1125 kips) at the abutments and

10,000 kN (2250 kips) at each pier, the ratio between tributary mass and stiffness is equal to $M/K = 0.95$ in all cases. Note that despite the considerable variation in stiffness of the piers, a regularized response has been obtained.

The equivalent damping of the bridge involving the average ductility demand of the isolators was estimated in the design phase as equal to 22%, considering an average ductility demand of the pier-isolator systems equal to 2 and using Eq. (6.13) factorized by 0.7 because of the assumed reduced energy dissipation capacity with respect to an elastic–perfectly plastic system:

$$\xi_{DE} = \frac{0.7 \times 2(1 - 1/2)}{\pi} = 0.22$$

The bridge models with and without isolators are shown in Fig. 6.30, together with frequencies of vibration, participating masses, and modes of vibration. The insertion of the isolating system significantly changed the response modes, mainly shifting the period (the first mode is shifted from 0.75 s to 1.96 s) and increasing the damping. The nonisolated bridge response is dominated by the first mode of vibration because the equal strength assumed for the piers tends to regularize the response. The insertion of the isolating system further regularizes the response, imposing equal stiffness to the piers. The participating mass of the first mode increases further, to 99.97%. Other frequencies of vibration of some importance, not represented in the figure, could be the local modes of each single pier, particularly in the case of tall piers. These local modes are usually well separated from the first (global) mode, and this large separation justifies direction addition of the individual mode response quantities rather than the application of more complex modal combination rules. The addition of higher modes could be important when the local modes of some pier become significant and always has to be recommended as appropriate when dealing with isolated bridges.

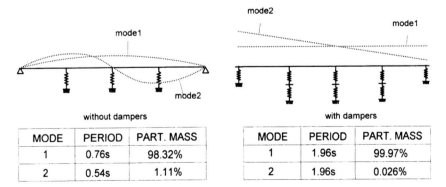

MODE	PERIOD	PART. MASS
1	0.76s	98.32%
2	0.54s	1.11%

MODE	PERIOD	PART. MASS
1	1.96s	99.97%
2	1.96s	0.026%

FIG. 6.30   Model of the bridge of Example 6.1, with and without dampers.

A modal analysis of the two systems, performed using a standard analysis program and a spectrum obtained combining the 5% and 22% spectra of Fig. 6.1, with the step between the two at 1.6 s, gave the following displacements of the deck at the pier locations:
Nonisolated bridge:

$$\Delta_1 = 0; \quad \Delta_2 = 114 \text{ mm, (4.5 in.)}; \quad \Delta_3 = 153 \text{ mm, (6.0 in.)};$$
$$\Delta_4 = 150 \text{ mm; (5.9 in.)}; \quad \Delta_5 = 0$$

Isolated bridge:

$$\Delta_1 = 201 \text{ mm, (7.9 in.)}; \quad \Delta_2 = 204 \text{ mm, (8.0 in.)}; \quad \Delta_3 = 203 \text{ mm, (8.0 in.)};$$
$$\Delta_4 = 202 \text{ mm, (7.9 in.)}; \quad \Delta_5 = 191 \text{ mm, (7.5 in.)}$$

In the case of the isolated bridge, the pier top displacements are: $\Delta_2 = 69$ mm (2.7 in.), $\Delta_3 = 19$ mm (0.75 in.), and $\Delta_4 = 154$ mm (6.1 in.). Obviously, the displacements obtained in the nonisolated case imply an unrealistic linear response and are given for comparison. It is interesting to observe that the increased damping significantly reduces the effects of the period elongation for what concerns the total displacements, resulting in a displacement increase on the order of 25%. The corresponding reduction of shear force on the central pier is on the order of 10 times (again with respect to a nonrealistic linear response for the case of the nonisolated bridge). The expected response of the isolated bridge used in this example is discussed in Example 6.3.

(c)   Time-History Analysis. A time-history analysis is the only feasible method for a nonlinear dynamic analysis, and it should be used in all cases when dealing with important bridges where I/D devices are used. The nonlinear behavior, however, could be limited to the isolation system and be character-ized by a bilinear force–displacement relation that can be simulated by several commercially available computer codes. As discussed in the preceding section, only when the response to extreme earthquakes has to be investigated does the nonlinear response of other elements need to be considered. A three-dimensional time-history analysis will generally be needed in all cases.

A specific problem could be encountered in the case of friction sliding devices for the difficulty of modeling a rigid–plastic behavior. A simple solu-tion is to add the pier or abutment flexibility, to obtain an equivalent elastic–plastic behavior. Since the evaluation of a reliable friction coefficient is in most cases impossible, two simulations should be run using conservative upper and lower limit values.

To perform a time-history analysis a number of generated accelerograms should be used, according to the expected motion, as discussed in Sections 2.4.5 and 4.5.4. Since the structure will vibrate with quite long natural periods, rather nonuniform responses should be expected from different input motions, because the earthquake power is distributed among fewer cycles. The maximum response obtained for the different accelerograms is therefore more significant than the average, as an indicator of design response. The input motion is usually assumed to be synchronous at all the pier bases, but in consideration of the flexibility of the isolation system, the influence of pseudo-static displacement equivalent to the possible effects of nonsynchronism should be checked (see Section 2.5) or nonsynchronous input considered.

Highly nonlinear programs (see, e.g., [I1]), with fiber elements of the kind described in Section 4.4.1, allow the most refined simulation of the response of isolated bridges, considering cracking and yielding when appropriate. A linear modal analysis is accepted in Japan [M18] and in the United States [A4], while in Europe [E9], and specifically in Italy [A11], a nonlinear time-history analysis is required in most cases. More specifically, the Eurocode 8 allows a multimodal equivalent analysis only when the following requirements are met:

- The distance of the bridge from all known active faults is greater than 15 km (approximately 10 miles).
- The soil is not soft.
- The isolated period of the structure does not exceed 3 s and is greater than three times the elastic fixed-base period of the structure.
- The bridge is approximately straight and the total mass of the piers is less than one-fifth of the mass of the deck.
- The effective stiffness of the isolating system at the design displacement (Δ_d) is at least 50% of the secant stiffness at 0.2 times Δ_d.
- The response of the isolating system does not depend on axial force and rate of loading.
- The isolation system produces a restoring force such that its increase between $0.5\Delta_d$ and Δ_d is at least equal to 0.025 times the total gravity load above the isolating system.
- The bridge is fully isolated.

The last condition refers to the possibility of isolating only a few piers or of intentionally designing an isolated bridge where the piers respond nonlinearly to the design earthquake. This appears to be a rather academic situation, since the same force has to pass through isolator and corresponding pier and either one of the two elements has to yield first, then protecting the other one. Document [A11], which applies to the Italian Autostrade network, allows a modal analysis only in the case of a 50- to 100-year-return-period earthquake, in which case the response is prescribed to be essentially elastic; in all other cases a nonlinear dynamic analysis is required, using at least eight accelerograms generated from a common response spectrum, which is a function of the soil type.

Example 6.2. For comparison, the bridge of Example 6.1 has been analyzed with a nonlinear dynamic program [G4] specifically developed for the analysis of isolated bridges, using a number of generated accelerograms. The program allows the simulation of soil with linear springs, of piers with degrading-stiffness nonlinear member models, of isolators with nonlinear springs, and of the deck with linear beam elements. Some of the results obtained with the accelerogram of Fig. 6.31 are shown in Figs. 6.32 and 6.33. As expected, the deck is moving almost perfectly in phase, with a maximum displacement equal to 245 mm, which is approximately 20% larger than the displacement predicted in the preliminary design. This is due essentially to the approximation of the viscous damping equivalent to the hysteretic behavior, which should have taken into account that only a limited number of cycles were reaching the maximum expected displacement. It would be easy to remain on the safe side, factoring Eq. (6.13) appropriately. It is interesting to observe that the maximum displacements obtained at the pier top are still slightly smaller than the yield displacements and that the isolator's response is therefore producing the required protection.

Using other accelerograms, generated from the same spectrum, differences of up to 40% in the maximum drift of the dampers have been obtained, confirming that the results depend largely on the specific accelerograms. Appropriate safety factors should be used to take this fact into account, and when a number of acclerograms are used, maximum rather than average response should be considered for design.

6.3.3 Design Principles

The first topic to be addressed when designing an isolated bridge is the existence of the basic conditions under which it may be appropriate to use I/D

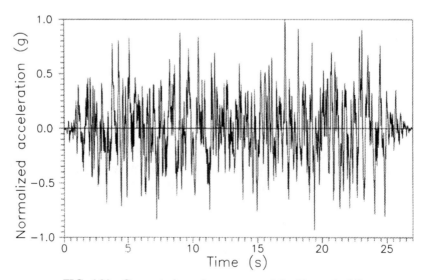

FIG. 6.31 Generated accelerogram used for Example 6.2.

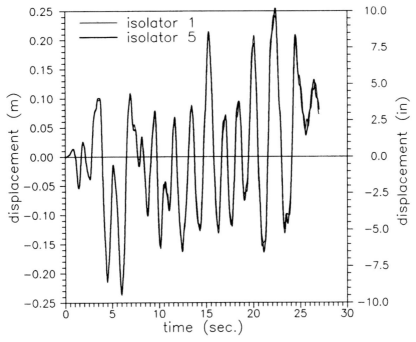

FIG. 6.32 Displacement time histories of the isolators used at the abutments of the bridge of Example 6.2, according to a nonlinear dynamic analysis program [G4].

devices. There are at least three conditions that, alone or together, could support the idea of an isolated bridge:

- The bridge has stiff piers, with a low period of vibration.
- The bridge is highly nonregular, for example with piers of significantly different height, and it therefore has a high potential for concentration of ductility demands.
- The nature of the expected motion is very well characterized for the given site, with high dominant frequencies and low energy at large periods of vibration: in most cases this means shallow earthquake, near fault, and foundation on rock.

The isolation system can also have quite different objectives, such as:

- To shift the main periods of vibration of the structure to values for which the power of the earthquake is low
- To increase the energy-dissipation capacity of the structure, or, in other words, to reduce shear forces and horizontal displacements, increasing the effective equivalent damping

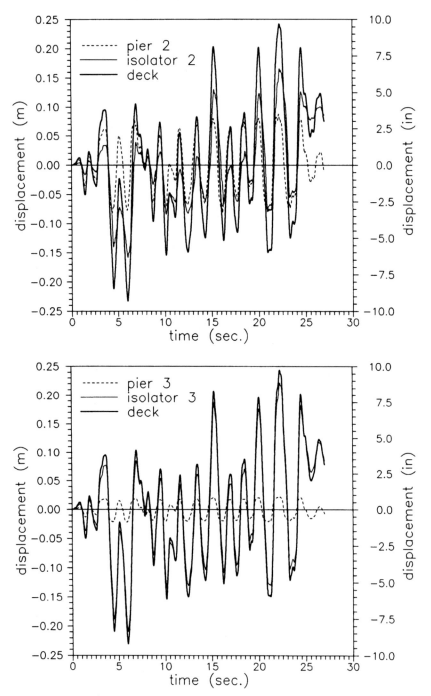

FIG. 6.33 Displacement time histories of piers, isolators, and deck of the bridge of Example 6.2, according to a nonlinear dynamic analysis program [G4].

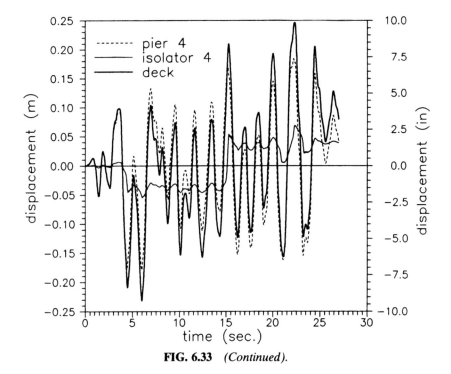

FIG. 6.33 *(Continued)*.

• To regularize the structural response, making the effective stiffness and expected displacements of the piers more similar to each other.

The first objective should be pursued per se only when the second and third conditions are met, while in most cases a shift of the period of vibration will result as a by-product of a correct application of the other two principles when regularization and increased damping are the main objectives. The design of a bridge provided with I/D devices can be simpler than the design of a standard bridge, since the elastic response of all members can be assured, while on the contrary, it is usually more demanding in the analysis phase, since a nonlinear analysis is in most cases to be recommended.

6.3.4 Capacity Design Principles and Protection Factors

When designing an isolated bridge it is appropriate to assume that the greater part of the energy dissipation takes place in the isolating system, and only in extreme cases is some ductility induced into the piers. This kind of response can easily be obtained, applying to the system the same principles used to avoid undesired modes of failure (i.e., protecting against pier yielding by designing the isolating system yielding at a lower force). When the response of the isolating system shows some strain hardening, the force corresponding to the expected displacement rather than the yielding force should be considered in setting the pier design force levels.

The protection factor to be applied obviously depends on the reliability of the mechanical characteristics of the isolating system; in most cases it is required that the actual strength of an I/D device does not differ by more than 10% from its design strength and it is therefore felt appropriate to require that the strength of the device at the expected displacement be equal to 85% of the design nominal strength of the pier. The protection factors applied to calculate the design strength will therefore protect the pier against yielding. Some initial yielding could still take place in the most external reinforcement because of the bilinear idealization of the actual force–displacement curve, which implies that yielding starts at a significantly lower force (see Fig. 5.28).

The capacity design principles discussed in Section 5.3 to avoid undesired failure modes in the pier and in the foundation system still have to be applied, although it will no longer be necessary to ensure, for example, that column shear strength exceeds column flexural strength. Capacity protection factors obviously have also to be applied to the strengths of supports, connections, and abutments.

The ultimate displacement capacity to be required of each device will be calculated applying a safety factor to the expected maximum displacement obtained from a preliminary analysis. The value of the safety factor will depend on the reliablity of the expected displacement (i.e., on the reliability of the input motion and of the method of analysis) and on the consequence of larger-than-expected displacement demand (i.e., on the properties of the pier and of the isolating system). For example, if the device shows increasing strain hardening at large deformation, as in the case of some steel dampers, a larger-than-expected displacement demand will result in some ductility demand in the piers. This should not be considered an undesirable event in case of an extremely large earthquake, and a significant safety margin in the ultimate displacement may not be necessary. On the other hand, for other kinds of devices a larger displacement demand could result in a local failure that will completely change the response of the bridge to the remaining part of the seismic excitation. This is obviously not acceptable and will require larger safety margins.

It is therefore important to underline that significant strain hardening, possibly increasing with deformation, is a desirable property of I/D devices that will favor exploitation of the structural resources when needed. Unfortunately, the contribution of the eventual pier ductility to the global ductility of the system could be significant only if the pier was rather flexible with respect to the isolating system and will be negligible for very stiff piers, as discussed in Section 6.3.1 with reference to Fig. 6.23.

6.3.5 Design Methods

As for all structures, the conceptual design of an isolated bridge is more art than science and cannot be obtained by the blind application of equations or rules. Two extreme cases can be considered to make the point: A good designer could produce the bridge design simply following his instinct and his experience, then proving with a series of time-history analyses that the response is acceptable; on the other hand, a large number of time-history analyses could be performed on bridges where the properties of the isolators are randomly selected until a bridge with an acceptable response is obtained. The actual design process is always intermediate. A designer will use instinct, experience, and simplified analysis methods to produce what he believes should respond acceptably, then will run a more detailed analysis, and where the response is found to be unsatisfactory, will use the results of the analysis to correct the preliminary design before rerunning the detailed analysis. The iteration process obviously converges faster if the designer is experienced. The design methods that will be discussed in the following are therefore an attempt to rationalize possible lines of thinking to design an isolated bridge. Several oversimplified assumptions will be made, and the resulting bridge will not necessarily respond as expected. It is therefore fundamental to temper the results with general experience and specific observation of the case to be studied.

When designing an isolated bridge, the bridge geometry and pier and deck sections have usually already been defined, resulting from functional and aesthetic requirements and non-seismic-load conditions. The pier reinforcement and the characteristics of the isolating system are therefore the design unknown. When dealing with the problem of retrofitting an existing bridge, the pier reinforcement is also known and the isolating system becomes the only unknown.

Two design approaches are presented here, reflecting the two situations described: reinforcement to be designed, and given reinforcement. The elastic response spectrum recommended in EC8 [E5] for medium-stiff soil is used in all the numerical examples. The acceleration and displacement spectra are shown in Figs. 6.1 to 6.3, where the two most relevant characteristics are evident. The structure displacement is assumed not to increase for periods longer than 3 s, and the spectral ordinate can be corrected by multiplying the reference spectrum ($\xi = 5\%$) by a factor η, a function of the global structural damping according to the following relation:

$$\eta = \left(\frac{7}{2 + \xi}\right)^{1/2} \tag{6.15}$$

(a) Design of I/D System and Pier Reinforcement. A comprehensive design approach for an isolated bridge will start from the design of the I/D devices, to allow full exploitation of their technical capabilities, and the pier reinforcement will then be selected to obtain a flexural strength higher than the strength of the damper at the expected maximum displacement, according to the capacity design philosophy discussed in Section 6.3.4. As discussed in Section 5.3.1,

two different approaches will be possible, either starting from acceleration (force) or from displacement.

In the case of a *displacement-based design approach* the procedure may go through the following steps:

1. The accepted maximum displacement (Δ_e) to be expected during the design earthquake will be decided. This maximum displacement will in principle apply to all piers and abutments, because a theoretically perfectly regular structure will be designed.

2. The yielding displacement of each pier (Δ_{Py}) will be calculated from the following relation (i.e., integrating the curvature along the column height and assuming a cantilever bending moment diagram):

$$\Delta_{Py,i} = \phi_{Py,i} \frac{H_i^2}{3} \tag{6.16}$$

where $\phi_{Py,i}$ can be obtained from Eq. (7.15) and H_i is the height of the *i*th pier. Soil–foundation deformation can easily be included if appropriate. The modifications required in the case of a nonzero moment at the pier top are also obvious.

3. The effective displacement of each isolator will be obtained from the following equation, which assumes that the force in each isolator at the expected displacement will be 85% of the yield force of each pier:

$$\Delta_{DE,i} = \Delta_e - 0.85\Delta_{Py,i} \tag{6.17}$$

4. The ductility demand desired for the isolators at the effective displacement (μ_{DE}) will be decided, applying an appropriate factor to the isolator ductility capacity to avoid collapse in case of an extreme seismic event. The yield displacement of each device ($\Delta_{Dy,i}$) will be calculated dividing the effective displacement by the ductility:

$$\Delta_{Dy,i} = \frac{\Delta_{DE,i}}{\mu_{DE}} \tag{6.18}$$

5. The effective ductility demand of each (foundation)–pier–isolation system will be calculated as

$$\mu_{E,i} = \frac{\Delta_e}{\Delta_{Dy,i} + 0.85\Delta_{Py,i}} \tag{6.19}$$

and the corresponding effective damping (ξ_i) will be calculated from the appropriate form of Eq. (6.13), considering the reduced dissipation with respect to an elastic–perfectly plastic response and the probability of several reduced amplitude cycles during the seismic response.

6. The global effective damping of the structure can be computed approximately from the following equation, which expresses a weighted average of the different damping ratios (M_i is the tributary mass of pier i and M_d is the total mass of the deck). This equation takes empirically into account that the equivalent yield displacement is shifted from the first yield displacement according to the ratio of the last yield force to the first yield force, but the area of the hysteresis loop is also reduced progressively. A better expression could easily be derived using the appropriate force–displacement curves for each foundation–pier–isolator system coupled in parallel.

$$\xi_b = \frac{\Sigma(\xi_i M_i)}{M_d} \tag{6.20}$$

Entering the spectrum of Fig. 6.3 with displacement and damping, the period of vibration (T_b), and the effective acceleration (S_a) of the first period of vibration of the bridge will be computed.

7. The equivalent stiffness of the bridge (K_b) will then result from the following equation, where M_d is the total mass of the deck:

$$K_b = \frac{4\pi^2 M_d}{T_b^2} \tag{6.21}$$

and the stiffness of each (foundation)–pier–isolator system can be calculated assuming proportionality between stiffness and tributary mass:

$$K_i = \frac{M_i K_b}{M_d} \tag{6.22}$$

8. In accordance with the previous discussion of capacity design principles, it will also be assumed that the damper effective force is equal to 85% of the yielding force of the pier. The proportionality between displacement and stiffness, used in conjunction with Eq. (6.10), will then imply the following relations:

$$K_{DE,i} = K_i\left(1 + 0.85\,\frac{\Delta_{Py,i}}{\Delta_{DE,i}}\right) \tag{6.23}$$

$$K_{P,i} = K_{DE,i}\,\frac{\Delta_{DE,i}}{0.85\Delta_{Py,i}} \tag{6.24}$$

9. The design forces for each pier can be obtained multiplying stiffness by

displacement, and the pier reinforcement will finally be designed in the usual way.

As already mentioned, the procedure above does not assure that a consistent and effective design will result. For example, the stiffness and strength obtained for the piers could be incompatible with a reasonable reinforcement ratio, or the effective coupling of the deck could cause a response significantly different from the response expected. A more refined analysis should therefore always be performed to check the response of the bridge, and some iteration may be necessary. The design procedure outlined is intended to serve essentially as a guidance to the way of reasoning and it is aimed at obtaining a regular structure, where all the (foundation)–pier–isolator systems should vibrate in phase, with similar periods of vibration and amplitudes and stiffnesses proportional to the tributary masses.

Example 6.3. Consider again the bridge of Example 6.1. Geometry, masses, and pier sections are given; the isolating system and the pier reinforcement have to be designed. It is assumed that the design earthquake has a frequency content corresponding to the spectra of Figs. 6.1 to 6.3 and that the peak ground acceleration is equal $0.5g$. The soil–foundation interaction is neglected and the abutments are assumed to be rigid.

1. The equivalent displacement (Δ_e) to be used during the design process is assumed to be 200 mm (7.9 in.).
2. The yield curvature of all piers results from Eq. (7.15), assuming that $\varepsilon_y = 2.35 \times 10^{-3}$:

$$\phi_{Py,i} = 1.23 \times 10^{-3} \text{ m}^{-1} \ (0.37 \times 10^{-3} \text{ ft}^{-1})$$

   and the yield displacement of each pier from Eq. (6.16):

$$\Delta_{Py,2} = 81 \text{ mm } (3.19 \text{ in.}), \ \Delta_{Py,3} = 20 \text{ mm } (0.79 \text{ in.}),$$
$$\Delta_{Py,4} = 182 \text{ mm } (7.17 \text{ in.})$$

3. The effective displacement required from each isolator is obtained from Eq. (6.17):

$$\Delta_{DE,1} = 200 \text{ mm}, \ \Delta_{DE,2} = 131 \text{ mm}, \ \Delta_{DE,3} = 183 \text{ mm},$$
$$\Delta_{DE,4} = 45 \text{ mm}, \ \Delta_{DE,5} = 200 \text{ mm}$$
$$(7.88, 5.16, 7.20, 1.77, \text{ and } 7.88 \text{ in.})$$

4. The ductility demand of the isolators at the effective displacement is assumed to be equal to 4. The yield displacements of the devices are obtained from Eq. (6.18) as

$$\Delta_{Dy,1} = 50 \text{ mm}, \ \Delta_{Dy,2} = 33 \text{ mm}, \ \Delta_{Dy,3} = 46 \text{ mm},$$
$$\Delta_{Dy,4} = 11 \text{ mm}, \ \Delta_{Dy,5} = 50 \text{ mm}$$
$$(1.97, \ 1.30, \ 1.81, \ 0.43, \ \text{and} \ 1.97 \text{ in.})$$

5. The effective ductility demand of each pier–isolator system can be calculated from Eq. (6.19):

$$\mu_{E,1} = 4, \quad \mu_{E,2} = 1.75, \quad \mu_{E,3} = 3.03, \quad \mu_{E,4} = 1.04, \quad \mu_{E,5} = 4$$

and the corresponding effective damping from Eq. (6.13). Assuming hysteresis loops with area equal to 70% of that of an elastic–perfectly plastic system, the equivalent damping ratios are

$$\xi_{E,1} = 0.33, \quad \xi_{E,2} = 0.19, \quad \xi_{E,3} = 0.30, \quad \xi_{E,4} = 0.01, \quad \xi_{E,5} = 0.33$$

6. From Eq. (6.20) the global equivalent damping is obtained as $\xi_b = 0.21$, and entering the spectrum of Fig. 6.3 with displacement and damping, the following period of vibration and effective acceleration for the first mode of vibration of the bridge are obtained: $T_b = 1.95$ s and $S_{ab} = 0.21g$.

7. The equivalent stiffness of the bridge (K_b) will then result from Eq. (6.21):

$$K_b = 42{,}290 \text{ kN/m} \ (2918 \text{ kips/ft})$$

and the stiffnesses of each (foundation)–pier–isolator system are obtained from Eq. (6.22), assuming a tributary weight of 10,000 kN (2250 kips) for each pier and of 5000 kN (1125 kips) for the abutments:

$$K_1 = 5286 \text{ kN/m}, \ (365 \text{ kips/ft}), \quad K_2 = 10{,}573 \text{ kN/m}, \ (730 \text{ kips/ft}),$$
$$K_3 = 10{,}573 \text{ kN/m}, \quad K_4 = 10{,}573 \text{ kN/m}, \quad K_5 = 5286 \text{ kN/m} \ (\text{kips/ft})$$

8. Imposing the design requirement that the damper effective force must be equal to 85% of the yielding force of the pier, the effective stiffness of dampers and piers can be obtained from Eqs. (6.23) and (6.24) as

$$K_{DE,1} = 5286 \text{ kN/m} \ (365 \text{ kips/ft}), K_{DE,2} = 16{,}130 \text{ kN/m} \ (1113 \text{ kips/ft}),$$
$$K_{DE,3} = 11{,}555 \text{ kN/m} \ (797 \text{ kips/ft}), K_{DE,4} = 46{,}921 \text{ kN/m} \ (3238 \text{ kips/ft}),$$
$$K_{DE,5} = 5286 \text{ kN/m} \ (365 \text{ kips/ft})$$

$$K_{P,2} = 30{,}690 \text{ kN/m} \ (2118 \text{ kips/ft}),$$
$$K_{P,3} = 124{,}386 \text{ kN/m} \ (8583 \text{ kips/ft}),$$
$$K_{P,4} = 13{,}648 \text{ kN/m} \ (942 \text{ kips/ft})$$

9. The minimum yield shear strength of each pier can be computed multi-

plying stiffness by displacement and the minimum yield bending strength multiplying shear strength by height, obtaining the following values:

$$V_{Py,2} = 2486 \text{ kN (559 kips)}, \quad V_{Py,3} = 2488 \text{ kN (560 kips)},$$
$$V_{Py,4} = 2484 \text{ kN (559 kips)}$$

$$M_{Py,2} = 34{,}804 \text{ kNm (25,720 kip-ft)},$$
$$M_{Py,3} = 17{,}416 \text{ kNm (12,870 kip-ft)},$$
$$M_{Py,4} = 52{,}163 \text{ kNm (38,548 kip-ft)}$$

It can be noted that in a nonisolated bridge the shorter piers tend to attract the major part of the seismic force, while according to the design procedure proposed above, in the case of an isolated bridge the forces are forced to be proportional to the tributary mass. As a consequence, larger flexural strengths have to be provided to taller piers if the tributary masses are approximately equal. This could, of course, be adjusted in the design process. The properties of piers and isolators obtained for the bridge of Example 6.3 are shown in Fig. 6.28. Note that the same bridge has been analyzed in Examples 6.1 and 6.2.

In the case of a more traditional *force approach*, the desired global period of vibration (usually in the range 1 to 3 s) and the acceptable damper ductility (μ_D) will be decided first; the design procedure could then be described in the following steps.

1. The equivalent damping (ξ_{DE}) of the isolating system will be calculated from the appropriate form of Eq. (6.13), as already discussed.
2. The modest contribution of the pier response to the global damping will be neglected in this preliminary design phase, and the appropriate acceleration will then be read from the response spectrum.
3. The design force will be calculated entering the acceleration spectrum with period and damping and multiplying acceleration by tributary mass. This force will be assumed to be the effective strength of the isolator at the expected maximum displacement; to obtain the design strength of the pier it will be factored by 1/0.85.
4. The pier reinforcement will be designed, enabling yield stiffness and yield displacement to be calculated.
5. The required isolator displacements (at yielding and at the expected displacement) will be calculated, therefore designing the devices.
6. The procedure will be repeated to convergence if the resulting system should be technically unacceptable (i.e., the dampers required are impossible or too expensive to manufacture) or when the effective resulting damping would be significantly different from the assumed damping.
7. Iterative corrections could also be required when the required pier reinforcements are not acceptable (too low or too high with respect to the limits discussed in Section 5.3.1(d)(ix)). It may then be necessary to

adopt different strength or stiffness for different bents, in which case a tributary mass concept will not be applicable. In this case it will be preferable to perform a series of multidegree-of-freedom modal analyses with each pier–isolator system modeled with the appropriate equivalent stiffness.

Example 6.4. Consider the bridge of Example 6.3, assuming a design global period of vibration of 2 s [i.e., global stiffness equal to 40,200 kN/m (26,894 kips/ft)] and 20% equivalent viscous damping. The corresponding spectral ordinate on the response spectrum of Fig. 6.1 is $S_a = 0.21$, from which the total horizontal force is found to be

$$V_H = 8400 \text{ kN (1890 kips)}$$

Assuming a distribution of this total force proportional to the tributary masses of piers and abutments, the following forces will be adopted as design forces for the isolators:

$$V_{D,1} = V_{D,5} = 1050 \text{ kN (236 kips)}, \quad V_{D,2} = V_{D,3} = V_{D,4} = 2100 \text{ kN (472 kips)}$$

The design yield strength of the piers will be obtained multiplying these forces by 1/0.85, obtaining

$$V_{P,2} = V_{P,3} = V_{P,4} = 2470 \text{ kN (556 kips)}$$

The effective displacement of each pier–isolator system should then be equal to

$$\Delta_E = \frac{8400}{40,200} = 0.208 \text{ m (8.19 in.)}$$

The pier and isolator displacements can then be calculated similarly to what was done in steps 2 to 4 of Example 6.3. Clearly, the end result is the same as with the displacement-based design.

(b) Given Reinforcement Design. When the reinforcement of the piers of a bridge is already given, for example when it is required to design the retrofit of a bridge or when it is desired to keep the reinforcement obtained from nonseismic constraints, only the I/D system has to be designed, but it will be difficult to obtain a regular response in the sense used in the previous examples. Since the strength of the piers will be known, the strength of the isolators at design response could be set at 85% the strength of the corresponding pier. In this case it will be impossible to have both similar stiffness and similar expected displacement for different piers if the strengths are different. A time-history analysis to check the design is in this case fundamental, since the real response could be significantly different from that obtained with a simplified model, because the coupling effect of the deck could have important effects. Another possible choice is to design the isolator using as a target the strength of the weakest pier. In this case the design process becomes easier and the bridge response regular, but the global low strength could imply the necessity of very large damping and low stiffness. In general, a compromise should be pursued, iterating the design process and verifying the effectiveness of the solution using a time-history analysis.

For the reasons given above, the design steps suggested below attempt to suggest a way of reasoning rather than giving impossible design recipies.

1. The first step will be to assume the yield strength to be assigned to the isolators, based on proportion of the pier strength, or applying more stringent capacity design principles. At the abutments there is usually more freedom, because they have been designed more conservatively. It could be convenient to have rather high yield strength in the devices used at the abutments, to reduce the importance of the rotational mode in the horizontal plane. For the same reason it is appropriate to try to have a rotationally equilibrated system of forces when all the isolators have yielded. The easiest choice could be to set all yield forces at the same level, considering the weakest pier.

2. Entering the design spectra with the total force, a reasonable couple of damping and period of vibration will be selected, obtaining a value for the effective displacement of the equivalent system, which should be kept in a range of 1 to 5% of the height of all piers, if possible.

3. The pier yield displacement and the pier displacement corresponding to yield in the isolators will be computed.

4. The effective displacement required from each isolator will be obtained as the difference between global displacement and pier displacement.

5. The accepted maximum ductility for the isolators will be decided, and the global ductility of each foundation–pier–isolator system will be computed. The equivalent damping of the bridge will then be computed from Eq. (6.20).

6. If the value of the global damping is not compatible with that assumed in step 2, some iteration is needed, either accepting larger ductility demand in the isolators or accepting a larger displacement (i.e., a larger period of vibration) at step 2.

If the foundation–pier–isolator systems all have similar effective force, displacement, and stiffness, it is very likely that the time-history analysis will confirm the expected response, which assumed that the total force was shared according to a tributary mass principle. If this should not be the case, it is likely that the coupling effect of the deck will be significant and the time-history analysis will become part of the iteration process. The higher frequency response modes of the piers themselves will also be checked at the end of the design process, since it is very likely that they will be uncoupled from the global response modes of the bridge.

Example 6.5. Consider again the bridge of Example 6.1. The reinforcement of the piers is now known. For this example it is assumed that they have a similar flexural yield strength, equal to 37,000 kNm (27,000 kips-ft) and therefore their yield shear forces have been computed as follows:

$$V_{py,2} = 2650 \text{ kN (596 kips)}$$

$$V_{py,3} = 5200 \text{ kN (1170 kips)}$$

$$V_{py,4} = 1750 \text{ kN (394 kips)}$$

The strength of the corresponding dampers could be assumed as 85% or less of these values. As a first trial, the yield strength of the three isolators will be computed assuming a tributary mass concept, as follows:

$$V_{Dy,2,3,4} = 1750 \times 0.85 = 1490 \text{ kN (335 kips)}$$

$$V_{Dy,1,5} = \frac{1490}{2} = 745 \text{ kN (168 kips)}$$

The total reaction force will therefore be $V_b = 5960$ kN (1341 kips). Assuming a peak ground acceleration equal to 0.5g and a total deck weight of 40,000 kN (9000 kips), the spectral acceleration should be

$$S_a = \frac{5960}{40,000} = 0.15$$

and entering the spectra of Figs. 6.1 to 6.3, scaled to the appropriate PGA, a possible combination of period and damping could be

$$T_b = 3 \text{ s}, \; \xi_b = 18\%$$

The global effective displacement corresponding to this couple is

$$\Delta_b = 335 \text{ mm (13.2 in.)}$$

The pier stiffnesses have been computed as follows, using the yield curvature computed in example 6.3 and assuming, in this case, a double bending condition for the piers as a result of superstructure torsional stiffness:

$$K_{p,2} = 62{,}000 \text{ kN/m (4278 kips/ft)}$$

$$K_{p,3} = 485{,}000 \text{ kN/m (33,465 kips/ft)}$$

$$K_{p,4} = 19{,}000 \text{ kN/m (1311 kips/ft)}$$

The pier displacement corresponding to the isolator yield forces will be computed dividing forces by stiffness, obtaining

$$\Delta_{P,2} = 24 \text{ mm (0.94 in.),} \quad \Delta_{P,3} = 3 \text{ mm (0.12 in.),} \quad \Delta_{P,4} = 78 \text{ mm (3.07 in.)}$$

The effective displacement required from each isolator is, therefore,

$$\Delta_{DE,1} = 335 \text{ mm (13.2 in.),} \quad \Delta_{DE,2} = 311 \text{ mm (12.2 in.),}$$
$$\Delta_{DE,3} = 332 \text{ mm (13.1 in.),} \quad \Delta_{DE,4} = 257 \text{ mm (10.1 in.),}$$
$$\Delta_{DE,5} = 335 \text{ mm (13.2 in.)}$$

The available ductility of the isolators at the effective displacement is assumed to be equal to 4. The yield displacements of the devices are, therefore,

$$\Delta_{Dy,1} = 84 \text{ mm (3.31 in.),} \quad \Delta_{Dy,2} = 78 \text{ mm (3.07 in.),}$$
$$\Delta_{Dy,3} = 83 \text{ mm (3.27 in.),} \quad \Delta_{Dy,4} = 64 \text{ mm (2.52 in.),}$$
$$\Delta_{Dy,5} = 84 \text{ mm (3.31 in.)}$$

The effective ductility of each pier–isolator system can be calculated from Eq. (6.19) as

$$\mu_{E,1} = 4, \ \mu_{E,2} = 3.28, \ \mu_{E,3} = 3.90, \ \mu_{E,4} = 2.36, \ \mu_{E,5} = 4$$

and the corresponding effective damping results from Eq. (6.13), assuming hysteresis loops with area equal to 70% of that of an elastic perfectly plastic system:

$$\xi_{E,1} = 0.33, \ \xi_{E,2} = 0.31, \ \xi_{E,3} = 0.33, \ \xi_{E,4} = 0.26, \ \xi_{E,5} = 0.33$$

From Eq. (6.20) the global equivalent damping is obtained as

$$\xi_b = 0.30$$

Since this value is significantly larger than the 0.18 that was needed, the response of the bridge should be less than assumed, and thus acceptable. A second couple of period and damping could be selected to optimize the design, choosing a smaller period and a larger damping. For example,

$$T_b = 2.4 \text{ s}, \qquad \xi_b = 28\%$$

The global effective displacement corresponding to this couple is equal to

$$\Delta_b = 217 \text{ mm } (8.54 \text{ in.})$$

and following the same procedure of the first iteration yields

$\Delta_{DE,1} = 217$ mm (8.54 in.), $\Delta_{DE,2} = 193$ mm (7.60 in.), $\Delta_{DE,3} = 214$ mm (8.43 in.),

$\Delta_{DE,4} = 139$ mm (5.47 in.), $\Delta_{DE,5} = 217$ mm (8.54 in.)

$\Delta_{Dy,1} = 54$ mm (2.13 in.), $\Delta_{Dy,2} = 48$ mm (1.89 in.), $\Delta_{Dy,3} = 53$ mm (2.09 in.),

$\Delta_{Dy,4} = 35$ mm (1.38 in.), $\Delta_{Dy,5} = 54$ mm (2.13 in.)

$\mu_{E,1} = 4, \quad \mu_{E,2} = 3.01, \quad \mu_{E,3} = 3.87, \quad \mu_{E,4} = 1.92, \quad \mu_{E,5} = 4$

$\xi_{E,1} = 0.33, \quad \xi_{E,2} = 0.30, \quad \xi_{E,3} = 0.33, \quad \xi_{E,4} = 0.21, \quad \xi_{E,5} = 0.33$

$\xi_b = 0.29$

This result shows that the energy-dissipation capacity of the bridge is dominated by the isolating system, and there is little variation of the global equivalent damping provided that a constant ductility of the isolators is kept.

A more effective design strategy could be studied using the coupling capacity of the deck and differentiating between the strength of the isolators. A multidegree-of-freedom substitute structure should be used in the preliminary design phase, and more detailed nonlinear dynamic analyses would be appropriate.

A model for the bridge of this example, with and without dampers, is shown in Fig. 6.34. Note that the elastic response of the original bridge is dominated by the second mode of vibration, because of the large stiffness (and strength) of the central pier, while the first mode dominates the response of the isolated

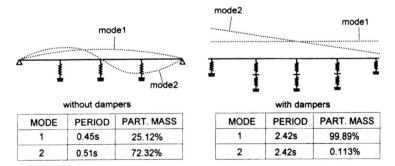

MODE	PERIOD	PART. MASS
1	0.45s	25.12%
2	0.51s	72.32%

MODE	PERIOD	PART. MASS
1	2.42s	99.89%
2	2.42s	0.113%

FIG. 6.34 Model of the bridge of Example 6.5, with and without dampers.

bridge. The force–displacement curves of piers and isolators are shown in Fig. 6.35.

The response obtained from the bridge, using the same analysis program as that used for Example 6.2, is shown in Figs. 6.36 and 6.37. The deck is vibrating without significant deformation, with the points corresponding to the piers moving perfectly in phase. The maximum displacements are approximately 235 mm, with a difference of less than 8% with respect to the 217 mm predicted. Similarly to Example 6.2, the relative displacements of isolators and piers are very close to the values predicted.

6.4 FOUNDATION ROCKING

6.4.1 Introduction

It has been observed after several earthquakes that a number of structures had responded to seismic excitation by rocking on their foundation, and in some cases this enabled them to avoid failure. Typically, this had been observed for elevated water or storage tanks, characterized by large masses at some distance from the ground and comparatively narrow bases. For example,

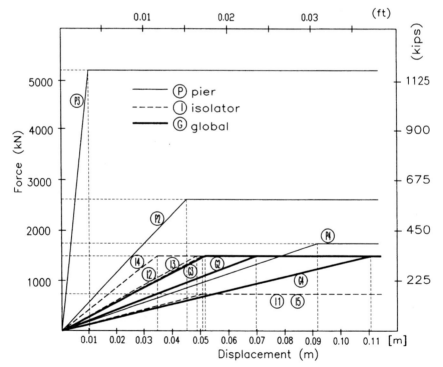

FIG. 6.35   Force–displacement curves for the piers and dampers of Example 6.5.

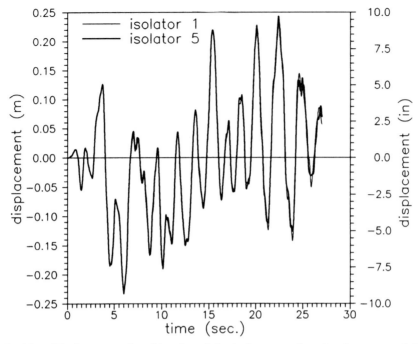

FIG. 6.36 Displacement time histories of the isolators used at the abutments of the bridge of Example 6.5, according to a nonlinear dynamic analysis program [G4].

after the Arvin Tehachapi earthquake (California, July 1952) it was found that a number of tall petroleum-cracking towers had escaped more serious damage by stretching their anchor bolts and rocking on their foundation pads [H6]. It can be observed that the response of slender structures is usually governed by the high overturning moment at the base, and if rocking and uplift is possible, this moment is then limited by the moment needed to lift the weight of the structure against the stabilizing moment due to gravity. It is therefore obvious that all the internal forces and deformations throughout the structure will be limited correspondingly.

The rocking phemomenon could be of interest for bridge piers, which often present geometry, mass distribution, and foundation characteristics that could favor a controlled rocking response, as discussed in Section 6.4.2. The possibility of allowing some rocking in the transverse direction should obviously be considered together with the capacity of the superstructure to accommodate the corresponding deformation. Appropriate details for rocking foundations are discussed in Section 5.6.2. It should also be noted that the mechanism of rocking will often be considered as a satisfactory response in assessment of existing bridges (see Section 7.4.10). It is not essential that rocking has to take place at the foundation level; on the contrary, a bridge has been designed and constructed in New Zealand with the pier feet free to uplift under severe

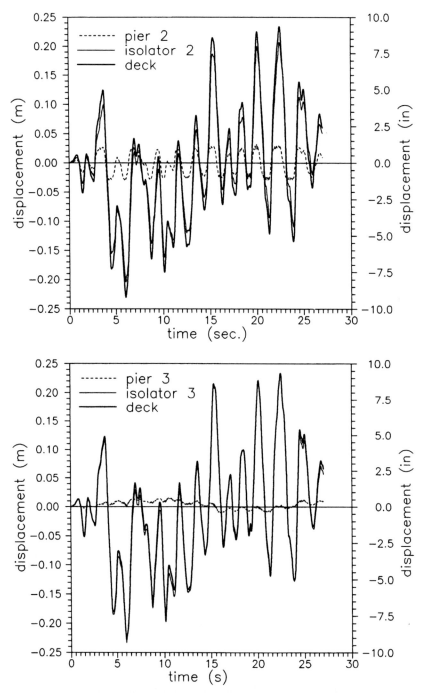

FIG. 6.37 Displacement time histories of piers, isolators, and deck of the bridge of Example 6.5, according to a nonlinear dynamic analysis program [G4].

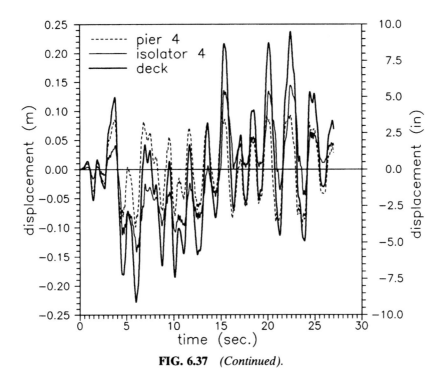

FIG. 6.37 *(Continued)*.

seismic loads. The bridge piers are up to 70 m (230 ft) tall, and each pier is a two-leg frame; the deck has a prestressed concrete box section with six spans, for an overall length of 315 m (1033 ft) [C21]. Each leg of each pier has a rocking pad close to its base, as shown in Figs. 6.38 and 6.39; a torsional-beam steel damper connects the upper and lower parts, limiting the lateral movements under rocking, and the maximum uplift of the legs is also limited to 125 mm (4.9 in.) by stops.

It can be noted that the seismic response of a rocking bridge presents many similarities with the response of a bridge isolated by means of a friction pendulum system (Section 6.2.1), which is in fact based on the same inverted pendulum concept. The response is approximately rigid plastic, with a substantial centering force, given by the uplift force itself.

6.4.2 Rocking Response of Bridges

A fundamental study of the rocking response of a rigid block is presented in [H6], where it is shown that the period of vibration of the rocking response increases with increasing amplitude of the motion, and it is recognized that

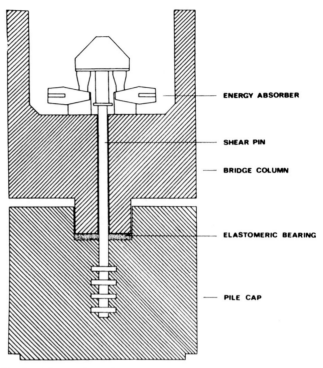

— ENERGY ABSORBER

— SHEAR PIN

— BRIDGE COLUMN

— ELASTOMERIC BEARING

— PILE CAP

FIG. 6.38 Rocking pad at the base of a stepping pier [C21].

FIG. 6.39 South Rangitikei viaduct under construction [C21].

520

much of the advantage to a rocking structure comes from the efficiency of the mechanism in dissipating energy. Assuming that the impacts are purely inelastic collisions (i.e., the only dissipated energy takes the form of radiation to the soil half-space), by equating momentum before and after impact a kinetic energy reduction factor (r) may be obtained, related to the block dimensions by the expression

$$r = \left[1 - \frac{MR^2(1 - \cos 2\alpha)}{I_0}\right]^2 \tag{6.25}$$

where M is the total mass, R the distance between the mass centroid and the center of rotation, α the angle between a vertical line and the line connecting the centroid and the center of rotation, and I_0 the mass moment of inertia of the block about the point of rotation. Using this coefficient the peak nondimensional displacement after n impacts (Δ_n, equal to the actual displacement divided by the width of the foundation) during natural decay of rocking is predicted as a function of the nondimensional initial displacement (Δ_0) as

$$\Delta_n = 1 - \{1 - r^n[1 - (1 - \Delta_0)^2]\}^{1/2} \tag{6.26}$$

If we now consider a single-degree-of-freedom oscillator with viscous damping, it is well known that the relative amplitude of different displacement peaks is related to the viscous damping (as percent of critical) by the expression

$$\xi = \frac{\ln(\Delta_0/\Delta_m)}{2\pi m} \tag{6.27}$$

where m is the number of complete cycles between Δ_0 and Δ_m. Considering that there are two impacts per cycle, and substituting Eq. (6.26) into Eq. (6.27), the following expression can be derived to express the equivalent viscous damping of a rigid rocking system [P27]:

$$\xi = \frac{\ln[\Delta_0/(1 - \{1 - r^n[1 - (1 - \Delta_0)^2]\}^{1/2})]}{\pi n} \tag{6.28}$$

A number of trial solutions of the equation above, all with $\Delta_0 < 0.5$ and the number of impacts $n = 2m < 16$, are shown in Fig. 6.40, demonstrating that under these hypotheses the relation between equivalent viscous damping and energy reduction factor is rather insensitive to the value of the initial displacement and the number of cycles; the interpolating line shown in the same figure can therefore be used for practical applications:

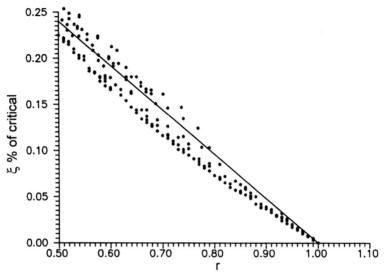

FIG. 6.40 Approximate relationship between equivalent viscous damping and energy reduction factor [P27].

$$\xi = 48(1 - r) \tag{6.29}$$

In the case of bridge piers, base rocking can occur when the foundations are spread footings or pile-supported footings with limited tension capacity of the piles. The basic geometric relationships are shown in Fig. 6.41. Note that the height H now refers to the distance between the center of the seismic weight (W_s) and bottom of the footing and that the total weight (W) at the base of the pier includes the portion of the pier weight (W_c) not included in the seismic weight, and the footing weight (W_f).

An approximate relation between seismic force and horizontal displacement will be derived from equilibrium consideration, assuming that the compression and tension forces R_c and R_t acting at the soil–footing interface have a rigid perfectly plastic interface pressure distribution (p_c and p_t), resulting in rectangular stress blocks. Equilibrium of vertical forces along the soil–footing interface requires that $R_c = R_t + W$, from which the depth of the compression stress block (a) can be determined as

$$a = \frac{BLp_t + W}{B(p_c + p_t)} \tag{6.30}$$

Equilibrium of moments results in a lateral overturning force V_E at the centroid of the seismic weight W_s as a function of the total displacement Δ of the general form

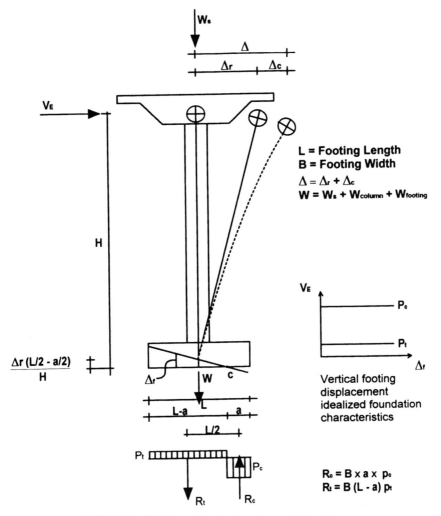

FIG. 6.41 Rocking of a single-column bridge bent.

$$V_E(\Delta) = \frac{R_t L/2 + W(L - a)/2 - W_s \Delta}{H + \Delta_r(L - a)/2H} \qquad (6.31)$$

where Δ_r is the displacement due to the rocking motion, Δ_c is the structural displacement, and $\Delta = \Delta_r + \Delta_c$.

Equation (6.31) can be further simplified in several ways:

- When there is no tension between soil and footing, $R_t = p_t = 0$
- When the pier is tall, $H \gg \Delta_r(L - a)/2H$, and therefore it is possible to assume $\Delta_r(L - a)/2H = 0$

- When the pier is stiff, $\Delta_c \ll \Delta_r$, and therefore it is possible to assume $\Delta_c = 0$

From the appropriate form of Eq. (6.31) the effective stiffness of the rocking pier can be evaluated as the ratio between force and displacement, and the period of vibration can be calculated from mass and stiffness. When a stiff superstructure connects several bents, rocking of the entire frame is more likely than rocking of a single bent. In this case, effective bent stiffnesses for n bents can be combined to an effective frame stiffness as

$$K_{\text{frame}} = \Sigma_n \frac{V_{E,n}}{\Delta} \tag{6.32}$$

and the characteristic rocking period can be obtained as

$$T_i = 2\pi \left(\Sigma_n \frac{W_{s,n}}{g K_{\text{frame}}} \right)^{1/2} \tag{6.33}$$

Equation (6.25) can also be simplified in the case of bridge piers. For example, neglecting the small contribution of pier and foundation, the mass moment of inertia takes the form

$$I_0 = MR^2 + \frac{M(bh^3 + b^3h)}{12bh} \tag{6.34}$$

where b and h are the width and height of the deck. In most case $b \gg h$, and therefore

$$I_0 \simeq M\left(R^2 + \frac{b^2}{12} \right) \tag{6.35}$$

Equation (6.25) therefore becomes

$$r = \left(1 - \frac{R^2(1 - \cos 2\alpha)}{R^2 + b^2/12} \right)^2 \tag{6.36}$$

The simplified relationships expressed by Eqs. (6.29), (6.31), and (6.36) allow a preliminary design (or assessment) of a rocking bridge, as discussed in the next section. It has to be noted, however, that the drastic simplification of the response could lead to large errors. A nonlinear time-history analysis has therefore always to be recommended, as discussed in Section 6.4.4.

6.4.3 Response Spectra Design Approach

The results discussed in Section 6.4.2 can be applied to design (or assess) a rocking bridge using a traditional trial-and-error procedure based on the substitute structure design method. In a preliminary phase, a bridge bent can be modeled as a rigid single-degree-of-freedom oscillator with constant damping and period of vibration proportional to the amplitude of rocking. Similar to the case with isolating devices, it is also assumed that the response will depend only on the equivalent elastic characteristics (period and damping) at peak response. The design process can then proceed as follows.

1. Using the no-rocking natural period and damping of the structure and the acceleration response spectra of the design earthquake, the elastic response acceleration is calculated to check that this will in fact induce rocking.

2. Using Eqs. (6.36) and (6.29), the viscous damping (ξ_e) of the rocking system equivalent to the soil radiation damping is calculated. Other forms of damping should also be considered if appropriate; for example, in the case of the South Rangitikei viaduct, the damping equivalent to the hysteretic response of the dampers should be introduced. The correct evaluation of the equivalent damping to be considered is the most critical step in the procedure.

3. The maximum rocking displacement at the center of mass of the rocking system is guessed to be Δ_1, the corresponding lateral force V_E is computed from Eq. (6.31), and a response period T_1 is calculated from Eq. (4.39).

4. A new displacement (Δ_2) of the equivalent elastic system (T_1, ξ_e) is calculated from the design displacement response spectrum, and the corresponding force and period will be calculated as in step 3. The iteration process will continue until convergence is obtained on a stable couple of period and displacement. In some cases no stable response can be achieved and rocking cannot be used as a seismic protection mechanism. This is more likely to happen when linearly increasing displacement response spectra are used, which do not correspond to reality, as discussed in Section 2.4.1.

5. The structure is then designed to behave essentially linearly until rocking takes place and to be able to accommodate the expected displacement.

Example 6.6. The procedure described above is illustrated below for the example of a rather tall column bent, shown in Fig. 6.42, which is part of a uniform bridge structure that can be expected to have the same rocking response for all piers. The pier is assumed to be 27 m (90 ft) tall from the base of the footing to the centroid of the seismic mass, the deck to have height

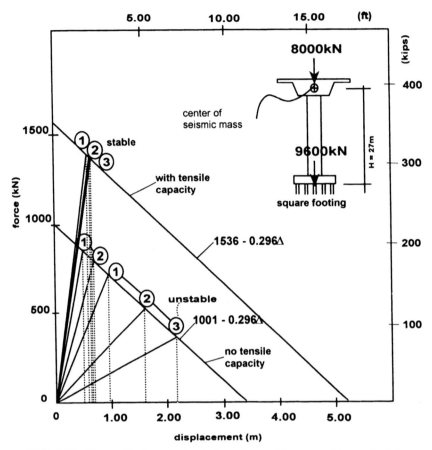

FIG. 6.42 Force–displacement relations for rocking piers (Example 6.6).

h equal to 2.4 m (8 ft) and width b equal to 14 m (46 ft), and the footing to be square with 7-m (23-ft) sides. The seismic weight is 8000 kN (1800 kips), and the total weight of the system at the base of the footing, 9600 kN (2116 kips). Seismic excitation is the EC8, Soil B spectrum for 0.7g PGA.

A spread footing is assumed with no tension capacity and with a plastic compression capacity of the soil equal to 1.0 MPa (142 psi). The depth of the compression block is then found from Eq. (6.30) as

$$a = \frac{9,600,000}{7000 \times 1.0} = 1371 \text{ mm (54 in.)}$$

and a linear relation between V_E and Δ is obtained from Eq. (6.31) as (V_E in kN, Δ in mm)

$$V_E(\Delta) = \frac{9600(7000 - 1371)/2 - 8000\Delta}{27,000} = 1001 - 0.296\Delta$$

[with V_E in kips, Δ in inches: $V_E(\Delta) = 221 - 1.66\Delta$], where it has been assumed that $R_t = 0$ and $\Delta_r(L - a)/2H = 0$. This linear relation is shown in Fig. 6.42.

The acceleration at the pier top needed to induce rocking can be evaluated as approximately equal to 0.125g, since the maximum moment at the base is

$$M_r = 9600\left(\frac{7 - 1.371}{2}\right) = 27,019 \text{ kN} \cdot \text{m} \ (19,856 \text{ kip-ft})$$

and the acceleration-inducing rocking

$$a_r = \frac{27,019}{27 \times 8000} = 0.125g$$

Rocking will therefore certainly be induced, and the equivalent damping is calculated from Eqs. (6.36) and (6.29):

$$R = \left[27,000^2 + \left(\frac{7000}{2} - 1371\right)^2\right]^{1/2} = 27,083 \text{ mm} \ (90 \text{ ft})$$

$$\cos 2\alpha = \cos 2 \tan^{-1} \frac{7000/2 - 1371}{27,000} = 0.988$$

$$r = \left[1 - \frac{27,083^2(1 - 0.988)}{27,083^2 + 14,000^2/12}\right]^2 = 0.977$$

$$\xi_e = 48(1 - 0.977) = 1.12\%$$

This damping has to be added to the structural viscous damping, which is taken as equal to 5%. The low value of the added damping is due to the small dimensions of the foundation with respect to the height of the pier. The rocking response of the bridge is therefore governed by the rocking period of vibration rather than by increased energy dissipation capacity. The iteration procedure is then started, using the displacement response spectrum of Fig. 6.2, considering the case of 0.7g ground acceleration and assuming a tentative displacement equal to $\Delta_1 = 500$ mm (19.7 in.), obtaining

$$V_{E1} = 1001 - 0.296 \times 500 = 852.7 \text{ kN} \ (192 \text{ kips})$$

$$K_{E1} = \frac{852.7}{0.5} = 1705 \text{ kN/m} \ (118 \text{ kips/ft})$$

$$T_1 = 2\pi\left(\frac{8000}{9.81 \times 1705}\right)^{1/2} = 4.34 \text{ s}$$

$$\Delta_2 = 722 \text{ mm (28.4 in.)}$$

$$V_{E2} = 1001 - 0.296 \times 722 = 787 \text{ kN (177 kips)}$$

$$K_{E2} = \frac{787}{0.722} = 1090 \text{ kN/m (75 kips/ft)}$$

$$T_2 = 2\pi\left(\frac{8000}{9.81 \times 1090}\right)^{1/2} = 5.43 \text{ s}$$

$$\Delta_3 = 722 \text{ mm (28.4 in.)}$$

If a more traditional but less realistic displacement spectrum should be used, with displacements increasing in proportion to T at long period of vibration instead of being bounded for T > 3s, the bridge bent would have become unstable according to the following iteration steps, which start again from a tentative displacement of 500 mm (19.7 in.):

$$V_{E1} = 1001 - 0.296 \times 500 = 852.7 \text{ kN (192 kips)}$$

$$K_{E1} = \frac{852.7}{0.5} = 1705 \text{ kN/m (118 kips/ft)}$$

$$T_1 = 2\pi\left(\frac{8000}{9.81 \times 1705}\right)^{1/2} = 4.34 \text{ s}$$

$$\Delta_2 = 995 \text{ mm (39.2 in.)}$$

$$V_{E2} = 1001 - 0.296 \times 995 = 706 \text{ kN (159 kips)}$$

$$K_{E2} = \frac{706}{0.995} = 710 \text{ kN/m (49 kips/ft)}$$

$$T_2 = 2\pi\left(\frac{8000}{9.81 \times 710}\right)^{1/2} = 6.73 \text{ s}$$

$$\Delta_3 = 1543 \text{ mm (60.7 in.)}$$

$$V_{E3} = 1001 - 0.296 \times 1543 = 202 \text{ kN (45 kips)}$$

$$K_{E3} = \frac{202}{1.543} = 131 \text{ kN/m (9 kips/ft)}$$

$$T_3 = 2\pi\left(\frac{8000}{9.81 \times 131}\right)^{1/2} = 15.7 \text{ s}$$

$$\Delta_4 = 3600 \text{ mm (142 in.)}$$

The results would change significantly if the foundation would be able to offer some tensile capacity. For example, assume that the footing is supported on piles, offering the same total compression capacity as in Example 6.5, and a tensile pull out capacity equal to 15% of the compression capacity. The depth of the compression block would again result from Eq. (6.30) as

$$a = \frac{7000 \times 7000 \times 0.15 + 9{,}600{,}000}{7000(1.0 + 0.15)} = 2106 \text{ mm (83 in.)}$$

It follows that

$$R_t = 7000 \times (7000 - 2106) \times 0.15 = 5{,}138{,}700 \text{ N (1133 kips)}$$

A linear relation between V_E and Δ is obtained as (V_E in kN, Δ in mm):

$$V_{E(\Delta)} = \frac{5138.7 \times 7000/2 + 9600(7000 - 2106)/2 - 8000\Delta}{27{,}000}$$

$$= 1536 - 0.296\Delta$$

[with V_E in kips, Δ in inches: $V_E(\Delta) = 338 - 1.66\Delta$]. This linear relation is also shown for comparison in Fig. 6.42. The damping would also be affected, not only because of the variation in R and α, but because of the energy dissipated in the pile–soil interaction.

Assuming a linearly increasing displacement response spectrum and a 10% additional damping, the response would be stable, as shown in the following calculation, where the first trial displacement is taken equal to 600 mm (23.6 in.).

$$V_{E1} = 1536 - 0.296 \times 600 = 1359 \text{ kN (306 kips)}$$

$$K_{E1} = \frac{1359}{0.6} = 2264 \text{ kN/m (156 kips/ft)}$$

$$T_1 = 2\pi \left(\frac{8000}{9.81 \times 2264} \right)^{1/2} = 3.77 \text{ s}$$

$$\Delta_2 = 628 \text{ mm (24.7 in.)}$$

$$V_{E2} = 1536 - 0.296 \times 628 = 1350 \text{ kN (304 kips)}$$

$$K_{E2} = \frac{1350}{0.628} = 2150 \text{ kN/m (148 kips/ft)}$$

$$T_2 = 2\pi \left(\frac{8000}{9.81 \times 2150} \right)^{1/2} = 3.87 \text{ s}$$

$$\Delta_3 = 646 \text{ mm (25.4 in.)}$$

$$V_{E3} = 1536 - 0.296 \times 646 = 1345 \text{ kN (303 kips)}$$

$$K_{E3} = \frac{1345}{0.646} \, 2082 \text{ kN/m (144 kips/ft)}$$

$$T_3 = 2\pi \left(\frac{8000}{9.81 \times 2082} \right)^{1/2} = 3.93 \text{ s}$$

$$\Delta_4 = 656 \text{ mm (25.8 in.)}$$

6.4.4 Time-History Analysis

As clearly stated in the discussion above, many assumptions have been made in order to develop a simple design procedure. It is therefore again of fundamental importance to check the structural response using a nonlinear dynamic approach. Unfortunately, an appropriate simulation is often difficult because of the limited knowledge of phenomena such as radiation damping and because of modeling problems when monolateral restraints are involved. For example, in [P29] it has been shown that the apparent damping of a rocking system on a Winkler foundation decreases when the amplitude of motion increases, and this fact could constitute a source of unconservatism.

An appropriate simulation is easier and more reliable when the rocking motion does not involve interaction with soil. The main source of equivalent damping is provided by the well-known hysteretic response of some artificial device, and some tensile reaction is offered to the system, as in the case of the South Rangitikei viaduct, mentioned in Section 6.4.1. If interaction with soil is involved, conservative assumptions on the damping should be adopted. The nonlinear behavior can generally be confined to the elements simulating the rocking motion. The linear deformability of the structure will overcome the assumption of a rigid-body motion, adopted in the preliminary design phase.

Example 6.7. A rocking bridge pier similar to that designed in Example 6.6, with 5% viscous damping, a foundation compression capacity of 1.0 MPa (145 psi), and negligible foundation tensile capacity (Fig. 6.43) has been analyzed. The compression spring was capable of dissipating energy as shown in Fig. 6.43. The same input acceleration used for the previous examples (Fig. 6.31) has been used, scaling the amplitude to 0.7g, obtaining the displacement and acceleration response shown in Figs. 6.44 to 6.46. In Fig. 6.44 the total horizontal displacement at the top is plotted, while in Fig. 6.45 only the displacement due to the base rotation is shown. The difference between the two diagrams

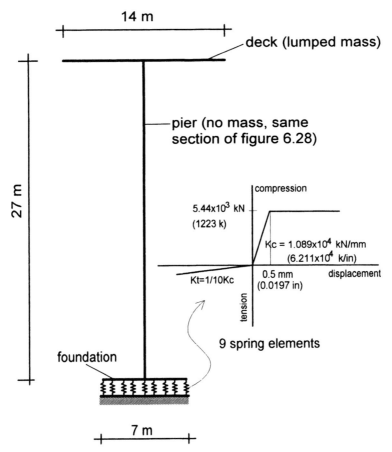

FIG. 6.43 Example 6.7: model used for the time-history simulation of the rocking response and properties of the springs used to model the soil behavior.

therefore represents the structure displacement. Note that the rocking motion is dominating the response even with this rather flexible pier, that the rocking mode is not continuously excited during the time history, and that the rocking period is not far from the predicted 5.4 s. The peak displacement is larger than predicted in the preliminary design phase by approximately 30%, but only one cycle reaches this limit. The limitation of the acceleration response provided by the rocking response is evident in Fig. 6.46, where the peak acceleration during rocking is approximately constant, around 0.1g.

6.5 ACTIVE CONTROL

Active control involves sensors to measure the response of the structure, real-time computer analysis of the response to calculate the optimum forces to be

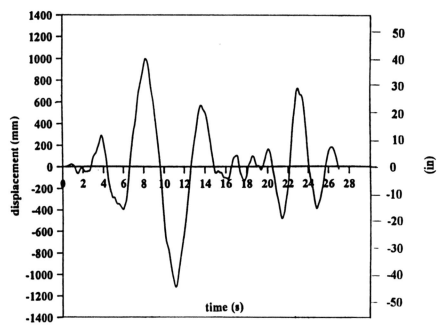

FIG. 6.44 Total displacement time history of the bridge pier of Example 6.7.

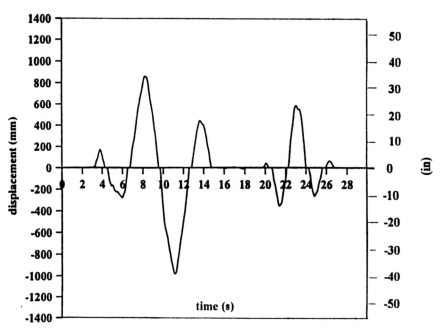

FIG. 6.45 Displacement time history of the bridge pier of Example 6.7. Only the rocking contribution is shown.

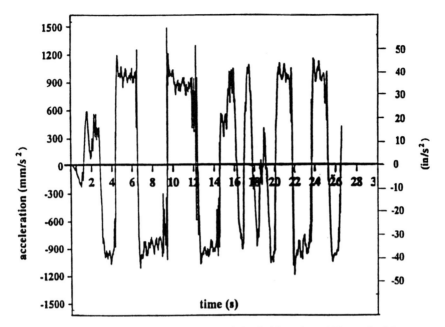

FIG. 6.46   Acceleration response of the bridge pier of Example 6.7.

applied, and the capacity for applying the forces required to counteract the seismic attack. The concept is clearly fascinating, and the technology to transform the concept into reality has been proved to exist. Nevertheless, active control is not mature enough to be applied extensively in the earthquake protection of civil engineering structures, is far from being economically competitive with more traditional techniques, and its potential has not yet been fully exploited even in the research field. For these reasons the right places to discuss active control are in scientific journals and at conferences rather than in a book on seismic design of bridges. On August 1994 the First World Conference on Structural Control was held in Los Angeles. As soon as the proceedings become available (only abstracts have been distributed [F2]) they will constitute the most recent reference on the subject. Very little material is present on active seismic protection of bridges, and no practical application is known to the authors of this book.

7

SEISMIC ASSESSMENT
OF EXISTING BRIDGES

7.1 INTRODUCTION: PROCEDURES FOR SEISMIC ASSESSMENT

The purpose of carrying out a seismic assessment analysis of an existing bridge is to determine the level of risk associated with loss of serviceability, severe damage, or collapse. With this risk quantified, rational decisions can be made as to whether the bridge should be retrofitted or replaced, or to accept the risk and leave the bridge in the existing state. Frequently, the choice to retrofit will be relative: Insufficient funds may be available to retrofit all deserving bridges, and hence the risk analysis will be used to aid in selection of the bridges on which the limited resoures will be spent.

There are generally two stages to a seismic assessment. The first involves a general screening and prioritization study to determine which bridges are most likely to pose the greatest risk. This is normally carried out based on generic indicators, such as age of bridge, soil conditions, structural type, site seismicity, and traffic density rather than on detailed analyses. The second stage involves a detailed structural analysis of the bridges identified as having high risk in the prioritization phase in relation to site seismicity and soil conditions.

7.1.1 Prioritization Schemes

The objective of a prioritization scheme is to identify and rank all high-risk bridges in a specified region so that an optimum allocation of resources for retrofit can be made. Ideally, this should be carried out within the constraints of a cost–benefit analysis so that absolute as well as relative (i.e., one bridge versus another) decisions can be made about retrofit. Although a full cost–benefit approach has recently been attempted in isolated prioritization exercises [M11], most prioritization schemes are developed in terms of relative risk without an attempt at quantification of cost or benefit.

Perhaps the best known prioritization schemes are those developed by ATC-6-2 [A2], Caltrans [G1], and the Japan Ministry of Construction [K1]. Although they consider the same general contributory factors for risk, the means for combining them differ. In all cases the contributory elements for assessing relative risk can be divided into three major categories: seismicity, vulnerability, and importance.

1. *Seismicity.* This includes site-specific information on the expected ground motion. Ideally, it should consist of a basic response spectrum shape, or shapes, with peak ground acceleration related to annual probability of exceedance, and should be appropriate to the soil conditions of the site.

2. *Vulnerability.* This represents the susceptibility to damage or collapse and is related primarily to structural aspects. The vulnerability will depend on the structural form: whether single or multispan; whether continuous over internal supports, or simply supported with movement joints; whether connections of superstructure to column are moment resisting or bearing supported; whether columns are tall or short; whether bents are single column or multicolumn; whether abutments are right or skewed. Vulnerability will also depend on the age of the bridge; in particular insofar as different details, such as amount of confinement and shear reinforcement in columns, depend on when the structure was designed. Soil conditions are considered in terms of liquefaction potential.

3. *Importance.* Importance relates to the consequences of damage or failure and typically includes consideration of traffic volume, type of crossing (a bridge crossing a freeway is more important than a bridge crossing a stream), length of detour resulting from bridge closure, and significance of the bridge to lifeline operations such as passage of emergency services vehicles to hospitals.

Many prioritization schemes involve ranking based on addition of suitably weighted values for seismicity (S), vulnerability (V), and importance (I) according to the relationship

$$R = w_s S + w_v V + w_I I \tag{7.1}$$

It will be seen that there are logical objections to such an additive relationship, since it may imply high seismic risk for a totally nonseismic region ($S = 0$) or for an invulnerable bridge ($V = 0$). As a consequence, some form of multiplicative relationship, such as

$$R = S^{w_s} V^{w_v} I^{w_I} \tag{7.2}$$

would appear more logical. Typically, the weighting factors in Eq. (7.2) would each be less than unity. Prioritization ranking based on Eq. (7.2) generally results in a much wider range of ranking than obtained from Eq. (7.1), making it easier to separate the prime candidates for retrofit from a large inventory of bridges.

No attempt to made in this book to quantify the elements of Eq. (7.1) or (7.2) or of their component constituents. Such quantification appears to depend rather critically on seismicity and design features, which may differ markedly from one seismic region (e.g., California) to another (e.g., Europe). Consequently, local information should be used to develop the appropriate coefficients. However, Fig. 7.1 shows weightings assigned by Caltrans in their additive model.

There appears to be real merit in developing a probabilistic risk analysis for guiding retrofit decisions [M11]. In such an analysis a crucial element is the relationship between damage and site seismicity for different levels of vulnerability, as expressed by the different structural options summarized above. If the site seismicity is expressed as a curve of intensity versus annual probability of exceedance, expected damage to each class of bridges can be expressed in dollar terms over a specified time frame. The influence of bridge importance is expressed by estimating the cost of bridge closure and for nonstructural aspects such as personal injury or death, also by equivalent dollar estimates. The cost of bridge damage and loss of use is then compared with the cost to reduce the risks and hence reduce the loss. Clearly, such an approach, although intellectually satisfying, requires incorporation of much "soft" data, particularly those relating seismic intensity to damage. However,

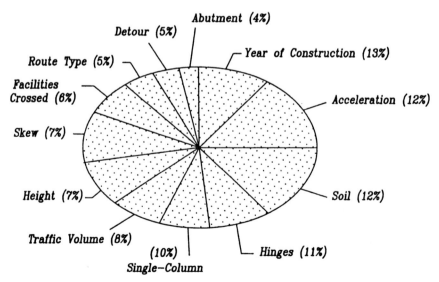

FIG. 7.1 Basic risk components in Caltrans prioritization scheme [G1].

as more data and improved structural assessments become available, the precision of the approach will naturally improve.

7.2 ASSESSMENT LIMIT STATES

Structure limit states were introduced in Section 1.3.6(b), where it was suggested that three limit states could be considered for design: the serviceability, damage control, and survival states. The same three limit states are appropriate for assessment, but less conservative estimates of conditions corresponding to the serviceability and damage control limits may be appropriate. This is not the case for the survival limit states: We should not accept higher probability of collapse and loss of life, even though higher probability of damage may be acceptable when assessing existing bridges.

7.2.1 Serviceability Limit State

For response to this limit state, the bridge would be expected to be serviceable immediately following an earthquake and should not need repair. Member flexural strengths could be achieved, and some limited ductility developed, provided that concrete spalling in plastic hinges did not occur and that residual crack widths remain sufficiently small so that remedial activity, perhaps in the form of epoxy injection of cracks, was not needed. For assessment, the recommendations of Section 5.3.2(d) for ideal strength can be assumed, and thus limit values of $\varepsilon_c = 0.004$ and $\varepsilon_s = 0.015$ are suggested.

The validity of these strain limits can be determined as follows. Typically, spalling of concrete is initiated at extreme fiber compression strains between $\varepsilon_c = 0.006$ and 0.01 [M12]. Thus the limit of 0.004 is a conservative estimate of the onset of concrete damage. The strain limit of $\varepsilon_s = 0.015$ was determined to ensure that residual crack widths would not exceed 1.0 mm (0.04 in.). If we consider a typical bridge plastic hinge region in a column where cracks form at an average spacing of 200 mm (8 in.), the crack width corresponding to an extreme reinforcement strain of 0.015 will be $0.015 \times 200 = 3.0$ mm. However, this crack width is the value at maximum response. For low ductility levels, residual displacements (and hence crack widths) are approximately one-third of maxima values, and hence a residual crack width of 1.0 mm (0.04 in.) can be expected. This is frequently taken as the maximum width than can be tolerated in normal environmental conditions without requiring remedial action.

It should be noted that the definition of serviceability as being the limit beyond which remedial action is required may be too severe in some cases. For example, a lifeline bridge may be required to be "serviceable" following a major earthquake, but this will normally mean in a state capable of transporting emergency services vehicles. Significant damage, more appropriate to the damage-control limit state, might hence be tolerable.

7.2.2 Damage-Control Limit State

The damage-control limit state represents the extreme level of seismic response, beyond which it would not be economically and technically feasible to repair the bridge. The damage-control limit state is probably the most important in terms of seismic assessment and, as such, is considered in detail in this chapter. It is taken to be the limit state beyond which lateral resistance diminishes with increasing displacement. In assessing existing bridges, an increased probability of attaining the damage-control limit state, compared with that for new bridges, may be acceptable.

7.2.3 Survival Limit State

Response to the survival limit state represents the extreme level of seismic response, beyond which collapse would occur. Conditions defining this limit state are discussed in some detail in Section 1.3.6(*b*). As discussed above, a higher probability of collapse under extreme earthquake intensity should not be accepted for existing bridges compared with new bridges.

7.3 ASSESSMENT ANALYSIS SCHEMES

The outcome of prioritization analysis will be to identify a group of bridges estimated to be at high risk and in probable need of retrofit. However, inevitably, the prioritization approach will provide only coarse estimates of true risk, which must be refined by more precise analysis. The purpose of the prioritization scheme is to reduce the number of bridges that need to be subjected to detailed analysis to manageable proportions.

There has been a tendency in the past to use rather rudimentary analytical tools for this phase of analysis. Considering that the results of the analysis may be a decision to retrofit or *not* to retrofit a bridge, it is appropriate to spend rather more analytical effort to reduce the retrofit cost. Our experience in this area has indicated that the results of increased analytical effort are very cost-effective in identifying where retrofit cost savings can safely be made. Given that resources for retrofit are likely to be less than required for fully upgrading all bridges in a given area, these cost savings translate into improved public safety.

As a consequence, it is important that the assessment analyses be adequately sophisticated. This, however, should not be taken to mean that very large and complex analytical tools, such as three-dimensional multimode analyses or inelastic time-history analyses should be adopted for all bridges. As explained in Chapter 4, these will not necessarily result in improved representation of response.

7.3.1 Capacity/Demand Ratio Analyses

An early and much used assessment approach is the capacity/demand ratio analysis developed in the mid-1980s by the Applied Technology Council [A2] and known as ATC-6-2. Forces resulting from elastic analysis demand of a bridge structure were compared with strength (capacity) to provide a demand/capacity ratio for actions at different parts of the structure. In its simplest form, a ratio greater than 1 implies failure. In its more complete form, the acceptable ratio is permitted to exceed unity, where ductile response is assured. Thus a demand/capacity ratio of 2 or 3 might be acceptable when moments at a given section are considered but not when shear is considered. In the ATC-6-2 approach, this is effected by modifying the capacity to allow for ductility, where appropriate.

Although not widely used until the 1989 Loma Prieta earthquake [E1] engendered increased awareness in bridge owners to the potential for collapse of their bridge stock in moderate earthquakes, it has been widely used since then. Simultaneously, the ATC-6-2 and similar capacity/demand procedures have been subjected to a more detailed examination in the light of recent advances in understanding of bridge seismic performance.

This reexamination has revealed some basic flaws in the procedure. To some extent these relate to details, such as methods for assessing lap-splice and anchorage competence, or shear strength of columns, and as such, could be corrected within the framework of the existing methodology. However, there are also some more basic problems with the overall concept of section capacity/demand ratios based on elastic analyses. Three examples are given below.

1. The inherent assumption of the capacity/demand approach for ductile action is that the moment demand/capacity ratio at a section can be equated to the section ductility demand, and hence an estimate of member ductility capacity can be compared to the moment demand/capacity ratio. It was established in Section 5.3.1(*b*) that although it is possible to relate structural displacement ductility to the force-reduction factor (which is just another name for moment-demand/capacity ratio), this could not be extended to individual sections, since the relationship between member and structure ductility demands depends on the structural geometry. In particular, with relation to the twin-column bent of Fig. 5.10, cap beam flexibility may increase the local ductility demand at a plastic hinge to many times the overall structure ductility demand. This was expressed in Eq. (5.21). Thus the moment demand/capacity ratio at a section is a poor estimate of member ductility demand.

2. Not all sections with demand/capacity ratios greater than unity are at risk. For example, if we again consider the two-column bent of Fig. 5.10, we may find that moment demand/capacity ratios of critical sections in the columns are 6, and in the beams are 4. Our assessment of ductility capacity supports

(say) a ratio of 2 in each case. The implication is that both columns and beams are at risk of failure, and both should be retrofitted. However, the *difference* in ratios implies that the columns will hinge first, at two-thirds of the moment capacity of the beams, and hence the beams are capacity-protected against developing plastic hinges by an overstrength factor of 1.50. Hence although the columns need retrofitting to increase ductility, the beams are satisfactory. Thus the demand/capacity approach can lead to incorrect decisions about the need to retrofit. Although for simple cases such as that discussed in this example, an experienced practitioner can make the necessary adjustments, these will not generally be obvious.

3. In assessing member strength and ductility, the axial force on the member is a critical consideration. The capacity/demand approach does not lead readily to establishing what the seismic axial load component should be. It has been common in such assessments for the analyst to use the axial force that comes out of the elastic seismic analysis as the value to compute the member moment capacities (and possibly the ductility capacities, if assessed directly). Again, taking the example of our two-column bent, this will be seen to be inappropriate. If the column moment capacities are one-sixth of the elastic demand moments, the axial loads resulting from seismic action will also be one-sixth of the axial loads from the elastic analysis. These should be used to determine the column moment capacities. It will thus be seen that the results are interactive and cannot be solved by a single elastic analysis. The adjustments for the column capacity can again be made by the experienced practitioner, but it will be recognized that the approach requires considerable manipulation. For example, the foundations may come out as having demand/capacity ratios of greater than unity, but the foundation strength will be critically affected by the axial load input from the column, which is again related to column capacity.

For these and related reasons, the capacity/demand ratio approach has lost favor in comparison with the more recently developed plastic collapse mechanism approach [P16].

7.3.2 Plastic Collapse Mechanism (Pushover) Analysis

As has been explained in Chapter 4, less emphasis is currently placed on results of multidegree-of-freedom elastic modal analysis for assessment of existing bridges than was the case perhaps 5 to 10 years ago. This has resulted from realization of the inability of elastic analysis to capture the modification of response caused by inelastic action of individual elements, as discussed above, or the behavior of movement joints, which respond with different opening and closing characteristics, and whose rotational characteristics about the vertical axis cannot be characterized as elastic. For long bridges with several movement joints, it is improbable that the excitation at all soil–structure boundaries will be coherent and synchronous (see Section 2.5), further adding to difficulties in simulation by elastic analysis.

More recently, the preferred analysis tool has been inelastic plastic collapse mechanism analysis, or push analysis, carried out on independent
stand-alone frames considered to be completely separated from adjacent
frames at the movement joints. In these analyses the superstructure is often,
though not exclusively, considered to be effectively rigid in the horizontal
plane.

An initial stage of the analysis is to carry out independent collapse analyses
of individual bents of columns in the direction considered (i.e., longitudinal
or transverse). For each bent, displacement is incremented, tracking the formation of plastic hinges, shear degradation, joint degradation, and plastic rotation.
Serviceability and ultimate (collapse) limit states are related to inelastic rotations of plastic hinges, onset of member or joint shear failure, or other degradation mechanisms. Details of member assessment are presented in the following sections.

With lateral force–deformation characteristics identified for each bent, the
frame is assembled, with bent stiffness represented by simple inelastic springs.
The initial center of rigidity, total transverse stiffness, and rotational stiffness
are calculated, and hence an effective stiffness at the center of mass is computed. The vector of bent displacements, including both translational and
rotational components, is then compared with bent displacement capacities,
in terms of yield and ultimate displacements, to identify the critical bent and
mode of failure.

The procedure outlined in Section 4.4.1 is summarized here for convenience.
Using the nomenclature of the generic frame of Fig. 7.2, the effective stiffness
K at the center of mass is given by

$$\frac{1}{K} = \frac{1}{\sum K_i} + \frac{\bar{x}^2}{\sum K_1 x_i^2} \tag{7.3}$$

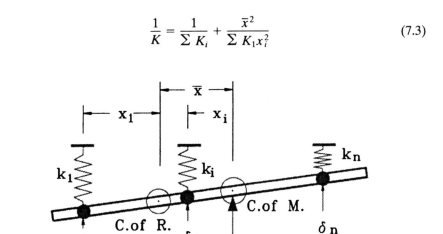

FIG. 7.2   Plan view of frame deformation under unit lateral force applied at center
of mass.

where $\bar{x}$ is the distance between the center of rigidity and the center of mass. The natural period of the frame is then

$$T = 2\pi \sqrt{\frac{M}{K}} \tag{7.4}$$

where M is the total mass of the frame. The vector of bent displacements for unit lateral response inertia force acting at the center of mass is thus

$$\delta_i = \frac{1}{\Sigma K_i} + \frac{\bar{x} x_i}{\Sigma K_i x_i^2} \tag{7.5}$$

If Δ_i is the vector of bent displacement capacities, calculated in accordance with the following sections, the value of

$$V_E = \min \left| \frac{\Delta_i}{\delta_i} \right| \, i \tag{7.6}$$

defines the equivalent elastic response inertia force, based on an equal-displacement approximation of response, and also identifies the critical bent.

The actual response force will be less than given by Eq. (7.6) if inelastic response results. In this case the frame push analysis is carried out incrementally, modifying the bent stiffness and center of rigidity sequentially, as inelastic action develops in one or another of the bents. An equivalent elastic response level is calculated whose magnitude depends on the calculated frame displacement ductility μ_Δ at the assessed limit state and the relationship between the frame fundamental period T and the period at peak spectral acceleration response T_o, as discussed in Section 5.3.1(b). The equivalent elastic response is thus given by

$$V_E^* = V_E \frac{Z}{\mu_\Delta} \tag{7.7}$$

where Z is found from Eq. (5.22) and V_E is given by Eq. (7.6) using the initial elastic vector of bent flexibilities. For $T > 1.5 T_o$, Eq. (5.22) implies that $Z = \mu_\Delta$, yielding $V_E^* = V_E$. For $T < 1.5 T_o$, the equivalent elastic response decreases with T and approaches $V_E^* = V_E/\mu_\Delta$ as $T \to 0$.

The equivalent elastic response acceleration $S_{ar(g)}$ at the limit state considered is then given by

$$S_{ar(g)} = \frac{V_E^*}{W} \tag{7.8}$$

where W is the effective weight of the frame considered. The procedure outlined above is represented schematically in Fig. 7.3.

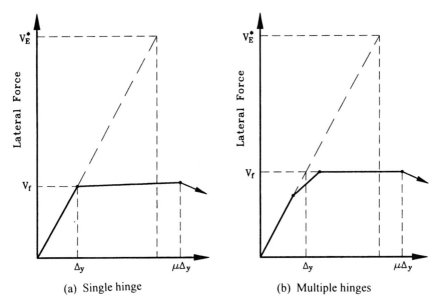

(a) Single hinge   (b) Multiple hinges

FIG. 7.3 Equivalent elastic lateral strength of a frame or bent.

The seismic risk associated with the equivalent elastic response acceleration $S_{ar(g)}$ is found by comparison with the spectral response of the assessment earthquake at the same period, as shown in Fig. 7.4(a), which has a known annual probability of exceedance, and by use of a relationship between relative intensity and annual probability of exceedance, shown in Fig. 7.4(b). Thus in Fig. 7.4(a), the calculated spectral response $S_{ar(g)}$ related to the assessment spectral response $S_{a(g)}$ indicates that an earthquake of lesser intensity, suggested by the dashed profile, will be sufficient to cause failure. Entering Fig. 7.4(b) with the calculated ratio $S_{ar(g)}/S_{a(g)}$ enables the risk of failure, in terms of annual probability, to be determined. This enables rational decisions to be made as to whether retrofit is required.

It will be apparent that relating the displacement response to equivalent elastic acceleration response is somewhat cumbersome, and does not allow consideration of differences in hysteretic-energy consideration between different modes of inelastic deformation. A more direct procedure uses the substitute structure approach described in Sections 4.4.3 and 5.3.1(c), relating the displacement capacity to the displacement demand through assessment displacement spectra for different damping ratios. The equivalent period chosen for comparison is that at displacement capacity of the frame.

The procedure outlined above is generally applicable but has limitations. For example, when adjacent frames have greatly different stiffnesses and adequate connection across movements joints is assured, response estimated by a stand-alone analysis may overestimate response of the more flexible

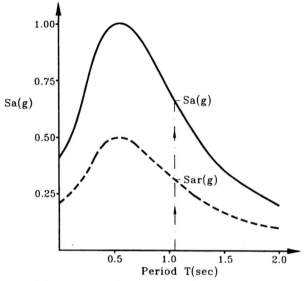

(a) Assessment and Capacity Acceleration
Response Spectra

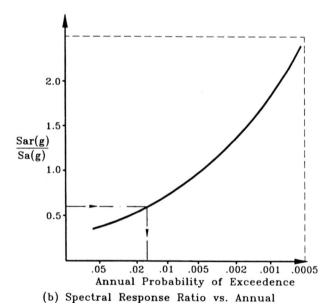

(b) Spectral Response Ratio vs. Annual
Probability of Exceedance

FIG. 7.4 Determination of seismic risk from limit-state spectral capacity.

frame and underestimate the response of the stiffer frames. Second, dynamic inelastic analyses indicate that push analyses tend to overestimate torsional response of frames [C6]. Third, it is difficult, though not impossible, to incorporate superstructure flexibility into a stand-alone push analysis. Finally, there are occasions when the bent stiffness calculated by independent bent push analysis is inappropriate when assembled into the full-frame stiffness model. In particular, this occurs with curved superstructures supported on single-column bents. A push analysis in the transverse direction would treat each bent as a simple cantilever, but the curved nature of the superstructure, combined with its torsional rigidity, results in moments being developed at the column top of opposite sign to those at the column base, as discussed in relation to Fig. 3.3.

Special-purpose computer programs such as SC-PUSH3D [S8] have been developed for inelastic push analyses for complete frames, capable of including curved bridge geometry, superstructure flexibility, and most of the deformation characteristics described in the following sections, and hence most of the limitations described above have been eliminated. Where interaction between frames is felt to be significant, however, it is advisable to carry out elastic modal analyses of a segment of bridge superstructure consisting of several frames, including the frame of interest. Comparisons between results from inelastic time-history analysis and elastic modal analysis indicate that provided that fundamental periods are larger than that corresponding to peak spectral acceleration response, bent displacements (but not forces) are predicted with adequate accuracy ($\pm 20\%$) by modal analyses. Bent-by-bent push analyses can then be used to determine the inelastic force–displacement response of each bent for comparison with demand displacement from the elastic modal analyses. If the ratio of limit-state displacement capacity Δ_c to elastic model demand displacement Δ_d is found for each bent, the minimum value of Δ_c/Δ_d may be used to enter Fig. 7.4(b) to obtain the annual probability of exceedance of the assessed limit state.

7.3.3 Inelastic Time-History Analysis

Although inelastic time-history analysis is clearly the most sophisticated method available for assessing bridge performance, there are currently still problems in application to assessment procedures. First, there are relatively few computer programs that are capable of realistic three-dimensional modeling, including interaction between biaxial inelastic response of columns, interaction of flexural and shear strength, modeling of joint characteristics, and modeling of degrading performance in terms of negative postyield stiffness and degraded hysteretic response characteristics. Without such modeling capabilities, there is little point in pursuing the complexities of inelastic time-history analysis. Further, to determine the response corresponding to different specified limit states, a series of analyses at different intensity of seismic input will be needed, since (for example) the plastic rotation of a specific hinge will not in general be linearly related to seismic intensity. This, together with uncertainty about if and how to model nonsynchronous support motion, makes the level of computational effort enormous and the precision of the results doubtful. As a consequence, the remainder of this chapter is built around the assumption of a plastic collapse analysis approach, and member characteristics are organized to be compatible with this. More complete information on inelastic time-history analysis is given in Section 4.5.4.

7.4 ASSESSMENT OF MEMBER STRENGTH AND DEFORMATION CHARACTERISTICS

7.4.1 Material Strengths

Member strengths and deformation characteristics should generally be calculated using the methods developed in Chapter 5. However, less conservatism should be adopted for assessment than for design, since the aim of the assessment analysis will be to obtain a best estimate of expected performance. For example, probable material strengths rather than nominal strengths will be used, and strength reduction factors, used to ensure dependable capacity, will generally not be used. Where possible, material strengths should be based on results of representative tests, including compression tests on concrete cores taken from representative areas of the bridge, supported by compression strength estimates at all critical areas from nondestructive methods such as impact hammer results and/or ultrasonic tests. Reliance on impact hammer or ultrasonic tests by themselves is not recommended. However, when correlated with the few selected core compression results, this enables a good estimate of the compression strength of the concrete to be obtained for all critical regions. Reinforcement strength should, where possible, be determined from mill certificates or from representative samples taken from the bridge.

Where testing is not feasible and mill certificates are not available, the following assessment strengths are suggested:

$$f'_{ca} = 1.5f'_c \tag{7.9a}$$

$$f_{ya} = 1.1f_y \tag{7.9b}$$

where f'_c and f_y are the specified material strengths. The 50% strength increase over specified 28-day compression strength accounts for typical conservative batching practice and strength gain with age and has been found to be a lower bound to actual compression strength of concrete of older bridges in California. In other regions it would be advisable to determine the appropriate strength increase from local data.

7.4.2 Relative Capacity Considerations

Modified capacity design considerations must be employed to determine which of alternative inelastic deformation mechanisms may develop. For example, initial calculations for strength of a critical member may indicate that shear strength exceeds flexural strength by (say) 10%. Both shear and flexural strength are based on probable member strengths and best-estimate analyses. A ductile flexural mode is thus indicated, and for the sake of example, the displacement ductility capacity is assessed to be $\mu_\Delta = 3$.

Assuming, for simplicity, that the fundamental period $T \geq 1.5T_o$, the equal displacement approximation applies, and from Section 5.3.1(c), the equivalent elastic response is $V_E = 3V_f$, where V_f is the strength of the flexural mechanism. The simplified elastoplastic response of the flexural mechanism is shown by line 1 in Fig. 7.5. However, the consequences of the flexural strength exceeding the probable value must be considered. It is conceivable that the longitudinal reinforcement has an unusually high yield strength and that the actual flexural strength exceeds $1.1V_f$. In this case the shear strength will be reached before flexural strength is achieved, and the brittle failure characteristic represented in Fig. 7.5 by line 2 is predicted. The ductility of this brittle mechanism must be assessed as $\mu_\Delta = 1$, and the equivalent clastic lateral strength is thus equal to the shear strength: $V_E = 1.1V_f$. Thus as a consequence of a small unanticipated *increase* in flexural strength, the equivalent elastic lateral

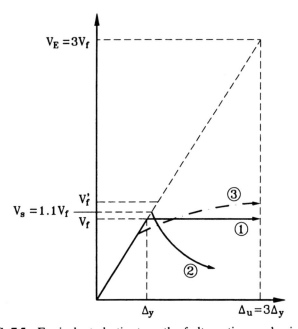

FIG. 7.5 Equivalent elastic strength of alternative mechanisms.

strength has been *decreased* by 63%, implying failure in an earthquake of greatly reduced intensity.

The example above is, of course, simplistic. A more realistic evaluation must consider the effects not only of excess yield strength, but the possibility of strain hardening and the fact that shear strength is not a constant value but depends on flexural ductility, as discussed in Section 5.3.4(*b*). Further, the assumption of elastoplastic flexural response is also unrealistic, implying strengths that are too high at low ductilities and too low (as a result of strain hardening) at high ductilities. A more realistic curve is shown in Fig. 7.5 by the dashed curve 3. The implication is that the equivalent lateral strength of $1.1V_f$ in the example above is excessively conservative, since the assessed flexural capacity will typically not be developed until displacement ductility factors of about $\mu_\Delta = 1.5$ to 2.

Consistent incorporation of the foregoing uncertainties in an assessment procedure would require sophisticated reliability analysis, involving a knowledge of the probability distributions of material strength, particularly the reinforcement yield strength. As discussed in the preceding section, uncertainty related to concrete compression strength can be minimized by in situ nondestructive testing, but information on the probability distribution of reinforcement yield strength is likely to be sketchy at best. Consequently, it is recommended that the deterministic plastic collapse analysis first be carried out for probable material properties, in accordance with Section 7.4.1. If the analysis indicates that flexural response is expected but with a small margin of strength over potential shear failure, the analysis should be rerun with amplified steel strength for longitudinal reinforcement in the plastic hinge region. In the absence of local information, it is recommended that an overstrength yield capacity of $f^\circ_{ya} = 1.15 f_{ya}$ be adopted. If the results from the revised analysis indicate a change in failure mode to shear failure, the reduced ultimate displacement should be adopted.

Note that if the initial analysis already indicated a shear failure was expected, this revised analysis would not be required. Although this example has been carried out with reference to member shear failure, it is clear that similar considerations apply to any brittle or undesirable failure mode.

7.4.3 Elastic Stiffness

The approach outlined above requires the estimation of ductility capacity by comparing ultimate and yield displacements. Thus, in addition to providing a best estimate of ultimate displacement, equal emphasis must be placed on providing a best estimate for yield displacements of ductile elements. Clearly, this involves consideration of the stiffness of cracked sections rather than the use of gross-member stiffnesses.

As discussed in detail in Chapter 4, at any early stage of the assessment procedure it should be determined which members are expected to remain essentially uncracked and hence which can be modeled by gross-section stiffness. Members with plastic hinges should be modeled by elastic stiffness properties appropriate for first yield, as given by Fig. 4.9, or from results from a moment–curvature analysis by

$$I_{\text{eff}} = \frac{M_{cm}}{\phi_{cm} E}$$ (7.10)

where M_{cm} and ϕ_{cm} are taken as moment and curvature at first yield. If the member or section considered is expected to develop moments significantly less than yield, the values for M_{cm} and ϕ_{cm} should represent the maximum expected response level. Appropriate techniques for modeling members with variation in reinforcement ratios along the length are presented in Section 4.4.2. Equation (7.15) presents a simplified expression for the yield curvature of columns.

7.4.4 Flexural Strength

To obtain the best possible estimate of flexural strength and deformation capacity, a moment–curvature analysis incorporating effects of confinement of the concrete core by transverse reinforcement (where appropriate) and strain hardening of longitudinal reinforcement should be used. Relevant details of the analysis are described in Section 5.3.1(d)(viii) and include definition of the flexural strength as the moment corresponding to an extreme compression fiber strain of $\varepsilon_c = 0.004$ or an extreme reinforcement bar tensile strain of $\varepsilon_s = 0.015$, whichever occurs first. The latter limit ($\varepsilon_s = 0.015$) is necessary for situations, such as may occur in cap beams, where the tension reinforcement ratio is very low and less than that of the compression reinforcement. Under such circumstances, tensile strains in the reinforcement may be excessive when $\varepsilon_c = 0.004$ is reached.

7.4.5 Flexural Strength of Column Sections
with Lap-Spliced Longitudinal Reinforcement

With bridges designed and constructed between 1940 and 1970, it is common to find lap splices at critical sections, such as the column base, as illustrated in Fig. 5.75(b). In new bridge design, these details are avoided since lap-splice failure is probable at low to moderate ductilities unless very large amounts of transverse reinforcement are provided.

In Section 5.5.4 the mechanism of lap-splice failure was discussed and the maximum force T_b that could be transferred through a lap splice given by Eq. (5.114). If the stress corresponding to T_b is less than the yield stress of the bar, the section will not be able to develop its flexural strength, and a reduced moment capacity M_s corresponding to a maximum tension stress of

$$f_s = \frac{T_b}{A_b} = \frac{f_t p l_s}{A_b}$$ (7.11)

is appropriate.

Even if Eq. (5.114) indicates that the yield strength of the lap-spliced reinforcement can be attained, it is probable that the flexural strength of the splice will degrade under cyclic loading to moderate ductility levels. This is because the tension strength of the concrete in the region of the lap splice is decreased by cyclic response, putting the concrete into compression. If the compression strains reach $\varepsilon_c = 0.002$, longitudinal microcracking develops, reducing the competence of the concrete for compression, and particularly tension, resistance. On reversal of moment direction, the previously compressed concrete in the lap splice is subjected to tension, and the reduced tension strength f_t results in reduced strength of the lap splice, from Eq. (7.11). As the cyclic response continues to higher levels of curvature ductility, the competence of the lap splice completely degrades, and no stress can be transferred across the splice.

The response of a circular column with a base lap splice under cyclic inelastic displacements is shown in Fig. 7.6. The column [C2] was able to sustain a peak moment close to the theoretical flexural strength calculated on the basis of measured material strengths, using the methods outlined in Section 5.3.1(d)(viii) but degraded under cyclic response rapidly, as is apparent from Fig. 7.6(b). It will be noted from this figure that the response at high ductility levels appears to be stabilizing at a residual moment capacity M_r. For columns with no effective transverse confinement reinforcement, this residual capacity is that moment which can be sustained by the axial compression force on the column, with no contribution from longitudinal reinforcement, using a reduced section size taken to be bounded by the inside of the longitudinal reinforcement cage, as indicated in Fig. 7.7 for rectangular and circular columns. For the rectangular section, the residual moment capacity based on axial force alone is thus

$$M_r = P \left(\frac{h' - a}{2} \right) \qquad (7.12a)$$

where $a = P/0.85 f'_{ca} b')$ and b' and h' are the dimensions of the residual core of the section. For a circular column, the corresponding strength is

$$M_r = P \left(\frac{D'}{2 - x} \right) \qquad (7.12b)$$

where x defines the centroid of the curved compression zone (see Fig. 7.7(b)).

If the lap splice is effectively confined by transverse reinforcement, the residual strength of the section will be increased accordingly. A circular column with properly detailed confinement reinforcement satisfying Eq. (5.118) will be capable of developing the full flexural strength, even when considerable strain hardening is taken into consideration. For lesser amounts of transverse reinforcement, the residual capacity may be linearly interpolated between

(a)

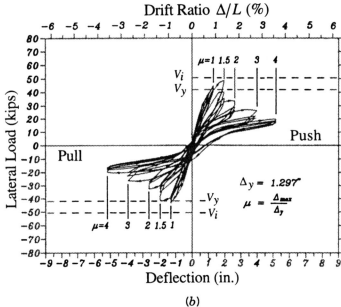

(b)

FIG. 7.6 Lap splice failure at the base of a circular column [C2]. (a) Fully degraded lap splice; (b) Lateral force–displacement response (1 kip = 4.45 kN).

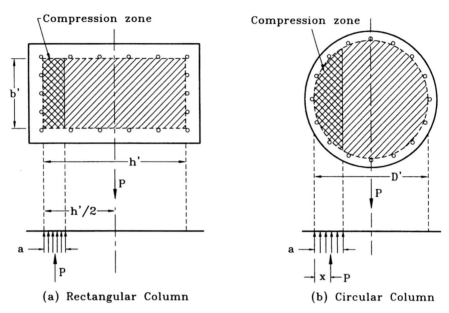

(a) Rectangular Column   (b) Circular Column

FIG. 7.7   Residual moment capacity of columns after lap-splice failure.

the value given by Eq. (7.12) and the overstrength capacity for the level of confinement provided by Eq. (5.118). However, it should be noted that the confinement reinforcement will be ineffective unless it is properly detailed, with welded hoops or continuous spirals welded at laps, or when ends of hoops or spirals are bent back into the core region by standard 135° hooks. It should also be noted that the strength enhancement provided by confinement cannot be considered to be additive to the initial strength provided by concrete tension capacity and governed by Eq. (7.11). The concrete dilation required to mobilize the confinement action is sufficiently large to ensure that all initial tension capacity of the concrete is dissipated completely.

7.4.6   Deformation Capacity of Plastic Hinges

(a)   Sections Without Lap-Spliced Reinforcement. Plastic rotation capacity may be found using the approach outlined in Section 5.3.2(*b*) and the equation for ultimate compression strain, Eq. (5.14). However, for assessment of existing bridges, a lower limit of $\varepsilon_{cu} = 0.005$ should be adopted. This is because examination of a large body of experimental data indicates that incipient spalling of cover concrete does not initiate before this strain when the plastic hinge forms against a supporting member, such as a footing or a cap beam [M12]. Tests on poorly confined columns indicate that theoretical ultimate displacements based on $\varepsilon_{cu} = 0.005$ provide a conservative estimate of actual response [C7,S9].

On the other hand, the beneficial effects of confinement implied by Eq. (5.14) should not be relied on unless the confinement is properly anchored by welding or by hooks into the core, since hoops lap spliced in the cover concrete will lose their integrity once the cover concrete spalls.

In general, critical elements will be potential plastic hinges in columns, and a moment–curvature analysis should be carried out to determine the appropriate inelastic response to incorporate in the plastic collapse analysis. The aim is to produce a simple bilinear moment–curvature approximation of response, as shown, for example, in Fig. 7.8. For poorly confined sections with $\varepsilon_{cu} = 0.005$, $M_u \approx M_n$ and an elastoplastic response is appropriate. Figure 7.8 includes the possibility of reduced flexural ductility capacity resulting from premature shear failure, to be discussed subsequently.

During the plastic collapse analysis, elastic stiffness of the member will be based on the effective stiffness $EI_{\text{eff}} = M_n/\phi_y$, which may also be obtained from the graphs of Fig. 4.9. After the plastic hinge has formed in the member, a reduced effective stiffness

$$EI_p = \frac{M_u - M_n}{\phi_u - \phi_y} \qquad (7.13)$$

is adopted. For elastic–perfectly plastic response this stiffness will be zero, and the behavior for incremental rotations is that of a perfect hinge. Once the hinge [whether perfect or with residual stiffness given by Eq. (7.13)] forms, the analysis needs to track the plastic rotation θ_p developing in the hinge and

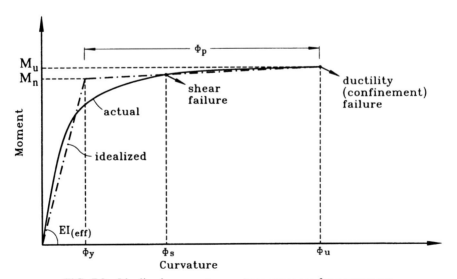

FIG. 7.8   Idealized moment–curvature response for assessment.

compare it with the plastic rotation capacity given by Eq. (5.40). It is thus of interest to examine the influence of the key parameters: longitudinal reinforcement ratio, axial load ratio, and section shape on the plastic curvature ϕ_p.

To this end, results of analyses of unconfined (i.e., $\varepsilon_{cu} = 0.005$) circular and rectangular sections are plotted in Fig. 7.9. The data in Fig. 7.9 are presented in dimensionless form. Thus the plastic curvature for any column

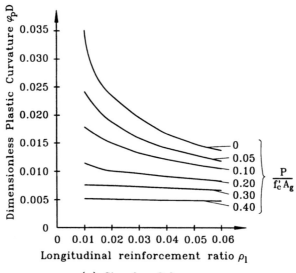

(a) Circular Columns

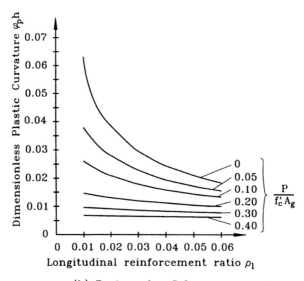

(b) Rectangular Columns

FIG. 7.9 Dimensionless plastic curvature capacity for unconfined column plastic hinges based on $\varepsilon_{cu} = 0.005$ [$f'_{ca} = 5$ ksi (34.5 MPa), $f_{ya} = 44$ ksi (30.3 MPa)].

may be found by dividing the dimensionless curvature by the section diameter D or depth h for circular or rectangular sections, respectively. Analyses were based on an assumed concrete compression strength of $f'_{ca} = 5.0$ ksi (34.5 MPa) and an assumed yield strength $f_{ya} = 44$ ksi (303 MPa). Other material strengths may be investigated with Fig. 7.9 using a modified longitudinal reinforcement ratio:

$$\rho'_l = \rho_l \frac{f_{ya}}{44} \frac{5}{f'_{ca}} \tag{7.14}$$

The adjustment represented by Eq. (7.14) is essentially exact for variations of f'_{ca} from the assumed value, but only approximate for variations of f_{ya} from the assumed value. Analyses with $f_{ya} = 66$ ksi (455 MPa) result in plastic curvatures lower than those of Fig. 7.9 by 10 to 15% for low axial load levels, with the difference decreasing at higher axial loads.

Results shown in Fig. 7.9 indicate a very strong dependency of plastic curvature on the axial load ratio $P/f'_{ca}A_g$, and a lesser but still significant influence of longitudinal reinforcement ratio, particularly associated with low axial load ratios. Figure 7.9 is sufficiently accurate to be used directly to estimate plastic rotation capacity of unconfined column hinges when multiplied by the plastic hinge length given by Eq. (5.39). It should be noted that despite the very large variation in plastic curvature indicated in Fig. 7.9, yield curvatures from the same, and further analyses for different yield strengths were rather insensitive to both axial load ratio and reinforcement ratio, with dimensionless yield curvatures of

$$\phi_y D = 2.45\varepsilon_y \pm 15\% \tag{7.15a}$$

$$\phi_y h = 2.14\varepsilon_y \pm 10\% \tag{7.15b}$$

being found for circular and rectangular sections, respectively, where ε_y is the longitudinal reinforcement yield strain.

(b) Sections with Lap-Spliced Reinforcement. As discussed in Section 7.4.5, the flexural strength of columns with base lap splices will degrade from the initial strength M_s to a residual strength M_r as cyclic inelastic response develops. For sections where splice failure initiates before the full nominal flexural strength is achieved, experimental results support an effective curvature ductility capacity of $\mu_\phi \approx 8$ for attainment of the residual capacity M_r [C7,S9]. Where lap-splice failure occurs after nominal capacity M_n is achieved, the residual capacity is developed at a higher curvature ductility.

Based on these observations, a model describing the moment–curvature response of different categories of plastic hinge is suggested in Fig. 7.10. Four different conditions are depicted. All have the same initial elastic (cracked-section) stiffness. Line 1 is a bilinear representation of response of a comparatively well-confined section. The nominal moment capacity M_n is reached at

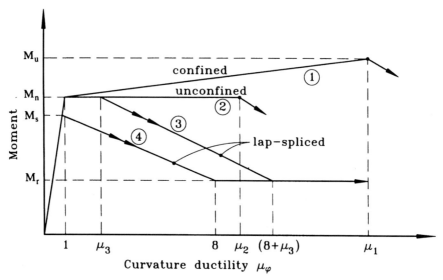

FIG. 7.10 Curvature ductility of different column sections.

a curvature ductility capacity of $\mu_\phi = 1$, and an enhanced ultimate moment capacity of M_u and ductility capacity μ_1, resulting from strain hardening and confinement effects, is found in accordance with Eqs. (5.14), (5.42), and (5.43). Line 2 represents a poorly confined column without lap splices in the plastic hinge region. Ultimate strength M_u = nominal strength M_n, and the maximum curvature ductility capacity of μ_2 corresponds to $\varepsilon_c = 0.005$ and is found from Eqs. (5.42) and (5.43) or from Fig. 7.9 and the average yield curvatures defined in the preceding section. When the ductility limit for lines 1 or 2 is reached, strength degrades rapidly due to crushing of core concrete and buckling of longitudinal reinforcement, and the damage-control limit state is reached.

Lines 3 and 4 represent different possibilities for lap-splice response. Line 4 represents degradation where Eq. (7.11) indicates that the nominal moment capacity M_n will not be reached. Strength starts degrading at less than $\mu_\phi = 1$ from a maximum strength M_s, based on the reduced effective reinforcement stress of Eq. (7.11), to the residual flexural capacity M_r calculated in accordance with Section 7.4.5 at a curvature ductility of $\mu_\phi = 8$. Line 3 represents degradation of a column with lap splices, where Eq. (7.11) indicates that the yield stress of the longitudinal reinforcement can be developed. The nominal moment capacity M_n is reached and degradation begins when a curvature ductility μ_3 corresponding to an extreme fiber compression strain of $\varepsilon_c = 0.002$ is reached. The capacity degrades to the residual strength M_r at a curvature ductility of $\mu_4 = 8 + \mu_3$.

Although experimental evidence indicates that the residual capacity M_r can be sustained while curvature ductilities increase beyond $\mu_\phi = 8$ (or 8 +

μ_3), the data base is small, and it is recommended that this be considered the ultimate curvature for assessment of the damage-control limit state. Larger ductilities could be accepted in assessing the survival limit state. Although the moment–curvature characteristics of lines 3 and 4 in Fig. 7.10 have negative stiffnesses during lap-splice failure, it is quite possible that the structure as a whole has positive stiffness as a consequence of increasing moment at other critical sections, and hence the onset of lap-splice failure does not necessarily imply that the damage-control limit state, defined in Section 7.2, has been achieved.

7.4.7 Flexural Strength and Ductility of Cap Beams

Although the assessment of flexural strength and ductility of cap beams follows the same general principles as for columns, there are specific conditions existing for cap beams which require special consideration. In particular, these relate to the influence of the ratio of positive to negative flexural reinforcement at critical sections and consequences of termination of longitudinal reinforcement within the span length. In assessing cap beam flexural strength and stiffness, full allowance for the contribution of superstructure deck and soffit slabs constructed monolithically with the cap beam should be made, in accordance with the recommendations of Section 4.4.2(c).

(a) Effects of Low Positive Moment Capacity. It is common to find critical cap beam sections at column faces with greatly reduced areas of bottom (positive moment) reinforcement compared to the top reinforcement. This results from the elastic design philosophy prevalent before the 1970s, with consequences that have been discussed in relation to Fig. 1.1. In such cases, use of an ultimate compression strain of $\varepsilon_{cu} = 0.005$ will result in very large plastic curvature, because of the very small neutral-axis depth corresponding to ultimate moment conditions. Steel tensile strains corresponding to $\varepsilon_{cu} = 0.005$ may be found to exceed the ultimate strain ε_{su}. In the assessment of existing bridge cap beams, with poor detailing of transverse reinforcement, it is recommended that the tensile strain in the reinforcement be limited to $\varepsilon_s = 0.04$. Although this is well below the ultimate strain, this limitation is necessary to minimize the possibility of buckling of the plastically strained reinforcement under moment reversal, which places the bottom steel in compression. Generally, the plastic curvature associated with $\varepsilon_s = 0.04$ will still be found to be very large, and positive-moment plastic hinges in cap beams, in consequence, have a high plastic rotation capacity. Note that this discussion implies that cap beam ductility, which is considered unacceptable for design of new bridges, should be permitted in assessment of existing bridges.

(b) Effects of Termination of Longitudinal Reinforcement. A typical layout of reinforcing steel in a cap beam designed in the 1950s or 1960s is shown

in Fig. 7.11(*a*). Features include premature termination of negative (top) reinforcement and comparatively little positive moment reinforcement anchored into the joint cores at the ends of the span. The first stage of the assessment analysis will involve computation of the distribution of flexural strength along the beam, shown by the solid lines marked "capacity" in Fig. 7.11(*b*).

Consider first the negative-moment capacity envelope. Over the central portion of the span, only nominal top reinforcement, to support transverse reinforcement, has been provided. The moment capacity over this region is thus equal to the cracking moment M_{cr}. Flexural strength provided by reinforcement increases gradually from zero at the end of the bar extending farthest into the span as bond stress increases the reinforcement stress. The rate

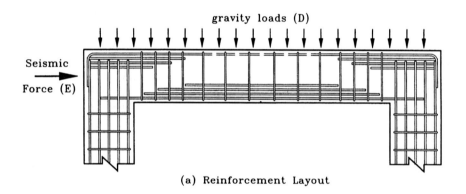

(a) Reinforcement Layout

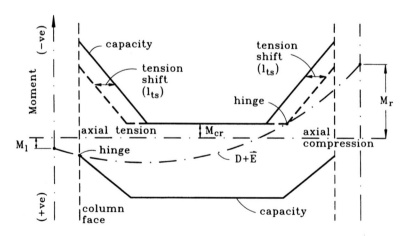

(b) Moment Capacity and Demand

FIG. 7.11 Seismic moments and capacity of a cap beam.

of increase of stress in the reinforcement may be represented approximately by
the relationship

$$f_s = \begin{cases} \dfrac{2x}{d_b} & \text{(ksi)} \leq f_{ya} & (7.16a) \\[2ex] \dfrac{13.8x}{d_b} & \text{(MPa)} \leq f_{ya} & (7.16b) \end{cases}$$

where x is the distance from the free end of the bar. Equation (7.16), which is
based on Eq. (5.108), is only approximate but appears reasonable for members
where the termination of longitudinal reinforcement is staggered, as shown
in Fig. 7.11(a). Where a large number of bars are terminated at the same
location, bond conditions will deteriorate and lower coefficients in Eq. (7.16)
than the value of 2 (13.8) will be appropriate.

Use of Eq. (7.16) enables the capacity envelope to be calculated, for comparison with the applied moments. Note that the flexural strength should include
consideration of axial forces in the cap beam [see Section 5.3.3(a)]. However,
when shear stresses in the critical region exceed that required to induce
diagonal cracking [as given by Eq. (5.63)], the effects of tension shift must
be considered [see Section 5.3.4(a)(ii)]. This is normally effected by lateral
displacement of the applied moment envelope, but for assessment purposes
it is more convenient to displace the *capacity* envelope, as suggested in Fig.
7.11(b) by the dashed lines of the negative-moment capacity envelope. The
amount l_{ts} of lateral shift depends on the proportion of shear carried by
transverse reinforcement and the angle of shear cracking ($\theta \geq 30°$) that is
needed to provide shear resistance. If adequate shear strength may be assumed
with a 45° mechanism, $l_{ts} = 0.5h_b$, where h_b is the cap beam depth, should be
satisfactory. If a flatter angle is needed to develop the required shear strength,
l_{ts} should be increased accordingly.

The capacity envelope for positive moments is found using considerations
similar to those for the negative moment. However, anchorage of bottom
reinforcement inside the column cage is likely to be more effective than for
top reinforcement development, and it is recommended that the coefficients
in Eqs. (7.16a) and (7.16b) be increased to 3 and 21, respectively, for that
portion of reinforcement anchored within the column cage. Shear stresses
within the potential plastic hinge region for positive moment are likely to be
low, and it is probable that no tension shift effects need be considered.

Having established the moment capacity envelopes, the lateral force can
be increased until plastic hinges form. In Fig. 7.11, the positive-moment hinge
forms at the column face, but the negative-moment hinge forms some distance
inside the span, in this case at the location of final termination of top reinforcement, a fairly typical occurrence. The cap beam moments M_l and M_r extrapolated back to the column centerlines can now be compared with moments

corresponding to formation of column hinges to determine whether hinges will form in the columns or cap beams.

7.4.8 Shear Strength

Shear strength of members may be assessed using the equations developed in Section 5.3.4(*b*), where a distinction was made between equations for assessment and design. The additive form of Eq. (5.74) is recommended for both beams and columns, using the best estimate values. For convenience these are summarized in the following:

$$V_n = V_c + V_s + V_p \tag{7.17}$$

where

$$V_c = k\sqrt{f_c'}A_e \tag{7.18}$$

$$V_s = \begin{cases} \dfrac{\pi}{2} \dfrac{A_h f_{yh} D' \cot \theta}{s} & \text{(circular columns)} & (7.19a) \\[3mm] \dfrac{A_v f_y D' \cot \theta}{s} & \text{(rectangular columns and beams)} & (7.19b) \end{cases}$$

$$V_p = P \tan \alpha \tag{7.20}$$

In Eq. (7.18) the effective shear area $A_e = 0.8A_{\text{gross}}$, and k may be expressed as a function of the curvature ductility factor μ_ϕ, as shown in Fig. 7.12. Outside the end region, taken to extend for a distance of $2D$ (or $2h$) from the critical section, the value of k applicable for $\mu_\phi = 1$ may be adopted. The definition for D' of Eq. (7.18) is given in Fig. 5.42, and the angle θ may be taken as $30°$ unless the corner-to-corner diagonal of the member subtends a larger angle with the member axis, in which case this larger angle should be used. Figure 5.44 defines the angle α of the axial force contribution to shear strength given by Eq. (7.20).

In Fig. 7.12(*b*) the concrete component for beams is expressed by ratio to the nominal strength given by the ASCE/ACI 426 approach of Eq. (5.63). Where the tension and compression reinforcement have equal areas, or the compression reinforcement has a larger area than the tension reinforcement, wide full-depth flexural cracks may develop within plastic hinge regions, and the shear strength of the concrete component degrades to zero at $\mu_\phi = 8$. However, if the tension reinforcement area exceeds the compression reinforcement area by a significant amount (say, $A_s > 1.2A_s'$), a concrete compression zone will always exist in the plastic hinge region and some residual capacity of the concrete contribution will exist. Consequently, in such cases, a residual capacity of $v_c = 0.25v_b$ may be adopted.

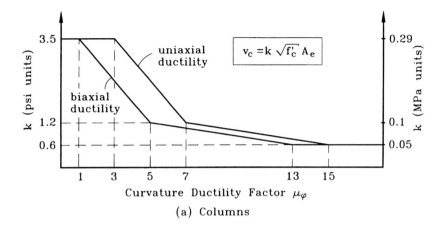

(a) Columns

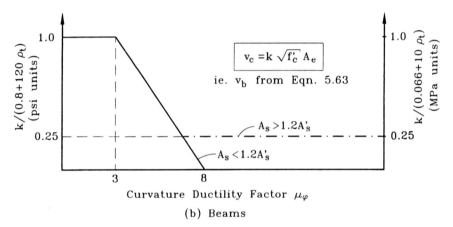

(b) Beams

FIG. 7.12   Relationship between strength of concrete component of shear resistance and curvature ductility for assessment purposes.

The form of Eq. (7.17) and Fig. 7.14 is suitable for incorporation in a plastic collapse analysis, since the total shear strength may be expressed in terms of curvature ductility, and compared with the flexural strength–ductility relationship. Figure 7.13 shows this comparison where the shear strength–curvature ductility relationship is compared with three flexural strength–ductility relationships such as might be applicable to three different levels of longitudinal reinforcement ratio. For convenience, the flexural moment–curvature relationships are expressed as equivalent shear force–curvature relationships.

Relationship 1, corresponding to the maximum longitudinal reinforcement ratio, develops a shear force at flexural strength of $V_1 > V_i$, the initial nominal shear strength, and hence a brittle shear failure results. Relationship 3, corresponding to the minimum longitudinal reinforcement ratio, develops a maxi-

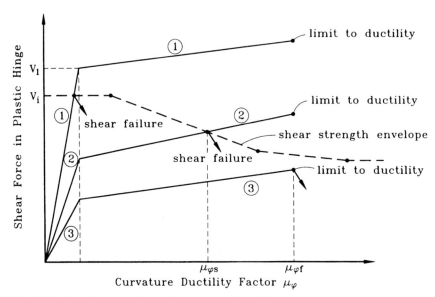

FIG. 7.13 Ductility at failure of columns with different longitudinal reinforcement ratios.

mum shear force corresponding to full ductile response that is inside the shear strength envelope, and hence shear failure does not occur, and the column fails when the flexural ductility capacity $\mu_{\phi f}$ is reached. Relationship 2, representing an intermediate reinforcement ratio, has a shear force corresponding to ideal flexural strength lower than the shear strength envelope, but as ductility develops the increasing shear force intersects the decreasing shear capacity. Limited ductile shear failure thus occurs at $\mu_{\phi s}$, less than the limit to flexural ductility. The ultimate curvature corresponding to this premature shear failure can be found by multiplying $\mu_{\phi s}$ by the yield curvature, and thus used to limit the plastic rotation of a hinge during plastic collapse analysis, as suggested in Fig. 7.8.

7.4.9 Joint Strength and Deformation Characteristics

The behavior and design of joints were examined in some detail in Section 5.4. The principles outlined in that section can also be used for assessment of existing joints. Initial joint cracking can be expected when the principal tension stress, calculated in accordance with Eq. (5.89), exceeds $3.5\sqrt{f_c'}$ psi ($0.29\sqrt{f_c'}$ MPa). If members framing into the joint remain elastic, higher tension stresses can be sustained as a more complete crack pattern develops in the joint. At nominal principal tension stress levels of about $5.0\sqrt{f_c'}$ psi ($0.42\sqrt{f_c'}$ MPa), a full diagonal pattern of cracks develops. Unless secondary mechanisms for joint force transfer in the cracked state can be identified, as discussed in Section 5.4.4, joint degradation is then anticipated.

If plastic hinging develops adjacent to the joint at $3.5\sqrt{f_c'} \leq p_t \leq 5\sqrt{f_c'}$ psi $(0.29\sqrt{f_c'} \leq p_t \leq 0.42\sqrt{f_c'}$ MPa), yield penetration into the joint increases with ductility, and joint failure eventually occurs. Currently, in the absence of more definitive data, it may be assumed that a principal tension stress of $5.0\sqrt{f_c'}$ psi $(0.42\sqrt{f_c'}$ MPa) may be sustained up to an adjacent member curvature ductility of $\mu_\phi = 3$, with tension strength degrading to $3.5\sqrt{f_c'}$ psi $(0.29\sqrt{f_c'}$ MPa) at $\mu_\phi = 7$, as shown in Fig. 7.14.

It will be recognized that Fig. 7.14 shows similarities to the model for degradation of plastic hinge shear strength shown in Fig. 7.12, on which it has been based. In particular, the dashed relationship for biaxial ductility is inferred from Fig. 7.12, by similarity, and is not based on test data. Nevertheless, it is felt advisable to adopt this conservative relationship in the absence of hard data.

If the principal tension stress remains below the strength envelope, the joint will not limit the ductility capacity of adjacent members. If, however, the principal tension stress reaches the strength envelope, joint strength degradation occurs, as may be seen from the force–deformation response of the knee-joint and tee-joint test units, shown in Figs. 5.58 and 5.65, respectively. In these figures it is noticeable that the strength of the knee joint, which failed under closing moment, degraded much more rapidly than did the strength of the tee joint. This is because there is no additional support to the outside of the knee joint once failure is initiated, whereas support from the adjacent cap beam, incorporating mechanisms described in Section 5.4.4(b), slows the rate of degradation. Similarly, for opening moments in knee joints, degradation after initial failure is more similar to that shown in Fig. 5.65 for tee joints than to the closing-moment failure of Fig. 5.58.

Joint degradation in itself is unlikely to cause catastrophic failure, in that gravity-load support should remain after joint failure, at least for single-level bridges. As noted in Chapter 1, joint shear failure of double-decker bridges

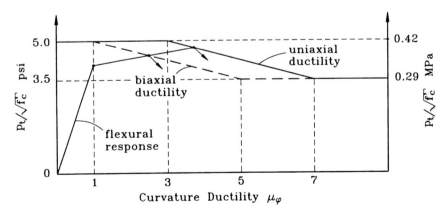

FIG. 7.14 Joint principal tension strength as a function of curvature ductility demand of connecting member.

can lead to catastrophic failure, as occurred with the I-880 Cypress viaduct in the Loma Prieta earthquake. Clearly, this could also occur with single-level bridges if joint failure results in a lateral collapse mechanism of greatly reduced strength. However, for assessment at the survival limit state, and even at the damage-control limit state, characterization of the rate of strength degradation is important to be able to define the full lateral force–displacement response. In consequence, the approximate relationship for principal tension strength versus joint shear strain presented in Fig. 7.15 may be used to determine the rate of degradation. Note that the shear strain in Fig. 7.15 may be considered to be a joint rotation, similar to the rotation occurring in a plastic hinge. Initial shear stiffnesses are characteristic of uncracked and cracked joint states for the situation up to and beyond tension stresses of $3.5\sqrt{f'_c}$ psi ($0.29\sqrt{f'_c}$ MPa), respectively.

Clearly, the information in Fig. 7.15 is not in a convenient form to use in plastic collapse analysis, since the principal tension stress is not likely to be calculated automatically during the analysis. However, it is relatively straightforward to relate this information back to the moment in the critical member framing into the joint. Equation (5.89) may be rearranged to yield the shear stress corresponding to a given tension stress p_t as

$$v_j = \sqrt{p_t^2 - p_t(f_v + f_h) + 2f_vf_h} \qquad (7.21)$$

where the sign convention is compression positive (i.e., p_t is negative). Equations (5.85) and (5.86) can then be used to determine the corresponding moment in the critical member framing into the joint. In this way the moment–

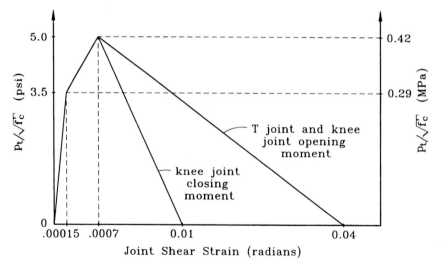

FIG. 7.15 Degradation of effective principal tension strength with joint shear strain (not to scale).

curvature relationship of the critical member framing into the joint may be adjusted to include the joint failure condition. This reduces the work load associated with the plastic collapse analysis and allows a preliminary calculation to be carried out to determine whether, for example, joint failure or column hinging will develop.

Equation (7.21) indicates that unless both horizontal and vertical direct stresses in the joint are compressive (i.e., $f_h > 0$ and $f_v > 0$), the joint shear capacity will eventually degrade to zero, since $p_t \to 0$ as the rotation increases. However, except in the case of closing-moment failure of a knee joint, it is probable that residual capacity will result from one of the alternative mechanisms described in Section 5.4.4(*b*). In this case the residual capacity should be calculated and used as the base for the moment–curvature response. In particular, the mechanism involving external shear reinforcement, shown in Figs. 5.60(*c*) and 5.66(*c*), should be investigated.

Example 7.1. To illustrate the procedure developed above, consider the column/cap beam tee joint shown in Fig. 7.16 under transverse seismic response. The column supports an axial load of 900 kips (4000 kN) and is reinforced with 28 No. 14 (D43) longitudinal bars that extend 42 in. (1067 mm) into the joint. Column height from a pinned base to the cap beam soffit is 50 ft (15.2 m),

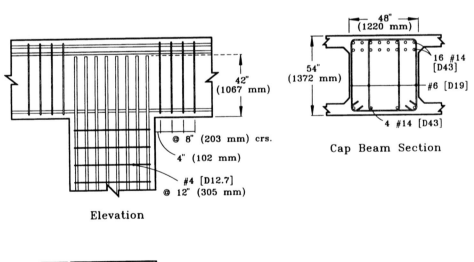

Elevation

Cap Beam Section

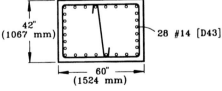

Column Section

FIG. 7.16 Details for joint assessment example.

and transverse reinforcement, shown in the column section of Fig. 7.16, is No. 4 at 12 in. (D12.7 at 305 mm).

Cap beam reinforcement at the column face consists of 16 No. 14 (D43) top bars in two layers and 4 No. 14 (D43) bottom bars, as shown in the cap beam section. Adjacent to the column, cap beam transverse reinforcement consists of sets of 4 No. 6 (D19) legs at 8-in. (203-mm) centers, with the first set 4 in. (102 mm) from the column face. Dead-load moments in the cap beam at the column centerline are 2000 kip-ft (2712 kNm); and column dead-load moment is negligible. It may be assumed that the cap beam reinforcement is carried sufficiently far into the cap beam before termination that the critical section is the column face for both positive and negative moments.

Specified material strengths for the bridge were f'_c = 3250 psi (22.4 MPa) and f_y = 40 ksi (275 MPa) for all reinforcement. Based on the recommendations of Section 7.4.1 and limited testing, assessment strengths are taken to be f'_{ca} = 5000 psi (34.5 MPa) f_{ya} = 44 ksi (303 MPa). Based on preliminary analyses, nominal moment capacities of the column and cap beam are found to be $M_{n,col}$ = 7030 kip-ft (9530 kNm), $M_{n,cap}$ = 1580 kip-ft (2140 kNm), −6000 kip-ft (8140 kNm). Taking dead-load moments into account, these indicate that the column should form a plastic hinge unless limited by joint strength.

The joint principal tension stress is first checked at the nominal flexural strength of the column. Using the simplified approach of Eq. (5.83), the horizontal joint shear force is

$$V_{jh} = \frac{M_{n,col}}{h_b} = \frac{7030}{4.5} = 1563 \text{ kips (6950 kN)}$$

and the joint shear stress is

$$v_j = \frac{V_{jh}}{h_c b_j} = \frac{1563}{60 \times 48} = 543 \text{ psi (3.74 MPa)}$$

In accordance with Section 5.4.3, the effective axial stress on the joint is found assuming a 45° spread from the column into the cap beam. Thus

$$f_v = \frac{900}{48(60 + 54)} = 164 \text{ psi (1.13 MPa)}$$

The axial force in the cap beam is approximately zero; therefore, f_h = 0. From Eq. (5.89) the principal tension stress is thus

$$p_t = \frac{164 + 0}{2} - \sqrt{\left(\frac{164 - 0}{2}\right)^2 + 543^2} = -467 \text{ psi} \quad \text{(tension)}$$

For f'_{ca} = 5000 psi this is equivalent to $p_t = 6.6\sqrt{f'_{ca}}$ psi ($0.55\sqrt{f'_{ca}}$ MPa), which exceeds the upper limit of $5\sqrt{f'_{ca}}$ psi $(0.42\sqrt{f'_{ca}}$ MPa) of Fig. 7.14. Joint shear failure will thus occur at less than column nominal capacity.

We thus determine the column moment at which joint failure occurs as follows: From Eq. (7.21), with $p_t = -5.0\sqrt{5000} = -353.5$ psi (2.44 MPa),

$$v_j = \sqrt{(-353.5)^2 - (-353.5)164} = 428 \text{ psi (2.95 MPa)}$$

The corresponding horizontal joint shear force [Eq. (5.87)] is

$$V_{jh} = 428 \times 60 \times 48 = 1232 \text{ kips (5480 kN)}$$

From Eq. (5.83),

$$M_{\text{col}} = 1232 \times 4.5 = 5540 \text{ kip-ft (7510 kNm)}$$

This is the maximum moment that is expected to develop in the column when joint failure is initiated. If no mechanisms are available to provide residual strength, the moment capacity would decrease to zero as the joint shear strain increased to 0.04 rad. However, it is evident that the beam stirrups adjacent to the column will enable the mechanism illustrated in Fig. 5.66(c) to develop, at least partially. Within a distance of $h_b/2$ from the column face, there are three sets of cap beam stirrups available to provide the tie force T_s in the mechanism involving external joint reinforcement.

Under positive cap beam moments, this reinforcement is not utilized because the dead load and seismic shears effectively cancel out, and the small residual shear can be carried by concrete shear-resisting mechanisms. The full tensile capacity of the three stirrup sets can thus be utilized for the external joint reinforcement mechanism. Thus

$$T_s = 3 \times 4 \times 0.44 \times 44 = 232 \text{ kips (1032 kN)}$$

This implies that the column reinforcement tension force that can be stabilized is $T = 2T_s = 464$ kips (2064 kN). Because no transverse reinforcement is provided within the joint and the column bars are terminated 12 in. (305 mm) below the deck surface, it is conservatively decided that no further tension force can be provided. The corresponding column moment capacity is then found to be

$$M_{r,\text{col}} \approx 464(54 - 4) + 900(30 - 4) = 46,600 \text{ kip-in.}$$
$$= 3880 \text{ kip-ft (5265 kNm)}$$

Note that for this mechanism to develop, an additional cap beam tension capacity of $T_{br} = 0.5T_s = 116$ kips (516 kN) is necessary, which requires $116/44 = 2.64$ in^2 (1700 mm^2). The cap beam bottom reinforcement area is

A_s = 4 × 2.25 = 9.0 in² (5800 mm²) and is hardly utilized for flexure since the seismic and dead-load moments effectively cancel ($M_b \approx -2000 + 3880/2$). The joint rotation characteristic, expressed in terms of applied column moment, is thus as shown in Fig. 7.17(*a*). In an assessment analysis it will often be convenient to include a special member between the joint centroid and bottom of the cap beam with these characteristics. However, a simpler and more direct approach is to modify the column moment–curvature relationship to include the joint deformation, as shown in Fig. 7.17(*b*).

The "virgin" moment–curvature relationship for assessment is shown in Fig. 7.17(*b*) by the solid line. This would normally be produced by a specific moment–curvature analysis but can also be generated with adequate accuracy using the procedure of Section 7.4.6(*a*) and Fig. 7.9. From Eq. (7.15), the yield curvature can be approximated as

$$\phi_y h = \frac{2.14 \times 44}{29,000} = 0.0033$$

Therefore,

$$\phi_y = \frac{0.0033}{60} = 55 \times 10^{-6} \text{ in}^{-1} \ (0.00217 \text{ m}^{-1})$$

Now the axial load ratio is $P/f'_{ca}A_y$ = 900/[5(60 × 42)] = 0.071, and the reinforcement ratio is ρ_l = 28 × 2.25/(60 × 42) = 0.025. Since the material strengths f'_{ca} and f_{ya} correspond to the values used to generate Fig. 7.9, the adjustment provided by Eq. (7.14) is not needed. By interpolation in Fig. 7.9(*b*), the dimensionless plastic curvature is found to be

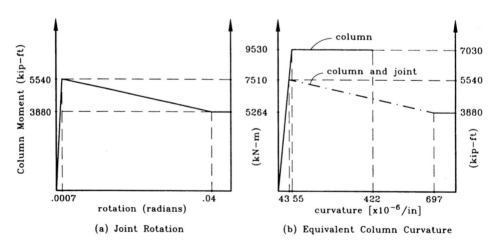

(a) Joint Rotation   (b) Equivalent Column Curvature

FIG. 7.17   Joint degradation for example of Fig. 7.16.

$$\varphi_p h = 0.022$$

Therefore,

$$\varphi_p = \frac{0.022}{60} = 367 \times 10^{-6} \text{ in}^{-1} \ (0.0144 \text{ m}^{-1})$$

and hence the ultimate curvature is found to be $\varphi_u = \varphi_y + \varphi_p = 422 \times 10^{-6}$ in^{-1} (0.0166 m^{-1}).

The virgin moment–curvature relationship can be adjusted by using the strength limit of Fig. 7.17(a), converting the rotation to equivalent curvature, by dividing by the plastic hinge length, and adding the result to the column elastic curvatures. From Eq. (5.39), the plastic hinge length is

$$L_p = 0.08(50 \times 12) + 0.15 \times 44 \times 1.69 = 59.2 \text{ in. } (1503 \text{ mm})$$

Ignoring the elastic joint rotation as insignificant compared with column elastic flexibility, the initial stiffness of the moment curvature curve is unchanged. Because of the reduced column moment at joint failure, the yield curvature is adjusted to

$$\varphi_y' = \frac{55 \times 5540}{7030} = 43.3 \times 10^{-6} \text{ in}^{-1} \ (0.0017 \text{ m}^{-1})$$

The curvature corresponding to a joint rotation of 0.04 is

$$\varphi_p = \frac{0.04}{59.2} = 675 \times 10^{-6} \text{ in}^{-1} \ (.0266 \text{ m}^{-1})$$

Taking into account the reduction in elastic curvature caused by moment degradation, the ultimate curvature is thus

$$\varphi_u = 675 + \frac{55 \times 3880}{9530} \times 10^{-6} \text{ in}^{-1} = 697 \times 10^{-6} \text{ in}^{-1} \ (0.0275 \text{ m}^{-1})$$

This relationship is shown in Fig. 7.17(b), which can now be used directly in an incremental collapse analysis.

It will be noted that the modified moment–curvature response incorporating joint failure has a higher equivalent ultimate curvature than the column. This is reasonable, since a ductility failure in the effectively unconfined region of the plastic hinge at the column top could result in rapid physical degradation and loss of gravity-load support, which is unlikely to occur in a joint principal tension failure, where confinement is provided by the cap beam.

7.4.10 Footing Strength and Deformation Characteristics

General seismic response of footings and pile caps was examined in Section 5.6 which also included specific design recommendations for flexure, shear, and joint design. Most of this information is also relevant to the assessment of footing performance of existing bridges. However, as with other areas of bridge performance, it is appropriate to be less conservative during assessment than during design. As a consequence, it is worthwhile spending more time in detailed assessment of expected footing performance to reduce unnecessary retrofit costs. In this context it is worth reiterating the point made in Section 5.6.2(a) that there are very few, if any, reports of footing failures except where caused by liquefaction or sliding of ground on sloping surfaces.

(a) Stability. A less conservative approach to assessing stability under the column plastic moment input (where appropriate) may be used than that advocated for design in Section 5.6.2(a). In particular, stability should first be checked under nominal input from the column using a strength reduction factor $\phi = 1.0$ in the stability Eqs. (5.120) to (5.124). If only marginal stability is available, or if the footing is clearly unstable, the consequences of rocking should be carefully considered. In many cases, rocking, either of spread footings or of pile-supported footings without tension connection between piles and footing, should be considered an acceptable form of response. The peak displacement can be estimated using the method suggested in Section 6.4.3. If the footing strength is sufficient to sustain the rocking mode without severe damage, damage to column and superstructure is likely to be minimal, since the rocking acts as a form of seismic isolation.

An example is afforded by the as-built footing shown in Fig. 5.86, which failed in joint shear. This footing was held down to prevent rocking. As is evident in Fig. 5.86(b), the joint failure is extreme. Prior to this phase of testing, the footing unit had been tested without uplift restraint, on a flexible foundation provided by rubber pads, simulating a spread footing on stiff soil. In this state of testing, the column sustained maximum moments of approximately $0.75M_i$ while rocking to drift angles of 0.05, with only minor cracking being sustained in the column and footing. Figure 7.18 shows conditions at maximum displacement and the very stable lateral force–displacement response of the test unit, which displayed the expected nonlinear elastic hysteresis characteristic. This behavior is clearly much more satisfactory than the retrofit solution of restraining uplift to enforce stability, shown in Fig. 5.86.

(b) Flexural Strength. Information contained in Section 5.6.2(c) and (d) should be reviewed when assessing footing flexural strength. However, for assessment, it is reasonable to increase the effective width b_{eff} given by Eq. (5.133). When both top and bottom reinforcement mats are provided, it is recommended that the effective width be increased to

(a)

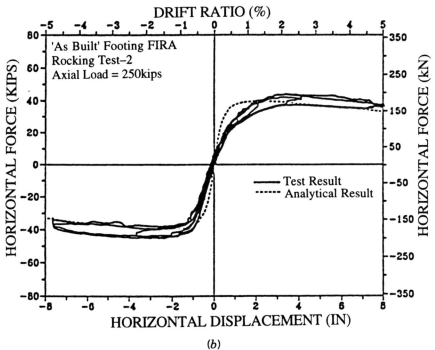

(b)

FIG. 7.18 Rocking response of unit shown in Fig. 5.86 prior to restraint of uplift.
(a) Condition at 5% drift; (b) Lateral force–displacement response.

$$b_{\text{eff}} = \begin{cases} D_c + 3d_f & \text{(circular columns)} & (7.22a) \\ B_c + 3d_f & \text{(rectangular columns)} & (7.22b) \end{cases}$$

When there is no top reinforcement, it would seem reasonable to assume the full width of the footing to be effective in flexure placing the bottom in tension unless the pad width is unusually wide, since moment reversal will not occur over the length of footing corresponding to the column depth.

Note that when diagonal struts between piles and the column compression zone are relied on to transfer shear force in the footing (see Fig. 5.85), extra reliance is placed on the anchorage of the footing bottom reinforcement because of tension shift effects caused by the inclined flexure-shear cracking. Although footing bars passing immediately above the piles will have improved anchorage provided by the compressive reaction in the piles, those midway between the piles will not be so confined, and anchorage failure could result, if the bars are developed without 90° hooks at the end of the footing, as is likely to be the case in older footings.

As with stability equations, it is appropriate to check footing flexural capacity against nominal input from the column, using a strength reduction factor of $\phi_f = 1$ to determine the most probable inelastic mode of deformation. In the event that footing flexural hinging is predicted, some limited ductility associated with this inelastic mode can be accepted. The maximum curvature ductility, and hence plastic rotation capacity of a footing plastic hinge, can be estimated based on a maximum compression strain of $\varepsilon_c = 0.005$ or a maximum reinforcement tensile strain of $\varepsilon_s = 0.04$. When the footing has no top reinforcement, there will be nothing acting to remove the residual displacement, and a tendency for displacements to increase under successive cycles of inelastic response must be expected, as shown in Fig. 7.19(b). As a consequence it is recommended that the effective maximum plastic displacement corresponding to footing plastic hinging be estimated to be 50% of that predicted from the footing hinge plastic rotation capacity θ_p. Thus, with respect to the nomenclature of Fig. 7.19(a), the dependable plastic displacement capacity should be limited to

$$\Delta_p = 0.5\theta_p H \tag{7.23}$$

in the anticipation that the actual displacement will increase to twice this level under multiple cycles of excitation.

When tension connection between the piles and footing is assessed to be competent, analyses may show that pile uplift is expected under the column plastic moment capacity. Again, this should not necessarily be considered to be undesirable, as the response will be similar to that of a rocking foundation with increased lateral strength (due to the pile tension capacity) and increased damping (due to the Coulomb friction associated with pile friction). The

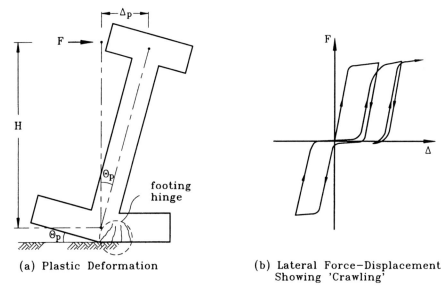

(a) Plastic Deformation

(b) Lateral Force−Displacement
Showing 'Crawling'

FIG. 7.19 Plastic displacement amplification due to successive inelastic cycles of a one-way footing hinge.

expected hysteretic characteristic, which can easily be incorporated within the rocking analysis approach described in Section 6.4.2, is shown in Fig. 7.20.

*(c) **Shear Strength.*** Shear strength of footings and pile caps is discussed in Section 5.6.3(*b*). For assessment purposes, a greater effective footing width

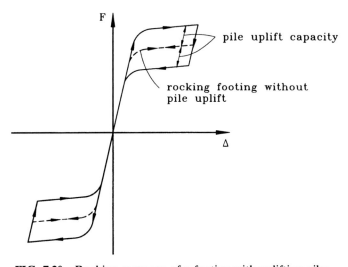

FIG. 7.20 Rocking response of a footing with uplifting piles.

may be assumed, in accordance with Eq. (7.22), as was the case for flexural strength. More reliance can also be placed on the formation of diagonal struts between the column compression zone and the footing compression reaction.

(d) Column–Footing Joint Strength. Procedures outlined in Section 7.4.9 for determining strength and deformation characteristics of column–cap beam connections can be applied directly to the assessment of column–footing joint behavior, with the following limitations. First, the effective width of the joint region for principal tension stress calculations should be given by Eq. (5.88). This will normally be considerably wider than with cap beams. Second, the joint shear force should be calculated in accordance with Eq. (5.135), as the simplified approach of Eq. (5.83) is likely to be excessively conservative when there is considerable holddown force R_t.

(e) Footing Failure as an Accepted Mechanism of Response. In all cases of footing assessment, a prime concern must be that regardless of what damage is expected in the foundation and which may be shown to be acceptable in terms of overall structural response, the footing must remain capable of supporting the column gravity load during and after an earthquake. In the event of severe footing damage associated with footing plastic hinging, the central region must then be capable of supporting the entire load transferred from the column.

When this gravity load support is assured, the designer may deliberately choose to allow considerable damage at the base of some or all columns and treat them as hinged in global plastic collapse analyses. This will be adopted when it is found that the cost of footing retrofit is excessive and the survival of the structure can be assured with pinned base (i.e., degraded footing) conditions. The cost of repairing the footings after a major earthquake may be little different from the initial retrofit cost. However, when performing the global assessment it is important to realize that the pinned-base condition cannot be relied upon when computing (say) column shear forces. The column shear force should be assessed on the basis of the probable footing strength *before* degradation. Failure to do this could lead to a serious underestimation of the column shear force and an incorrect decision not to retrofit the column for shear.

(f) Strength and Rotation Capacity of Column-Base Hinges. As discussed in Chapter 3, a common design detail for multicolumn bents in California involves a hinged base detail, to reduce the input forces to footing and foundation structure and hence reduce substructure costs. It is common in design to consider these details to be perfect hinges, with zero moment capacity. When determining shear forces for capacity protection or when assessing existing bridges, it is important, however, to realize that this detail, shown in Fig. 7.21, may transmit significant moment and is required to transmit significant shear.

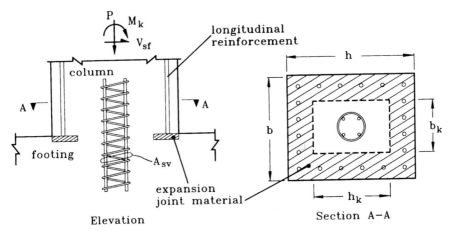

FIG. 7.21 Force transfer at a hinged column base.

The hinged base is typically provided by stopping all column longitudinal reinforcement at the column base and providing an area A_{sv} of vertical hinge reinforcement, placed at the center of the column. Expansion joint material is placed around the perimeter of the column to reduce the size of the key to dimensions h_k and b_k. However, since this must be designed conservatively for the maximum anticipated column axial force P, the key dimensions are still quite large.

The moment capacity of the hinge can readily be determined for the axial load P, central steel area A_{sv}, and key dimensions. Assuming that the hinge steel yields, this is conservatively given for the strong direction as

$$M_k = (P + A_{sv}f_{ya})\frac{h_k - a}{2} \tag{7.24}$$

where

$$a = \frac{P + A_{sv}f_{ya}}{0.85f'_{ca}b_k}$$

Since the key concrete is likely to be well confined by the less highly stressed adjacent concrete in the footing and column, a high value of f'_{ca} is appropriate, and it is suggested that $f'_{ca} = 2f'_c$ be adopted. This also allows for aging effects, noted in Section 7.4.1.

Plastic rotation capacity can be estimated by assuming a plastic hinge length allowing for strain penetration into both footing and column, which is thus given by the lower limit of Eq. (5.39). A very high ultimate compression strain is appropriate as a result of the excellent confinement of the key concrete. Assuming a permissible steel strain of (say) 5%, it will normally be found that high plastic rotations, of at least $\theta_p = 0.05$, will be attainable, assuming that the equivalent plastic hinge length is given by the lower limit of Eq. (5.39). Note that the limit to rotation is imposed by steel strains, not concrete strains, in this condition. Strain hardening should be considered in determining ultimate moment capacity. It should, however, be noted that at very high rotations, the expansion joint material may become fully compressed, resulting in significant compression force being transferred through it, with consequent increase in the moment capacity. This condition should be considered to exist when the expansion joint material is compressed to 25% of initial thickness at its edge.

Shear force is transferred through shear friction, and thus the base shear capacity is given by

$$V_{SF} = \mu(P + A_{sv}f_{ya}) \qquad (7.25)$$

Under seismic tension forces in the column it may be found that the key shear friction capacity is less than the applied shear, limiting bent shear capacity. In line with code recommendations [A5], a value of $\mu = 1.4$ may be adopted where the key concrete is poured monolithically with column and footing, causing a natural, rough crack to form at the shear key. If a roughened construction joint was provided at the key, a reduced value of $\mu = 1$ should be used.

(g) Pile Capacity. When assessing the lateral strength of foundations incorporating piles or foundation cylinders, lateral resistance provided by pile bending and by end bearing of the footing should be considered. Pile bending moment distributions should be established from a soil–structure interaction model, as discussed in Section 4.4.2(d). The first aim of the assessment will be to determine whether the pile system has sufficient strength to force plastic hinging into the column or to sustain the base shear force for a hinged base detail.

The possibility of shear failure can readily be assessed simply by distributing the total shear uniformly between the piles and checking the capacity of one pile. Although this approach will not give accurate results for any one pile, since the influence of axial force variations on shear strength will be considerable, the total resistance of the pile group will be reasonably accurate. Since limited ductility in shear is possible, the shear redistribution implied by the averaging process is acceptable. Further, the stiffness of the piles with reduced axial force, and hence reduced shear strength, will be less than those with increased axial force, so the redistribution implied by the averaging process will be a reality.

Pile moments must be calculated with due respect to the fixity of the footing–pile connection detail, which may vary in practice from pinned, through partially fixed, to fully fixed. An attempt should be made to model this detail as closely as possible. Moments developed in-ground will depend on the soil stiffness and the shear force to be transferred. A simple initial check may be provided using the average shear per pile, as above. If this indicates that pile hinging is likely, a more detailed analysis using distributed winkler springs to model the foundation (as shown in Fig. 4.28) should be adopted.

Although ductility of the potential plastic hinge at the footing base may be limited by poor confinement details (in the case of a reinforced or prestressed pile), the in-ground hinge will have a large plastic rotation capacity because of its long hinge length (see Fig. 5.30) and the additional confinement provided to the compression face from soil pressure. Even if unconfined by transverse reinforcement, a comparatively high ultimate compression strain will be appropriate (say, $\varepsilon_{cu} = 0.01$).

Steel shell piles filled with concrete have excellent ductility capacity. In the case of a steel shell confining a reinforced concrete core, the ductility at the footing interface will be extremely large and can be estimated by considering the shell as equivalent confinement reinforcement. The ductility of the in-ground plastic hinge is likely to be limited by "elephant's foot" buckling of the steel casing, which occurs at a curvature ductility factor of approximately $\mu_\phi = 10$, regardless of shell thickness [P17]. It is very important in the plastic collapse analysis that the flexibility of the foundation be incorporated in the analysis, either by provision of appropriate spring stiffnesses or, preferably, by proper soil–structure modeling as discussed in Section 4.4.2(d).

7.4.11 Superstructure Strength and Deformation Characteristics

Design of superstructures for longitudinal seismic effects was considered in Section 5.7. When monolithic superstructure–substructure connections exist, it will often be found that superstructure flexural capacity will be inadequate to force plastic hinging into columns, particularly when superstructure movement joints are close to the column considered, as illustrated in Fig. 5.89.

As with cap beam assessment, some liberalization, including greater effective contributory width of superstructure compared with new design, and limited ductility of the superstructure should be permitted. Consequently, it is recommended that the effective width of superstructure contributing to seismic resistance on each side of a column be increased by 50% above the guidelines of Section 5.7. Also, when it is determined that even with this increased effective width, superstructure hinging is predicted, limited ductility, corresponding to the serviceability limit state (i.e., $\varepsilon_c \leq 0.004$, $\varepsilon_s \leq 0.015$), should be permitted regardless of the limit state considered.

Since limited ductility is permitted, moment redistribution capacity across the superstructure is thus implied. The check for whether column hinging will develop is thus made on the sum of superstructure moments at the superstructure–column joint center. Thus, referring again to Fig. 5.89, if it can be shown that in terms of nominal moment capacities, $M_{lc} + M_{rc} > M_{cc}$, plastic hinging will develop in the column even if analysis predicts that (say) superstructure positive moment hinging will precede this. The amount of plastic rotation required of the superstructure hinge in such cases will be small.

Care is needed in the assessment of joint B, Fig. 5.89, adjacent to the hinge. Equilibrium consideration along the span containing the hinge will generally indicate that the maximum moment M_{lb} that can be developed will be governed by the nominal moment capacity M_{ra} and will be considerably less than the nominal moment capacity for M_{lb}. This will make it more difficult to force plastic hinging into the column.

7.4.12 Abutment Assessment

Seismic response of abutments has been considered in some detail in Section 4.4.2(e). In overall bridge assessment, perhaps the most critical item to be quantified will be the strength and stiffness of the abutment–superstructure connection. This will have a critical influence on forces and deformations developed in internal bents.

The assessment of the strength of abutment elements loaded by soil–structure interaction effects will inevitably be less precise than possible with simpler parts of the bridge. Areas that will frequently be found deficient are the connection between the wing wall and the back wall under opening moment, the strength of shear keys locating the superstructure on the abutment seat against transverse response, the back wall capacity against passive pressure generated when the superstructure moves into the back wall, and the capacity of abutment piles when the superstructure moves away from the back wall or soil slumping of abutment back wall fill occurs. Many of these aspects have been discussed in Chapters 1 and 3 and will not be discussed further. It should be noted that although severe abutment damage is rather common in earthquakes, it rarely results in catastrophic failure because of the relatively small displacements involved. Repair may also be relatively inexpensive. Temporary access for emergency services vehicles can typically be provided within a matter of days, if not hours, by placing new fill material over slumped approaches or settlement slabs. Consequently, a greater degree of damage is often accepted for abutments than for other elements of bridges assessed to the damage-control limit state.

7.5 MISCELLANEOUS DETAILS

In the assessment of expected seismic performance of any bridge, there will always be a number of details requiring special attention and whose behavior may be more difficult to predict with any certainty than the more fundamental components of the bridge. These will require special and careful study, as their performance may critically affect the modeling assumptions that are appropriate for determining actions developed in the rest of the bridge, even if the failure of these components or details might not in itself be critical. A few of these details are discussed below. However, the list is incomplete, and these critical details should be identified and evaluated carefully at the earliest stage of the assessment process.

7.5.1 Shear Keys

Shear keys between superstructure and abutment and across superstructure movement joints critically affect the seismic response of bridges, in particular the distribution of seismic inertia forces and displacements between bents and abutments. The lateral strength of shear keys was considered in Section 5.8.2. For assessment purposes, the shear strength reduction factor ϕ_s should be omitted from Eq. (5.140). The shear strength of the keys can thus be assessed with reasonable confidence. If the strength of the key is developed and the clamping reinforcement that passes across the potential sliding surface yields, strength and stiffness degrade rapidly. The residual plastic strains developed in the reinforcement imply dilation of the crack surface, which at low displacements means that the frictional restraint is lost. Although the lateral resistance may pick up again as displacements approach the previous maximum value, the sliding that will have occurred on the crack surface reduces the effectiveness of the aggregate interlock and significant strength degradation will occur.

As a consequence of this rapid degradation in performance, it is common initially to run analyses based on the assumption that the shear keys are competent. If as a result of these analyses it is found that forces larger than the strength are developed in the shear keys, a second analysis is run, with the overloaded shear keys assumed to have failed and removed from the analysis. However, it should be pointed out that elastic analyses of multiframe bridge segments, in which inelastic response of some or all of the bents is expected, may overestimate the forces developed in the shear keys. A method for estimate shear key forces is presented in Section 5.8.2. This method is also suitable for assessment purposes, although full overstrength of bents need not be assumed for a capacity-based assessment analysis.

7.5.2 Strut Action in Wide Bridges

The idealization of bridges containing internal movement joints by linear centerline models is analytically convenient but not always appropriate, since the simulation of the abutment and center movement joints as perfect hinges is simplistic. As shown in Fig. 7.22, with wide bridges, lateral resistance can be provided by compression struts under lateral response. When the lateral displacement is sufficient to close movement joints at the corners, struts C form between the closed corners and the inertia forces F_L and F_R. As idealized in Fig. 7.22, these struts are straight, but because of the distributed nature of the inertia forces F_L and F_R, they will actually be curved compression arches. Analysis to include this additional resistance will require simulation of the

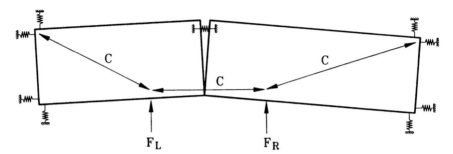

FIG. 7.22 Lateral resistance from strut action in wide bridges.

deck as a two-dimensional element or grid or by including the appropriate rotational stiffnesses in hinges of the centerline model.

7.5.3 Bearings

Bearings and their connections to superstructures and supports will often be found to have inadequate strength to transmit the expected seismic forces. At abutments, rocker bearings are typically restrained laterally by keeper plates bolted to superstructure and abutment. This provides a stiff but weak lateral restraint that initially attracts significant force but fails early in the seismic response by shearing of the keeper bolts, as shown in Fig. 7.23. The lateral resistance of this form of support can be estimated as

$$V_A = \mu_s P_A + nA_b f_s \tag{7.26}$$

where P_A is the axial load supported by the rockers, μ_s the steel-to-steel

FIG. 7.23 Abutment rocker keeper plate bolt fracture.

coefficient of friction ($\mu_s \approx 0.25$), A_b the keeper plate bolt area with shearing strength f_s, and n the number of effective bolts resisting lateral displacement. The value of n will depend on the connection detail, particularly on whether the keeper plates completely surround the rocker, making the bolts on both sides of the rocker fully effective, or if they are in two halves, making only the bolts on one side effective. Since a gap is normally provided between the keeper plates and rockers, it is probable that not all bolts of multiple bearings will engage simultaneously, and hence n should be reduced from the full potential number of effective bolts. Judgment and measurement of the variation of gap sizes will be required.

It is also common on older bridges to find steel-to-steel sliders or pin supports to have rusted solid, providing support conditions considerably different from those intended by the designer. Analyses should be based on current assessment of the support conditions rather than on the intentions of the original designer.

7.5.4 Restrainer and Movement Joint Details

Analysis and design of restrainers have been considered in some detail in Sections 5.8.1 and 5.8.3. Assessment for seismic performance of frames should generally be on a frame-by-frame basis, but special consideration will be required when frames of considerably different stiffness are connected across movement joints. Also, to ensure satisfactory performance of stand-alone frames, it is necessary to determine that unseating will not occur. Consequently, analyses must be carried out to consider the interaction between frames. As discussed in Section 5.8.3, only inelastic time-history analyses can be expected to yield realistic estimates of relative displacements across movement joints and hence of restrainer forces. The analyses described in that section indicate that unless restrainer stiffness are at least of similar value to that of the more flexible of the frames connected by the restrainers, the restrainers will have little influence on the magnitude of movement joint opening. A reasonable estimate of movement joint opening, Δ_{lr} under longitudinal response, can be obtained from the difference between the absolute value of maximum displacement estimated from the response of the two frames acting independently as stand-alone units. That is,

$$\Delta_{lr} = |\Delta_{l1}| - |\Delta_{l2}| \tag{7.27}$$

where $|\Delta_{l1}|$ and $|\Delta_{l2}|$ are the absolute values of the more flexible and less flexible frames, respectively. Whether the full magnitude of longitudinal displacement represented by Eq. (7.27) can develop may depend on the competence of the abutment details, which may restrain the movement joint opening to smaller values.

Movement joints opening under transverse response should be estimated using a model that represents the constraints to transverse response implied

by the strut or arch action of Fig. 7.22. Further information relevant to estimation of movement joint opening is given in Section 5.8.1.

Restrainer details should be checked to ensure that they are capable of
sustaining the anticipated displacements without failure. Early restrainer designs [S7] were often based more on intuition than on relevant analysis, and
as discussed in Section 5.8.1, the factors governing response are complex.
Connection and anchorage details were often such that brittle failure can be
expected when, or before, yield strength is reached [Fig. 7.24(a)].

Potential punching shear failures of movement joint diaphragms under the
anchorage forces imposed by the restrainers should also be checked. Several
examples of this type of failure occurred in the 1989 Loma Prieta earthquake
and the 1994 Northridge earthquake [Fig. 7.24(b)]. However, it is not clear
that these failures would have occurred if other bridge components had not
failed first, placing unusual demands on the restrainers.

7.5.5 Steel Superstructure Assessment

Principles for seismic assessment of steel superstructures should be the same
as for concrete bridges. It will be expected that the superstructure should
remain elastic at inertia force levels corresponding to either elastic response
of the bridge, or ductile response of the substructure. Initial analyses should
seek to provide a simplified representation of the stiffness of the superstructure
so that substructure response may reasonably be estimated, since this will
normally be more critical than the superstructure. Section 4.4.1 provides guid-

(a)   (b)

FIG. 7.24 Failures of restraint systems across movement joints. (a) Restrainer fracture; (b) Diaphragm punching failure.

ance in this matter. The superstructure should then be checked under the
inertia force levels corresponding to a high estimate of probable substructure strength.

Particular importance needs to be placed on the force transfer from the
superstructure into the substructure at the bents. It is common to find the
lateral bracing systems at the bents to be inadequate to transfer full inertia
forces without buckling. Similarly, the lateral bracing along the span collecting
lateral interia forces and transferring these back to the supports should be
assessed carefully, taking into account additional bracing provided by the
concrete deck surface, if present.

Some of the potential problem areas are identified in Fig. 7.25 for a steel
girder bridge. If the concrete deck has transverse joints at regular intervals
along the length, as will often be the case, the bracing system, shown in the
plan in Fig. 7.25(a), must be checked for its ability to carry the forces back
to the supports. Bracing in the end bays will be critical.

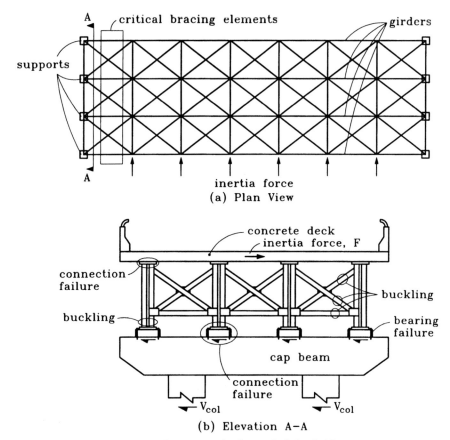

(a) Plan View

(b) Elevation A–A

FIG. 7.25 Force transfer in steel girder bridges.

In the elevation of Fig. 7.25(b), the path from the inertia force F down to the columns must be fully checked. If the concrete deck is continuous rather than with shrinkage joints as above, force transfer will be primarily within the plan of the end bracing. The system shown has potential problems with connection shear failure between deck and girder, with buckling of the bracing system elements, with lateral buckling of the girders below the bottom chord of the bracing system [see, e.g., Fig. 1.29], and with displacement capacity of bearings or connection strength of the bearings to the girders or cap beam.

When assessing steel truss or arch superstructures, the seismic force investigated should again correspond to the full elastic response or the capacity of the substructure whichever is smaller. Tests on riveted connections [A8] indicate that the stiffness and strength may be substantially less than that appropriate for welded connections.

8

RETROFIT DESIGN

8.1 INTRODUCTION: RETROFIT PHILOSOPHY

There are two fundamental decisions to be made at the start of a bridge seismic retrofit. The first, based on the results of the detailed seismic assessment carried out in accordance with the principles of Chapter 7, is whether the calculated risk of damage or failure warrants retrofit. This decision will generally be relative, depending on comparison with other bridges being considered for seismic retrofit and on the financial resources available. Following a decision in favor of retrofitting, the second decision will be the level to which the bridge should be retrofitted. Ideally, this should be based on a cost–benefit analysis. It may be possible to reduce seismic risk considerably by, say, extending seat widths at movement joints. The additional reduction in risk provided by a full bridge retrofit involving column, footing, and perhaps superstructure retrofit may be small and achieved only at considerable expense.

There is, however, a case to be made for full retrofit to current standards of seismic performance expected of new bridge designs. First, cost–benefit analysis is a rather imperfect science, and the possibility of considerable error in assessment of the return period of the retrofit-level earthquake should be considered. Second, although cost–benefit analysis may be perceived to be relatively sophisticated and useful as a generalized statistical tool when applied to a large inventory of bridges, the general public does not share the same level of sophistication when disasters are concerned. The failure in an earthquake of a bridge that has been seismically upgraded to a moderate risk level will be seen by the public, and the media in particular, as unacceptable when it is established that the bridge could have been upgraded to a level where survival is assured. It can also be argued that current seismic design provisions for new bridges are not based on cost–benefit analyses, which might in themselves justify much lower levels of seismic protection. The public expects that it should be protected against large-scale loss associated with infrequent catastrophes even if the amortized cost of such losses is small in comparison to that corresponding to more frequent events such as road accidents.

These considerations imply that the probability of exceedance of the survival limit state should be the same for a retrofitted and a new bridge. There may, however, be a case for accepting a higher probability of achieving the damage-control limit state for retrofitted structures than for new structures, on the basis of a cost–benefit analysis. As a consequence, there has been a change in philosophy of retrofit in recent years, particularly in California. Following the 1971 San Fernando earthquake, the California Department of Transportation embarked on a retrofit program that involved placing restrainers across movement joints of a large number of bridges to reduce the potential for unseating [G1]. Similar activities took place in Japan [K1]. Although not specifically based on a cost–benefit analysis, this was clearly an attempt to get the most effective reduction in risk for the minimum cost, while accepting that seismic risk had not been reduced to that considered acceptable for new bridges. Following the failures of restrained bridges in the 1987 Whittier Narrows [P1], the 1989 Loma Prieta [E1], and the 1994 Northridge earthquakes [E2], Caltrans has required that bridge retrofits provide survival limit-state protection at seismic intensities appropriate for new bridges.

If retrofit designs are based on rational analysis rather than general guidance rules, it is comparatively straightforward to design for a specified level of seismic risk, whether it corresponds to new bridge standards or a lesser level. The procedures adopted should be such that the risk after retrofit should be capable of being assessed using the procedures developed in Chapter 7. As a consequence, the design information for retrofitting included in this chapter is deliberately formulated to be compatible with the plastic collapse mechanism analyses espoused in Chapter 7.

8.2 RETROFIT OF CONCRETE COLUMNS

Concrete columns are commonly deficient in flexural ductility, shear strength, and flexural strength when affected by lap splices in critical regions or by premature termination of longitudinal reinforcement. Flexural strength of columns without lap splices is generally adequate, as a result of conservative flexural design equations adopted prior to the 1970s, as discussed in Section 1.2.3.

A number of column retrofit techniques have been developed and tested, with a rather smaller number implemented in actual retrofit design at the time of writing this book. Column retrofit techniques include steel jacketing, active confinement by wire prestressing, use of composite materials jackets involving fiberglass, carbon fiber, or other fibers in an epoxy matrix, and jacketing with reinforced concrete. Of these, the most common retrofit technique implemented to date has been steel jacketing, with a small amount of retrofit involving reinforced concrete jackets or composite materials jackets. These three approaches are discussed in the following.

8.2.1 Column Retrofit Techniques

(a) Steel Jacketing. The procedure was originally developed for circular columns [C2, C7]. Two half shells of steel plate rolled to a radius of 0.5 to 1.0 in. (12.5 to 25 mm) larger than the column radius are positioned over the area to be retrofitted and are site-welded up the vertical seams to provide a continuous tube with a small annular gap around the column. This gap is grouted with a pure cement grout, after flushing with water. Typically, a space of about 2 in. (50 mm) is provided between the jacket and any supporting member (footing or cap beam), to avoid the possibility of the jacket acting as compression reinforcement by bearing against the supporting member at large drift angles. This is to avoid excessive flexural strength enhancement of the plastic hinge region, which could result in increases in moments and shears in footings and cap beams under seismic response.

The jacket is effective in passive confinement. That is, lateral confining stress is induced in the concrete by flexible restraint as the concrete attempts to expand laterally in the compression zone as a function of high axial compression strains, or in the tension zone as a function of dilation of lap splices under incipient splice failure, as described in Section 5.5.4. The level of confinement induced depends on the hoop strength and stiffness of the steel jacket.

A similar action occurs in resisting the lateral column dilation associated with development of diagonal shear cracks. In both cases—confinement of flexural hinges or potential shear failures—the jacket can be considered equivalent to continuous hoop reinforcement. Fig. 8.1(a) shows details of the jacket.

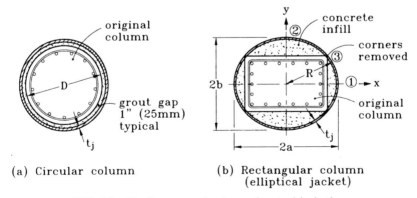

(a) Circular column   (b) Rectangular column (elliptical jacket)

FIG. 8.1 Confinement of columns by steel jacketing.

For rectangular columns, the recommended practice is to use an elliptical jacket that provides a continuous confining action similar to that for a circular column but with a confining stress potential that varies around the circumference because of the continuously changing curvature of the jacket. The space between the jacket and column is filled with normal unmodified concrete rather than grout.

Rectangular steel jackets on rectangular columns are not recommended. Although these can be expected to be fully effective for shear strength enhancement, a column retrofitted for shear will normally also require enhanced flexural ductility, which will not be provided by the steel jacket except at the corners, since there will be little restraint of lateral dilation of the core provided by bending of the jacket. Thus confinement for enhanced compression strain capacity or improved lap-splice performance is unlikely to be effective. Tests on various designs of stiffened rectangular jackets have shown them to be significantly less effective than elliptical jackets [S9].

Thin rectangular steel jackets epoxy-bonded to concrete columns over regions of premature termination of reinforcement have been used in Japan [K1] to locally augment the flexural and shear capacity of the columns, ensuring that inelastic action occurs only at the intended plastic hinge at the column base. Steel jacketing has been widely used in Calfornia as the major retrofit technique for bridge columns, with several hundred bridges thus retrofitted by 1994. During the 1994 Northridge earthquake, some 50 bridges with steel-jacketed columns were subjected to peak ground acceleration of 0.3g or higher. None of these bridges suffered damage to columns requiring subsequent remedial work.

(b) Concrete Jacketing. Addition of a relatively thick layer of reinforced concrete in the form of a jacket around the columns can be used to enhance flexural strength, ductility, and shear strength of columns. Although this technique has been used more frequently for building columns than for bridge columns, it has been used in some Japanese bridge retrofits. By doweling the longitudinal reinforcement of the jacket into the footing with sufficient anchorage length to develop the reinforcement strength, the column flexural strength can be enhanced, although this must generally be accompanied by footing retrofit measures to enhance footing flexural and shear strength sufficiently to ensure that plastic hinging develops in the column.

Enhanced confinement of circular columns is relatively easy to achieve with a concrete jacket, by use of close-spaced hoops or a spiral of small pitch, as shown in Fig. 8.2(a). However, unless the concrete jacket is made of elliptical or circular shape, it is difficult to achieve effective confinement by a rectangular concrete jacket. Longitudinal bars in the midregion of each face will be susceptible to buckling, and only the concrete near the corners will be effectively confined. The situation can be improved by chipping the corners of the existing concrete column back to the corner bars and using hoops for the concrete jacket which include 45° corner bends, alternately with full peripheral hoops, as shown in Fig. 8.2(b). The use of midside links in holes drilled through the

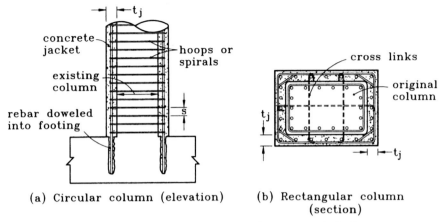

(a) Circular column (elevation)   (b) Rectangular column
                                            (section)

FIG. 8.2 Confinement of columns by concrete jacketing.

core of the existing column, as suggested by the dashed lines in Fig. 8.2(*b*), is likely to be expensive, and the links will be difficult to place because of the need to bend the hook at one end in place after threading through the hole.

(c) Composite-Materials Jackets. There has been a considerable body of research performed to establish the effectiveness of column retrofit using jackets of composite materials such as fiberglass, carbon fiber, and kevlar, generally bonded together and to the column with epoxy. Figure 8.3 shows

(a)   (b)

FIG. 8.3 Retrofitting with composite-materials jackets. (*a*) High-strength fiberglass and epoxy: hand-layup; (*b*) Carbon fiber and epoxy: machine winding.

two forms of application, one involving the hand layup of a jacket made from layers of epoxy-impregnated fiberglass fabric, and the second, machine winding of a column with carbon fiber tows, also impregnated with epoxy. Because of the greater strength and stiffness of the carbon fiber, lesser thicknesses are needed than for the less expensive but more flexible and weaker fiberglass jacket. Both techniques have proven effective in laboratory tests, as discussed subsequently. Fiberglass jacketing has been implemented fairly extensively for building columns but has seen only trial applications for bridges so far. Carbon-fiber retrofitting is still in the developmental stage.

In both cases, the techniques are most suitable for circular columns, since obtaining full confinement for rectangular columns requires placing concrete bolsters or other means of section shape modification to enable the jacket to be placed over a continuously curved surface. However, it has been found that reasonable enhancement of ductility has been achieved for rectangular carbon fiber or fiberglass–epoxy jackets on rectangular columns [P7,P18], as discussed subsequently.

8.2.2   Column Retrofit Design Criteria

(a)   Confinement for Flexural Ductility Enhancement. With poorly confined columns that are expected to sustain large inelastic rotations in plastic hinges, a prime concern will be retrofit design to enhance the ductility capacity. The procedure adopted follows closely the approach developed in Section 5.3.2(*c*), which relates the volumetric ratio of confinement to the required plastic rotation θ_p. For convenience, these steps are summarized below.

1. On the basis of the plastic collapse analysis, the required plastic rotation θ_p of the plastic hinge being considered is established.
2. The plastic curvature is found from the expression

$$\phi_p = \frac{\theta_p}{L_p} \tag{8.1}$$

where the plastic hinge length is given by

$$L_p = \begin{cases} g + 0.3f_y d_{bl} & (f_y \text{ in ksi}) & (8.2a) \\ g + 0.044f_y d_{bl} & (f_y \text{ in MPa}) & (8.2b) \end{cases}$$

Equation (8.2) corresponds to the lower limit of Eq. (5.39), with the addition of g, the gap between the jacket and the supporting member. It is typically found [P6] that the effect of the jacket is to concentrate plasticity at the gap, with strain penetration on either side, resulting in Eq. (8.2).

3. The maximum required curvature is

$$\phi_m = \phi_y + \phi_p \tag{8.3}$$

where the equivalent bilinear yield curvature may be found from moment–curvature analysis, or from Eq. (7.15).

4. The maximum required compression strain is given by

$$\varepsilon_{cm} = \phi_m c \tag{8.4}$$

where c is the neutral-axis depth (from moment–curvature analysis or flexural strength calculations).

5. The volumetric ratio of confinement required, ρ_s, is given by

$$\rho_s = \Phi_j(\varepsilon_{cm}) \tag{8.5}$$

where Φ_j is a materials-dependent relationship between ultimate compression strain and volumetric ratio of jacket confinement. For a steel jacket, this may conservatively be taken as Eq. (5.14).

(i) Steel Jacket Retrofit. The effective volumetric ratio of confining steel for a circular steel jacket of diameter D is

$$\rho_s = \frac{4t_j}{D} \tag{8.6}$$

Hence, from Eq. (5.14), Eq. (8.5) becomes

$$\varepsilon_{cm} = 0.004 + \frac{5.6 t_j f_{yj} \varepsilon_{sm}}{D f'_{cc}} \tag{8.7}$$

where t_j is the jacket thickness, with yield stress f_{yj} and strain at maximum stress of ε_{sm}, and the compression strength of the confined concrete f'_{cc} is given by Eq. (5.6), which is expressed for convenience in Fig. 8.4 in terms of ρ_s and the ratio f_{yj}/f'_c. Equation (8.7) can then be solved for the jacket thickness t_j as

$$t_j = \frac{0.18(\varepsilon_{cm} - 0.004) D f'_{cc}}{f_{yj} \varepsilon_{sm}} \tag{8.8}$$

Design charts for required steel jacket thickness for circular columns are given in Fig. 8.5 as functions of column longitudinal reinforcement ratio and axial load ratio. The charts have been prepared for two longitudinal bar

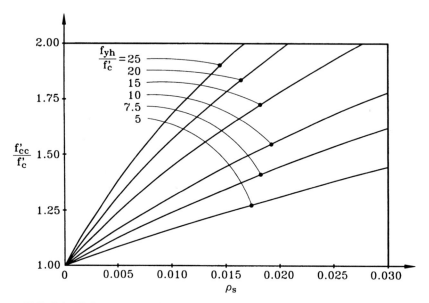

FIG. 8.4 Enhancement of concrete compression strength by confinement.

diameters, No. 11 (D35.8) and No. 18 (D57.2) grade 40 bars [f_{ya} = 44 ksi (303 MPa)] to provide a dependable plastic drift of θ_p = 0.045 or an approximate total drift of θ = 0.05, and thus represent extreme deformation requirements. Required jacket thicknesses for lesser plastic drift angles may conservatively be found by proportionate reduction. The jacket steel assumed was an A36 steel [f_{yj} = 36 ksi (248 MPa), ε_{sm} = 0.015], but results are not very

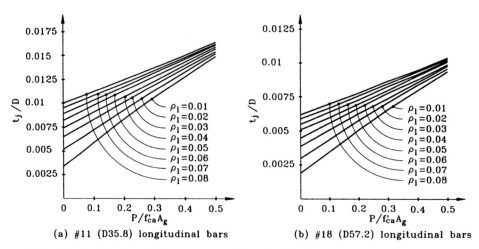

FIG. 8.5 Steel jacket thickness to provide a plastic drift of 0.045 rad [f_{yl} = 44 ksi (303 MPa), f'_{ca} = 5 ksi (34.5 MPa)].

sensitive to steel type since the product $f_{yj}\varepsilon_{sm}$ [see Eq. (8.7)] is approximately constant as yield strength varies.

Results for other bar sizes may be found by interpolation between (or extrapolation from) curves. Although results are given only for grade 40 longitudinal bars, they apply conservatively to grade 60 [$f_{ya} = 66$ ksi (455 MPa)] longitudinal bars, since the plastic hinge length [Eq. (8.2)] is proportionately increased, thus reducing the required plastic curvature while the neutral-axis depth c [Eq. (8.4)] is increased at a rate lower than proportional to f_{ya}.

Figure 8.1(b) shows an elliptical jacket retrofit of a rectangular column. The equation of the jacket circumference may be expressed as

$$\frac{x^2}{a^2} + \frac{y^2}{b^2} = 1 \tag{8.9}$$

A reasonable approximation to the confining effect may be taken using Eqs. (8.6) and (8.7) with an equivalent diameter $D = 2R$, where with reference to Fig. 8.1(b),

$$R = \begin{cases} \dfrac{R_1 + R_2}{2} & (x \text{ direction}) & (8.10a) \\[2mm] \dfrac{R_3 + R_2}{2} & (y \text{ direction}) & (8.10b) \end{cases}$$

and where $R_1 = b^2/a$, $R_3 = a^2/b$, and R_2 is found from Eq. (8.9) for the corner coordinates. Note that for rectangular sections of high aspect ratio, the radius for weak direction response will be very large, resulting in ineffectual confinement. However, it will generally be found that these sections have adequate displacement capacity in the weak direction (normally, this is the bridge longitudinal direction) without the need for confinement.

The design charts of Fig. 8.5 can also be applied to retrofit of rectangular columns with elliptical jackets, using the appropriate definition of equivalent diameter, as above.

To provide adequate protection against buckling of longitudinal reinforcement in the original column, the effective volumetric ratio of confining steel given by Eq. (8.6) should also satisfy Eq. (5.54). "As-built" columns of both circular and rectangular shape retrofitted in accordance with these provisions have performed very well in both laboratory tests [P6] and field exposure to earthquakes [P5]. Figure 8.6 shows lateral force–displacement response for typical examples.

For circular columns, flexural response of the columns is typically limited by the effective ultimate tension strain of the longitudinal reinforcement. As discussed in Section 5.2.3(a), this may be taken as $0.75\varepsilon_{su}$, where ε_{su} is the strain at maximum stress, unless a more detailed analysis in accordance with

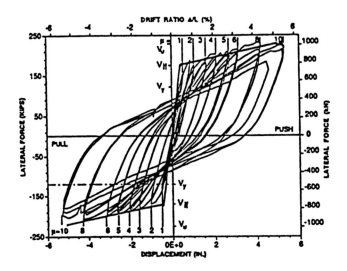

(a) circular column with circular jacket

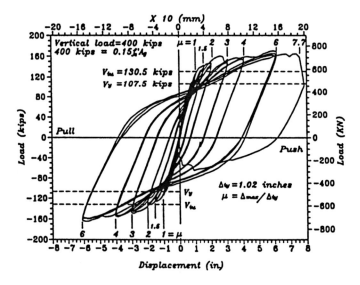

(b) rectangular column with
elliptical jacket

FIG. 8.6 Lateral force displacement response of columns retrofitted with steel jackets for enhanced ductility.

Fig. 5.6 is undertaken. The rectangular column of Fig. 8.6(b) had a longitudinal steel ratio of 5% and was confined with an elliptical jacket extending beyond the expected plastic end region. Failure, at a displacement ductility factor of 8 eventually occurred as a result of inadequate shear strength in the unconfined region beyond the jacket.

(ii) Concrete Jacket Retrofit. Design of concrete jacket retrofits to enhance flexural ductility of circular columns essentially follows procedures for assessing transverse reinforcement requirements for new columns. Thus, Eqs. (8.1) to (8.5) are used with Eq. (5.14), where

$$\rho_s = \frac{4A_h}{D's} \tag{8.11}$$

and D' is the diameter of the hoop or spiral provided in the concrete jacket, at vertical spacing s. The amount of confinement provided should also satisfy Eq. (5.54).

For a rectangular concrete jacket around a rectangular column with only peripheral and octagonal hoops, as shown in Fig. 8.2(b), the confinement effectiveness given by Eq. (5.13) will be reduced. Based on considerations of the volume of effectively confined core, it is suggested that a value of $K_e = 0.5$ be used and that the effective volumetric ratio of confinement be similarly taken to be 50% of the calculated value in assessing ultimate compression strain capacity. It should be appreciated that although reasonable confinement of the concrete may be achieved, the effectiveness of antibuckling restraint for longitudinal reinforcement at the center of each face will be small. Consequently, a rectangular jacket should not be used on tall columns with high aspect ratios or where high ductilities are expected under seismic response.

(iii) Composite-Materials Jackets. Tests on circular columns retrofitted with composite-materials jackets to improve ductility indicate that the confinement effectiveness is more efficient than with steel jackets. It is thought that this is a result of the elastic nature of the jacket material. With a steel jacket, yield under hoop tension may occur early in seismic response, particularly if designed to the limits permitted by Eqs. (8.1) to (8.5). On unloading, residual plastic strains remain in the jacket, reducing its effectiveness for the next cycle of response, and requiring increased hoop strains for each successive cycle. With materials such as fiberglass and carbon fiber, which have essentially linear stress-strain characteristics up to failure (see Section 5.2.6), there is no cumulative damage, and successive cycles to the same displacement result in constant rather than increasing hoop strain. Thus the experimentally derived expression for Eq. (8.5) for composite-materials jackets indicates greater efficiency than for steel jackets and is given by the following equation:

$$\varepsilon_{cu} = 0.004 + \frac{2.5\rho_s f_{uj} \varepsilon_{uj}}{f'_{cc}} \tag{8.12}$$

where ρ_s is given by Eq. (8.6) for circular columns, f'_{cc} by Eq. (5.6), and f_{uj} and ε_{uj} are the ultimate stress and strain of the jacket material. Equations (8.12) and (8.6) can be combined and solved for the required jacket thickness as

$$t_j = \frac{0.1(\varepsilon_{cu} - 0.004)Df'_{cc}}{f_{uj} \varepsilon_{uj}} \tag{8.13}$$

The volumetric ratio of confinement ρ_s should also satisfy Eq. (5.54), using the appropriate value for the modulus of elasticity E_t of the jacket material.

Tests on rectangular columns retrofitted with rectangular composite-materials jackets wrapped directly onto the existing column have been shown to provide a reasonable enhancement of ductility capacity. An example is shown in Fig. 8.7 of a column with a rectangular fiberglass–epoxy jacket placed to enhance shear strength of a short rectangular column. Although it was not expected that the jacket would provide significant enhancement to ductility, it is seen from Fig. 8.7(b) that the column sustained displacement ductilities up to $\mu_\Delta = 8$, corresponding to drift angles of 4% before jacket failure. On the basis of this and similar tests, it was found that a jacket effectiveness of 50% was reasonable, and hence Eq. (8.12) would be modified to

$$\varepsilon_{cu} = 0.004 + \frac{1.25\rho_s f_{uj} \varepsilon_{uj}}{f'_{cc}} \tag{8.14}$$

where, for a rectangular jacket, the volumetric ratio of confinement may be expressed as

$$\rho_s = 2t_j \left[\frac{b + h}{bh} \right] \tag{8.15}$$

and where b and h are the section dimensions of the column.

From Eq. (8.14), the required value of ρ_s is

$$\rho_s = \frac{0.8(\varepsilon_{cu} - 0.004)f'_{cc}}{f_{yj} \varepsilon_{uj}} \tag{8.16}$$

Because of the limited data base for rectangular jackets on rectangular columns, and also because of increased probability of longitudinal bar buckling, rectangular jackets should only be used for columns with axial load ratios of $P \leq 0.15f'_{ca}A_g$, longitudinal reinforcement ratios of $\rho_l \leq 0.03$, and column aspect ratios $M/Vh \leq 3$.

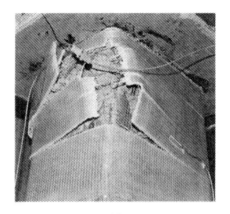

(a)

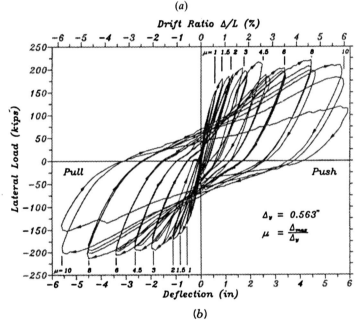

(b)

FIG. 8.7 Rectangular column with fiberglass–epoxy rectangular jacket. (a) Failure by jacket fracture; (b) Lateral force–displacement response.

(iv) Extent of Jacket for Confinement. The region of the column to be confined for provision of enhanced ductility may be taken as that given by Section 5.3.2(c)(iii) and Fig. 5.34 for new designs, except that it is recommended that the minimum length over which confinement be provided, of 20% and 30% of distance from the critical section to the point of contraflexure for $P/f'_{ce}A_g < 0.3$ and > 0.3, respectively, be increased to 25% and 37.5%. This is because the increased confinement and reduced plastic hinge length provided by jacketing typically results in higher moment enhancement as a consequence of increased concrete compression strength and strain hardening of reinforcement.

Because of reduced demand on the jacket as distance from the critical section increases, it is acceptable to reduce the jacket thickness by 50% for that half of the required jacket length farthest away from the critical section. For steel jackets, this will rarely be practical, but for composite materials jackets, the economies afforded by this relaxation may be worth considering. Extent-of-confinement requirements are summarized in Fig. 8.8.

(b) Confinement for Flexural Integrity of Column Lap Splices. In Section 5.5.4 it was shown that propensity for splice failure could be predicted by an assessment of the concrete tensile capacity across a potential splitting failure surface (see Fig. 5.78). After cracking develops on this interface, splice failure can be inhibited if adequate clamping pressure is provided across the fracture surfaces by confinement. As mentioned in Section 5.5.4, a coefficient of friction of $\mu = 1.4$ is appropriate, provided that the equivalent radial dilation strain is less than $\varepsilon_s = 0.0015$. This provision can be used to design the lap-splice retrofit. The confining stress necessary to inhibit lap-splice failure of a lapped bar of area A_b and transfer stress f_s is thus

$$f_l = \frac{A_b f_s}{\mu p l_s} \tag{8.17}$$

where p is the perimeter of the crack surface, given by Eq. (5.115) or (5.116), and l_s is the splice length. If the section is required to form a plastic hinge, as at the column base, the value of stress to be transferred should correspond to extensive strain hardening and should include the possibility of high material overstrength. Thus unless more detailed calculations are undertaken to determine the appropriate value for f_s, it is recommended that $f_s = 1.7 f_{yl}$ be assumed, where f_{yl} is the nominal yield strength of the longitudinal reinforcement. Substituting in Eq. (8.17) and taking $\mu = 1.4$ leads to

$$f_l = \frac{1.21 A_b f_{yl}}{p l_s} \tag{8.18}$$

Note that the requirements for confinement of plastic hinges and for lap-splice containment should not be considered additive when the lap slice occurs within the plastic hinge, as will be the case, for example, with a column–base lap splice. This is because at any instant the lap-splice containment and concrete confinement will be occurring on opposite sides of the column. The more stringent of the two requirements will thus apply.

(i) Steel Jacket Retrofit. As noted above, the dilation strain should not exceed $\varepsilon_s = 0.0015$. Thus noting that for a circular jacket $f_l = 0.5 \rho_{sj} f_{sj}$, where $f_{sj} = 0.0015 E_{sj} \le f_{yj}$, Eq. (8.18) is equivalent to requiring that

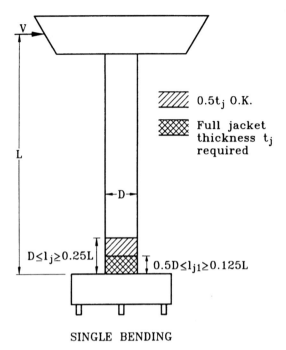

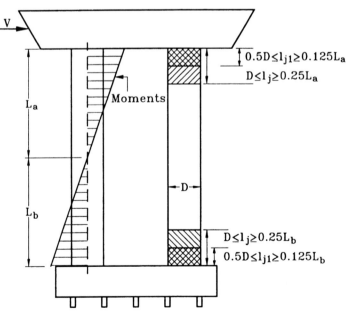

FIG. 8.8 Extent of jacket required for plastic hinge confinement when $P/f'_{ce}A_g \leq 0.3$ (increase by 50% when $P/f'_{ce}A_g \leq 0.3$).

$$\rho_{sj} = \frac{2.42 A_b f_{yl}}{p l_s (0.0015 E_{sj})} \tag{8.19a}$$

but not less than

$$\rho_{sj} = \frac{2.42 A_b f_{yl}}{p l_s f_{yj}} \tag{8.19b}$$

The required jacket thickness is then given by Eq. (8.6).

For elliptical jackets on rectangular columns, Eqs. (8.17) and (8.19) still apply with the appropriate value for p [Eq. (5.116)]. The required jacket thickness is found from Eq. (8.6), with an average value for the jacket diameter $D = 2R$, where R is given by Eq. (8.10).

Appropriate relationships for steel jacket retrofit of lap splices in circular columns, based on Eq. (8.19b), are given for two bar sizes in Fig. 8.9. Data are based on a longitudinal yield stress of $f_{yl} = 44$ ksi (303 MPa) and a jacket yield stress of 36 ksi (248 MPa). Cover was conservatively assumed to be 2 in. (51 mm) to longitudinal steel. As a consequence of the low jacket yield stress, which is appropriate for the commonly used A36 steel, the strain limit implied in Eq. (8.19a) is not reached, and Eq. (8.19b) applies. The lower limit of $t_j/D = 0.0052$ in Fig. 8.9 applies for a failure mechanism involving 45° splitting, as shown in Fig. 5.78(c).

The data in Fig. 8.9 apply to a lap splice of $l_s = 20 d_b$. Required jacket thickness for other splice lengths may be found from inverse proportional variation. Similarly, different longitudinal steel yield stress may be accommodated by proportional variation in required jacket thickness. Finally, different jacket yield strength can be accommodated by inverse proportional variation of jacket thickness, except that the jacket yield stress must not be taken larger than $0.0015 E_{sj}$, as noted above. Because of the large number of possible reinforcement configurations in rectangular sections, it is not feasible to develop design charts for elliptical retrofit of rectangular columns, and the basic design approach, as outlined above, should be used.

Figures 8.10 and 8.11 show steel jacket retrofits of lap-spliced column reinforcement in circular and rectangular columns and the corresponding force–displacement response [P6]. The column of Fig. 8.10 was a duplicate of the column shown in Fig. 7.6, which was tested as-built. Comparison of the as-built and retrofitted force–displacement response indicates the great improvement in behavior of lap splices possible with retrofitting. Both of the columns shown in Figs. 8.10 and 8.11 were capable of sustaining high ductility and drift levels, with the circular column eventually failing by tensile fracture of the longitudinal reinforcement. The energy absorbed by the retrofitted column prior to failure was more than 100 times that absorbed by the as-built column. The rectangular column of Fig. 8.11 eventually suffered gradual degradation of performance due to splice failure. This occurred because of the

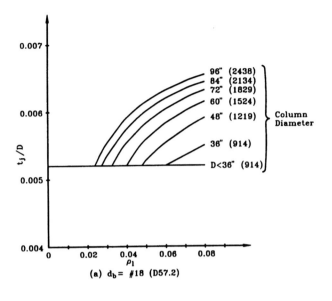

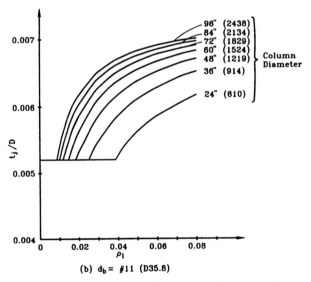

FIG. 8.9 Steel jacket thickness for lap-splice retrofit of circular columns [f_{yl} = 44 ksi (303 MPa), f_{yj} = 36 ksi (248 MPa), l_s = 20d_b].

gradually developing incompatibility of the deformation of the stiff elliptical concrete-filled jacket and the more flexible column, which opened up a vertical crack surface between the column and infill concrete at the base of the column, reducing the clamping pressure in this region. Despite this, the performance must be considered to be extremely satisfactory.

(a)

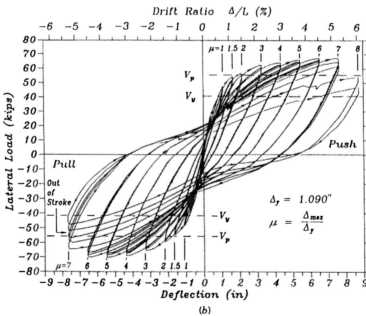

(b)

FIG. 8.10 Circular column–base lap splice retrofitted with a steel jacket. (a) Column under test; (b) Lateral force–displacement response (1 kip = 4.45 kN).

(a)

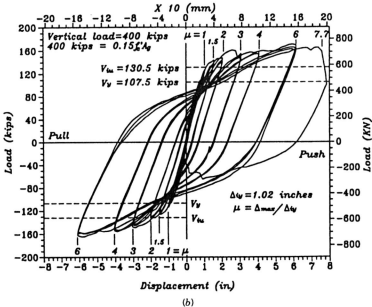

(b)

FIG. 8.11 Rectangular column–base lap splice retrofitted with a steel jacket. (*a*) Column test; (*b*) Lateral force–displacement response.

(ii) Concrete Jacket Retrofit. Although we are not aware of tests on column lap splices retrofitted with concrete jackets, it seems evident that these should be adequate provided that the volumetric ratio of hoop reinforcement in the jacket satisfies Eq. (8.17) or (8.18), as appropriate. Rectangular columns should be jacketed with elliptical concrete jackets or rectangular jackets where the confining reinforcement is of elliptical shape. However, it would appear that if the jacket is thick enough, a rectangular jacket should be effective, provided that an appropriately low value of the effectiveness coefficient, $k_e = 0.5$, is used in Eq. (5.13). The jacket thickness would need to be at least one-sixth of the length of the jacketed side to confine the central region of that side effectively [M4] and therefore is of limited practical interest.

(iii) Composite-Materials Jackets. Retrofit of lap splices with composite-materials jackets follows the same principles and equations as those for steel jackets. The design charts of Fig. 8.9 may be used with appropriate modification for effective jacket stress. However, because of the low modulus of elasticity of many composite materials, particularly fiberglass-based composites, confining stress developed at a dilation strain of 0.0015 will be low, and very large jacket thicknesses will be necessary. An alternative is to provide a degree of active confinement by either winding the jacket material onto the column under tension or by prestressing the jacket by pressure grouting between the jacket and the column after placing the jacket. In such cases the jacket thickness required will be reduced, and referring to the general form of Eq. (8.17), will be given by

$$\rho_{sj} = 2 \left(\frac{A_b f_s / \mu p l_s - f_a}{0.0015 E_{sj}} \right) \tag{8.20}$$

where f_a is the active confining stress, after losses, provided by prestressing the jacket.

This technique of lap-splice retrofit has been tested successfully using prestressed fiberglass–epoxy composite jackets [P19], as shown by the hysteresis response of Fig. 8.12(*b*). However, this approach should be adopted only with a thorough knowledge of the long-term strength characteristics under sustained stress of the jacket material. Strength of many composites degrades under sustained stress, and thus it is important that the level of stress induced in the jacket by the combined effects of active and passive confinement be low. In the absence of specific test data, it is recommended that the maximum jacket stress, when active confinement forms a significant component, should not exceed 25% of the short-term tensile strength.

(iv) Extent of Jacket. When the only reasons for jacketing is to enhance the lap-splice strength, there is no need to extend the jacket beyond the extent of the lap splice. This condition might occur when the lap splice is not located at the potential plastic hinge region. When the lap splice is located at the

(a)

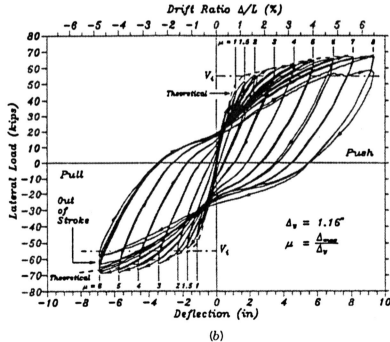

(b)

FIG. 8.12 Circular column–base lap splice retrofitted with a fiberglass–epoxy jacket. (*a*) Column test; (*b*) Lateral force–displacement response.

critical section, as in Figs. 8.10 to 8.12, the jacket thickness and extent will be the more stringent of the requirements of this and the preceding section, governing confinement.

(v) Minimum Splice Length. It is clear that if the lap splice is too short, effective clamping cannot be achieved and the failure mechanism will change from the splitting mode of Fig. 5.74(*a*) to the sleeving failure mode of Fig. 5.74(*b*). Under these circumstances, confinement is unlikely to be very benefi- cial. Consequently, it is recommended that full strength of the retrofitted lap splice be relied upon only when

$$
l_s \geq \begin{cases} \dfrac{0.025 d_b f_{ya}}{\sqrt{f'_{ca}}} & \text{in. (psi units)} & (8.21a) \\[4mm] \dfrac{0.3 d_b f_{ya}}{\sqrt{f'_{ca}}} & \text{mm (MPa units)} & (8.21b) \end{cases}
$$

It will be noted that this is considerably less than the minimum recom- mended splice length for new design given by Eq. (5.117) and is permissible only because of the high degree of confinement provided by the jacket. Equa- tion (8.21) conforms to the minimum anchorage requirements of Eq. (5.111) except that the assessment material strengths f'_{ca} and f_{ya} are used rather than nominal strengths. With shorter laps, the moment capacity should be reduced proportionately. Because of the sustained clamping stress, the reduced mo- ment capacity can be expected to degrade only slowly under cyclic response.

(c) Shear Strength Enhancement. In Sections 5.3.4(*b*)(iii) and 7.4.8 it was suggested that less conservative estimates of shear strength could be adopted for assessment of existing structures than for design. This was primarily to ensure that limited financial resources for retrofitting could better be allocated to the most needy structures, while recognizing that the better knowledge of material properties for existing, as opposed to intended (i.e., under design) columns, results in reduced uncertainty for strength.

Since the costs of retrofitting columns for shear strength will not be greatly affected by jacket thickness, it is appropriate to adopt the same conservatism for retrofit shear design as for new design. Hence the design shear force should be based on conservatively high estimates of column plastic hinge flexural capacity, in accordance with Section 5.3.3, and conservatively low estimates of shear strength, in accordance with recommendations of Section 5.3.4(*b*)(iii). Where the dependable shear strength $\phi_s(V_c + V_s + V_p)$ is less than the maximum feasible shear force $V°$, corresponding to overstrength capacity of plastic hinges, the additional shear strength required to be imparted by the column retrofit will be

$$\phi_s V_{sj} \geq V^\circ - \phi_s(V_c + V_s + V_p) \qquad (8.22)$$

(i) Steel Jacket Retrofit. The shear resistance of a circular passive steel jacket may be found by analogy to hoop reinforcement. The jacket may be considered as equivalent to hoop reinforcement of area $A_h = t_j$ at a spacing of $s = 1$. This conservatively ignores the additional strength imparted by membrane shear flow in the jacket. Thus Eq. (5.76a) may be modified to yield

$$V_{sj} = \frac{\pi}{2} t_j f_{yj} D \cot \theta \qquad (8.23)$$

In accordance with the recommendations of Section 5.3.4(b)(iii), an angle of $\theta = 35°$ should be used in Eq. (8.23) unless the column corner-to-corner diagonal subtends a larger angle with the column axis. In conjunction with Eq. (8.22), Eq. (8.23) implies a required jacket thickness of

$$t_j \geq \frac{V^\circ/\phi_s - (V_c + V_s + V_p)}{0.5\pi f_{yj} D \cot \theta} = \frac{V_{sj}}{2.24 f_{yj} D} \qquad (8.24)$$

The relationship implied by Eq. (8.24) is plotted in Fig. 8.13 in terms of $t_j D$ against V_{sj} for a variety of typical jacket yield strengths. It will be seen that only comparatively small jacket thicknesses are required to achieve large shear strength enhancement. For example, a shear strength deficit of $V_{sj} = 2624$ kips (11,680 kN) can be supplied by A40 steel ($f_{yj} = 40$ ksi $= 275$ MPa) by a value of $t_j D = 29.3$ in^2 (18,900 mm^2). For a column diameter of $D =$

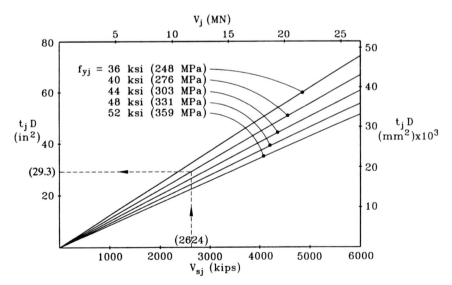

FIG. 8.13   Circular column shear strength enhancement by circular steel jacket.

96 in. (2440 mm), this requires a steel jacket thickness of only 0.305 in. (7.75 mm). In all cases it should be checked that the nominal shear stress in the concrete, V°/A_e, after retrofitting does not exceed the limit of $0.2f'_{ca}$ recommended in Section 5.3.3(b)(iv).

Enhancing the shear strength of a rectangular column with a rectangular steel jacket bonded to the concrete surfaces should be effective, and the strength enhancement, by analogy to rectangular hoops of area t_j and spacing $s = 1$, can be found by manipulating Eq. (5.76b) as

$$V_{sj} = 2t_j f_{yj} h \cot \theta \tag{8.25}$$

where h is the overall column dimension parallel to the shear force applied. However, installation of rectangular jackets on large bridge columns can cause problems. If the jacket is epoxy bonded to the surface, this will need to be done one side at a time, to ensure good contact between column and jacket, with the corners connected by welding or bolting. Grouting a gap between the column and jacket by pressure grouting, as for circular columns, will cause excessive bowing of the steel sides even under low pressures. Also, as has been discussed earlier, rectangular steel jackets will provide little enhancement to flexural ductility capacity, which will commonly be required as well as shear strength enhancement.

For these reasons, rectangular steel jacket retrofits have been used very infrequently in bridge retrofit, although they have found some application in building column retrofit when flexural ductility was not a concern [J2]. On the other hand, the elliptical jacket retrofit discussed in the preceding section for flexural ductility retrofit of rectangular columns has also been widely implemented for shear strength enhancement. For an elliptical jacket, shear strength enhancement may be determined by consideration of equilibrium of forces parallel to the applied shear in a fashion similar to that of a circular jacket and as shown in Fig. 8.14 [P20]. Considering the jacket to be equivalent to a series of elliptical hoops stacked vertically, which are each assumed to yield in hoop tension as a consequence of resisting shear, the total shear force capable of being resisted by the jacket in the strong direction is

$$V_{sj} = \int D_i f_{yj} t_j \cos \delta \, dx \cot \theta \tag{8.26}$$

where the angle δ is defined in Fig. 8.14. Results from the integration of Eq. (8.26) are given in Fig. 8.15, and may be approximated by the following linearizations

$$V_{sj} = \begin{cases} 2f_{yj}t_j D_j \left[1 - \left(1 - \dfrac{\pi}{4}\right)\dfrac{B_j}{D_j}\right] \cot \theta & \text{(strong direction)} \quad (8.27a) \\[4mm] 2f_{yj}t_j B_j \left[1 - \left(1 - \dfrac{\pi}{4}\right)\dfrac{D_j}{B_j}\right] \cot \theta & \text{(weak direction)} \quad (8.27b) \end{cases}$$

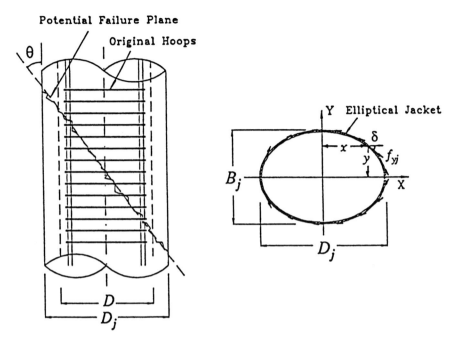

FIG. 8.14 Shear strength enhancement by an elliptical jacket.

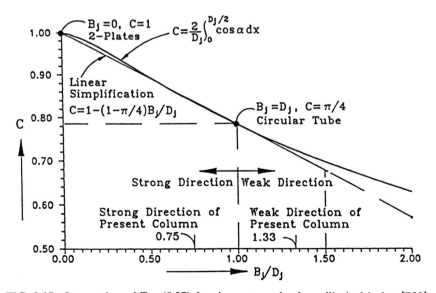

FIG. 8.15 Integration of Eq. (8.27) for shear strength of an elliptical jacket [P20].

Note that Eq. (8.27) degenerates to Eq. (8.23) when $B_j = D_j$ (a circular jacket), and to Eq. (8.25) when $B_j/D_j = 0$ and $D_j = h$, simulating two parallel surfaces (rectangular column). However, when $B_j/D_j > 1.5$, Eq. (8.27) will be rather conservative for evaluating the weak direction strength.

Steel jackets have proved in extensive testing to be extremely effective in inhibiting shear failure of columns. Figures 8.16 and 8.17 compare lateral force–displacement response of as-built and retrofitted columns tested under cyclic reversals of deformation. The as-built circular column of Fig. 8.16(a) achieved its ideal flexural strength indicated by V_{if} but failed in shear at a

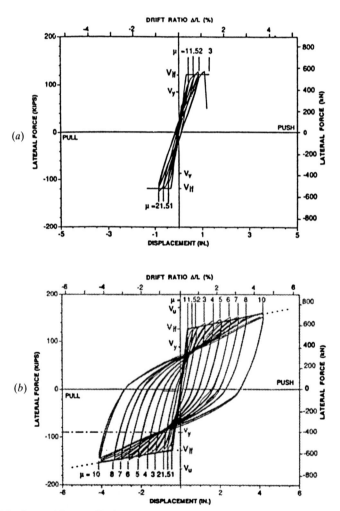

FIG. 8.16   Lateral force–displacement response of as-built and steel-jacketed circular columns with high shear force [P11]. (a) As-built column; (b) Column retrofitted with a steel jacket.

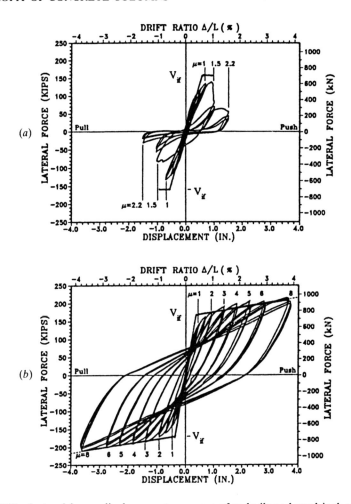

FIG. 8.17 Lateral force–displacement response of as-built and steel-jacketed rectangular columns with high shear force [P11]. (*a*) As-built column; (*b*) Column retrofitted with elliptical steel jacket.

flexural ductility of less than $\mu_\Delta = 3$, corresponding to behavior described by line 2 in Fig. 7.13. As may be seen, strength and stiffness degradation following shear failure were extremely rapid. The companion retrofitted column, which was cast from the same batch of concrete as the as-built column and retrofitted with a circular steel jacket, performed extremely well, as is apparent in the lateral force–displacement response of Fig. 8.16(*b*). Hysteresis loops are very stable at a given displacement ductility level, and the column was able to sustain drift levels of more than 4% without strength or stiffness degradation.

Similar behavior is apparent in Fig. 8.17 for rectangular columns. However, the as-built column of Fig. 8.17(*a*) failed in shear before reaching its ideal flexural strength V_{if}, corresponding to behavior described by line 1 of Fig. 7.13. The companion column, retrofitted with an elliptical jacket, whose response is shown in Fig. 8.17(b), has again performed extremely well, with stable hysteresis loops under multiple cycles to high ductility and drift levels.

(ii)   Concrete Jacket Retrofit. Shear strength enhancement by concrete jackets around existing concrete columns requires little comment, as the principles are clearly very similar to those of new column design, covered in Section 5.3.4(b). However, because of the possibility of poor bond between old and new concrete, particularly under large deformations within the plastic hinge region, it is recommended that the strength of the concrete shear resisting mechanism, V_c, be based only on the original concrete, and that the enhanced shear strength imparted by the jacket be considered to be applied solely by the transverse reinforcement within the jacket. However, in Eq. (5.76), the core dimension D' (see Fig. 5.42) for the jacket may be taken as the center-to-center dimension of the jacket hoop reinforcement, thus providing additional strength as a result of the increased section dimensions.

(iii)   Composite-Materials Jackets. Composite-materials jackets have proved, like steel jackets, to be very effective in enhancing the shear strength of concrete columns and inhibiting shear failures. However, design equations for strength enhancement by steel jacketing, such as Eqs. (8.23) and (8.27), require minor modification since composite materials typically do not exhibit a yield stress. Use of the ultimate jacket strength f_{uj} would imply large dilation strains and hence degradation of aggregate interlock action, which is essential to the concrete shear resisting mechanism, and would not provide adequate protection against the possibility of jacket failure under unexpected overload. Consequently, composite-materials jackets have generally been designed for jacket stress levels corresponding to a jacket strain of $\varepsilon_j = 0.004$. Equations (8.23) to (8.27) can thus be used with f_{yj} substituted by

$$f_j = 0.004E_j \qquad\qquad (8.28)$$

Tests have been carried out on columns retrofitted with carbon fiber–epoxy and fiberglass–epoxy composites [P18,S11] designed to this criterion, with results comparable to those of steel-jacketed columns. Figure 8.18 shows results from a fiber-wrapped circular column, which was a further companion to the column shown in Fig. 8.16(a) but was constructed at a later date. The lateral force–displacement response of Fig. 8.18(a) is comparable to that of the steel-jacketed column of Fig. 8.16(b), although the fiber-wrapped column shows higher elastic flexibility (see Section 8.2.2(d)). Hoop strain measurements on the jackets from gauges on the generator at 90° to the loading axis (i.e., shear-induced strains) are shown in Fig. 8.18(b). It will be seen that these increase in magnitude as the ductility increases as a result of the decreased competence of the concrete shear-resisting mechanisms, and spread from the

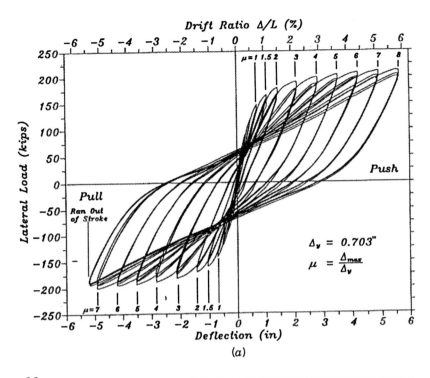

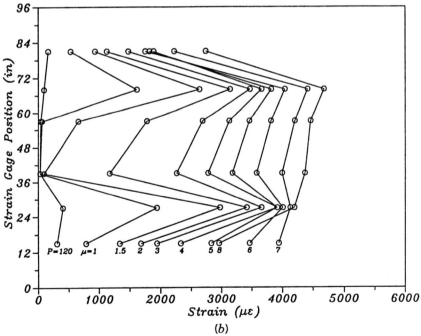

FIG. 8.18 Cyclic response of a circular column retrofitted with a fiberglass–epoxy jacket (height = 4 D; M/VD = 2). (*a*) Lateral force–displacement response; (*b*) Shear-induced hoop strains.

plastic hinge zones at top and bottom of the column into the central region at high displacement ductility factors.

It will be seen that at maximum response, jacket strains are in close agreement with the design value of 0.004 and are almost constant over the central region of the column. As is apparent from Fig. 8.7, rectangular composite jackets are also effective in enhancing shear performance of rectangular columns. This column was a retrofitted companion of the as-built rectangular column whose lateral force–displacement response is shown in Fig. 8.17(a).

(iv) Extent of Jacket. As discussed in Section 5.3.4(b)(v), the plastic end region over which reduced strength of the concrete shear resisting mechanisms applies should be taken as $2D$ (or $2h$ for a rectangular column). Within this region a thicker jacket will be required for shear retrofit than in regions farther from the critical section, where the full concrete capacity, appropriate for μ_ϕ ≤ 1, may be assumed. Jacket thickness calculated for shear need not be added to those required for confinement or lap-splice restraint, since the resisting actions occur at 90° to each other.

(d) Column Stiffness Considerations

(i) Stiffness Enhancement by Jacketing. Placing jackets on columns will increase the column stiffness, thus attracting more force to the retrofitted columns and less to unretrofitted columns (if any) in the elastic range of response. The extent of stiffness enhancement depends on the jacket material, original column section shape, and extent of retrofit. The values given in Table 8.1 give general guidance. However, it should be noted that the stiffening provided by concrete jackets varies greatly depending on the height, thickness, and

TABLE 8.1 Stiffness Enhancement (%) of Columns by Jacketing, at First Yield

Retrofit Case	Steel Jacket	Concrete Jacket	Composite Material Jacket
Circular column			
Plastic hinge retrofit (partial column height)	10–20	20–50	0–5
Shear retrofit (full column height)	20–40	25–75	0–5
	Elliptical Jacket		Rectangular Jacket
Rectangular column			
Plastic hinge retrofit (partial column height)	20–40	20–50	0–10
Shear retrofit (full column height)	40–70	25–75	0–5

amount of longitudinal reinforcement and should be calculated from first principles.

With very flexible bents it may be desirable to reduce structural displacements to minimize $P-\Delta$ effects and relative displacements between frames. For single-column bents this can be achieved, to a modest extent, by steel or concrete jacketing, as indicated by the stiffness enhancements listed in Table 8.1, although it will be rare that jacketing will be used solely to increase member stiffness.

(ii)   Stiffness Enhancement by Link Beams. With tall multicolumn bents, an alternative is to consider connecting the columns of the bent by link beams as shown in Fig. 8.19(*a*). If the link beam is located at column midheight, lateral strength of the bent will be doubled and elastic displacements halved. The link beam may be designed to force plastic hinges into the columns above and below the beam, or by reducing its strength, to force column hinges into the ends of the link beam, thus protecting the column from inelastic action at midheight. Plastic hinges will still, however, be required to form at column top and bottom to complete a plastic collapse mechanism. It should be appreciated that if the link beam is strengthened sufficiently to force column hinges at midheight, higher mode effects may cause a soft-story mode of inelastic response to develop, with the majority of plastic deformation concentrated in the hinges either above or below the link beam, which should thus be designed for amplified plastic rotations. To ensure satisfactory behavior, it will frequently be necessary to jacket the columns, thus reducing displacements further if steel or concrete jackets are adopted. Also, a consequence of the increased lateral strength is increased shear force in the columns, which should be checked by capacity design principles to determine whether shear strength is adequate. Shear retrofitting may be necessary. A consequence of this form of stiffness retrofit is that seismic axial forces on the footing will be increased by an amount equal to the seismic shear force induced in the link beam. Figure 8.19 shows two further applications of link-beam retrofit, discussed in Section 8.3.

(e)   Column Repair by Jacketing. The methods of column jacketing discussed in previous sections can also be used to repair columns damaged in moderate earthquakes. Considerable judgment will need to be exercised in deciding whether the level of damage to the column is such that full reinstatement of strength is possible. Tests have indicated that spalling of cover concrete and yielding of either longitudinal or transverse reinforcement should not be considered to militate against retrofitting. However, if reinforcement has been fractured, buckled, or deformed significantly out of the straight, column replacement should be adopted rather than repair. Similarly, excessive crack widths and spalling of core concrete should indicate that column replacement is necessary.

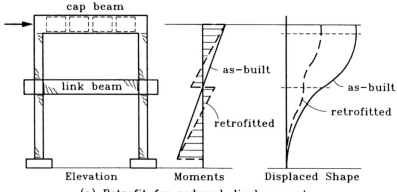

(a) Retrofit for reduced displacements.

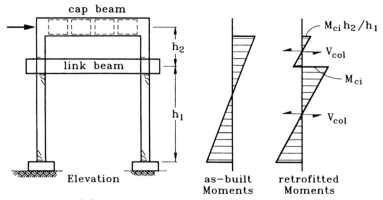

(b) Retrofit for reduced cap beam forces.

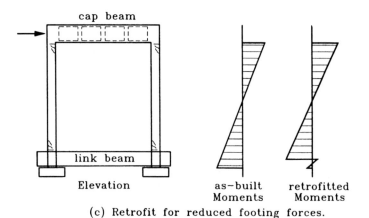

(c) Retrofit for reduced footing forces.

FIG. 8.19 Use of link beam to improve transverse seismic response of multicolumn bents.

With these provisions, damaged columns repaired by jacketing have performed well in tests. The column shown in Fig. 7.6, which suffered a lap-splice failure at the column base was repaired by placing a steel jacket around the lower one-third of the column height after removal of loose cover concrete, with grout being pumped into the space between the jacket and column, as for a standard column retrofit [C2]. Lateral force–displacement response for the repaired column is shown in Fig. 8.20. Although the response exhibits reduced energy dissipation compared to response of a retrofitted undamaged column, the repaired column was able to develop its theoretical flexural strength to large displacements without significant strength degradation.

A further example of column repair is provided in Fig. 8.21, where a circular column previously tested to failure in shear was repaired by wrapping with a fiberglass–epoxy composite jacket, with internal cracks and cover spalling reinstated by epoxy injection grouting. Condition of the column prior to repair is shown in Fig. 8.21(*a*). Lateral force–displacement response of Fig. 8.21(*b*) is essentially the same as for a retrofitted undamaged column (compare with Figs. 8.16 to 8.18).

When designing repair measures for columns that have failed in shear, it is recommended that both concrete (V_c) and truss (V_s) mechanisms be consid-

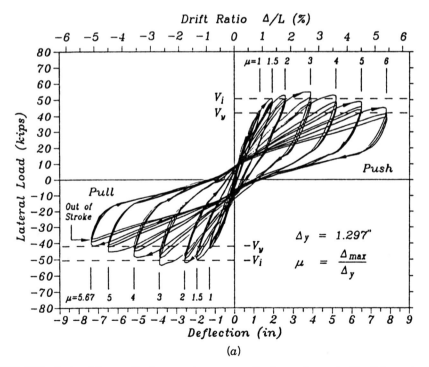

(*a*)

FIG. 8.20 Lateral force–displacement response of column shown in Fig. 7.6 repaired with a steel jacket.

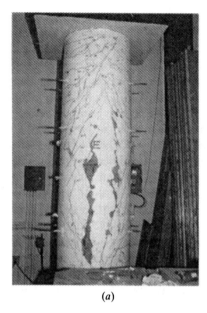

(a)

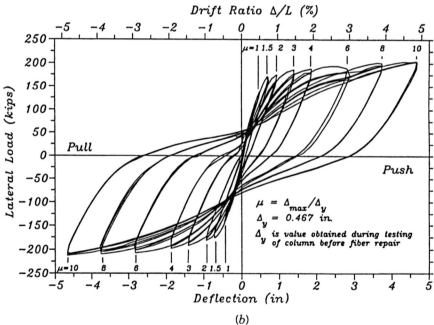

(b)

FIG. 8.21 Repair of a shear-failed circular column by composite-materials jacket. (a) Condition prior to repair; (b) Lateral force–displacement response of repaired column.

ered ineffective and that reliance for shear resistance be placed solely on the jacket capacity V_j and axial load component V_p.

(f) *Example 8.1: Column Steel Jacket Retrofit Design.* The 72-in. (1829-mm) circular column of Fig. 8.22(a) is part of a multicolumn bent with columns fixed at top and bottom. Longitudinal reinforcement consists of 54 No. 14 (D43) bars of yield strength f_{yl} = 50 ksi (345 MPa) bundled in pairs, with transverse reinforcement No. 4 (D12.7), f_{yh} = 44 ksi (303 MPa) at 12-in. (305-mm) centers. Axial load is 1000 kips (4.45 MN), and probable concrete strength is f'_{ca} = 6 ksi (41.4 MPa). A steel jacket is to be designed for the column to enable it to sustain a plastic displacement of Δ_p = 7.5 in. (190.5 mm). Jacket yield stress is f_{yj} = 40 ksi (275 MPa), and strain at maximum stress for the jacket steel is ε_{sm} = 0.15. Consideration will be given later to an alternative retrofit design in composite materials.

Solution. First consider confinement in the plastic hinge region. The plastic rotation is $\theta_p = \Delta/L = 7.5/(12 \times 18) = 0.0347$. From Eq. (8.2) with g = 2 in. (51 mm), $L_p = 2 + 0.3 \times 50 \times 1.69 = 27.4$ in. (696 mm). Therefore, from Eq. (8.1),

$$\phi_p = \frac{0.0347}{27.4} = 0.00127/\text{in. (0.05/m)}$$

From a preliminary moment–curvature analysis, $\phi_y = 67 \times 10^{-6}/\text{in.}$ (0.00236/m). Therefore, from Eq. (8.3),

$$\phi_m = 0.00133/\text{in. (0.0524/m)}$$

Also from the preliminary moment–curvature analysis the neutral axis depth at maximum response is c = 16.5 in. (419 m). Therefore, from Eq. (8.4),

$$\varepsilon_{cm} = 0.00133 \times 16.5 = 0.0219$$

The required jacket thickness is given by Eq. (8.8). This requires an estimate of the compression strength of the confined concrete, which is obtained from Fig. 8.4. With $f_{yj}/f'_{ca} = 40/6 = 6.67$ and an estimated $\rho_s = 0.015$, $f'_{cc} = 1.32 f'_{ca} = 7.92$ ksi (54.6 MPa). Hence, from Eq. (8.8),

$$t_j = \frac{0.18 \times (0.0219 - 0.004) \times 72 \times 7.92}{40 \times 0.15} = 0.306 \text{ in. (7.8 mm)}$$

The volumetric ratio of transverse reinforcement is thus, from Eq. (8.6),

$$\rho_s = 4 \times \frac{0.306}{72} = 0.017$$

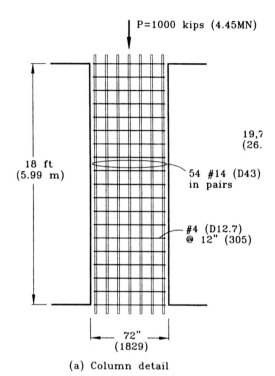

P=1000 kips (4.45MN)

18 ft
(5.99 m)

19,7
(26.

54 #14 (D43)
in pairs

#4 (D12.7)
@ 12" (305)

72"
(1829)

(a) Column detail

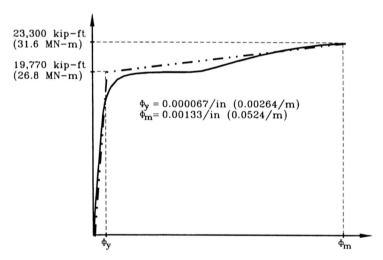

23,300 kip–ft
(31.6 MN–m)

19,770 kip–ft
(26.8 MN–m)

Φ_y = 0.000067/in (0.00264/m)
Φ_m= 0.00133/in (0.0524/m)

Φ_y

Φ_m

(b) Moment–curvature at overstrength

FIG. 8.22   Details for design example of Section 8.2.2(f).

This is close enough to the assumed value of $\rho_s = 0.15$ not to warrant a further iteration, since available plate thicknesses for the jacket would dictate a minimum thickness of $\frac{5}{16}$ in. [0.313 in. (7.94 mm)].

Alternatively, an approximate and conservative estimate of jacket thickness can be obtained from Fig. 8.5. The longitudinal steel ratio is $\rho_l = 54 \times 2.25/4069 = 0.03$. Hence, from Fig. 8.5(a), t_j/D for No. 11 bars (D35.8) is 0.0072, and from Fig. 8.5(b) for No. 18 bars (D57.2), $t_j/D = 0.0044$. Interpolating for No. 14 bars, the thickness/diameter ratio to provide a plastic drift of 0.045 rad is

$$\frac{t_j}{D} = 0.0072 - (0.0072 - 0.0044)\left(\frac{1.69 - 1.41}{2.25 - 1.41}\right) = 0.00626$$

The required plastic drift is 0.0347. The required jacket thickness is thus

$$t_j = 72 \times 0.00626 \times \frac{0.0347}{0.045} = 0.0348 \text{ (8.84 mm)}$$

This is 14% larger than the "exact" approach.

Check the antibuckling requirement of Eq. (5.56), which will be conservative since $f_y = 50$ ksi, while Eq. (5.55) was developed for grade 60 rebar ($f_y = 414$ MPa nominal):

$$\rho_s > 0.0002 \times 54 = 0.0108 < 0.017 \text{ provided (O.K.)}$$

Note that the column ratio is M/VD < 4, indicating antibuckling protection is not critical.

Shear requirements: To determine the shear requirements, a maximum longitudinal reinforcement yield stress of $f_{ylo} = 60$ ksi (414 MPa) and a maximum concrete compression strength (unconfined) of $f'_{co} = 8$ ksi (55.2 MPa) are assessed. A maximum axial load of $P = 1200$ kips (5.35 MN) is determined. Results of a moment–curvature analysis of the jacketed column using program *SEQ-Mϕ* [S12] are shown in Fig. 8.22(b). This indicates a maximum feasible moment of 23,300 kip-ft (31.6 MNm) at the maximum design curvature of $\phi_m = 0.00133/\text{in.}$ (0.0524/m). Hence the overstrength shear force is

$$V^\circ = 2 \times \frac{23,300}{18} = 2590 \text{ kips (11.52 MN)}$$

Check shear capacity of existing column:

V_c: maximum curvature ductility factor is $\mu_\phi = 0.00133/0.000067 = 19.9$.

Thus, from Fig. 5.46,

$$V_c = 0.5\sqrt{6000} \times 0.8 \times 4069 = 126 \text{ kips (560 kN)}$$

V_s: From Eq. (5.76a) with $\theta = 35°$ and $D' = 72 - 4 + 0.5 = 68.5$ in. (1740 mm),

$$V_s = \frac{\pi}{2} \times \frac{0.2 \times 44 \times 68.5 \times \cot 35°}{12} = 113 \text{ kips (501 kN)}$$

V_p: From Eq. (5.77b),

$$V_p = 0.85 \times \frac{1000 \ (72 - 16.5)}{12 \times 18} = 218 \text{ kips (970 kN)}$$

Therefore, existing strength

$$V_c + V_s + V_p = 126 + 113 + 218 = 457 \text{ kips (2033 kN)}$$

The required jacket thickness within the plastic hinge region is thus, from Eq. (8.24),

$$t_j \geq \frac{2590/0.85 - 457}{2.24 \times 40 \times 72} = 0.401 \text{ in. (10.2 mm)}$$

This exceeds the thickness required for confinement and hence governs the plastic hinge region. Using U.S. standard plate sizes, a thickness of $\frac{7}{16}$ in. would probably be chosen, although the metric size of 10 mm would be considered satisfactory, considering conservatism in the design. It will be noted that there is some apparent extra conservatism in the design in that the overstrength shear has been based on an axial load of 1200 kips (5.35 MN) and a concrete strength of $f'_{co} = 8$ ksi (55.2 MPa), while the shear strength has been based on the probable axial load and concrete strengths. It would be more consistent to check thickness requirements using the same values for input and strength. This would require checking for requirements corresponding to nominal and overstrength values for P and f'_c and choosing the more severe.

The total column height is only $3D$. Since the end region for shear is $2D$ from top and bottom hinge locations, the entire column height must be considered a hinge region for shear and the jacket thickness maintained over the full column height.

(g) Example 8.2: Alternative design with composite materials jacket: An option using a carbon fiber–epoxy composite jacket is to be considered. Design data for the jacket material are $E_j = 12 \times 10^6$ psi (82.8 GPa), $f_{uj} = 150$ ksi (1034 MPa) and $\varepsilon_{uj} = 0.0125$.

Confinement: From Eq. (8.13),

$$t_j = 0.1 \, (0.0219 - 0.004) \times 72 \times \frac{7.92}{150 \times 0.0125}$$

$$= 0.544 \text{ in. (13.8 mm)}$$

From Section 8.2.2(a)(iv) the end region for confinement is the greater of D (72 in. (1829 mm) and $0.25 \times 9 \times 12 = 27$ in. (686 mm). For half this length [i.e., 36 in. (915 mm)] full confinement is required, and for the next 36 in. (915 mm), the jacket thickness, for confinement, could be reduced by 50% to $t_j = 0.272$ in. (6.9 mm).

Check shear: Since the material has no yield, the usable jacket stress, from Eq. (8.28), is

$$f_j = 0.004 \times 12 \times 10^3 = 48 \text{ ksi (331 MPa)}$$

Hence, from Eq. (8.24),

$$t_j = \frac{2590/0.85 - 457}{2.24 \times 48 \times 72} = 0.334 \text{ in. (8.5 mm)}$$

This is less than the required confinement thickness in the end regions of the column but more than the reduced thickness in the outer half of the plastic hinge. The final jacket design would thus require $t_j = 0.544$ in. (13.8 mm) for 3 ft (915 mm) at the top and bottom of the column, and $t_j = 0.334$ in. (8.5 mm) over the central region of length 12 ft (3660 mm).

8.3   RETROFIT OF CAP BEAMS

Cap beams provide the link in force transfer between the superstructure and columns. Under transverse seismic response, the cap beams of multicolumn bents will be subjected to flexure and shear. As discussed in Section 7.4.7, deficiencies in flexural strength are common, particularly as a consequence of low positive reinforcement ratios at column faces and premature termination of negative reinforcement. Inadequate shear strength of cap beams is also common. Under longitudinal response, cap beams supporting superstructures via bearings [Fig. 8.23(a)] are unlikely to have problems, but monolithic superstructure/cap beams/column designs may develop cap beam torsional problems, particularly when the columns are located outside the superstructure [Fig. 8.23(b)].

Cap beam deficiencies can be difficult and expensive to alleviate. Two basic approaches may be adopted: The cap beam strength can be increased to the level required to sustain the column plastic hinges, or the seismic forces developed in the cap beam can be reduced by a number of means. These are discussed briefly in the following section.

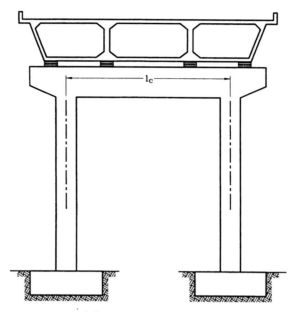

(a) Bearing–supported
superstructure

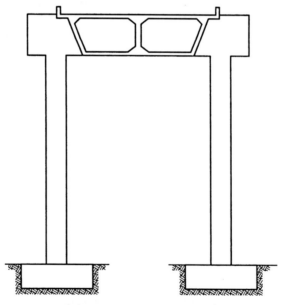

(b) Integral cap beam/
superstructure

FIG. 8.23 Cap beam–superstructure connection.

8.3.1 Reduction of Cap Beam Seismic Forces

When cap beam and particularly column/cap beam joint forces induced by seismic action severely exceed capacity, it may be worth considering means for cap beam force reduction. One effective method involves the use of the link beam concept presented in Fig. 8.19 and discussed earlier in relation to stiffening of bent lateral response [Fig. 8.19(a)].

As shown in Fig. 8.19(b), if located high on the columns, the link beam will also be effective in reducing seismic input to the cap beam. Column moments between the cap beam and the link beam will be small if plastic hinging is forced into the column section below the link beam and if the free height between the cap beam and the column is small compared with the column height below the link beam. If the link beam and column mass is small compared with the superstructure mass, as is generally the case, the column shear force V_{col} will be essentially constant over the column height. With the nomenclature used in Fig. 8.19(b), the column input moment at the cap beam midheight will thus be approximately $M_{ci}h_2/h_1$, where M_{ci} is the moment capacity of the column hinge below the link beam, extrapolated to the link beam midheight. If h_2/h_1 is small enough, no further retrofit of the cap beam should be needed. Note that the link beam strength must be determined by capacity principles to ensure that plastic hinging occurs in the column below the link beam and not in the link beam itself. If link beam hinges form [which were considered acceptable in Fig. 8.19(a)], a further hinge is required at the top of the column or in the cap beam to complete the plastic mechanism. In this case the link beam will have been ineffective in protecting the cap beam, although plastic rotations will be reduced, and lateral strength increased.

The link beam should be constructed by removing the column cover concrete over the height of the link beam and using a link beam width sufficient to place the longitudinal reinforcement outside the column core, as shown in Fig. 8.24(a). In accordance with Section 5.4.4, special joint reinforcement will need to be placed in the link beam in the joint regions [Fig. 8.24(a)]. Figure 8.24(b) shows link beam retrofits of tall bents of the I-10 freeway in Los Angeles. These retrofits, constructed shortly before the 1994 Northridge earthquake, enabled this section of the freeway to survive the earthquake without damage, despite collapse of other sections of the freeway some 10 km away [P5].

Cap beam forces may often also be decreased in bearing-supported bents such as that of Fig. 8.23(a) by replacement of existing bearings by low-friction bearings (i.e., PTFE/stainless steel sliders) or elastomeric bearings of low stiffness. Relevant details are discussed in Chapter 6.

8.3.2 Enhancement of Cap Beam Strength

(a) *Flexural Strength.* Generally, the retrofit philosophy will be to increase the cap beam flexural strength sufficiently to force plastic hinging into the

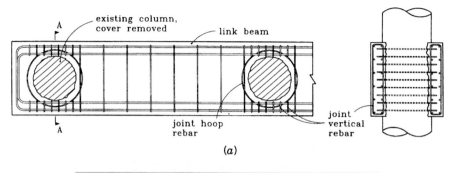

(a)

(b)

FIG. 8.24 Retrofit using link beams. (a) Link beam reinforcement details; (b) Link beam implementation on I-10 Santa Monica viaduct (Los Angeles).

columns. With a separate cap beam supporting the superstructure via bearings, as shown in Fig. 8.25(a), flexural strengthening can be achieved by adding reinforced concrete bolsters to the sides after roughening the interface. The new and old concrete should be connected by dowels, preferably passing right through the existing cap beam. If the amount of tension reinforcement in a bolster is A_{sb}, and assuming it to be stressed to yield at the face of the supporting column, the area $(\Sigma\ A_d)$ of dowel reinforcement required to transfer the force back into the existing cap beam and thus ensure composite action will be

$$\Sigma\ A_d = \frac{A_{sb}f_{yl}}{f_{yd}} \tag{8.29}$$

where f_{yl} and f_{yd} are yield strengths of bolster longitudinal and dowel reinforcement, respectively. This assumes a coefficient of interface friction of $\mu = 1$. To avoid excessive dilation of the interface between old and new concrete, it is recommended that f_{yd} not be taken larger than 60 ksi (414 MPa) unless the

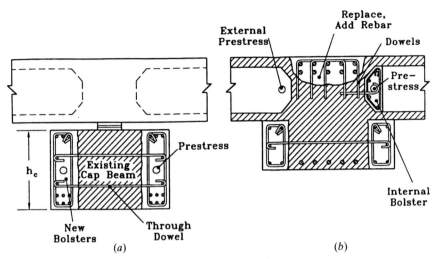

FIG. 8.25 Flexural and shear retrofit of cap beams. (*a*) Bearing-supported superstructure; (*b*) Integral cap beam.

dowels are through-bolts, stressed after the bolsters are cast. The dowels should be distributed over an area $h_c l_c/4$ [see Figs. 8.23(*a*) and 8.25(*a*)] consisting of the lower half of the cap beam (for positive moment reinforcement), from column face to cap beam centerline.

An alternative, or supplemental means of flexural strength enhancement is to prestress the cap beam using strong-backs at the cap beam ends. The prestress may be inside bolsters, as shown in Fig. 8.25(*a*), or using external prestressing without bolsters. In this case the prestress essentially acts as an additional axial compression force on the cap beam, hence enhancing its flexural strength. Where external prestress is used, it should be protected against corrosion by being placed in grouted galvanized ducts, or stainless steel prestressing rods used, as might be the case in Europe. These have the advantage of not requiring corrosion protection and being accessible for inspection and replacement, if necessary.

Enhancing the flexural capacity of integral cap beams [Fig. 8.25(*b*)] is more difficult because of physical constraints imposed by the existing superstructure. Bolsters may be added at the bottom to enhance positive moment capacity, and negative moment capacity can be increased by removing the top concrete and adding additional reinforcement [see Fig. 8.25(*b*)]. Clearly, this will involve disruption of traffic. Prestressing, either in external grouted galvanized ducts or in holes cored through the length of the cap beam, will generally be the most economical means of enhancing both positive and negative moment capacity.

(b) Shear Strength. Full or partial-depth bolsters [see Fig. 8.25(*a*) and (*b*), respectively] can be reinforced with transverse reinforcement to enhance cap beam shear strength, as indicated in Fig. 8.25. As with column retrofit, it is recommended that retrofit shear design be based on the design recommendations of Section 5.3.4(b)(iii). Prestressing will also enhance the cap beam shear strength, as discussed in relation to Fig. 5.47, and can be incorporated directly in the design approach recommended in Section 5.3.4(b)(iii) and given by Eq. (5.78). Tests on a large-scale model of a retrofit concept for the San Francisco double-deck viaducts following the Loma Prieta earthquake [P13,P14], which incorporated cap beam prestressing to enhance flexural and shear strength of the existing cap beam, indicated that the approaches for flexure and shear enhancement described herein were dependable and conservative. Figure 8.26 shows cap beam retrofit details and compares maximum flexural and shear response with predicted capacities based on the assessment procedures of Chapter 7 rather than the more conservative retrofit design procedures recommended in this chapter. It will be seen that peak flexural and shear response

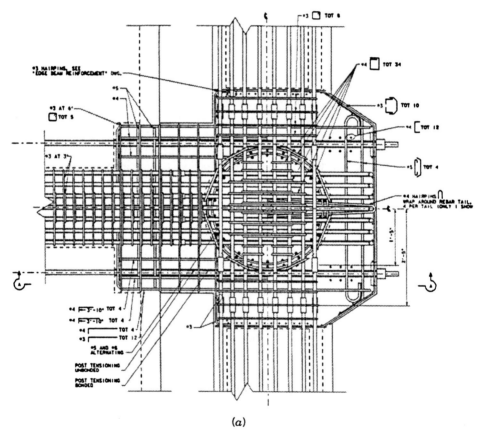

(a)

FIG. 8.26 Seismic response of a cap beam for a retrofit concept for the San Francisco double-deck viaducts (retrofitted to provisions of this chapter). (a)Cap beam and joint retrofit details; (b) Cap beam flexural strength and moment envelopes; (c) Cap beam shear strength and moment envelopes.

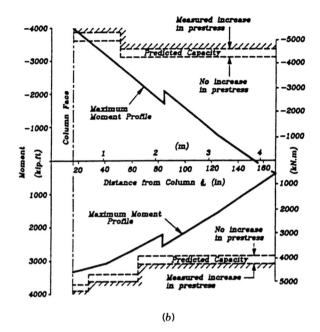

(b)

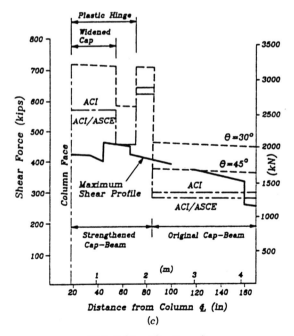

(c)

FIG. 8.26 *(Continued).*

was very close to predicted capacity without development of excessive cracking due to either flexural or shear response. Further details of this test are given in Section 8.5.2(*b*).

Cap beam shear strength may also be enhanced by composite materials bonded to the sides of the cap beam. This will be most effective when the composite layer can be wrapped around the top and soffit of the cap beam, as will generally be the case for bearing-supported superstructures. Where the composite material cannot be turned over onto the top or soffit surface, the efficiency of the retrofit will be reduced unless a positive means of clamping the fabric as close as possible to the top and soffit is provided.

If the adhesion between the composite material matrix and the existing concrete is the sole means of bonding, it is recommended that the efficiency of the retrofit layer, to be used in conjunction with Eq. (5.76*b*), be taken as 60%. Thus for a total effective composite layer thickness, measured in the direction perpendicular to the cap beam axis of t_j (this is the net thickness of two sides if both sides of the cap beam are retrofitted), the shear strength enhancement is

$$V_{sj} = 0.6t_j(0.004E_j)h_j \cot 35° \qquad (8.30)$$

where $f_j = 0.004E_j$ is the jacket stress corresponding to a strain of 0.004 and h_j is the height of the composite layer perpendicular to the cap beam axis.

(c) *Torsional Strength.* Enhancement of torsional strength by transverse reinforcement requires that complete closed hoops be added. This will require concrete jacketing around the entire section, which will be possible for bearing-supported superstructures with independent cap beams [Fig. 8.23(*a*)]. Composite-materials fabric wrapped around the full section will also be effective.

For integral cap beams, such as Fig. 8.23(*b*), it can be extremely difficult to provide effective transverse reinforcement to enhance torsional strength. In these cases, reliance should be placed on axial prestressing to inhibit torsional cracking and enhance torsional strength. It should be noted, however, that provided that the superstructure remains elastic under longitudinal response, it is difficult to envisage torsional distress of cap beams enclosed within the superstructure box girders, as the torsional rotations of the cap beam would require significant distortions and warping of the superstructure, which will be resisted by in-plane membrane forces in the deck and soffit slabs.

8.4 RETROFIT OF CAP BEAM–COLUMN JOINT REGIONS

The behavior and design of beam–column joints for bridges were examined in considerable detail in Section 5.4. The recommendations discussed in that section can be used to guide retrofit design. As with cap beam retrofit, a number of options are available, including joint force reduction, damage acceptance with subsequent repair, joint prestressing, jacketing, and joint replacement.

8.4.1 Joint Force Reduction

The means discussed in Section 8.3.1 to reduce cap beam forces will also, naturally, reduce column–cap beam joint forces. The benefits of reduced retrofit effort for cap beams and joints will often outweigh the costs of link-beam construction, which is comparatively straightforward, and can often be achieved without traffic disruption.

8.4.2 Damage Acceptance with Subsequent Repair

If it can be shown in the assessment phase of the as-built structure that joint failure will not lead to bent collapse under the design earthquake, an option to be considered is to accept the probability of damage in a major earthquake, with subsequent joint repair or replacement. In such cases there must be certainty that the joint failure will not jeopardize the gravity-load capacity of the structure, and the possibility of punch-through failures, where columns have penetrated the deck surface as occurred with the Struve Slough bridge in the Loma Prieta earthquake [E1], carefully considered.

This mechanism of failure is illustrated for a box-girder superstructure in Fig. 8.27. As is evident from Fig. 5.65(a), failure of a joint can result in extreme physical degradation of the joint region, to the extent that shear transfer from superstructure into the column can only be relied upon from the relatively undamaged concrete above the joint region. This can conservatively be as-

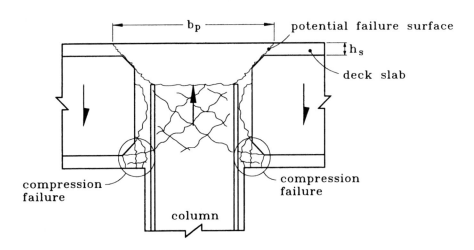

FIG. 8.27 Potential punching shear failure following joint failure.

sumed to be the deck thickness. Using the nomenclature of Fig. 8.27, the effective area for punching shear resistance may be taken as

$$A_p = b_o h_s \qquad (8.31)$$

where $b_o = 4b_p$.

This assumes that the column is square or circular. The modification to b_o for rectangular columns should be obvious. Allowing a shear stress of $4\sqrt{f'_c}$ psi $(0.33\sqrt{f'_c}$ MPa) in accordance with ACI 318 [A5], the ultimate capacity of the slab will be

$$P_u = \begin{cases} \phi_s\, 4\sqrt{f'_c}\, b_o h_s & \text{psi units} & (8.32a) \\ \phi_s\, 0.33\sqrt{f'_c}\, b_o h_s & \text{MPa units} & (8.32b) \end{cases}$$

The test unit shown in Fig. 5.65 was found able to sustain an axial load exceeding that corresponding to Eq. (8.32) after joint failure without developing a punching failure.

8.4.3   Joint Prestressing

In many cases unacceptable joint performance can be improved by addition of prestressing. As well as increasing the flexural and shear strength of the cap beam, it will reduce the tendency for joint cracking because of the increase in horizontal stress f_h in Eq. (5.89). An example of the effectiveness of cap beam prestressing was given in Section 5.4.3(c). However, the prestress increases the principal compression stress, which should be checked to ensure that the limit of $0.3f'_c$ suggested in Section 5.4.3(b) is not exceeded. Figure 8.28 shows the results of a knee-joint test where a duplicate of the as-built knee joint of Fig. 5.58 was retrofitted by external prestressing anchored to a strong-back at the back of the joint. The prestressing was designed to increase the cap beam flexural strength sufficiently to ensure that column plastic hinges developed for both opening and closing moments. Although joint principal tension stresses were reduced by the prestress, they still exceeded the crack-initiation stress of $3.5\sqrt{f'_c}$ psi $(0.29\sqrt{f'_c}$ MPa).

As can be seen from the joint condition of Fig. 8.28(a) and the force–displacement response of Fig. 8.28(b), the prestress did not, in this case, inhibit joint failure. However, strength, ductility, and energy absorption capacity were greatly improved compared to the response exhibited in Fig. 5.58. Failure occurred by crushing of the joint region under closing moments. Peak nominal compression strength in the joint, given by Eq. (5.89), exceeded the recommended limit of $0.3f'_c$. The effectiveness of cap beam prestressing in improving joint performance was discussed further in Section 5.4.4(b) for tee joints.

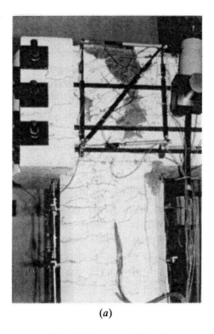

(a)

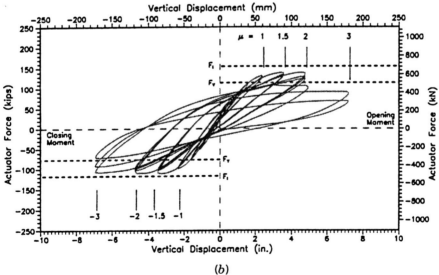

(b)

FIG. 8.28 Seismic response of a joint retrofitted by prestressing, with high principal compression stress (compare with 'as-built' Fig. 5.58). (a) Joint crushing at failure; (b) Force displacement response.

8.4.4 Jacketing

Prestressing of joints is likely to be most effective when principal tension stresses are reduced below $3.5\sqrt{f_c'}$ psi ($0.29\sqrt{f_c'}$ MPa). With higher levels of tension stress, joint degradation is likely at moderate drift angles, as is apparent from Fig. 8.28(b). Although there will be cases where the comparatively gradual degradation of the force–displacement response of Fig. 8.28(b) is acceptable, as a consequence of redundancy in the lateral resistance load path, there will often be a need for more effective retrofitting measures. In such cases, jacketing of the joint by concrete, steel, or composite-materials jackets should be considered.

With concrete jacketing, the necessary reinforcement to satisfy the load-path mechanisms developed in Section 5.4.4 is placed in the concrete jacket, which is connected to the existing concrete by dowels and by roughening the surface of the existing joint. Typically, the jacket will extend beyond the original joint dimensions into the cap beam and column, utilizing a haunch, as indicated in Fig. 8.29(a), which shows a retrofit example for an outrigger knee joint. Jacketing also increases the joint thickness, thus reducing joint stress levels. Extending the size of the joint into the existing column and cap beam also increases development length for the column reinforcement and creates new critical sections for moment capacity at the edge of the jacket.

As will be seen from the lateral force–displacement response of Fig. 8.29(b) [I6], this form of retrofit can be completely effective. In this knee-joint example, plastic hinging formed in the column, and the limit to response was caused by a confinement failure in the column at moderately high curvature ductility factors. Similar improvements in joint behavior have been obtained with steel jackets epoxy bonded to the concrete surface connected through the joint thickness with through-bolts to assist in the transfer of the joint force resistance mechanism from the outer steel plates to the beam and column stress resultants [T2]. It would appear that composite materials could also be used to enhance shear strength, although we are not aware of any tests in support of such a retrofit.

In all cases the jacket should encompass the joint completely to ensure adequate force transfer. Although this is possible for exposed outrigger knee joints of the type shown in Fig. 8.29(a) and will also generally be possible where the cap beam supports the superstructure by bearings, this will not be the case for monolithic column–cap–superstructure designs such as are common in California. In such cases more reliance will need to be made on cap beam prestressing. As shown in Fig. 8.30, addition of reinforced side bolsters prestressed through the cap beam using high-strength through-bolts is a possible solution for longitudinal response. The combined effects of transverse and longitudinal prestressing of the cap beams are likely to be fully effective in solving joint problems of this kind.

8.4.5 Joint Replacement

In some cases none of the options discussed above will be acceptable—the only satisfactory solution will be joint replacement. This might be the case

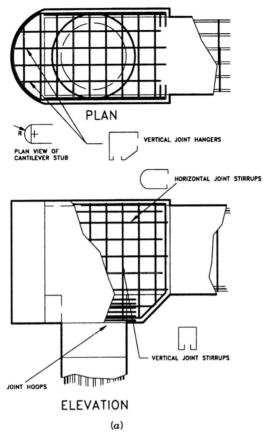

PLAN

PLAN VIEW OF
CANTILEVER STUB

VERTICAL JOINT HANGERS

HORIZONTAL JOINT STIRRUPS

VERTICAL JOINT STIRRUPS

JOINT HOOPS

ELEVATION

(a)

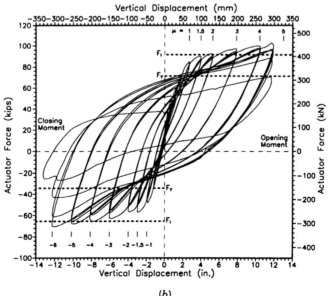

(b)

FIG. 8.29 Concrete jacketing of a deficient knee joint.

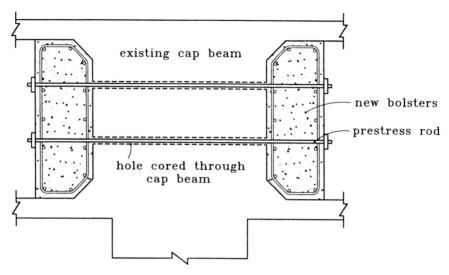

FIG. 8.30 Retrofitting monolithic cap–superstructure for joint force transfer under longitudinal response.

where strengthening by prestress or jacketing was technically unfeasible and where damage under seismic response would be unacceptable because of the resulting loss of lateral capacity or because the bridge was an essential part of post-earthquake lifeline networks. Joint replacement will generally require temporary falsework support of the superstructure, and because of the reduction to bent lateral capacity during the replacement, very few joints in a frame should be replaced simultaneously. Joint replacement will also be an option following damage to a joint in an earthquake where the design option of Section 8.4.2 is adopted or where damage occurs before retrofit could be implemented.

An example of joint replacement is shown in Fig. 8.31. The joint shown in Fig. 8.31(a) is a model of the actual repair and replacement of a knee joint damaged in the Loma Prieta earthquake and shown in Figs. 1.27 and 1.42, on which the model tests of Figs. 5.58 and 8.28 were based. The joint redesign, based on the principles of Section 5.4.4(a), involved complete removal of the joint concrete and a portion of column and cap beam framing into the joint to allow a haunched joint detail, increasing the amount of horizonal joint reinforcement confining the interlocking circles of the column longitudinal reinforcement (see Fig. 1.27), widening and extending the joint region and reinforcing this with both horizontal and vertical reinforcement to provide joint force transfer and to restrain the tails of the bent-down beam reinforcement, and partial reliance on existing cap beam shear reinforcement for force transfer under opening moments. The damaged test unit of Fig. 5.58 was repaired similarly [I5] by joint replacement, with details closely approximating the prototype repair details and based fully on the provisions of Section 5.4.4(a). Results of testing the repaired unit are shown in Fig. 8.31(b). The

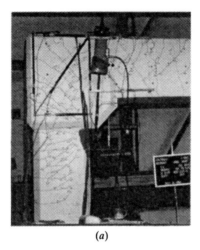

(a)

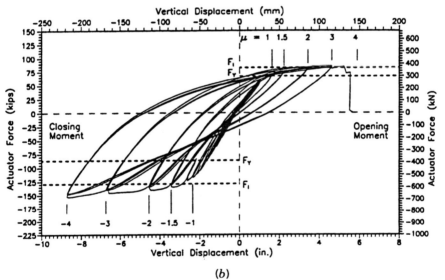

(b)

FIG. 8.31 Seismic response of a damaged knee joint unit after joint replacement and retesting (joint as-built behavior is shown in Fig. 5.58) [I5]. (a) Crack condition at failure; (b) Lateral force–displacement response.

repaired unit performed in close agreement with predictions, reaching theoretical flexural strength in both opening and closing directions, and finally failing due to fracture of the cap beam bottom reinforcement at the column–joint interface.

The joint performed very well throughout the test, with strain gauge readings on the new joint reinforcement indicating strain readings slightly less than yield strain. Figure 8.32 shows a comparison of the force–displacement envelopes of the as-built, repaired, and retrofitted unit, and also of a redesigned

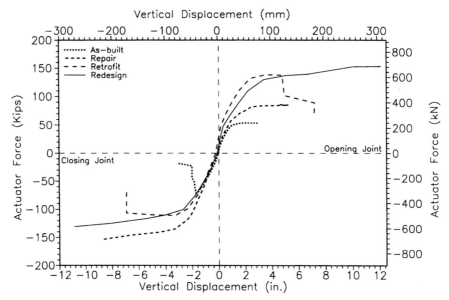

FIG. 8.32 Comparison of as-built, repaired, retrofitted, and redesigned test units for the I-980 joint failure of Fig. 1.24 [15].

unit based on the full design principles of Section 5.4.4, where the cap beam as well as the joint were redesigned to ensure formation of column plastic hinges. This unit eventually suffered a column flexural ductility failure under closing moments at a high drift angle of response.

8.5 SUPERSTRUCTURE RETROFIT

Superstructure deficiencies are normally associated with insufficient seating length at movement joints, and inadequate flexural capacity, particularly for positive moments, to force plastic hinges into columns.

8.5.1 Movement Joint Retrofit

There are two possible courses of action when displacements at movement joints are judged to be excessive. Restrainers may be placed across the joint in an attempt to reduce the relative displacements, or the displacement capacity of the movement joint can be increased. Often, both actions are taken.

(a) Restrainers. As well as being used to restrain displacements, restrainers may be placed in order that longitudinal seismic force can be transferred between adjacent frames. As discussed in Section 7.5.4, analysis and design for the interaction of inelastic frames connected by restrainers are complex

and cannot be achieved by simple elastic analyses. Results from dynamic inelastic analysis incorporating relatively sophisticated modeling of the movement joints indicate that maximum longitudinal displacements can be estimated from Eq. (7.27) and that restrainers are not likely to have a significant effect in reducing seismic relative displacements across movement joints unless the stiffness of the restrainer system is at least as high as that of the more flexible of the two frames connected by the restrainers.

Determination of the appropriate strength for the restrainers and reduction in displacement caused by stiff restrainers are difficult unless dynamic inelastic analyses are carried out. However, in many bridges the appropriate treatment for movement joints will be to use restrainers to lock the movement joints so that essentially no relative movement can occur. Older bridges frequently have movement joints at much closer spacing than would be provided for new bridges, yet it will generally be reasonable to assume that all creep and shrinkage movements of the bridges will have taken place by the time that retrofit is considered. Thus only thermal movements should need to be considered. It follows that many movement joints of existing bridges could be locked without causing column distress, particularly when realistic stiffnesses based on cracked-section analyses are adopted for determining column moments induced by thermal effects.

If internal movement joints are to be locked, it is relatively straightforward to determine the maximum seismic forces to be transferred across the joints. With reference to Fig. 8.33, the maximum longitudinal inertia force $F°$ that can be developed is the sum of the resisting forces at overstrength of the resisting elements:

$$F° = \mu P_1 + \sum_2^6 V°_{coli} + \mu P_7 \qquad (8.33)$$

where P_1 and P_7 are the axial loads at the abutments and μ is the appropriate coefficient of friction. The seismic inertia force per unit length is thus approximately

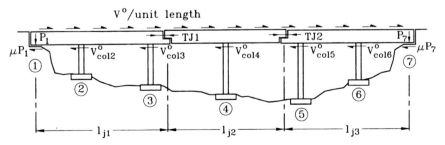

FIG. 8.33 Seismic forces across locked movement joints.

$$v^° = \frac{F^°}{l_{j1} + l_{j2} + l_{j3}} \tag{8.34}$$

For the direction of seismic response shown, the maximum tension force that can be developed across the movement joint between columns 3 and 4 is thus

$$TJ_1 = \mu P_1 + V_{col2}^° + V_{col3}^° - v^° l_{j1} \tag{8.35a}$$

and across the movement joint between columns 4 and 5 is

$$TJ_2 = \mu P_1 + \sum_{2}^{4} V_{coli} - v^°(l_{j1}+l_{j2}) = v^° l_{j3} - V_{col5}^° - V_{col6}^° - \mu P_7 \tag{8.35b}$$

Additional forces will be developed due to lateral response. Those can be estimated from the moments induced in the superstructure at maximum displacement. Again, these cannot be adequately calculated from an elastic analysis which will tend to overestimate the superstructure curvature unless the analysis is based on a substitute structure approach, reducing column stiffness in proportion to ductility demand, as outlined in Section 4.4.3. Longitudinal and transverse response should be combined in accordance with Section 4.3.2.

The length of bridge deck that can be locked will depend on temperature ranges, the coefficient of thermal expansion of the concrete, which can vary between $(6 - 14) \times 10^{-6}/°C$ [$(3.3 - 7.8) \times 10^{-6}/°F)$] [P21], and column flexibility. Locking of joints, which may involve grouting of the existing gap as well as placing and stressing restrainers should occur with the bridge at average temperature condition. Generally, movement joints at abutments should not be locked, to reduce seismic and thermal forces. Provision of seat extension at abutments is preferable and inexpensive [see Section 8.5.1(b)]. There may be concern that locking internal movement joints may induce excessive displacements in stiffer columns as a result of force transfer across joints. However, it should be pointed out that under the reversed direction of seismic response, placing the movement joint in compression, these forces will be inevitable anyway, and the columns will need appropriate retrofit.

Where superstructure movement joints are to be locked, this will normally be effected by prestressing the frames together across the joint. Another possibility is the use of oil viscous dampers connecting the frames across the joint. Under thermal slip rates these provide little effective restraint, but under the relative displacement rates typical of seismic response, the response is effectively rigid, limiting displacements and providing some damping. Further details are given in Section 6.2.1(e).

Flexible restrainers used in California typically use high-strength steel cables anchored to the diaphragms or webs of concrete bridges (Fig. 8.34) or to the bottom flange of steel girders (Fig. 8.35). Looping cables around the

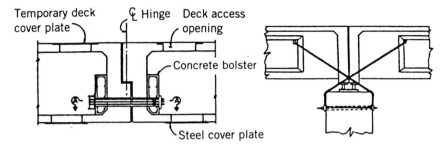

Temporary deck cover plate — ℄ Hinge — Deck access opening — Concrete bolster — Steel cover plate

FIG. 8.34 Cable restrainers for concrete superstructure movement joints. (*a*) Box-girder bridges; (*b*) I-girder bridges.

bent cap anchors the spans to the bent cap as well as to the adjacent spans but may induce undesirable force levels in the bents. The bent cables are susceptible to fracture at lower strains than with straight cables. Japanese practice has used a variety of restrainer types, including chains, rigid links, and knee-joint links, which provide limited displacement before locking up [P6].

(b) Seat Extenders. Where locking movement joints is undesirable or impractical, more reliance should be placed on extending the effective seating length of the movement joint than on limiting displacements with flexible restrainers unless nonlinear time-history analyses are carried out to determine the effectiveness of the design. Seat extenders are generally comparatively simple and inexpensive to install. Support lengths at abutments or under simply supported spans may be increased by corbels or brackets added to the sides of abutments or bents, as shown in Fig. 8.36(*a*). With internal movement joints, where the seating is provided within the depth of the superstructure section, direct seat extension is not possible. Figure 8.36(*b*) shows two alternatives that have been developed by Caltrans for use in California. The first involves thick-walled pipe seat extenders which are connected to the diaphragm on one side of a movement joint and slide freely through the other diaphragm. The pipes are designed to have sufficient strength to support the superstructure if unseating occurs. With the second approach, a series of

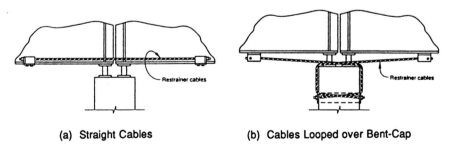

(a) Straight Cables   (b) Cables Looped over Bent-Cap

FIG. 8.35 Cable restrainers for steel superstructure movement joints.

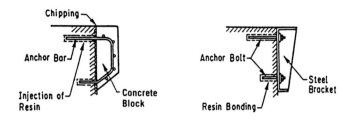

(a) corbels and brackets for bearing supported superstructures

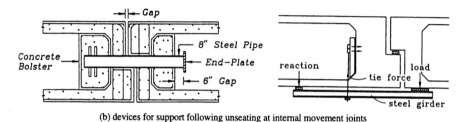

(b) devices for support following unseating at internal movement joints

FIG. 8.36 Seat extension to accommodate large longitudinal displacements.

underslung beams is bolted to the diaphragm of the supporting half of the movement joint and designed to carry the span by cantilever action if unseating occurs. Although less expensive than the pipe seat-extender option, this solution is rather unsightly and involves considerable displacement before the beams can support the load. Dynamic impact effects must be considered in design of underslung beams, which will have maximum instantaneous force levels at least twice the gravity load without considering additional effects from vertical seismic response.

In retrofit assessment of movement joints, it should be determined that it is physically possible for the joints to open to the extent that unseating can occur. If the abutments are assessed to be competent at the design level of seismic response, restraint to joint opening by the physical constraints of the abutments may mean that unseating is extremely unlikely. In very long bridges it is inconceivable that the full bridge length will be subjected to in-phase ground displacements, and as a consequence, dynamic response may be eliminated or considerably reduced. In such cases the deck may remain essentially stationary in the longitudinal direction while the ground moves relative to the superstructure. Maximum joint openings can then be computed from considering out-of-phase displacement components of the traveling wavefront (see Section 2.5.1).

8.5.2 Superstructure Flexural Capacity

Where superstructure flexural capacity, determined in accordance with the recommendations of Section 7.4.11, which are significantly more liberal than those suggested for new bridges in Section 5.7, is found to be inadequate, retrofit will need to be considered. Before a decision to retrofit is made, all other avenues should be explored since longitudinal superstructure retrofit will be expensive. It should be determined whether the levels of displacement required to develop superstructure strength can actually develop within the constraints to longitudinal response imposed by abutments and nonsynchronous seismic input, and if so, whether the superstructure has adequate ductility capacity to sustain the expected displacements without excessive response. It is recommended that this be assessed as corresponding to the serviceability limit state. When a decision to retrofit is made, similar options to those explored for cap beams will be available, notably to increase strength or to reduce forces in the superstructure.

(a) Strength Increase. External prestressing of the superstructure may be a feasible choice for superstructure strength enhancement. Generally, however, it will be a comparatively small region of the superstructure immediately adjacent to the support that is deficient in strength, particularly for positive moment capacity. This can be enhanced by coring through the cap beam above the soffit slab and placing renforcement bonded to the soffit slabs on either side of the cap beam, as shown in Fig. 8.37. This may be in the form of mild steel reinforcement, providing integral action with the existing soffit through roughening of the top surface and connection by dowels, as shown, or by high-strength bars connected to end brackets bolted into the soffit slab (not shown). The alternative of full superstructure prestress, which will enhance both positive and negative capacity, is also shown in Fig. 8.37. The length *l* the soffit strength enhancement needs to be taken will depend primarily on the positioning of terminated soffit slab reinforcement. Normally, this will be found to increase rapidly with distance from the cap beam, resulting in quite small values of *l*. It should also be recognized that the width of superstructure effective in resisting the input seismic moment from the column will increase with distance from the column in accordance with a 45° spread.

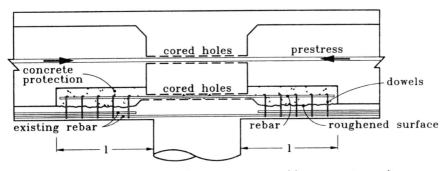

FIG. 8.37 Enhancement of superstructure positive moment capacity.

Even without pickup of terminated soffit slab reinforcement, this will often result in rapid increase in resisting moment capacity.

(b)   Force Reduction. Use of low-friction or low-damping bearings to limit longitudinal response will not be an option, since high superstructure moments will develop only for monolithic superstructure–column designs. Thus seismic force reduction can be achieved only by use of longitudinal link beams, in a fashion similar to that discussed in relation to cap beam retrofit, with reference to Fig. 8.19. These can be placed either parallel with the deck or below the deck, depending on locations of columns with respect to the superstructure width and physical constraints under the bridge. The two options are shown in Fig. 8.38.

When the superstructure is supported with cap beam outriggers, as shown in Fig. 8.38(*a*), it will be practical to place the link beam at the same level as the superstructure. In this case, the link beams *add* to the existing longitudinal capacity of the superstructure. However, because of their location in the plane of the columns, they attract a greater proportion of the seismic force than the superstructure, which is thereby protected from damage. A three-dimensional analysis should be carried out to determine the proportion of longitudinal seismic force carried by the link beams and superstructure.

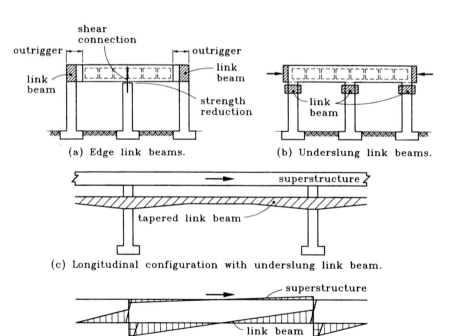

(a) Edge link beams.   (b) Underslung link beams.

(c) Longitudinal configuration with underslung link beam.

(d) Longitudinal bending moments with underslung link beam.

FIG. 8.38   Reduction of superstructure longitudinal moments with link beam.

A problem with this solution is also illustrated in Fig. 8.38(a). Unless extensive removal of superstructure is undertaken, it will not be feasible to place a link beam above the central column and superstructure damage may thus still occur at this location. It may be considered desirable to avoid this by reducing the column strength at the superstructure soffit. This can be achieved by cutting the longitudinal reinforcement at the column top and providing additional shear connection drilled through from the deck surface.

An alternative solution is shown in Fig. 8.38(b), utilizing underslung link beams in very similar fashion to the cap beam retrofit of Fig. 8.19. Figure 8.38(c) shows this detail in elevation. In this case link beams can be placed around all columns, reducing seismic moments in the superstructure to nominal values, as shown in Fig. 8.38(d). The link beam can be tapered, to reflect the seismic moment demand and to minimize self-weight moments and unnecessary additional seismic mass.

It will be recognized that the solutions suggested in Fig. 8.38 are extremes, to be avoided if at all possible. Following the collapse of the I-880 Cypress viaduct in Oakland in the 1989 Loma Prieta earthquake [E1], link beams were incorporated in the retrofit strategies for remaining double-deck viaducts in the San Francisco area. These retrofit designs incorporated edge link beams, which in some cases were connected to the existing superstructure, rather than underslung link beams. Figure 8.39 shows views of retrofit of these bridges.

Prior to accepting the concept of the edge beam retrofit extension experimental studies were carried out on large-scale models. Figure 8.40 shows the half-scale model of the retrofit concept using an isolated edge beam under test at the University of California–San Diego [P13,P14]. This test made use of symmetry to reduce the extent of structure modeled and an elaborate system of 13 computer-controlled actuators. The test unit modeled the structure from midspan to midspan on either side of a typical bent and from one column to

(a)   (b)

FIG. 8.39 Retrofit of the San Francisco double-deck viaducts, using edge link beams.

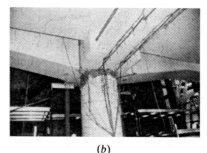

(a)   (b)

FIG. 8.40 Half-scale model of retrofit concept for San Francisco double-deck viaducts [P13, P14]. (a) General view of model under test; (b) Column hinge condition at twice design displacement.

midspan of a two-span bent in the transverse direction. The upper deck was not modeled because its pinned connection to the upper column made it possible to fully simulate its influence on structural response through applied forces and displacements.

This structural test, which was carried out under simulated transverse, longitudinal, and diagonal (i.e., transverse + longitudinal) response, established that the retrofit concept, which also included cap beam prestressing, as dicussed in Section 8.3.2(a), could be expected to be fully effective in reducing damage in a major earthquake to a level where continued traffic flow after the earthquake would not be impeded by repair measures. However, the cost of the repairs was of similar magnitude to complete replacement, and it could be argued that the longitudinal retrofit involving link beams, which formed a major part of the cost, could have safely been eliminated because of considerations discussed at the start of this section.

8.6 FOOTING RETROFIT

Retrofit of footings is potentially the most expensive aspect of bridge seismic design, and as a consequence, careful consideration should be given to the extent of damage expected in the as-built condition, and whether this may be sustained without detrimental effects on the response of the full structure. As discussed in Section 5.6.2(a), there are few, if any, reports of footing failure in earthquakes, except where caused by liquefaction or sliding of ground on sloping surfaces. The high cost of footing retrofit and the relative absence of damage to footings may together point toward less conservatism in retrofit design compared with other bridge components.

8.6.1 Stability Enhancement

It has been discussed in Sections 6.4 and 7.4.10(a) that rocking of footings may be considered as a form of seismic isolation, reducing input to bridge piers and superstructures. Rocking will occur when the stability calculations, carried out in accordance with Section 5.6.2(*a*), indicate that overturning capacity is less than the column moment capacity. This does not, however, necessarily indicate a potential for collapse any more than does formation of a plastic hinge at the base of a column.

When rocking response is unacceptable, perhaps because of excessive displacements induced in the superstructure, remedial actions will be required. Three main alternatives exist. The first involves the use of soil anchors or positive connection of footing to existing concrete piles by tie rods placed in rods drilled through the footing into the piles, as shown in Fig. 8.41(*a*). This might be adopted when existing piles are not connected by reinforcement anchored into the footing, or where a spread footing exists. Although overturning resistance will be enhanced, a consequence will be development of negative bending moments in the footing with the potential for flexural crack propagation from the top surface. Since such footing designs will not normally possess a top mat of reinforcement, failure will occur unless a reinforced footing overlay is added.

The second option is illustrated in Fig. 8.41(*b*) and (*c*). In this case, footing stability is enhanced by increasing the footing plan dimensions, placing new additional peripheral piles, properly connected to the new footing concrete, and providing a reinforced concrete overlay.

A third option, which may be possible for superstructures continuous over the full bridge length, with strong abutments, may be to permit rocking but to limit rocking displacements by the use of damping devices placed between the superstructure and abutments. Principles developed in Chapter 6 can be employed for the design of such a retrofit system. If footing retrofit of this magnitude is undertaken, it should be designed in accordance with provisions for new bridges, given in Section 5.6.

Finally, footing seismic forces can be greatly reduced and stability enhanced by placing a link beam between adjacent columns immediately above the footings, as illustrated in Fig. 8.19(*c*).

8.6.2 Flexural Enhancement

Retrofitting to enhance footing flexural strength will involve an overlay of reinforced concrete doweled to the existing footing, as shown in Fig. 8.41(*b*) and (*c*), or more infrequently, coring horizontally through the existing footing and prestressing the footing. With recent reductions in the cost of concrete coring, at least in California, the latter option has become more competitive. Where a top mat is placed in an overlay, the bulk of the top reinforcement should be located within a distance equal to the retrofitted footing depth from the column sides in accordance with the design recommendations of Section 5.6.2(*c*). The surface of the existing footing should be roughened, to improve shear transfer, and dowels set into the top and sides of the existing footing capable of transferring the interface shear stress using a coefficient of friction of $\mu = 1.0$.

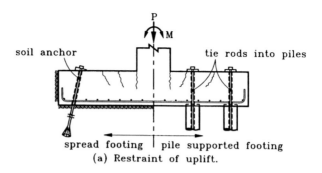

spread footing | pile supported footing
(a) Restraint of uplift.

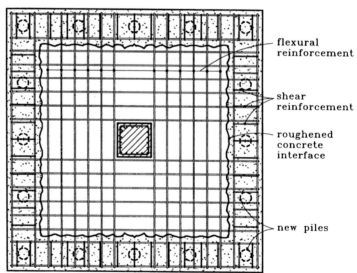

(b) Footing overlay and extension.

(c) Plan view of footing overlay and extension
— new top reinforcement mat shown.

FIG. 8.41 Footing retrofit measures.

Increasing the depth of the footing will increase the positive moment capacity as a result of the increased section depth. Further increase can be achieved by placing bottom reinforcement parallel to the footing in the side extensions. However, as noted in Section 5.6.2, rebar placed farther than a distance equal to the retrofitted footing depth from the column face will have reduced efficiency in resisting column flexure unless significant inelastic deformations are accepted in the footing adjacent to the column. Longitudinal reinforcement perpendicular to the existing footing boundaries should be exposed by chipping back the bottom edge of the footing to enable extensions to be welded or coupled to the existing reinforcement.

8.6.3 Shear Enhancement

Footing shear strength deficiencies are more problematical than flexural deficiencies. Since shear retrofit within the confines of the existing footing dimensions will be difficult, alternative mechanisms of shear transfer, including diagonal compression struts between compression piles and the column compression zone, as shown in Fig. 5.85, should be investigated. Some shear enhancement will accrue from the increase in depth resulting from footing overlay, and dowels connecting the footing overlay to the existing footing will increase shear strength if carried down to the base of the footing. This enhancement will be less than for an equivalent area of properly anchored transverse reinforcement, and an efficiency factor of 50% is tentatively suggested.

Prestressing of the footing in the horizontal direction, suggested as a mechanism for flexural strength enhancement, will also clearly enhance shear strength of the footing by providing an axial force component in the shear strength model proposed in Section 5.3.4(b)(iii).

Although closed stirrups can be placed within the full depth of footing extensions, these will act primarily to assist shear transfer from corner piles back into the existing footing and will do little to enhance shear strength in the critical region of the footing close to the column compression force resultant. An example of footing retrofit employing most of the details illustrated in Fig. 8.41(b) was presented in Fig. 1.43.

8.6.4 Footing Joint Shear Force Enhancement

It was shown in Section 5.6.4(a) that joint failure underneath the column could be expected when footing joint principal tension stress levels exceeded capacity. As shown by the example of Fig. 5.86, failure can be catastrophic. Although expensive to remedy, a number of options are possible. First, an overlay such as shown in Fig. 8.41(b) will increase the depth of the footing, thus reducing joint shear stress and hence principal tension stress given by Eq. (5.89). The option of a large footing depth increase, effected with a sloping top surface to minimize overlay cost and suggested in Fig. 8.42, could be considered. This will also have significant impact in flexural and shear strength enhancement and could be used to confine a column base splice, if present. Long dowels placed within the region given in Fig. 5.87(b) and extending as close as possible to the base of the footing will also be effective, by mobilizing the mechanisms described in Fig. 8.43 and discussed subsequently, when joint strength is still inadequate. Axial prestress, shown in Fig. 8.42(b), will inhibit joint cracking by reducing principal tension stresses in the joint region. Finally, high-strength through-bolts placed in holes drilled through the footing within the area defined by Fig. 5.87(b) and anchored against the footing base could be considered, although this would require excavation under the footing, which is expensive.

Tests on retrofitted footings at the University of California–San Diego have indicated that retrofit designs based on the concepts presented in Fig. 8.41(b) can successfully improve the seismic response of footings deficient in all of the foregoing areas. Figure 8.43 shows a model footing under retrofit implementation, and subsequent lateral force–displacement response. This footing consisted of an as-built footing and column duplicating the design of the footing in Fig. 5.86, which failed in the as-built configuration. Retrofit consisted of footing overlay, footing extension, and the addition of perimeter piles. Although the retrofit design was sufficient to increase footing flexural and shear strength to the extent that column hinging was expected, joint principal tensions stresses still exceeded the limits of $3.5\sqrt{f'_c}$ psi $(0.29\sqrt{f'_c}$ MPa) recommended for fully ductile columns. Despite this, the column achieved its theoretical flexural strength, indicated in Fig. 8.43(b) by the sloping line H_{if}, which includes the P–Δ reduction, up to displacement ductility factors of $\mu_\Delta = 3$, although the hysteresis loops displayed poor hysteretic energy absorption capacity.

Observations during the test [XI] indicated that a fan-shaped pattern of radial cracks was developing on the upper surface of the footing, as shown in Fig. 8.44(a), and that the upper surface was developing significant convex curvature, with the concrete adjacent to the column displaced upward by about 0.75 in. (19 mm) relative to the sides of the footing. This indicated a horizontal crack plane of significant width within the footing, verified after testing completion by cutting through the footing.

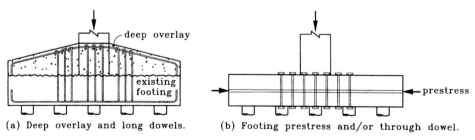

(a) Deep overlay and long dowels.   (b) Footing prestress and/or through dowel.

FIG. 8.42   Retrofit measures for joint shear stress deficiencies.

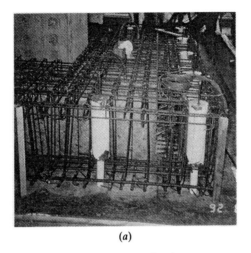

(a)

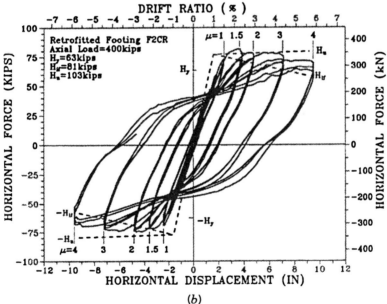

(b)

FIG. 8.43 Retrofit and testing of a model footing. (a) Retrofit measures; (b) Lateral force–displacement response.

Analyses carried out after the test [X2] indicated that the formation of this crack plane, which developed at the level of the base of the dowels connecting the overlay to the existing footing, was essential to the satisfactory test result. Joint shear failure had indeed occurred during the test, and resistance to failure was provided by the mechanism suggested in Fig. 8.44(b). Pull-out of the column tension reinforcement was resisted by struts to the column compression zone and outward into the footing. The latter component required

(a) fan cracks on top surface

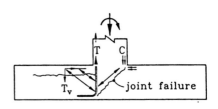

(b) restraint of tension force after joint failure

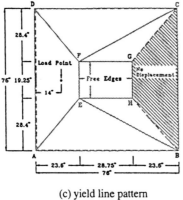

(c) yield line pattern

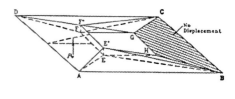

(d) displacement of yield slabs

FIG. 8.44 Mechanism for force transfer after joint failure of footing.

a vertical force $T_v \approx 0.5T$ for stability. Since no vertical reinforcement passed through to the base of the existing footing, the only mechanism available to resist T_v was from the formation of a yield line pattern, involving the overlay reinforcement and the depth of concrete bonded by the connecting dowels. This mechanism is illustrated in schematic form in Fig. 8.44(c) and (d) and is apparent in the crack pattern of Fig. 8.44(a). Using standard yield line theory [P22] it was possible to obtain close agreement between the predicted resistance to the uplift force T_v and the capacity developed by the test unit.

Clearly, although this test unit survived, the degree of damage was rather extreme and undesirable, as it would require extensive post-earthquake remedial action. A second footing, incorporating longer dowels to mobilize a greater effective slab depth for the yield line action, performed significantly better [X2]. Although not tested to date, it is felt that the retrofit options suggested in Fig. 8.42 are likely to result in even better response.

8.7 SEISMIC ISOLATION IN BRIDGE RETROFIT

In many cases of seismic retrofit of bridges supported by bearings on cap beams it will be necessary to replace the bearings. An attractive retrofit concept in this case will be to use isolation and damping devices for the replacement bearings, designed in accordance with the principles developed in Chapter 6. This approach has been used in several cases in the United States [M3] and in many cases in Italy [D4].

When the bridge has a continuous superstructure and has bents of different heights, as is commonly the case for bridge crossings over navigable rivers or harbors, there will often be severe seismic demands on the shorter, stiffer bents. In such cases the bearings can be designed with different stiffnesses so that the stiff bents support the superstructure through laterally flexible bearings, while the taller, more flexible bents use stiffer bearings to support the superstructure. In this fashion, as discussed in more detail in Chapter 6, the demands on the individual bents can be better distributed in accordance with the bent capacities.

REFERENCES

A1   Ang, B. G., M. J. N. Priestley, and T. Paulay, *Seismic Shear Strength of Circular Bridge Piers*, Report 85-5, Department of Civil Engineering, University of Canterbury, Christchurch, New Zealand, July 1985, 146 pp.

A2   ATC, *Seismic Retrofitting Guidelines for Highway Bridges*, Report ATC-6-2, Applied Technology Council, Palo Alto, Calif., August 1983.

A3   Nilson, A. H., *Design of Prestressed Concrete*, 2nd ed., Wiley, New York, 1978.

A4   *AASHTO Standard Design Specifications for Highway Bridges*, American Association of State Highway and Transportation Officials, Washington, D.C., 1992.

A5   *ACI Building Code Requirements for Reinforced Concrete* (ACI 318-89) and *Commentary* (ACI 318R-89), American Concrete Institute, Detroit, Mich., revised 1992, 347 pp.

A6   Ang, B. G., M. J. N. Priestley, and T. Paulay, "Seismic Shear Strength of Circular Reinforced Concrete Columns," *ACI Structural Journal*, Vol. 86, No. 1, January/February 1989, pp. 45–59.

A7   ASCE/ACI Task Committee 426, "The Shear Strength of Reinforced Concrete Members," *ASCE Journal of Structural Engineering*, Vol. 99, No. ST6, June 1973, pp. 1091–1187.

A8   Astaneh-Asl, A, and Cho, S. W., "Compressive Ductility of Critical Members of the Golden gate bridge". Report No UCB/CE-STEEL 93/01, University of California, Berkeley, 1993.

A9   Arias, A., "A Measure of Earthquake Intensity," in *Seismic Design for Nuclear Plants*, MIT Press, Cambridge, Mass., 1970, pp. 438–469.

A10   Parker, D. R., and R. A. Cameron, *Strength and Ductility Analysis of Typical Concrete Piers and Steel Laced Members for the Richmond San Rafael Bridge*, ANATECH Report, San Diego, Calif., November 1994.

A11   Autostrade, *Guidelines for Seismic Design of Bridges with Isolator/Dissipator Devices*, Autostrade S.P.A., Roma, Italy.

B1   Bazant, Z. P., and P. D. Bhat, "Endochronic Theory of Elasticity and Failure of Concrete," *Proceedings, ASCE*, Vol. 102, No. EM4, April 1976, pp. 701–702.

B2   Berrill, J. B., M. J. N. Priestley, and R. Peek, "Further Comments on Seismic Design Loads for Bridges," *Bulletin of the New Zealand National Society for Earthquake Engineering*, Vol. 14, No. 1, March 1981, pp. 3–11.

B3   Budek, A., G. Benzoni, and M. J. N. Priestley, "In-Ground Plastic Hinges in Column/Pile Shaft Design," *Proceedings, 3rd Annual Caltrans Seismic Research Workshop*, California Department of Transportation, Division of Structures, Sacramento, Calif., June 1994, 9 pp.

B4   Bekingsale, C. W., *Post-elastic Behavior of Reinforced Concrete Beam–Column Joints*, Research Report 80-20, Department of Civil Engineering, University of Canterbury, Christchurch, New Zealand, August 1980, 398 pp.

B5   Bernal, D., "Amplification Factors for Inelastic Dynamic P–Δ Effects in Earthquake Analysis," *Earthquake Engineering and Structural Dynamics*, Vol. 15, 1981, pp. 635–651.

B6   Bartlett, S. F., and T. L. Youd, *Empirical Analysis of Horizontal Ground Displacement Generated by Liquefaction-Induced Lateral Spreads*, Technical Report 92-0021, National Center for Earthquake Engineering Research, Buffalo, N.Y., 1992.

B7 Banon, H., and D. Veneziano, "Seismic Safety of Reinforced Concrete Members and Structures," *Earthquake Engineering and Structural Dynamics*, Vol. 10, 1982, pp. 179–193.

B8 Benzoni, G., and M. J. N. Priestley, Lateral Seismic Forces in Bridge Abutment Shear Keys', Structural Systems Research Report SSRP 95/15, University of California, San Diego, Oct. 1995.

B9 Berg, V. B., and J. L. Stratta, *Anchorage and the Alaska Earthquake of March 27, 1964*, American Iron and Steel Institute, New York, 1964, 63 pp.

C1 Caltrans, *Bridge Design Specifications*, California Department of Transportation, Sacramento, Calif., 1993.

C2 Chai, Y. H., M. J. N. Priestley, and F. Seible, "Seismic Retrofit of Circular Bridge Columns for Enhanced Flexural Performance," *ACI Structural Journal*, Vol. 88, No. 5, September/October 1991, pp. 572–584.

C3 Calvi, M. C., and G. R. Kingsley, "Displacement Based Seismic Design of Multi-Degree-of-Freedom Bridge Structures," *Proceedings, 2nd International Workshop on the Seismic Design of Bridges*, Queenstown, New Zealand, August 1994.

C4 Vecchio, V. J., and M. P. Collins, "The Modified Compression-Field Theory for Reinforced Concrete Elements Subjected to Shear," *ACI Structural Journal*, Vol. 83, No. 2, March–April 1986, pp. 219–231.

C5 Collins, M. P., and D. Mitchell, *Prestressed Concrete Structures*, Prentice Hall, Englewood Cliffs, N.J., 1991.

C6 Carvalho, E. C., D. Coelho, and A. V. Azevedo, "Seismic Assessment and Retrofit of an Existing Bridge," *Proceedings, 2nd International Conference*, Queenstown, New Zealand, 1994, pp. 777–794.

C7 Chai, Y. H., M. J. N. Priestley, and F. Seible, *Flexural Retrofit of Circular Reinforced Concrete Bridge Columns by Steel Jacketing*, Structural Systems Research Project, Report SSRP-91/06, University of California, San Diego, October 1991, 151 pp.

C8 Calvi, G. M., "Evaluation of Energy Spectra for Seismic Design of Structures," *Proceedings, International Meeting on Earthquake Protection of Buildings*, Ancona, Italy, 1991, pp. B65–74.

C9 Celebi, M. (Ed.), *Seismic Site-Response Experiments Following the March 3, 1985 Central Chile Earthquake*, Geophysical Survey Report 86-90, U. S. Department of the Interior, Menlo Park, Calif., 1986.

C10 Clough, R. W., and J. Penzien, *Dynamics of Sructures*, 2nd ed., McGraw-Hill, New York, 1994.

C11 Calvi, G. M., M. Ciampoli, and P. E. Pinto, "Guidelines for Seismic Design of Bridges: Background Studies, Part 1" p 114: European Earthquake Engineering, Vol. 2, 1989, pp. 3–16.

C12 Caltrans, *Seismic Design References*, California Department of Transportation, Sacramento, Calif., 1992.

C13 Caltrans, *Seismic Design of Highway Bridge Foundations: Training Course Manual*, California Department of Transportation, Sacramento, Calif., 1994.

C14 Cheney, R. S., and R. G. Chassie, *Soils and Foundations Workshop Manual*, FHWA-HI-88-009, NHI Course 13212, Federal Highways Administration, Washington, D.C., 1982.

C15 Carrubba, P., and M. Maugeri, "In-Situ and Laboratory Evaluation and Initial Shear Modulus of a Clay," *Proceedings, 10th European Conference on Soil Mechanics and Foundation Engineering*, Discussion Session 1b, Florence, 1991.

C16 Committee on Earthquake Engineering, National Research Council, *Liquefaction of Soils During Earthquakes*, National Academy Press, Washington, D.C., 1985.

C17 Carr, A. J., *Ruaumoko Inelastic Analysis Program*, University of Canterbury, Christchurch, New Zealand, 1992.

C18 Carter, D. P., *A Nonlinear Soil Model for Predicting Lateral, Soil Response*, Report 359, Civil Engineering Department, University of Auckland, New Zealand, 1984.

C19 Cook, R. D., D. S. Malkus, and M. E. Pleska, *Concepts and Applications of Finite Element Analysis*, 3rd ed. Wiley, New York, 1989.

C20 Carr, A. J., "Dynamic Analysis of Structures," *Bulletin of the New Zealand National Society for Earthquake Engineering*, Vol. 27, No. 2, June 1994, pp. 129–146.

C21 Cormack, L. G., "The Design and Construction of the Major Bridges on the Mangaweka Rail Deviation," Transactions, Institution of Professional Engineers, New Zealand, 15, I/CE, 1988 pp 16–23.

C22 Ciampoli, M., and P. E. Pinto, "Effects of Soil-structure Interaction on Inelastic Seismic Response of Bridge Piers," *Journal of Structural Engineering*, ASCE, Vol. 121, No. 5, 1995, pp. 806–814.

D1 Dodd, L. L., "The Dynamic Behavior of Reinforced Concrete Bridge Piers Subjected to New Zealand Seismicity." University of Canterbury, Department of Civil Engineering, Christchurch, New Zealand, Research Report No. 92-04, 1992.

D2 Der Kiureghian, A., and A. Neuenhoferf, "Response Spectrum Method for Multisupport Seismic Excitation," *Earthquake Engineering and Structural Dynamics*, Vol. 21, 1992, pp. 713–740.

D3 Darwin, D., and D. A. Pecknold, "Nonlinear Biaxial Stress–Strain Law for Concrete," *Journal of the Engineering Mechanics Division, ASCE*, Vol. 103, No. EM2, April 1977.

D4 Dolce, M., "Retrofitting-Europe," *Proceedings, International Workshop on Seismic Design and Retrofitting of Reinforced Concrete Bridges*, Bormio, April 1991, pp. 441–468.

E1 EERI, "Loma Prieta Earthquake Reconnaissance Report," *Earthquake Spectra*, Special Supplement to Vol. 6, May 1990, 448 pp.

E2 EERI, "Northridge Earthquake Reconnaissance Report," *Earthquake Spectra*, Special Supplement to Vol. 11, Feb 1995, 116 pp.

E3 EERI, "The Chile Earthquake of March 3, 1985," *Earthquake Spectra*, Vol. 2, No. 2, 1986, 513 pp.

E4 EERI, "Costa Rica Earthquake Reconnaissance Report," *Earthquake Spectra*, Special Supplement to Vol. 7, October 1991, 127 pp.

E5 *Eurocode 8 Design Provisions for Earthquake Resistance of Structures, Part 1: General Rules*, ENV 1993-2, Comité Europeen de Normalization, Brussels, 1994.

E6 Izzuddin, B. A., C. G. Karayannis, and A. S. Elnashai, "Advanced Nonlinear Formulation for Reinforced Concrete Beam-Columns," *ASCE Journal of Structural Engineering*, Vol. 120, No. 10, October 1994, pp. 2913–2934.

E7 Eligihausen, R., E. P. Popov, and V. V. Bertero, *Local Bond Stress–Slip Relationships of Deformed Bars Under Generalized Excitation*," Report UCB/EERC 83/23, Earthquake Engineering Research Center, University of California, Berkeley, 1983, 178 pp.

E8 Esteva, L., and E. Rosenblueth, "Espectras de Temblores a Distancias Moderadas y Grandes," *Bulletin Sociedad Mexicana Ingenieria Sismica.*, Vol. 2, No. 1, 1964, pp. 1–18.

E9 *Eurocode 8 Design Provisions for Earthquake Resistance of Structures, Part 2: Bridges*, ENV 1998-2, Comité Europeen de Normalization, Brussels, 1994.

E10 Esteva, L. "Bases para la Formulucion de Decisións de Diseño Sismico," Doctoral thesis, National University of Mexico, 1968.

E11 EERC "The Hyogo-ken Nambu Earthquake-Preliminary Reconnaissance Report," Report No. 94-04, University of California, Berkeley, 1995.

F1 Fung, G. G., R. J. Lebeau, E. D. Klein, J. Belvedere, and A. F. Goldschmidt, *Field Investigation of Bridge Damage in the San Fernando Earthquake*, Technical Report, Bridge Department, Division of Highways, California Department of Transportation, Sacramento, Calif., 1971, 209 pp.

F2 "Final Program and Abstract," 1st World Conference on Structural Control, Los Angeles, 1994.

G1 Gates, J., and I. G. Buckle, "Basic Design Concepts," *Proceedings, International Workshop on Seismic Design and Retrofitting of Reinforced Concrete Bridges*, Bormio, Italy, April 1991, pp. 7–15.

G2 Gulkan, P., and M. Sozen, "Inelastic Response of Reinforced Concrete Structures to Earthquake Motions," *ACI Journal*, December 1974.

G3 Ghali, A., and A. M. Neville, *Structural Analysis*, 3rd ed., Chapman & Hall, London, 1989.

G4 Giannini, R., G. Monti, G. Nuti, and T. Pagnoni, *ASPIDEA: A Program for Nonlinear Analysis of Isolated Bridges Under Non-synchronous Seismic Action*, Report 5/92. Dipartimento di Ingegneria Civile Délle Acque e del Terréno, Universitá del' Aquila, 1992.

H1 Housner, G. W., "Limit Design of Sructures to Resist Earthquakes," *Proceedings, First World Conference on Earthquake Engineering*, Berkeley, Calif., Vol. 5, 1956, pp. 5.1–5.13.

H2 Hambley, E. C., *Bridge Deck Behavior*, Chapman & Hall, London, 1976.

H3 Housner, G., "Competing Against Time," report to Governor George Deukmejian from the Governor's Board of Inquiry on the 1989 Loma Prieta Earthquake, May 1990.

H4 Hilber, H. M., T. J. R. Hughes, and R. L. Taylor, "Improved Numerical Dissipation for Time Integration Algorithms in Structural Dynamics," *Earthquake Engineering and Structural Dynamics*, Vol. 5, 1977, pp. 283–292.

H5 Hughes, T. J. R., *The Finite Element Method*, Prentice Hall, Englewood Cliffs, N.J., 1987.

H6 Housner, G. W., "The Behavior of Inverted Pendulum Structures During Earthquakes," *Bulletin of the Seismic Society of America*, Vol. 53, No. 2, 1963, pp. 403–417.

H7   Hagiwaza, R., Earthquake Ground Motion Characteristics, *Proceedings, 1st Japan–Italy Workshop on Seismic Design of Bridges*, Public Works Research Institute, Tsukuba, Japan, 1995.

I1   Izzuddin, B. A., and S. A. Elnashai, *ADAPTIC: A Program for Adaptive Large Displacement Elastoplastic Dynamic Analysis of Steel, Concrete, and Composite Frames*, Report ESEE 7/89, ESEE, Imperial College, London, 1989.

I2   Ingham, J. M., M. J. N. Priestley, and F. Seible, "Outrigger Bent Knee Joint Retrofit and Repair," *Proceedings, 3rd NSF Workshop on Bridge Engineering Research in Progress*, La Jolla, Calif., November 16–17, 1992, pp. 297–300.

I3   *Proceedings, World Congress on Joint Sealing and Bearing Systems for Concrete Structures*, Publication SP-70, American Concrete Institute, Detroit, Mich., 1981.

I4   *Proceedings, 2nd World Congress on Joint Sealing and Bearing Systems*, Publication SP-94, American Concrete Institute, Detroit, Mich., 1986.

I5   Ingham, J., M. J. N. Priestley, and F. Seible, *Seismic Performance of Bridge Knee Joints*, Vol. 1, Sructural Systems Research Project, Report SSRP-94/12, University of California, San Diego, June 1994, 277 pp.

I6   Ingham, J. M., M. J. N. Priestley, and F. Seible, "*Seismic Performance of Bridge Knee Joints*, Vol. 2, *Circular Column/Cap Beam Experimental Results*, Structural Systems Research Project, Report SSRP-94/17, University of California, San Diego, October 1994, 324 pp.

I7   Iwan, W. D., "Estimating Inelastic Response Spectra from Elastic Spectra," *Earthquake Engineering and Structural Dynamics*, Vol. 8, 1980, pp. 375–388.

I8   Idriss, I. M., and H. B. Seed, "An Analysis of Ground Motions During the 1957 San Francisco Earthquake," *Bulletin of the Seismological Society of America*, Vol. 58, No. 6, 1968, pp. 2013–2032.

I9   Idriss, I. M., J. Lysmer, R. Hwang, and H. B. Seed, *QUAD-4: A Computer Program for Evaluating the Seismic Response of Soil Structure by Variable Damping Finite Element Procedures*, Report 73-16, Earthquake Engineering Research Center, Berkeley, Calif., 1973.

I10 Ingham, J., M. J. N. Priestley, and F. Seible, *Seismic Performance of Bridge Knee Joints: Design Examples,* Structural Systems Research Project, Report SSRP95-14, University of California, San Diego, 1995.

I11 Idriss, I. M., "Evaluating Seismic Risk in Engineering Practice." Theme Lecture No. 6, *Proceedings, XI International Conference on Soil Mechanics and Foundation Engineering*, San Francisco, August 1985, Vol. 1, pp. 255–320.

J1   Jirso, J. O., and K. A. Woodward, *Behavior Classification of Short Reinforced Concrete Columns Subjected to Cyclic Deformations*, PMFSEL Report 80-2, Department of Civil Engineering, University of Texas at Austin.

J2   Jirso, J. O. (Editor) "Progress Reports, 1995—Repair and Rehabilitation Research for Seismic Resistance of Structures" Report R/R 1995, University of Texas, Austin, February 1995.

J3   Jennings, P. C., G. W. Housner, and N. C. Tsai, "Simulated Earthquake Motions for Design Purposes," *Proceedings, 4th World Conference on Earthquake Engineering*, Santiago, 1960, pp. 145–160.

K1   Kawashima, K., H. Ichimasu, and H. Ohuchi, "Retrofitting," *Proceedings, International Workshop on Seismic Design and Retrofitting of Reinforced Concrete Bridges*, Bormio, Italy, April 1991, pp. 471–501.

K2 Kent, D. C., and R. Park, "Flexural Members with Confined Concrete," *Proceedings, ASCE*, Vol. 97, No. ST7, July 1971, pp. 1969–1990.

K3 Kowalsky, M., M. J. N. Priestley, and G. A. MacRae, "Displacement-Based Design of RC Bridge Columns," *Proceedings, 2nd International Workshop on the Seismic Design of Bridges*, Vol. 1, Queenstown, New Zealand, August 9–12, 1994, pp. 138–163.

K4 Thorkildsen, E., M. Kowalsky, and M. J. N. Priestley, "Use of Lightweight Concrete in Seismic Design of California Bridges," *Proceedings, International Symposium on Structural Lightweight Aggregate Concrete*, Sandefjord, Norway, June 1995.

K5 Kawashima, A. T., K. Aizawa, and K. Takahashi, "Attenuation of Peak Ground Acceleration, Velocity and Displacement Based on Multiple Regression Analysis of Japanese Strong Motion Records," *Earthquake Engineering and Structural Dynamics*, Vol. 14, 1986, pp. 199–215.

K6 Kurkchubasche, A., and F. Seible, *PCYCO: Program for Cyclic Analysis of Concrete Structures: Users Reference Guide*, Structural Systems Research Project, Report SSRP-94/07, University of California, San Diego, January 1994, 55 pp.

K7 Kang, Y. J., *Nonlinear Geometric, Material and Time-Dependent Analysis of Reinforced and Prestressed Concrete Frames*, UC-SESM Report No. 77-1, University of California, Berkeley, 1977.

K8 Kowalsky, M. J., M. J. N. Priestley, and G. A. MacRae, "Displacement-based Design. A Methodology for Seismic Design applied to Single Degree of Freedom Reinforced Concrete Structures," *Report SSRP 94/16*, University of California, San Diego, 1994, 131 pp.

L1 Leonhardt, F., *Bridges: Aesthetics and Design*, Deutsche Verlagsanstalt, Stuttgart, Germany, 1982.

L2 Leslie, P. D., "Durability of Reinforced Concrete Bridge Piers," M. E. thesis, Department of Civil Engineering, University of Canterbury, Christchurch, New Zealand, 1974, 147 pp.

L3 Lysmer, J., T. Ukada, C. F. Tsai, and H. B. Seed, *FLUSH: A Computer Program for Approximate 3-D Analysis of Soil Structure Interaction Problems*, Report 75-30, Earthquake Engineering Research Center, University of California, Berkeley, 1975.

L4 Luco, J. E., and H. L. Wong, "Response of a Rigid Foundation to a Spatially Random Ground Motion," *Earthquake Engineering and Structural Dynamics*, 1986, pp. 891–908.

L5 Lin, L. F., "Back Analysis of Lateral Load Tests on Piles," M.E. thesis, Civil Engineering Department, University of Auckland, New Zealand, 1988.

L6 Lam, I. P., *ATC-32*, "Seismic Design of Bridges, Resource Document, Foundations." Applied Technology Council, Palo Alto, Calif., 1995 (Draft).

L7 Lam, I. P., "Soil Structure Interaction Related to Piles and Footings," *Proceedings, 2nd International Workshop on the Seismic Design of Bridges*, Vol. 1, Queenstown, New Zealand, August 1994.

M1 "Bridge Manual: Design and Evaluation." *Proceedings, International Workshop on Seismic Design and Retrofit of Reinforced Concrete Bridges*, Bormio, Italy, April 1991, pp. 49–81.

M2 MacGreggor, J., *Fundamentals of Reinforced Concrete Mechanics and Design*, Prentice Hall, Englewood Cliffs, N.J., 1988.

M3 Mayes, R. L., *Seminar Notes on the Seismic Isolation of Bridges*, Dynamic Isolation Systems, Berkeley, Calif., 1993.

M4 Mander, J. B., M. J. N. Priestley, and R. Park, "Theoretical Stress–Strain Model for Confined Concrete," *Journal of the Structural Division, ASCE*, Vol. 114, No. 8, August 1988, pp. 1804–1826.

M5 Mahin, S., F. Zayati, S. Mazzoni, J. Moehle, and C. Thewald, "Evaluation of a Seismic Retrofit Concept for Double Deck Viaduct," *Proceedings, 3rd NSF Workshop on Bridge Engineering Research in Progress*, La Jolla, Calif., November 1992, pp. 317–320.

M6 Mander, J. B., M. J. N. Priestley, and R. Park, *Seismic Design of Bridge Piers*, Report 84-2, Department of Civil Engineering, University of Canterbury, Christchurch, New Zealand, February 1984, 442 pp.

M7 Mattock, A. H., and Z. Wang, "Shear Strength of Reinforced Concrete Members Subject to High Axial Compressive Stress," *ACI Structural Journal*, Vol. 81, No. 3, pp. 287–298.

M8 MacRae, G. A., M. J. N. Priestley, and F. Seible, *Santa Monica Viaduct Retrofit: Large-Scale Column–Cap Beam Joint Transverse Test*, Technical Report 94/02, University of California, San Diego, August 1994, 117 pp.

M9 Mahin, S. A., and R. L. Baroschek, "Influence of Geometric Nonlinearities on the Seismic Response of Bridge Structures," *Proceedings, 3rd NSF Workshop on Bridge Engineering Research in Progress*, La Jolla, Calif., November 1992, pp. 317–320.

M10 MacRae, G., M. J. N. Priestley, and J. Tao, *P–Δ Design in Seismic Regions*, Structural Systems Research Project, Report SSRP-93/05, University of California, San Diego, June 1993, 114 pp.

M11 Maffei, J., and R. Park, "A New Method of Prioritising Bridges for Seismic Upgrading," *Proceedings of the New Zealand National Society for Earthquake Engineering, Annual Conference*, 1995, Rotorua, New Zealand, 11 pp.

M12 Mander, J. B., M. J. N. Priestley, and R. Park, "Observed Stress–Strain Behavior of Confined Concrete," *Journal of the Structural Division, ASCE*, Vol. 114, No. 8, August 1988, pp. 1827–1849.

M13 Monti, G., C. Nuti, P. E. Pinto, and I. Vanzi, "Effects of Non-synchronous Seismic Input on the Inelastic Response of Bridges," *Proceedings, 2nd International Workshop on Seismic Design and Retrofitting of R.C. Bridges*, Queenstown, New Zealand, 1994, pp. 95–112.

M14 Mahin, S. A., and J. Lin, *Construction of Inelastic Response Spectra for Single-Degree-of-Freedom Systems*, Report 83/17, Earthquake Engineering Research Center, Berkeley, Calif., 1983.

M15 Martin, P. P., and H. B. Seed, *MASH: A Computer Program for the Nonlinear Analysis of Vertically Propagating Shear Waves in Horizontally Layered Deposits*, Report 78-23, Earthquake Engineering Research Center, Berkeley, Calif., 1978.

M16 Maroney, B. H., and Y. H. Chai, "Bridge Abutment Stiffness and Strength Under Earthquake Loadings," *Proceedings, 2nd International Workshop on the Seismic Design of Bridges*, Queenstown, New Zealand, 1994.

M17 *Manual for Menshin Design of Highway Bridges*, Ministry of Construction, Japan (English version: EERC, Berkeley, Calif., Report 94/10).

N1 Newmark, N. M., and E. Rosenblueth, *Fundamentals of Earthquake Engineering*, Prentice Hall, Englewood Cliffs, N.J., 1971.

N2 Newmark, N. M., and W. J. Hall, *Procedures and Criteria for Earthquake Resistant Design, Building Practice for Design Mitigation*, Building Science Series 45, U.S. Department of Commerce, National Bureau of Standards, Washington, D.C., 1973, pp. 209–236.

N3 Newmark, N. M., "Effects of Earthquakes on Dams and Embankments," *Geotechnique*, Vol. 15, No. 2, 1965, pp. 139–160.

N4 Nuncio-Cantera, J. A., and M. J. N. Priestley, *Moment Overstrength of Circular and Square Bridge Columns*, Structural Systems Research Project, Report SSRP-91/04, University of California, San Diego, September 1991, 128 pp.

O1 Oliveira, C. S., H. Hao, and J. Penzien, "Ground Modelling for Multiple-Input Structural Analysis," *Structural Safety*, 1991, pp. 79–93.

O2 Ohta, Y., and N. Goto, "Empirical Shear Wave Velocity Equation in Terms of Characteristic Soil Indices," *Earthquake Engineering and Structural Dynamics*, Vol. 6, 1978, pp 167–187.

O3 Ordox, M., and E. Facciola, "Site Response Analysis in the Valley of Mexico: Selection of Input Motion and Extent of Non-Linear Soil Behavior," *Earthquake Engineering and Structural Dynamics*, Vol. 23, 1993, pp 895–908.

P1 Priestley, M. J. N., "Damage of the I-5/I-605 Separator in the Whittier Earthquake of October 1987," *Earthquake Spectra*, Vol. 4, No. 2, 1988, pp. 389–405.

P2 Priestley, M. J. N., F. Seible, and G. MacRae, *The Kobe Earthquake of January 17, 1995: Initial Impressions from a Quick Reconnaissance*, Structural Systems Research Project, Report SSRP-95/03, University of California, San Diego, February 1995, 71 pp.

P3 Park, R., and T. Paulay, *Reinforced Concrete Structures*, Wiley, New York, 1975, 769 pp.

P4 Paulay, T., and M. J. N. Priestley, *Seismic Design of Reinforced Concrete and Masonry Buildings*, Wiley, New York, 1992.

P5 Priestley, M. J. N., F. Seible, and C. M. Uang, *The Northridge Earthquake of January 17, 1994: Damage Analysis of Selected Freeway Bridges*, Structural Systems Research Project, Report SSRP-94/06, University of California, San Diego, February 1994, 260 pp.

P6 Priestley, M. J. N., F. Seible, and Y. H. Chai, *Design Guidelines for Assessment Retrofit and Repair of Bridges for Seismic Performance*, Structural Systems Research Project, Report SSRP-92/01, University of California, San Diego, August 1992, 266 pp.

P7 Priestley, M. J. N., and F. Seible, "Design of Seismic Retrofit Measures for Concrete and Masonry Structures," invited contribution to *Construction and Building Materials*, Special Issue: *Application of Polymeric Materials to the Construction Industry*, 1995.

P8 Pussegoda, L. N., "Strain Age Embrittlement in Reinforcing Steel," Ph.D. thesis report, Department of Mechanical Engineering, University of Canterbury, Christchurch, New Zealand, 1978.

P9   Priestley, M. J. N., F. Seible, and G. Benzoni, *Seismic Response of Columns with Low Longitudinal Steel Ratios*, Structural Systems Research Project, Report SSRP-94/08, University of California, San Diego, June 1994, 79 pp.

P10   Zahn, F., R. Park, M. J. N. Priestley, and H. C. Chapman, *Development of Design Procedures for Flexural Strength and Ductility of Reinforced Concrete Columns*, Pacific Concrete Conference, Auckland, New Zealand, November 1988, 12 pp.

P11   Priestley, M. J. N., F. Seible, Y. Xiao, and R. Verma, "Steel Jacket Retrofitting of Reinforced Concrete Bridge Columns for Enhanced Shear Strength, Part II: Test Results and Comparison with Theory," *ACI Structural Journal*, Vol. 91, No. 5, September/October 1994, pp. 537–551.

P12   Priestley, M. J. N., R. Verma, and Y. Xiao, "Seismic Shear Strength of Reinforced Concrete Columns," *ASCE Journal of Structural Engineering*, Vol. 120, No. 8, August 1994, pp. 2310–2329.

P13   Priestley, M. J. N., F. Seible, and D. L. Anderson, "Proof Test of a Retrofit Concept for the San Francisco Double-Deck Viaducts, Part 1: Design Concept, Details and Model," *ACI Structural Journal*, Vol. 90, No. 5, September/October 1993, pp. 467–479.

P14   Priestley, M. J. N., F. Seible, and D. L. Anderson, "Proof Test of a Retrofit Concept for the San Francisco Double-Deck Viaducts, Part 2: Test Details and Results," *ACI Structural Journal*, Vol. 90, No. 6, November/December 1993, pp. 616–631.

P15   Pardoen G. C., M. A. Haroun, R. Shepard, N. H. Flynn, R. P. Kazanjy, and S. A. Mourad, "Strong Axis and Weak Axis Strength of Pier Walls," *Proceedings, First Annual Seismic Research Workshop*, California Department of transportation, Sacramento, Calif., December 1991.

P16   Priestley, M. J. N., and F. Seible, "Seismic Assessment of Existing Bridges," *Proceedings, 2nd International Workshop on the Seismic Design of Bridges*, Vol. 2, Queenstown, New Zealand, August 1994, pp. 46–70.

P17   Park, R. J. T., M. J. N. Priestley, and W. R. Walpole, "The Seismic Performance of Steel Encased Reinforced Concrete Piles," *Bulletin of the New Zealand National Society for Earthquake Engineering*, Vol. 16, No. 2, June 1983, pp. 123–140.

P18   Priestley, M. J. N., F. Seible, and E. Fyfe, "Column Seismic Retrofit Using Fibreglass/Epoxy Jackets," *Proceedings, ACMBS-1 Conference*, Quebec, Canada, October 1992, pp. 287–297. Republished, *Proceedings, 3rd NSF Workshop on Bridge Engineering Research in Progress*, La Jolla, Calif., November 1992, pp. 247–251.

P19   Priestley, M. J. N., F. Seible, and E. Fyfe, "Column Retrofit Using Prestressed Fiberglass/Epoxy Jackets," *Proceedings, '93 FIP Symposium*, Kyoto, Japan, October 1993, pp. 147–160.

P20   Priestley, M. J. N., F. Seible, Y. Xiao, and R. Verma, "Steel Jacket Retrofitting of Reinforced Concrete Bridge Columns for Enhanced Shear Strength, Part I: Theoretical Considerations and Test Design," *ACI Structural Journal*, Vol. 91, No. 4, July/August 1994, pp. 394–405.

P21   Priestley, M. J. N., "The Thermal Response of Concrete Bridges," in *Concrete Bridge Engineering: Performance and Advances*, R. J. Cope (Ed.), Elsevier Applied Science, New York, 1988, pp. 143–188.

P22 Park, R., and W. L. Gamble, *Reinforced Concrete Slabs*, Wiley, New York, 1980, 618 pp.

P23 Pender, M. J., "Aseismic Pile Foundation Design Analysis," *Bulletin of the New Zealand National Society for Earthquake Engineering*, Vol. 26, No. 1, March 1993, pp. 49–160.

P24 Priestley, M. J. N., and M. J. Stockwell, "Seismic Design of South Brighton Bridge: A Decision Against Mechanical Energy Dissipators," *Bulletin of the New Zealand National Society for Earthquake Engineering*, Vol. 11, No. 2, June 1978, pp. 110–120.

P25 Prakash, V., G. H. Powell, and S. Campbell, *Drain-3 DX: Base Program User Guide*, Report UCB/SEMM-92/36, Department of Civil Engineering, University of California, Berkeley, 1992.

P26 Park, R. (Ed.), *Proceedings, 2nd International Workshop on the Seismic Design of Bridges*, Vol. II, Queenstown, New Zealand, August 1994.

P27 Priestley, M. J. N., R. J. Evison, and A. J. Carr, "Seismic Response of Structures Free to Rock on Their Foundations," *Bulletin of the New Zealand National Society for Earthquake Engineering*, Vol. 11, No. 3, September 1978, pp. 141–150.

P28 Pender, M. J., and T. W. Robertson (Eds.), "Edgecombe Earthquake: Reconnaissance Report," *Bulletin of the New Zealand National Society for Earthquake Engineering*, Vol. 20, No. 3, 1987, pp. 201–249.

P29 Psycharis, I. N., and P. Jennings, "Rocking of Slender Rigid Bodies Allowed to Uplift," *Earthquake Engineering and Structural Dynamics*, Vol. 11, 1983, pp. 57–76.

P30 Priestley, M. J. N., "Design of Single Level Bridges for Joint Shear," Structural Systems Research Report SSRP 93/02, University of California, San Diego, February 1993.

R1   Rosenblueth, E. (Ed.), *Design of Earthquake Resistant Structures*, Wiley, New York, 1981.

R2   Robertson, R. K., R. G. Campanella, D. Gillespie, and A. Rice, "Seismic CPT to Measure In-Situ Shear Wave Velocity," *Journal of Geotechnical Engineering*, Vol. 112, No. 8, 1986.

R3   Richards, R., and D. G. Elms, *Seismic Behaviour of Retaining Walls and Bridge Abutments*, Report 77-10, Department of Civil Engineering, University of Canterbury, Christchurch, New Zealand, 1977.

R4   Reese, L. C., W. R. Cox, and F. D. Koop, "Analysis of Laterally Loaded Piles in Sand," *Proceedings, 6th Offshore Technology Conference*, Houston, Texas, 1974, pp. 473–483.

R5   Robinson, W. R., and L. R. Greenbank, "An Extrusion Energy Absorber Suitable for the Protection of Structures During an Earthquake," *Earthquake Engineering and Structural Dynamics*, Vol. 4, 1976, pp. 251–259.

R6   Robinson, W. H., "Lead-Rubber Hysteretic Bearings Suitable for Protecting Structures During Earthquakes," *Earthquake Engineering and Structural Dynamics*, Vol. 10, 1982, pp. 593–604.

R7   Ricles, J., M. J. N. Priestley, F. Seible, R. Yang, R. Imbsen, and D. Liu, *The Whittier Narrows 1987 Earthquake: Performance, Analysis, Repair and Retrofit of the I-5/I-605 Separator*, Structural Systems Research Project, Report SSRP-91/08, University of California, San Diego, December 1991, 358 pp.

S1   Sheikh, S. A., and M. Uzumeri, "Strength and Ductility of Confined Concrete Columns," *Proceedings, ASCE,* Vol. 106, No. ST5, May 1980, pp. 1079–1102.

S2   Standards Association of New Zealand, *Concrete Structures Standard, Part 1: Code; Part 2: Commentary,* NZ S3101, Wellington, New Zealand, 1995.

S3   Seible, F., G. A. Hegemier, and V. Karbhari, "Advanced Composites for Bridge Infrastructure Renewal," National Research Council, Transportation Research Board, *Proceedings, 4th International Bridge Engineering Conference,* San Francisco, August 1995.

S4   Seible, F., S. Zunling, and G. Ma, *Glass Fiber Composite Bridges in China, Report ACTT 93/01,* University of California San Diego, April 1993, 36 pp.

S5   Seible, F., G. A. Hegemier, and G. Nagy, *The Aberfeldy Glass Fiber Composite Pedestrian Bridge, Report ACTT 94/01,* University of California San Diego, Jan. 1994, 52 pp.

S6   Seible, F., M. J. N. Priestley, C. T. Latham, and P. Silva, *Full-Scale Bridge Column/ Superstructure Connection Tests Under Simulated Longitudinal Seismic Loads,* Structural Systems Research Project, Report SSRP-94/14, University of California, San Diego, June 1994, 183 pp.

S7   Selna, L. G., L. J. Malvar, and R. J. Zelinski, "Box Girder Bar and Bracket Seismic Retrofit Devices," *ACI Structural Journal,* September/October 1989.

S8   *SC-Push 3D: Manual and Program Description,* SC Solutions, San Jose, Calif., 1995.

S9   Sun, Z., F. Seible, and M. J. N. Priestley, *Flexural Retrofit of Rectangular Reinforced Concrete Bridge Column by Steel Jacketing,* Structural Systems Research Project, Report SSRP-93/07, University of California, San Diego, February 1993, 215 pp.

S10 Seible, F., M. J. N. Priestley, and Z. Sun, *San Francisco Flexural Retrofit Validation Tests on Rectangular Columns,* Structural Systems Research Project, Report SSRP-90/07, University of California, San Diego, December 1990, 47 pp.

S11 Seible, F., G. A. Hegemier, M. J. N. Priestley, D. Innamorato, J. Weeks, and F. Policelli, *Carbon Fiber Jacket Retrofit Test of Circular Shear Bridge Column, CRC-2, Report ACTT 94/02,* University of California San Diego, September 1994, 49 pp.

S12 *SEQ-Mø Manual and Program Description,* Seqad Consulting Engineers, Solana Beach, Calif., 1995.

S13 Seed, H. B., and J. L. Alonso, "Effects of Soil–Structure Interaction in the Caracas Earthquake of 1967," *Proceedings, First Venezuelan Conference on Seismology and Earthquake Engineering,* 1974.

S14 Seed, H. B., and I. M. Idriss, *Ground Motion and Soil Liquefaction During Earthquakes,* Earthquake Engineering Research Institute Monograph, 1982.

S15 Schnabel, P. B., J. Lysmer, and H. B. Seed, *SHAKE: A Computer Program for Earthquake Response Analysis of Horizontally Layered Sites,* Report 72-12, Earthquake Engineering Research Center, Berkeley, Calif., 1972.

S16 Shibata, A., and M. A. Sozen, "Substitute Structure Method for Seismic Design in Reinforced Concrete," *ASCE Structural Journal,* Vol. 102, No. ST1, January 1976.

S17 Singh, S. K., E. Mena, and R. Castro, "Some Aspects of the Source Characteristics and the Ground Motion Amplification In and Near Mexico City from the Acceleration Data of the September 1995 Michoacan, Mexico Earthquake." *Bulletin, Seismological Society of America,* Vol. 78, 1988, pp. 451–477.

S18 Singh, S. K., J. Lermo, T. Dominguez, M. Ordox, J. M. Espinosa, E. Mena, and R. Quaas, "A Study of Amplification of Seismic Waves in the Valley of Mexico with Respect to a Hill Zone Site (CU)," *Earthquake Spectra,* Vol. 4, 1988, pp. 653–673.

S19 Seible, F., and M. J. N. Priestley, "Damage and Performance Assessment of Existing Concrete Bridges Under Seismic Loads," *Proceedings, 8th International Bridge Conference,* Pittsburgh, Pa., June 1991.

S20 Skinner, R. G., W. H. Robinson, and G. H. McVerry, *An Introduction to Seismic Isolation,* Wiley, New York, 1993.

T1 Tao, J. R., A. Krimotat, and V. Sobash, "An Analytical Model for Pounding Between Bridge Structures," *Proceedings, 3rd NSF Workshop on Bridge Engineering Research in Progress,* La Jolla, Calif., November 1992, pp. 231–235.

T2 Thewalt, C. R., and B. Stojadinovic, "Capacity Estimation and Retrofit of Outrigger Beam/Joint Systems," *Proceedings, 3rd NSF Workshop on Bridge Engineering Research in Progress,* La Jolla, Calif., November 1992, pp. 293–296.

T3 Trifunac, M. D., and A. G. Brady, "A Study on the Duration of Strong Earthquake Ground Motion," *Bulletin of the Seismological Society of America,* Vol. 65, 1975, pp. 581–626.

T5 Turner, M. J., R. W. Clough, H. C. Martin, and L. J. Topp, "Stiffness and Deflection Analysis of Complex Structures," *Journal of Aeronautical Science,* Vol. 23, No. 9, 1956.

T6 Tyler, R. G., "A Tenacious Base Isolation System Using Round Steel Bars," *Bulletin of the New Zealand National Society for Earthquake Engineering,* Vol. 11, No. 4, 1978, pp. 273–281.

T7 Tyler, R. G., "Dynamic Tests on PTFE Sliding Layers Under Earthquake Conditions," *Bulletin of the New Zealand National Society for Earthquake Engineering,* Vol. 10, No. 3, 1977.

T8 Tyler, R. G., "Rubber Bearings in Base-Isolated Structures: A Summary Paper," *Bulletin of the New Zealand National Society for Earthquake Engineering,* Vol. 24, No. 3, 1991, pp. 201–249.

T9 Terzaghi, K., "Evaluation of Coefficients of Subgrade Reaction," *Geotechnique,* Vol. 5, No. 4, 1955.

U1 Uang, C. M., and V. V. Bertero, "Evaluation of Seismic Energy in Structures," *Earthquake Engineering and Structural Dynamics,* Vol. 19, 1990, pp. 77–90.

U2 Universitá di Pavia "Certificato di Prova N.23414/184" Dipartimento di Meccanica Strutturale, June, 1993.

V1 Vallenas, J., V. V. Bertero, and E. P. Popov, *Concrete Confined by Rectangular Hoops Subjected to Axial Loads,* Report EERC 77/13, Earthquake Engineering Research Center, University of California, Berkeley, August 1977, 114 pp.

V2 Vecchio, F. J., and M. P. Collins, *The Response of Reinforced Concrete to In-Plane Shear and Normal Stresses,* Publication 82-03, University of Toronto, Canada, March 1982.

V3 Gasparini, D. A., and E. H. Vanmarcke, "SIMQKE: A Program for Artificial Motion Generation," Dept. of Civil Engineering, Mass. Institute of Technology, 1976.

W1   Watson, S. C., and M. K. Hurd (Eds.), *Esthetics in Concrete Bridge Design*, Publication MP1, American Concrete Institute, Detroit, Mich., 1990, 333 pp.

W2   Whitman, R. V., "Liquefaction: The State of the Knowledge," *Bulletin of the New Zealand National Society for Earthquake Engineering*, Vol. 20, No. 3, 1987.

W3   Wong, Y.-L., T. Paulay, and M. J. N. Priestley, "Response of Circular Reinforced Concrete Columns to Multi-directional Seismic Attack," *ACI Structural Journal*, Vol. 90, No. 2, March/April 1993, pp. 180–191.

W4   Wong, P. K. C., M. J. N. Priestley, and R. Park, "Seismic Resistance of Frames with Vertically Distributed Longitudinal Reinforcement in Beams," *ACI Structural Journal*, Vol. 87, No. 4, July/August 1990, pp. 488–498.

W5   Wilson, E. L., A. Der Kiureghian, and E. P. Bayo, "A Replacement of the SRSS Method in Seismic Analysis," *Earthquake Engineering and Structural Dynamics*, Vol. 9, 1981, pp. 187–194.

W6   Wakabayashi, M., *Design of Earthquake Resistant Buildings*, McGraw-Hill, New York, 1986.

W7   Whitman, R. V., "Seismic Design and Behavior of Gravity Retaining Walls," *Proceedings, Conference on Design and Performance of Earthquake Retaining Structures*, ASCE, Cornell University, Ithaca, N.Y., 1990, pp. 817–842.

W8   Wilson, E. L., *1 SADSAP: Static and Dynamic Analysis Programs*, Structural Analysis Programs, Inc., Berkeley, 1992.

X1   Xiao, Y., M. J. N. Priestley, and F. Seible, "Seismic Assessment and Retrofit of Bridge Column Footings." *ACI Structural Journal*, Vol. 93 No. 1, January-February 1996, pp. 1–16.

X2   Xiao, Y., M. J. N. Priestley, F. Seible, and N. Hamada, *Seismic Assessment and Retrofit of Bridge Footings*, Structural Systems Research Project, Report SSRP-94/11, University of California, San Diego, May 1994, 200 pp.

X3   Commission of the European Communities, Eurocode No. 8, *Structures in Seismic Regions, Part I: General and Building*, Report EUR 12266 EN, Brussels, 1989.

X4   Committee on Earthquake Engineering, National Research Council, *Liquefaction of Soils During Earthquakes*, National Academy Press, Washington, D.C., 1985.

Y1   Yang, R., M. J. N., Priestley, F. Seible, and J. Ricles, *Longitudinal Seismic Response of Bridge Frames Connected by Restrainers*, Structural Systems Research Project, Report SSRP-94/09, University of California, San Diego, August 1994, 50 pp.

Z1   Zahn, F. A., R. Park, and M. J. N. Priestley, "Flexural Strength and Ductility of Circular Hollow Reinforced Concrete Columns Without Confinement on Inside Face," *ACI Structural Journal*, Vol. 87, No. 2, March/April 1990, pp. 156–166.

Z2   Zahrah, T. F., and W. J. Hall, "Earthquake Energy Absorption in SDOF Structures," *ASCE Journal of Structural Engineering*, Vol. 110, No. 8, 1984, pp. 1757–1772.

Z3   Zerva, A., "Response of Multi-span Beams to Spatially Incoherent Ground Motions," *Earthquake Engineering and Structural Dynamics*, 1990, pp. 819–832.

Z4   Zayas, V. A., S. L. Low, and S. A. Mahin, "The FPS Earthquake Revising System," Report No. 87/01 UCB/EERC, University of California, Berkeley, 1987.

SYMBOLS LIST

A	=	seismic coefficient (a_o/g)
A	=	gross horizontal section area of elastomeric bearing
A'	=	overlap of top and bottom area of elastomeric bearing at maximum displacement
A_c	=	cross-sectional area of confined core of reinforced concrete section
A_e	=	effective shear area of cross-section
A_g	=	gross section area
A_h	=	area of one bar of shear reinforcement
A_h	=	hysteretic energy dissipated in one complete cycle of response to given displacement Δ_m
A_{jv}	=	cap beam vertical reinforcement for joint shear strength (Eq. (5.100))
A_l	=	total longitudinal mild steel area in section for torsion calculation
A_o	=	area inside perimeter p_0 of tubular section for shear flow
A_p	=	total prestressed reinforcement area in section for torsion calculation
A_r	=	total cross-section area of restrainers across a movement joint
A_{ST}	=	total area of longitudinal reinforcement in section
A_s	=	tension reinforcement area in section
A_s'	=	compression reinforcement in section
A_{sb}	=	longitudinal reinforcement area in a cap-beam retrofit bolster
A_{sc}	=	total area of longitudinal reinforcement in column section

A_{sh} = total area of transverse reinforcement within spacing s, perpendicular to section considered

A_{sp} = bar area of spiral reinforcement

A_{sv} = total longitudinal reinforcement area passing through a column-base hinge

A_v = area of transverse reinforcement in a layer in directin of applied shear force

A_{ve} = effective shear area

A_{vi} = cap beam/column joint vertical reinforcement required for joint shear strength (Eq. (5.102))

a = ground acceleration

a_b = depth of equivalent rectangular compresssion stress block in beam

a_c = depth of equivalent rectangular compression stress block in column

a_o = peak ground acceleration

a_s = acceleration coefficient

B = superstructure width

B_f = width of footing

b = width of rectangular section

b_{je} = effective column/cap beam joint width

b_{jeff} = footing effective width for joint shear

b' = rectangular section core width measured to inside of longitudinal reinforcement cage

b_b = beam width perpendicular to axis

b_c = column width perpendicular to axis

b_{eff} = footing effective width for flexure

b_{eff} = effective slab width contribution to section strength

b_k = column hinge section width (Fig. 7.21)

b_0 = perimeter length for punching shear resistance analysis

b_w = web width

C = compression force

C = soil cohesion

C_b = compression stress resultant in beam

C_c = compression stress resultant in column

C_N = factor for effective overburden pressure, (Fig. 2.19)

$\mathbf{c}$ = damping matrix for multidegree of freedom system

c = distance from neutral axis to extreme compression fiber of section

c = concrete cover to longitudinal reinforcement

c = viscous damping (force per unit velocity)

c_{cr} = critical damping coefficient

c_{eq} = equivalent viscous damping coefficient

c_u = value of neutral axis depth at ultimate strength of section

D = dead load effects

D	$=$	diagonal strut in joint
D	$=$	diameter of circular column
D'	$=$	core diameter of circular column
D_H	$=$	horizontal displacement in soil flow failure analysis
d	$=$	effective depth of rectangular section
d_b	$=$	reinforcement bar diameter
d_{bl}	$=$	longitudinal bar diameter
d_f	$=$	effective depth of footing
d_f	$=$	equivalent depth to fixity in pile shaft
d_m	$=$	depth to maximum moment in pile shaft
E	$=$	modulus of elasticity
E	$=$	energy
$\bar{E}, \tilde{E}$	$=$	directional seismic effects
E_c	$=$	concrete modulus of elasticity
E_d	$=$	damping energy of structure responding to an earthquake
E_{ds}	$=$	double modulus of longitudinal reinforcement for buckling calculation
E_h	$=$	hysteretic energy of structure responding to an earthquake
E_i	$=$	initial elastic modulus of longitudinal reinforcement
E_i	$=$	input energy of structure responding to an earthquake
E_j	$=$	modulus of elasticity of jacket material
E_k	$=$	kinetic energy of structure responding to an earthquake
E_r	$=$	modulus of elasticity of restrainers across a movement joint
E_s	$=$	modulus of elasticity of reinforcing steel
E_s	$=$	elastic energy of structure responding to an earthquake
E_{sec}	$=$	secant modulus of elasticity
E_t	$=$	modulus of elasticity of hoop or spiral reinforcement
E_{xx}	$=$	modulus of elasticity of fiber composite in direction x
F	$=$	force
$\mathbf{F}$	$=$	force matrix for multidegreee of freedom system
F_e	$=$	response force for an elastic system to an earthquake
F_H	$=$	horizontal force
F_p	$=$	axial prestress force
FS_E	$=$	seismic factor of safety of slope against failure
FS_l	$=$	factor of safety against liquefaction
FS_s	$=$	static factor of safety of slope against failure
F_y	$=$	yield force
f	$=$	frequency
f_{ya}^o	$=$	overstrength yield stress of reinforcement for seismic assessment of an existing bridge
f_c'	$=$	compressive strength of unconfined concrete
f_{ca}'	$=$	concrete compressive strength to be used in seismic assessment of an existing bridge

f'_{cc}	=	compressive strength of confined concrete
f'_{ce}	=	design compressive strength of unconfined concrete for plastic hinge
f'_{co}	=	maximum feasible compressive strength of unconfined concrete
f'_l	=	effective lateral confining stress
f'_{lx}	=	effective lateral confining stress in x direction
f'_{ly}	=	effective lateral confining stress in y direction
f'_t	=	concrete tensile strength
f_1	=	lowest theoretical frequency of vibration
f_a	=	'additional flexibility' coefficient $= \Delta_a/\Delta_y$
f_c	=	concrete compressive stress
f_d	=	damping force
f_h	=	nominal horizontal direct stress in joint
f_i	=	inertia force
f_j	=	hoop stress developed in jacket material
f_l	=	lateral confining stress
f_p	=	prestress in cap-beam
f_{pc}	=	average prestress in member
f_s	=	reinforcement stress
f_s	=	restoring force
f_s	=	required transfer stress in reinforcement in lap-splice
f_{sh}	=	permitted stress in hoop reinforcement for shear-friction calculations
f_{sh}	=	tension stress in hoop reinforcement
f_{su}	=	ultimate stress of reinforcement
f_{uj}	=	ultimate tensile strength of retrofit jacket material
f_v	=	nominal vertical direct stress in joint
f_v	=	shear flexibility of column
f_{xx}	=	strength of fiber composite in direction x
f_y	=	yield stress of reinforcement
f_{ya}	=	reinforcement yield stress to be used in seismic assessment of an existing bridge
f_{yd}	=	yield stress of dowel reinforcement
f_{ye}	=	design yield strength for longitudinal reinforcement in plastic hinge
f_{yh}	=	yield strength of transverse reinforcement
f_{yj}	=	yield stress of retrofit jacket material
f_{yo}	=	maximum feasible yield strength of longitudinal reinforcement
f_{yv}	=	yield strength of joint vertical reinforcement
G	=	shear modulus
g	=	acceleration due to gravity
g	=	gap between jacket and member supporting, or supported by, the column, measured parallel to column axis

H	=	height
H	=	soil reaction force against end of footing
H	=	column height to center of seismic force
H_c	=	column clear height
H_e	=	effective column height including strain penetration effects
H_s	=	superstructure depth
h	=	depth of rectangular section
h'	=	rectangular section core dimension measured to inside of longitudinal reinforcement cage
h_b	=	cap beam section depth
h_c	=	column section depth perpendicular to axis
h_f	=	footing thickness
h_k	=	column hinge section depth (Fig. 7.21)
I	=	importance factor in seismic prioritization (Eq. 7.1)
I/D	=	isolation/dissipation device
I_a	=	gross second moment of area of section
I_e	=	effective second moment of area of cracked section
I_f	=	second moment of area of footing/soil interface
J	=	torsional stiffness
j	=	rotational mass moment of inertia
K_A	=	static active earth pressure coefficient
K_{AE}	=	dynamic active earth pressure coefficient
K_b	=	bearing stiffness
K_{DE}	=	effective lateral stiffness of seismic I/D system at expected peak displacement
K_e	=	effective stiffness of column at peak displacement response
K_e	=	confinement effectiveness coefficient
K_f	=	rotational stiffness of footing
K_G	=	global effective stiffness of pier including effect of seismic I/D system flexibility
K_i	=	column stiffness at first yield
K_P	=	equivalent stiffness of pendulum system
K_{PY}	=	secant stiffness of pier at first yield of reinforcement, in direction considered
$\mathbf{k}$	=	stiffness matrix for multidegree of freedom system
k	=	factor for axial load variation due to vertical acceleration response
k	=	factor for shear strength of concrete shear resisting mechanisms (Eq. (5.75))
k	=	stiffness
k	=	rubber bulk modulus
k_e^L	=	bent longitudinal stiffness
k_e^T	=	bent transverse stiffness
k_b	=	bending stiffness of column

k_o	=	initial stiffness of Takeda degraded stiffness model
k_s	=	coefficient of subgrade reaction
k_s	=	soil reaction coefficient (force/length^3)
k_u	=	unloading stiffness of Takeda degraded stiffness model
k_v	=	seismic acceleration coefficient in vertical direction
k_v	=	shear stiffness of column
k_y	=	yield coefficient in sliding block analysis (Eq. 2.15)
L	=	length
L_c	=	clear distance between adjacent columns in a bent
L_{cap}	=	cap-beam length
L_f	=	length of footing
L_o	=	cap-beam overhang beyond end column
L_p	=	effective length of plastic hinge
L_r	=	length of restrainers across a movement joint
L_{sk}	=	length of shear key parallel to direction of shear force
l	=	length
l_a	=	length of anchorage of column longitudinal reinforcement provided in joint
l_b	=	distance from critical section to point of contraflexure
l_d	=	development length of reinforcement
l_{ts}	=	tension shift of moment diagram due to diagonal cracking
M	=	moment
M	=	Richter magnitude of earthquake
M_b	=	moment in beam
M_c	=	moment in column
M_{cr}	=	cracking moment
M_e	=	moment corresponding to elastic response to an earthquake
M_f	=	moment on footing
M_i	=	ideal flexural strength of section based on measured material properties
M_l	=	moment at left end of member, or left side of joint
M_n	=	nominal flexural strength
M^o	=	overstrength moment of plastic hinge
M_p	=	maximum moment capacity of plastic hinge
M_r	=	residual capacity of section, after failure by lap-splice
M_r	=	required flexural strength
M_r	=	moment at right end of member, or right side of joint
M_s	=	moment capacity of section, as limited by lap-splice strength
M_u	=	ultimate moment capacity
m	=	mass
$\mathbf{m}$	=	mass matrix for multi degree of freedom system
$\overline{m}$	=	mass per unit length

m^*	=	generalized mass characterizing contribution from distributed column mass to generalized displacement u^*
N	=	blow count in Standard Penetration Test
N_1	=	blow count corrected to eliminate influence of confining pressure
n	=	number of longitudinal bars contained within spiral or hoop
n	=	number of piles in a row
P	=	axial compressive force
P_b	=	axial compression force in beam
P_c	=	axial compression force in column
P_D	=	axial force due to dead load
PSA	=	peak pseudo acceleration
PSV	=	peak pseudo velocity
p	=	perimeter of crack surfaces around bar in lap splice failure
p	=	annual probability of exceeding a given limit state
p	=	pressure (in soil p-y curves)
p_t^{cr}	=	nominal principal tension stress at onset of torsional cracking of member
p_t^{deg}	=	nominal principal tension stress at onset of torsional degradation in member without special torsional reinforcement
p_c	=	nominal principal compression stress in joint
$\mathbf{p}_{eff}$	=	forcing function matrix
p_i	=	modal participation coefficient for mode i
p_o	=	perimeter of tubular section measured at centerline of tube thickness for shear flow
p_t	=	nominal principal tension stress in joint
p_u	=	ultimate soil pressure under footing
q	=	behavior coefficient representing ductility capacity in Eurocode seismic design
R	=	ranking value for seismic prioritization (Eq. 7.1)
R	=	radius of curvature of an elliptical jacket
R_c	=	radius of slip circle in embankment failure analysis
R_e	=	epicentral distance from a given site
$\mathbf{R}_i$	=	maximum spectral response for mode i
R_t	=	net tension force in footing tension piles
$\mathbf{R}^x$	=	response quantity from x-direction excitation
$\mathbf{R}^y$	=	response quantity from y-direction excitation
r	=	bilinear factor for second slope stiffness (Fig. 5.16(b))
r	=	live-load reduction factor
$\mathbf{r}$	=	ground motion influence coefficient for mode i
r	=	radius of curvature of pendulum motion

r = kinetic energy reduction factor for rocking response (Eq. 6.25)

r_d = stress reduction coefficient depending on deformability of soil column (Fig. 2.18)

r_o = post-elastic stiffness of idealized hysteretic response ignoring P-Δ effects (Fig. 5.97)

r_p = post-elastic stiffness of idealized hysteretic response including P-Δ effects (Fig. 5.97(b))

S = abutment skew in degrees

S = generalized action under seismic loading

S = ground slope (%)

S = shape factor of elastomeric bearing

S = seismicity factor in seismic prioritization (Eq. 7.1)

S_a = design (or assessment) elastic response acceleration corresponding to a specified limit state

S_{ar} = actual equivalent elastic response acceleration corresponding to a specified limit state

S_i = response spectrum ordinate for mode i

S_n = nominal strength

S^0 = action corresponding to M^0

S_R = residual undrained shear strength

S_r = required strength

s = longitudinal spacing of transverse reinforcement

T = natural period of vibration

T = tension force

T = torque

T_b = tension stress resultant in beam

T_b = maximum rebar force that can be transmitted through a lap splice

T_c = tension stress resultant in column

T_{cr} = torque at onset of torsional cracking of member

T_d = design period

TJ = maximum tension force capable of being developed across a locked movement joint

T_n = period of highest significant mode of vibration

T_o = period at peak elastic spectral acceleration response

T_p = natural period of pendulum system

t = time

t_{av} = average thickness of tube in shear flow calculation

t_d = average cyclic shear stress in liquefaciton analysis (Eq. 2.19)

t_j = thickness of retrofit jacket

t_{slab} = deck slab or soffit slab thickness

u = pore pressure in soil

$\mathbf{u}$ = displacement matrix

u^*	$=$	generalized displacement for system with distributed mass
u_g	$=$	ground displacement
$\ddot{u}_g$	$=$	ground acceleration
u_s	$=$	structural displacement
u_t	$=$	absolute displacement
$\dot{u}_s$	$=$	structural velocity
u_u	$=$	ultimate bond stress
u_x	$=$	displacement in x direction
u_y	$=$	displacement in y direction
u_z	$=$	displacement in z direction
V	$=$	shear force
V	$=$	vulnerability factor in seismic prioritization (Eq. 7.1)
V_E^*	$=$	equivalent elastic response force for a ductile system
V_B	$=$	applied shear force in beams
V_b	$=$	cap beam shear force
V_c	$=$	strength (force) of concrete shear resisting mechanisms
V_{col}	$=$	applied shear force in column
V_d	$=$	design shear strength
V_d	$=$	expected design lateral strength of vertical cantilever
V_{de}	$=$	elastic response base shear force for design earthquake
V_f	$=$	shear force corresponding to flexural strength
V_i	$=$	ideal shear corresponding to flexural strength of vertical cantilever based on measured material properties
V_{jh}	$=$	joint horizontal shear force
V_{jv}	$=$	joint vertical shear force
V_L	$=$	longitudinal shear force
V_m	$=$	average peak lateral force in complete cycle of response
V_{md}	$=$	ductile response base shear force for design earthquake
V_{me}	$=$	elastic response base shear force for moderate intensity earthquake
V_p	$=$	shear strength provided by axial force in member
V_r	$=$	required shear strength (force)
V_s	$=$	strength of shear resisting mechanisms involving transverse reinforcement
V_{sj}	$=$	shear strength enhancement provided to member by retrofit jacket
V_T	$=$	transverse shear force
V_V	$=$	vertical shear force
V_y	$=$	yield lateral strength of vertical cantilever
v	$=$	shear force per unit length
v_b	$=$	nominal shear stress of concrete shear resisting mechanisms (ASCE-ACI 426 approach)
v_c	$=$	strength (stress) of concrete shear resisting mechanisms
v_g	$=$	peak ground velocity in earthquake

v_{jh}	=	nominal horizontal shear stress in joint
v_{jv}	=	nominal vertical shear stress in joint
v_P	=	velocity of propagtion of P waves in earths crust
v_S	=	velocity of propagation of S waves in earths crust
W	=	seat width at superstructure internal movement joint
W	=	weight
W_f	=	weight of footing
W_s	=	equivalent weight for seismic analysis
w_S, w_V, w_I	=	weighting factors for seismicity, vulnerability and importance in seismic prioritization (Eq. 7.1)
x	=	distance
x	=	displacement
$\dot{x}$	=	velocity
$\ddot{x}$	=	acceleration
Z	=	adjustment factor for ductility and risk (Fig. 5.13)
z	=	depth of soil
α	=	coefficient representing influence of boundary conditions on column stiffness (Eq. 4.17)
α	=	unloading stiffness parameter for Takeda degraded stiffness model, (Eq. 4.98)
α	=	angle between tangent to spiral and direction of shear force; angle between column axis and axial force strut (Eq. (5.77)
Δ	=	displacement
Δ_a	=	contribution to yield displacement from mechanisms other than column flexure
ΔA_{sb}	=	additional area of cap beam reinforcement required for joint shear strength (Eq. 5.101)
Δ_B	=	bearing deformation
Δ_b	=	bearing shear deflection
Δ_c	=	limit-state displacement capacity
Δ_D	=	lateral displacement of simple verticale cantilever corresponding to V_{de}
Δ_d	=	design displacement
Δ_d	=	limit-state displacement demand
Δ_e	=	elastic strain energy stored in system at maximum force V_m and displacement Δ_m
Δ_m	=	average peak displacement in complete cycle of response
Δ_m	=	lateral displacement of simple vertical cantilever corresponding to V_{me}
Δ_p	=	plastic displacement
Δ_s	=	soil deformation
Δ_s	=	structural deformation
Δ_t	=	time step in time-history analysis
Δ_u	=	ultimate displacement

Δ_{ub}	=	bearing lateral displacement at onset of pier yield in seismic I/D system
Δ_y	=	yield displacement
ε_c	=	concrete compression strain
ε_{cc}	=	strain at peak stress for confined concrete
ε_{co}	=	strain at peak stress for unconfined concrete
ε_{cm}	=	maximum concrete compression strain
ε_{cu}	=	ultimate compression strain in concrete
ε_s	=	reinforcement strain
ε_{sh}	=	reinforcement strain at onset of strain hardening
ε_{sm}	=	strain at maximum stress in confining jacket material
ε_{su}	=	ultimate reinforcement strain (at $f_s = f_{su}$)
ε_t	=	tensile strain
ε_{uj}	=	ultimate tensile strain of composite-material jacket
ε_y	=	yield strain of reinforcement
ϕ	=	curvature
ϕ	=	friction angle of cohensionless soil
ϕ_f	=	flexural strength reduction factor
ϕ^0	=	capacity protection factor M^0/M^n
ϕ_p	=	plastic curvature
ϕ_s	=	strength reduction factor for action s, or for shear force
ϕ_u	=	ultimate curvature
ϕ_y	=	yield curvature
γ	=	unit weight of soil or concrete
$\lambda(M)$	=	probability of exceedence of earthquake of magnitude M occuring in volume V of earths crust in unit time
μ	=	ductility factor
μ	=	coefficient of friction
μ_D	=	displacement ductility in I/D system
μ_Δ	=	displacement ductility factor
$\mu_{\Delta c}$	=	displacement ductility factor for a column
μ_ϕ	=	curvature ductility factor
μ_s	=	steel-to-steel coefficient of friction
ν	=	Poisson's ratio
θ	=	angle between inclined diagonal tension cracking and member axis
θ	=	angle of incidence of seismic propagation to bridge axis
θ	=	rotation, angle
θ	=	stability index, defined in (Eq. 5.157)
θ_f	=	rotation of footing under applied moment
θ_p	=	plastic rotation in a plastic hinge
θ_x	=	rotation about x axis
θ_y	=	rotation about y axis
θ_z	=	rotation about z axis
$\ddot{\theta}$	=	rotational acceleration

ρ = material density

ρ_{ij} = cross-modal coefficient for complete quadratic modal combination

ρ_l = area ratio of longitudinal reinforcement

ρ_l' = longitudinal reinforcement ratio adjusted for yield strength and concrete strength (Eq. 7.14)

ρ_s = volumetric ratio of transverse reinforcement

ρ_{sj} = volumetric ratio of retrofit jacket

ρ_v = minimum area ratio of transverse reinforcement in footing

ρ_x = area ratio of transverse reinforcement in x direction

ρ_y = area ratio of transverse reinforcement in y direction

σ_0' = critical vertical effective stress in liquefaction analysis

ω = concrete unit weight

ω = dynamic amplification factor accounting for higher mode response

ω = angular frequency

ω_d = damped angular frequency

δ = displacement for unit lateral force

Φ_i = normalized mode shape for mode i

γ = principal tension stress coefficient (Eq. 4.72)

γ_{xz} = allowable shear strain in elastomeric bearing

η = correction factor applied to 5% damped spectral ordinates for system damping other than 5% (Eq. 6.15)

ξ = equivalent viscous damping ratio

ξ_{DE} = equivalent viscous damping ratio for non-viscous seismic I/D system

ξ_{DV} = viscous damping ratio for viscous seismic I/D system

ξ_{eq} = equivalent damping ratio

ξ_G = global effective damping of pier, including I/D system damping

ξ_p = equivalent viscous damping ratio for viscous seismic I/D system

Ψ = vibration deformation shape for system with distributed mass

INDEX

Abutments:
  analysis method, 218
  assessment of, 578
  damage to, 9
  modeling of, 216
  soil capacity, 218
  stiffness, 217
  superstructure connection, 139
  types, 139
Acceleration, vertical, effects of, 122, 304
Accelerograms:
  baseline correction of, 115
  characteristics of, 67
  soil modification of, 76
Amplification:
  dynamic, 36
  geographical, 59, 73, 152
Analysis:
  capacity/demand ratio, 539
  dynamic, 184
  frequency domain, 184
  linear elastic, 232
  methods, 227
    dynamic, 228
    response spectrum, 228, 241, 493
    static, 227
  moment-curvature, 298
  nonlinear time history, 184, 243, 327, 498, 530, 545
  plastic collapse (push), 232, 540
  risk, 536, 585
  for seismic isolation, 487, 490
  for spectral response, 185
  substitute structure, 223, 237
  tools, choice of, 160, 228
Anchorage:
  column bars, in cap beams, 395
  confinement, influence on, 393
  design for, 389, 392
  failure, 23, 28, 390
Arches, concrete, 153
Assessment:
  effective stiffness for, 548
  of existing bridges, 46, 57, 322, 534
  flexural strength, 549
  material properties, 546

Bearings:
  elastomeric, 133, 466
  influence on ductility, 135
  influence on vertical response, 136
  isolation, 57, 136, 466
  lead-rubber, 469
  pendulum, 473

period shift by, 134
PTFE, 133
to regularize bent response, 490
rocker:
  assessment of, 580
  failure of, 580
  sliders, damping of, 174, 472
Bent:
  configurations, 141
    linked column, 142, 143
    multi-column, 143
    single-column, 141
  modeling of, 203, 205
  stiffness, regularization of, 119
Bond, flexural, 400, 557
Bracing, of steel superstructures, 129,
  583
Bridges:
  long span, design considerations for,
    154
  wide, assessment of, 579
Buckling:
  of column longitudinal rebar, 314
  design of confinement against, 315

Cantilever construction, 153
Capacity design:
  basic equation, 36
  illustrative example, 36
  principles, 33, 265
  procedure, 280
  for seismic isolation, 503
Capbeam:
  damage, 22
  design forces, 324
  ductility, assessment of, 557
  effective width of, 207
  live load, for design of, 312
  modeling of, 206
  retrofit, 51, 623
  strength, assessment of, 557
  stiffness of, 207
Carbon fiber, 278, 589
Column:
  confinement of, 267, 313
  design axial force levels for, 304
  design examples, 435
  height, effective, 172
  modeling of, 203

reinforcement limits for, 301
  section shape alternatives, 144
Column/pile-shaft designs, equivalent
  depth to fixity of, 216, 284
Composite materials, advanced:
  material properties, 278
  retrofit using, 589, 595, 604, 612
Concrete material properties:
  compression strength:
    for assessment, 546
    confined, 267, 592
    unconfined, 266, 294
  influence of strain rate, 273
  modulus of elasticity, 266, 270
  tensile strength, 267
  ultimate strain, 272, 591
Confinement:
  design examples, 319, 320, 435
  influence of longitudinal rebar, 269
  of plastic hinges:
    required extent, 317
    required ratio (force-based), 313
    by transverse reinforcement, 267
    volumetric ratio of, 272
Connections:
  bent/foundation alternatives:
    fixed-base, 149
    pinned-base, 148
  superstructure/abutment alternatives:
    bearing-supported, 139
    monolithic, 137
    seismic isolated, 139
    transverse design of, 139
  superstructure/pier alternatives:
    bearing-supported, 133
    moment-resisting, 130
    seismic isolated, 136
Constraints:
  functional:
    alignment, 117
    bifurcations, 121
    horizontal curvature, 120
  geographical:
    liquefiable ground, 123
    unstable slopes, 123
    valley crossings, 152
Control, active, 531
Counterforts, 140

Cracking:
  of cap/column joints, 355
  extent in columns, 301
  limit state, 41
  strength, 302
  torsional, 200
Curvature:
  ductility, 33
  horizontal, of bridges, 120, 431
  yield, 555

Damage to bridges:
  abutment slumping, 9
  anchorage, 23, 28
  bearings, 580
  cap-beam, 22
  causes, 3
  column/cap-beam joint, 23
  column ductility, 16
  column, flexural, 12, 16
  column, lap-splice, 14
  column, shear, 18
  footing, 27
  pounding, 8
  restrainers, 582
  skew effects, 5
  span unseating, 5
  steel column, 27
  steel superstructure, 27
Dampers, seismic isolation, 446
  fatigue life of, 477
  hydraulic, 478
  lead extrusion, 480
  lead-rubber, 469
  pendulum, 473
  sliders, 472
  steel hysteresis, 475
  testing and performance
    requirements, 483
Damping, 460, 491
  coulomb, 176
  devices, 366, 458
  equivalent viscous, 162, 174, 457, 522
  general issues, 174, 457
  hysteretic, 175, 244
  influence on period of vibration, 179
  Rayleigh, 245
Design, seismic, 265
  conceptual, 55, 116, 136, 462

displacement-based, 289, 443, 506
  flexural, of plastic hinges, 293
  force-based, 281
  philosophy, 29, 462, 585
  for unstable ground, 123
Detailing, 56, 314, 317, 318
  of footings, 408, 409
Development, of reinforcement, 389,
  559
Diaphragms, superstructure, 127, 128
  modeling of, 202
Discretization, geometric, 159
Displacement:
  absolute, 162
  design level, 290
  generalized, 162
  of isolated structures, 463
  structural, 162
  yield, 289
Drift, total plastic, 290
Ductility:
  assessment of, 307
  curvature, 33
  definition, 32
  displacement, 33, 135, 310
  influence of bearings on, 135
  influence of capbeam flexibility on,
    281
  influence of foundation flexibility on,
    284
  required, 307
  vs. required strength, 99, 285
Dynamic amplification, 36
Dynamics, structural:
  fundamentals of, 160
  matrix formulation of, 181

Elements, structural:
  line, 193, 194
  plate and shell, 193, 196
  solid, 193, 196
Energy:
  considerations in design, 72
  dissipation of, 247, 457
Examples:
  analysis:
    capacity design principles, 36
    of MDOF model, 182
    push-over, 233

of SDOF model, 180
of spectral response, 186
substitute structure, 237
viaduct response assessment, 249
assessment, of cap/column joint, 565
design:
  of cantilever column and footing,
    431
  of cap beam, 452
  of cap/column joint principal
    stresses, 358
  for column flexure, 435, 444, 447
  of column/footing joint, 413
  for column shear, 437, 449
  of confinement reinforcement:
    displacement-based, 320
    force-based, 319
  displacement-based, 443
  footing, 441, 455
  foundation, 439
  of knee joint design, 374
  pile, 440
  of plastic rotation capacity, 310
  of tee joint design, 386
  of two-column bent, 443
retrofit:
  column composite jacket, 622
  column steel jacket, 619
seismic isolation:
  analysis, 494, 500
  design, 508, 511
  footing rocking, 525, 530
  retrofit design, using, 513
Exceedence, probability of, 66, 100, 544

Flexibility:
  flexural, 169
  shear, 170
Flexural strength:
  design for (plastic hinges), 293, 297
  reduction factors, 295
  required (capacity protected), 322
  required, of plastic hinges, 281
Fiberglass/epoxy composite, 278, 589
Footings:
  analysis of, 403
  damage, 27
  deformation characteristics, 570
  design of, 400

flexural strength of, 406, 570
force reduction by link beam, 616, 647
modeling of, 211
retrofit, 53, 646
rocking concept for, 403, 570, 516
shear strength of, 408, 573
stability, overturning of, 401, 570
strength of existing, 570
Force, axial, in columns, 304
Force-reduction factor, 286
Foundations:
  design alternatives, 149
    integral pile/column, 151
    pile, CIDH, supported, 152
    pile, driven, supported, 151
    rocking, 150
    spread footing, 149
  modelling of, 210
Freedom, degrees of,
  characteristics of multi-models, 181
  characteristics of single-models, 179

Ground:
  dislocation, 81
  improvement techniques, 49
  slumping, 81
Ground motion:
  acceleration, peak, 70
  acceleration, vertical, 70
  artificial, generation of, 106, 113
  attenuation laws, 59
  characterization, 59
  coherency, 114
  design, 94
  directionality effects, 69
  displacement, peak, 71
  duration, 67, 71
  energy considerations, 72
  geographical amplification of, 73
  intensity, 69
  magnitude, 69
  non-synchronous:
    analysis for, 108
    design approach, 110
    effects, 108, 122, 154
    generation of, 113
    nonlinear considerations, 110
    relative displacements, 111
  return periods, 65

risk evaluation, 65
soil modification of, 76
source mechanisms, 59
velocity, peak, 71
wave propagation, 61

Hinge, column base (pinned):
  conceptual aspects, 148
  strength and rotation capacity, 574
Hinge length, plastic:
  of columns, 308
  of column retrofits, 590
  of pile-shafts, 311
Hysteresis rules:
  bilinear, 247
  elasto-plastic, 247
  influence on dynamic response, 261
  Takeda, 247

Integration, Newmark, 243
Isolation, seismic, 457
  application, 464
  conceptual aspects, 136, 462, 465
  for design, 457, 487
  design principles for, 500
  devices, 466
  influence on mode shapes, 497, 515
  period shift by, 134, 458, 515
  preliminary design of, 487
  response in earthquakes, 464
  for retrofit, 57, 511
  retrofit using, 513, 653

Jackets:
  composite materials, 49, 589
  concrete, 588
  retrofit, steel, 49, 587
Joints:
  capbeam/column:
    cracking of, 357
    crushing of, 358
    damage, 23
    deformation characteristics, 562
    design of, 348
    design examples, 458, 374, 386
    design requirements, summary, 371
    force transfer in, 360
    knee joints, 352, 360
    longitudinal response, 384

nominal shear stress in, 355
principal stresses, 355, 564
retrofit, 51, 631
shear force in, 349, 353
strength, 348, 562
tee joints, 353, 377
column/footing:
  assessment of, 574
  damage to, 28, 410, 574
  design of, 410
  design example, 413
  design requirements, summary, 412
  force transfer in, 411
  principal stresses in, 411
  reinforcement for, 411
movement:
  force transfer across, 128
  influence on longitudinal response,
    416
  limiting devices (seismic isolation),
    481
  longitudinal displacements across:
    assessment of, 581
    retrofit against, 639

Limit states:
  acceptable probability of exceedence,
    100
  for assessment, 537
  cracking, 41
  damage control, 43, 538
  first yield, 41
  serviceability, 42, 537
  spalling, 42
  survival, 43, 538
  ultimate, 42
Limiting devices (seismic isolation), 481
Link beams, 142, 615, 631, 647
  retrofit using, 615, 631, 644
Liquefaction:
  analysis of, 88
  damage caused by, 7
  design, structural for, 124
  factors affecting, 88
  retrofit against, 49
Live load (cap beams), 325

Mass: 163
  centroid, 163

of columns, 166
distributed, 163
generalized, 166
lumped, 163
moment of inertia, 165
Modal combination, 185, 242
CQC, 185
direct, 185
SRSS, 185, 188
Mode shapes, 182
normalized, 184
Modes, natural, of vibration, 182
Modeling:
of abutments, 216
of columns, 203
of foundations, 210
general issues of, 189
introduction to, 56, 157
of joints, 208
of movement joints, 219
objectives, 157
of pile shafts, 213
of structural systems, 191
Models:
bent, 192
fiber, 195
finite element, 159, 202
frame (stand alone), 192
global, 158, 191
grillage, 201
linear, 198
lumped parameter, 159
member, 193
structural component, 159
superstructure, 197
Moment of inertia, effective, 171, 173
cracked-section, 171
torsional, 199
Moment redistribution (columns), 305
Motion, equations of, 160
Movement joints, modeling of, 219

Participation factors, modal, 184
Penetration strain, 171, 308
P-Δ effects:
collapse due to, 44
design against, 303, 427
influence of hysteresis rule on, 428

modeling of, 204
numerical example, 261
Philosophy:
design, 29
retrofit, 45
Piers, 144
Pile caps, design of, 400
Piles:
axial forces in, 402
capacity, assessment of, 576
conceptual aspects:
CIDH, 152
compression, 152
driven, 151
tension, 152
design of, 415
effective depth to fixity, 216, 284
lateral deformation characteristics of,
217
modeling:
of axial stiffness, 211
of lateral stiffness, 211
of pile shafts, 213
group effects, 212
shear failure of, 576
Pile shafts:
depth of plastic hinge, 312
equivalent depth of fixity, 216, 286
modeling of, 213
plastic hinge length, 311
Plastic end region, 317
Prestressing:
influence on joint design, 370, 374,
383
for retrofit, 51, 627, 632
steel material properties, 277
Prioritization, for retrofit, 534

Reinforcement:
anchorage and development, 389, 393
for cap beam/column joints, 360, 364,
370, 378, 385
for columns, 293, 298, 301, 312, 314
distributed, vs. lumped, 330
headed, 408
longitudinal, limits to, 301, 303
material properties:
for assessment, 546
cyclic loading, 275

influence of strain rate, 276
modulus of elasticity, 273
monotonic loading, 273
strain-age embrittlement, 276
strain-hardening, 276, 294
stress-strain curves, 273, 299
strain, ultimate, 273
yield strength variability, 275
spacing, maximum, 315
splicing, 389, 396
transverse:
  for antibuckling, 314
  for confinement (hoops), 269, 313,
    318
  for confinement (spirals), 268, 312,
    318
Repair, of columns:
  using composite jackets, 617
  using steel jackets, 615
Response combination:
  modal, 185
  of orthogonal components, 188
Response spectra:
  acceleration, elastic:
    characteristics of, 94
    design, 102
    soil modification of, 76
  combined acceleration/displacement,
    104
  displacement, 103
  ductility, 99
  examples of, 96
  inelastic design, 97, 105
Response, vertical, 122, 461
Restrainers, 48
  assessment of, 581
  design of, 425
  failure of, 582
  influence on movement joint
    displacement, 426
  retrofit with, 641
Retrofit, 585
  cap beam, 623
    force reduction in, 625
    strength enhancement of, 625
  cap/column joint, 630
    acceptable damage of, 631
    force reduction in, 631
    jacketing of, 634

by prestressing, 625, 632
replacement of, 634
column:
  by composite materials jacketing,
    589, 595, 604, 612
  by concrete jacketing, 588, 595, 604,
    612
  design example, 619, 622
  for flexural ductility, 590
  for lap-splices, 598
  for shear strength, 606
  by steel jacketing, 587, 590, 598, 607
  for stiffness enhancement, 634
concepts, 47
footing, 646
  force reduction in, 647
  joint strength, 649
  overlay, 647
of movement joints, 639
philosophy, 45, 585
prioritization, 45, 534
seismic isolation for, 653
superstructure, 638, 642
  force reduction in, 644
  strength increase in, 642
Rocking, foundation, 516
  analysis of response, 519
  applications, 520
  design approach, 525
  design example, 525
  time-history analysis of, 530
Rotation, plastic, 290, 307
  assessment of, 307, 552
  design example of, 310
  of unconfined plastic hinges, 554

Seat extension, for retrofit, 641
Seating:
  design minimum length of, 419
  for retrofit, 490
  retrofit of, 641
Seismicity, 58
  design, selection of, 55
Serviceability considerations, 321
Settlement:
  of abutment fill, 8
  slabs, 141
Shear
  design example, 437

design recommendations, 341
failures, of columns, 18
failures, of joints, 22
force, design level of, 36, 322
retrofit for, 627
strength, for assessment purposes,
    338, 560
strength, design, 331, 343
    of cap beams, 344
    of columns, 331
    influence of ductility on, 336, 338,
        343
Shear friction, 347, 424
Shear keys:
    at abutments, 140
    assessment of, 579
    design of, 424
    design forces in, 421
    at internal movement joints, 128
Slope stability:
    dynamic, 85
    static, 81
Slope, unstable, structural design for,
    123
Soil:
    layering, 80
    modification to response spectra, 76
    shear modulus, 78
    sliding analysis, 85
Soil-structure interaction, 212, 250
Solution strategies:
    for cyclic loads, 230
    for monotonic loads, 229
Splices, lap, 389, 396
    ductility of, 555
    flexural strength of, 549
    minimum length, for retrofit, 606
    residual capacity of, 550
    retrofit of, 598
Stiffness, 168
    center of, 164
    effective, 171, 541, 548, 553
    enhancement by retrofit jackets, 614
    flexural component of, 168
    post-yield, influence on response, 263
    shear component of, 140

of substitute structure, 218, 225, 290
torsional, 199
Strain, concerete:
    influence of confinement, 270
    at peak stress, 270
    at spalling, 271, 294, 322
    ultimate, 272
Strength:
    definitions, 38
    dependable, 39
    design, 39
    design, of capacity protected actions,
        326
    expected, 39
    extreme, 39
    flexural, for assessment, 549
    flexural, of columns, 297, 327
    flexural, of other members, 329
    ideal, 39
    nominal, 38
    overstrength, 39, 293, 296, 323
    reduction factors, 36
        flexural (plastic hinges), 293, 295
    required, 38, 322, 547
    shear, 322, 560
Subgrade reaction, modulus of, 214
Superstructure:
    arches, concrete, 153
    arches, steel, 584
    cable stayed, 154
    concrete, section types, 125
    deformation capacity of, 577
    design seismic of, 416
    effective width of, 417
    modeling of, 197
    precast, continuity of, 126
    retrofit of, 638, 642
    segmental construction, 154
    steel, assessment of, 582
    steel, section types, 129
    suspension bridges, 154

Tension shift, 329
Torsion, 345

Walls, wing, 138, 140
Wave propagation, 61